The Pattern of Change

One of life's most fundamental revelations is change. Presenting the fascinating view that pattern is the manifestation of change, this unique book explores the science, mathematics, and philosophy of change and the ways in which they have come to inform our understanding of the world. Through discussions on chance and determinism, symmetry and invariance, information and entropy, quantum theory and paradox, the authors trace the history of science and bridge the gaps between mathematical, physical, and philosophical perspectives. Change as a foundational concept is deeply rooted in ancient Chinese thought, and this perspective is integrated into the narrative throughout, providing philosophical counterpoints to customary Western thought. Ultimately, this is a book about ideas. Intended for a wide audience, not so much as a book of answers, but rather an introduction to new ways of viewing the world.

ROBERT V. MOODY is a mathematician and co-discoverer of the Kac–Moody algebras, significant in mathematics and physics. He is a co-winner of the Wigner Medal and a member of the Royal Society of Canada (1980). He was made an Officer of the Order of Canada in 1991 and was the Founding Scientific Director of the Banff International Research Station (2001). He has published more than 100 papers on Lie theory and long-range aperiodic order.

DENG MING-DAO is an author, artist, graphic designer, martial artist, and teacher. He is the author of twelve books on Daoism, including *365 Tao* (1992) and *The Living I Ching* (2006), a translation and commentary on the ancient Chinese classic, *The Book of Changes*. He brings a unique insight into traditional ideas of change and how patterns can be discerned using a combination of discrete and holistic approaches.

LI　理

LI — pattern (in wood or jade), inner essence, inner-pattern, principle, intrinsic order, structural coherence, science, truth

WULI　物理 — innate laws of things, physics

Calligraphy of Chungliang Al Huang

The Pattern of Change

A Mathematical and Philosophical Study of How We See the World

Robert V. Moody
University of Alberta

Deng Ming-Dao

Shaftesbury Road, Cambridge CB2 8EA, United Kingdom

One Liberty Plaza, 20th Floor, New York, NY 10006, USA

477 Williamstown Road, Port Melbourne, VIC 3207, Australia

314–321, 3rd Floor, Plot 3, Splendor Forum, Jasola District Centre, New Delhi – 110025, India

103 Penang Road, #05–06/07, Visioncrest Commercial, Singapore 238467

Cambridge University Press is part of Cambridge University Press & Assessment, a department of the University of Cambridge.

We share the University's mission to contribute to society through the pursuit of education, learning and research at the highest international levels of excellence.

www.cambridge.org
Information on this title: www.cambridge.org/9781009546515

DOI: 10.1017/9781009546508

When citing this work, please include a reference to the DOI 10.1017/9781009546508

First published 2025

Printed in the United Kingdom by CPI Group Ltd, Croydon CR0 4YY

Cover image: Japanese Garden at The Huntington Library, Art Museum, and Botanical Gardens in San Marino, California.

A catalogue record for this publication is available from the British Library

A Cataloging-in-Publication data record for this book is available from the Library of Congress

ISBN 978-1-009-54651-5 Hardback
ISBN 978-1-009-54653-9 Paperback

Additional resources for this publication at www.cambridge.org/9781009546515.

Contents

Preface

The only constant in this world of change is change itself. So it is often said. Already Heraclitus living in the fifth century BCE famously emphasized the flux that underlies all things, and change is the central feature of the famous *Yijing* (*I Ching, The Book of Changes*) that dates back some five centuries earlier. Life leaves us with no doubt about the reality of change. But is it really true that there are no other constants than change itself? Change is not entirely random—it is not without predictable effects. Within change there arise forms and patterns, and were it not for these, which we observe, predict, and depend on, there would be no world full of life with its amazing forms that arise, thrive, and regenerate in the great cycles of things.

To say that the only constant is change itself is too blunt. What emerges from change is pattern, and it is pattern that we can recognize and depend on. Is it not the case that rivers are rivers because something about them does not change, something that supports both fish and fishermen? A closer look at what Heraclitus said reveals that he understood more than we cannot step into the same river twice:

> As they step into the same rivers, other and still other waters flow upon them. [94, fragment L]

This suggests that although we cannot step into the same river *water* twice, what makes the river a river is its flowing water: no flow, no river. The river is defined as a pattern through its flow. Here we sense a deeper understanding of how sameness/difference turns out to be an important way in which we continuously reposition ourselves. The river stays the same by always consisting of new (other) water. It is the same river because it is in how it changes that it becomes itself. Thus sameness and difference are together what makes a river a river.

This is the underlying theme of this book: the world is a world of change, but change manifests itself in the form of *pattern*. It is through these patterns of change, which we recognize and act upon, that we are able to live in this world. We might even say that they are how we come to "make" our world; and so it is for all forms of life.

Still, as a word in English, pattern itself is a nebulous entity. There are weather patterns, dress patterns, holding patterns for planes, patterns of behavior, patterns of numbers, flocking patterns, and on and on *ad infinitum.* We have heard pattern described as being as information, as memory, as repetition, all of which have elements of truth, and mathematics has been called the science of patterns. Yet out of all this variety of ideas there is one that strikes us as fundamental, even universal—the one that we have already alluded to: the origin of pattern is dynamic, pattern is the signifier and identifier of change.

Put into a single sentence, our understanding is that

> *pattern is the means by which we experience change,*
> *change is what we are able to infer from pattern.*

In effect, change and pattern are aspects of the same thing. Important to this view is that it is *systemic* in conception, in the sense that neither change nor pattern stand alone, but each supports the other, and their unity is system based. *Patterning* may be the best way to think of it.

It doesn't seem that elaborating on this should be a book-length endeavor, but our intention is to develop a formalism that encompasses this concept and to show how deeply embedded it is in the mathematics and science of our time. The ideas that lie behind patterning—change, chance, systems, information, invariance, uncertainty, measures, laws, all of which we will meet—seem common enough, but formalizing them and giving them precise meanings is decidedly modern. In many ways their story parallels the history of science, and that is how we approach it.

This not a book about philosophy, at least not explicitly so, yet in dealing with ideas about how we see and function in the world philosophical questions are bound to arise, and it is our hope that these in themselves will stimulate the reader to exploring new ways of thinking. A central idea, reflected in its title, is that we do not live in a world of "things", but rather in a world of process and relation within change. Change is foundational, pattern is how it is manifested through our senses and minds. It is neither entirely deterministic nor entirely random yet there is a deep sense of relation that entwines chance and law. Forms emerge, they persist, though changing through time, and then return to the source from which they arose. Nature seems both bound by enduring principles and free to be endlessly creative.

These ideas are not new, but they have gathered extraordinary empirical evidence over the past few hundred years. To a great extent the scientific enterprise has been fueled by foundational philosophical ideas of the West, based on belief in an externally created universe, teleologically directed, with humanity as a supreme achievement and capable of revealing its internal design. Strangely, the more we have learned the more science itself has undermined these ideas. We don't seem so special as to consider ourselves some supreme achievement and there is no clear evidence that the evolution of life is teleological or directed to some anthropological ideal.

For many it has left empty the meaning of life—who are we, why are we, whither our journey? At the same time it has led to an understanding that is more in tune with one that arose in ancient China during a period of war and chaos much like our own today. There, a philosophy developed that is both remarkably consistent with what we have learned from science and deeply instructive about our entwined relationships with all the transient processes of life on this planet. We have tried to introduce some of this ancient philosophical thought into our development because we think it is hugely important. Vital, even. It is subtle and hard to grasp, our minds so implicitly bound to a culture that has for millennia defined itself through anthropocentric absolutes. But there is wonderful depth and unity to it. China's artists, poets, and intellectuals were attuned to Nature and found both meaning and repose within it, witnessing themselves as expressions in its weave of ever changing pattern.

The aim of the book is to color in all these concepts and see how they fit into a consistent framework which we might consider as a basis for a theory of pattern, or better a theory of *patterning*. At the same time we want it to read as a sort of history of science and show how it raises ideas that are important to our age. Yet a third intention is to give a gentle path through some of the history of mathematics and the ways, often surprisingly metaphorical, in which it is used to give meaning to such abstractions as change, randomness, and invariance. In all of this we hope that the reader can have better access to some of the remarkable revelations of modern science. A particularly challenging goal has been to conclude with a section of quantum theory, for it is astoundingly predictive, increasingly important in modern technology, and perhaps an important ingredient in consciousness, yet it too is pattern based and must be a necessary part of our story. But it is also fascinating because its domain is a land where ordinary language and thought fail to encapsulate its strange conclusions, and where philosophical debate has always been fierce.

So this book is held together with a thread of mathematical ideas. However, there is plenty of ordinary text too, and we hope that the book can be read at many levels and will be accessible by anyone interested in how science has come to address the idea of change. Mathematics is often seen as numerical and quantitative, technical, and difficult to manipulate. But at a deeper level it is *relational*. As such it can be used to represent any situation in which those relationships appear, and it can offer neutral ground where everything other than those relations are ignored. That's the key. Much of mathematics over the centuries has been developed precisely for this reason, and by now there is a great array of ideas and mathematical tools that are available to deepen our understanding.

The mathematics in this book has to do with the relational side of things. The difficulties that it presents are not technical. They are conceptual, and they have to do with the difficulties of finding ways in which abstract ideas can be formalized into a mathematical context. The difficulties that we encounter as readers may seem difficult, but that is not because we are no good at mathematics. The scientists and mathematicians who created these forms of expression had enormous difficulties with them too. It's like the invention of zero. We have all learned to be at ease with the concept, but its creation was a genuine step in the history of mathematics–and it did not come easily. A symbol for "nothing"? Really?

Often getting a new idea depends on grasping the right metaphor. We may not think of mathematics as having anything to do with metaphor but a great deal of it derives, not surprisingly, from attempts to model aspects of the world that we experience. We will see many examples of this.

In any case, in reading the book there is nothing wrong with breezing over the mathematics and getting some conceptual sense of what it is about. It is fine to read ahead and then go back if a concept is still vague. Perhaps the ideal reader is one with some sort of STEM background, but even a good background in high school mathematics should be sufficient. For the mathematically educated, there will be much that is familiar but, we hope, also things that are new. All the mathematics is subservient to the deeper idea that we want to present: that pattern is a fundamental attribute of change, and that it is through pattern, or patterning, that we see the world. We have tried to make the conceptual ideas clear, convincing, and readable, but at the same time to lay down a rigorous foundation for a theory of change and pattern.

The book naturally falls into three basic parts:

- Five chapters offering an expansive view of how pattern and change are related, including ideas around continuous change, discrete change, and random change and their historical development.
- Three chapters that lay down the precise mathematical concepts that are involved in formalizing pattern and change, as well as some of their consequences. Important in this part are ideas around information and what it means for pattern systems to interact and form larger integrated pattern systems: synthesis as opposed to the usual analysis.
- Seven chapters that show how universally these ideas of pattern and change apply. At the same time, starting at the chapter on waves, we see there is a new way of looking at pattern systems. We call it the *linearization* of pattern systems, and it is this that opens up the real power of the modern mathematics. It is also foundational in quantum theory.

Beyond the physical book itself there are two sets of cross-referenced online addenda that are freely downloadable (see the Table of Contents). The first is a set of additional chapters and the second a set of endnotes that arise out of the text and elaborate on details of further interest. Some of the harder or more extended parts of the mathematics are to be found in the addenda, for those who wish to see more details. For instance the mathematical insights into the structure of the periodic table are there. Most of the details are pitched at a level compatible with the idea of the book. Online sections and chapters are referenced with the symbol ‡§ or ‡Ch. Endnotes are referenced by superscript numbers, starting from 1 at each chapter.

The unlikely combination of a mathematician and a Chinese martial artist, writer, philosopher writing this book reflects the fact that pattern and change are universal. We felt the need for seeing them from the two sides of what is often seen as an east-west divide. Human beings are human beings, but cultural history leads us into seeing things from differing points of view. Nature is not divisible into right and wrong or good and bad cultures. It cannot be reduced to a single set of facts. It is in seeing from other sides that we understand how much we implicitly assume without being aware of it, and learn thereby what we have been missing.

The Universe is both mysterious and beautiful. Looking at it from a scientific point of view can do nothing to diminish this. Rather it can do much to inspire and deepen our sense of wonder, for it reveals again and again the extraordinary coherence and unity of Nature.

Yes, we do feel that Nature deserves its capitalization!

mathematic (n.)
*"mathematical science," late 14c. as singular noun, mathematik (replaced since early 17c. by mathematics, q.v.), from Old French mathematique and directly from Latin mathematica (plural), from Greek mathēmatike tekhnē "mathematical science," feminine singular of mathēmatikos (adj.) "relating to mathematics, scientific, astronomical; pertaining to learning, disposed to learn," from mathēma (genitive mathēmatos) "science, knowledge, mathematical knowledge; a lesson," literally "that which is learnt;" from manthanein "to learn," from PIE root *mendh- "to learn."* (*Online Etymology Dictionary*)

Acknowledgments

The writing of this book has involved the teachings of countless mentors throughout the years of my life. Remembered or not, I thank them all. However, I would like to give special thanks to the people who have directly been involved. First Michael Baake and Uwe Grimm who gave two weeks of their time to reading, correcting, and advising me on an earlier version of the book out of which the present version came to be. It is particularly sad that Uwe died abruptly, far too young, just two weeks after this reading. Michael has been and continues to be an inspiration. It was through my longtime co-researcher Jiří (George) Patera (1936–2022) that I was introduced to the world of mathematical physics, which ultimately is so important in this book.

I have had a long association with Jan Zwicky that involves photography and poetry, but it has been her depth and clarity as a philosopher that has been important here, especially giving me more nuanced views into meaning and gestalt. Many hours spent with neuroscientist Mark Bodner pointed me in the right directions in starting the book. David Robertson has been an enthusiastic reader and commenter over various drafts of this book, and pleasant company for many Wednesday afternoon meetings at local cafes around Victoria. It was Marjorie Senechal who introduced me to aperiodic order and she has been an important supporter of this book. Also I want to thank my former student, and now professor, Jeong-Yup Lee who read the entire book and made many corrections.

The friendship of Yves Meyer, along with his interest in the book and his careful reading of some of its more mathematical parts, has given me enormous encouragement.

It was through the teachings of Chungliang Al Huang that I, a very much in-the-head type person, truly learned what it means to be fully embodied, both body and mind, and in a deeper sense to be free. We are indebted to him for his contribution of the cursive calligraphy of the important Chinese character Li, which is the frontispiece of the book.

I would like to give thanks to the Natural Sciences and Engineering Research Council of Canada for five decades of research support without which this book would never have been written. On the production side, we are particularly grateful to our editor Roger Astley, who did so much to promote it.

Finally I owe deep thanks to my wife Wendy McKay, who has been my go-to sage for all things to do with the complex world of LaTeX as well as an endless source of patience, discussion, and support over the years of work on this book. RVM.

Chapter 1

Introduction

1.1 What is pattern?

Pattern is the manifestation of change, it is by pattern that we know the world, and it is only by pattern that we know the world. This applies not only to human life, but to all life. Indeed pattern underlies all aspects of the physical world, at least as we can describe them in conceptual terms.

Making such a grandiose statement involves considerable justification, not the least being exactly what we mean by pattern. In some sense we all know what pattern is, and can recognize it effortlessly. But if we ask around, or even look the word up in a dictionary, we realize that it has a diverse array of potential meanings. Over the last seven hundred years, it has developed at least ten major meanings, including ideas of design, recurrence, configuration of natural or chance events, discernible coherence or interrelationship of component parts, and that which serves as a model for imitation. It is the last of these that comes closest to the etymology of the word, deriving from the fourteenth-century Middle English *patron* and ultimately from *pater* meaning father, and by implication someone to follow. However, common usage goes well beyond that. We are quite happy talking about weather patterns, dress patterns, wallpaper patterns, patterns of behavior, cyclical patterns, and flight patterns, rhythmic patterns, rhyming patterns, migration patterns, cultural patterns, and pattern recognition both natural and artificial. We talk about a familiar pattern, a disturbing pattern, and an evolving pattern. Even the difficult concept of information is often described as being pattern. In all of these senses of pattern there is something that everyone understands and everyone can recognize. Yet no simple definition seems to embrace them all.

Nonetheless, the incredible number of phenomena that we can recognize and describe as instances of pattern seems to suggest that there really is something fundamental and even universal in our appreciation of it. This goes far beyond the human sphere: it is evident that all animals recognize and react to pattern. Think of their reactions to light, fire, human presence, thunder, or seasonal changes, and think of their mating rituals, nesting habits, predators, and prey. This is not restricted to conscious recognition of pattern either. The antennae of lunar moths contain sensilla, sense organs, which can detect individual molecules which are chemical signals or pheromones. Notably these are sex cues that are dispersed in minute quantities into the air by receptive females. These molecules are, in effect, patterns that are recognized as such by the sensilla of males and then transported within through the insect's neural system to effect an appropriate physical response. It has been reported that Indian lunar moths can respond to just one or two of these molecules, and at distances measured in miles. The induced response in the moth is then to fly into the wind, searching for further pheromones that will gradually lead it to the source. As we open our minds to the possibilities, we begin to recognize pattern everywhere. But that only opens the question wider. What is it that makes pattern pattern?

This book tries to give some answers to this question.

At the outset we should state that we think that pattern is really not about things, but about process. We can get some inkling of this from the fact that pattern can hardly be separated from recognition, and recognition itself is process. Part of what pattern involves is the concept of re-cognition, literally "knowing again", and knowing again involves process, typically a flow of time and engagement with pre-existing experience of this pattern. In other words memory. The pheromone that stirs a lunar moth might by itself be considered as bare pattern, and it is well established that similarly shaped molecules can be designed to have the same effect on the insect. Looking further though, we sense that the real pattern here lies in a much larger context. The molecular shape of the pheromone is pattern only because there is a complementary shape in the receptor that can accept it, and this acceptance, along with the subsequent conditioned response to it, is process. That process itself is based on genetic memory which itself has been handed down generation to generation by process. Most pattern is open-ended like that. In reality a better word for pattern is *patterning*, and that is how we shall often refer to it.

The little male pufferfish, only some 6 cm in length, will spend a week, nonstop, making the magnificent pattern shown in Fig.1.1 in the sandy seafloor. There is something profoundly about this. A tiny creature far removed from the human evolutionary tree deliberately creates a pattern of extraordinary symmetry, some 2m in diameter, that is immediately accessible to our own human sensibilities. But of course its efforts must be seen in light of its own existence. When a male goes to all that trouble to make this apparently unproductive piece of art, one can be sure that in the background there is a female to impress. Indeed this is the case—it is there to attract a female pufferfish. If she is sufficiently impressed she will swim into the pattern, whereupon mating, release

Figure 1.1: The amazing nesting circle worked into the sandy seabed by the tiny puffer fish, Torquigener sp. (Tetraodontidae). Kawase, H., Okata, Y. & Ito, K. With the kind permission of Hiroshi Kawase [95].

of her eggs, and spawning will take place. The sand pattern, having served its purpose, is left to disappear under the ceaseless motion of the sea.

Extraordinary though it is in its own right, it is even more enlightening to think of this beautiful pattern in the sand as part of the far larger pattern which is the dynamical life cycle of this species of puffer fish. Beyond its physical form, it is also a piece of information, ready to attract a female and stimulate her into releasing her eggs within its bounds. We soon become aware of a complex of overlapping pattern systems: the neural patterning in each fish; how their mating ritual is replicated in each subsequent generation; the sea-patterning of the sand itself; the lifecycle of both fish; and the continuity of the species. We could even consider how and why the local aquatic environment supports the existence and life of this unique species of puffer fish, which has only been found in the coastal waters off southern Amami-Oshima, Japan. That such patterning can be generated from the neural systems of such a small fish and beyond that, the fact that we, members of a species seemingly remote from puffer fish, are instantly drawn to it, suggests the profound ways in which pattern and pattern recognition must be woven into the fabric of reality.

The little puffer fish is a gentle reminder that pattern is not the private domain of humankind or of a species endowed with special mental powers. It is rooted in the motions of the planets, in the genetic code of all life, in the stature of trees, the shape of molecules, the spider's web, and the ice on the winter pond. With all this in mind, it is easy to augment our list without end, for pattern can be found everywhere. Yet, with all these forms, how can we articulate what pattern itself is? How do we move beyond this general gut feeling and turn it into a concept, or perhaps a family of related concepts, through which we can speak about pattern with precision and clarity? Is that even possible?

Our initial interest in pattern began from precisely this question, for we saw the many occasions in which scientists and philosophers, puzzling about information, complexity, the workings of the brain, the natural world, or social problems, would return to our inherent sense of pattern, taking that as a given that everyone understands. In fact, while there are many books written around the highly practical and extraordinarily difficult subject of pattern recognition, one is hard pressed to find within them a definition of pattern itself. We are hard-wired to detect it, and we are so wonderfully good at doing this that it is hard to gain perspective. That's part of the problem. We are too good at it to even realize that at every moment we are in the midst of a swirling ocean of pattern.

In trying to clarify what pattern is and to locate founding principles, we have often found ourselves looking more backwards than forwards. The marvels of patterning are wonderful, and we want to explore them, but what we are looking for are the fundamental ideas that lie at the basis of all our patterning. In doing this we will find inspiration from the natural sciences, the life sciences, mathematics, and more generally from the insights of philosophy, both from the West and from the East.

1.2 Grasping the idea

In writing this book, we have come to think about pattern in terms of a few fundamental ideas — words — that we think underlie it: context, difference and sameness, change and invariance, chance and determinism, order and disorder, meaning and metaphor, reductionism and synthesis, equilibrium and evolution, repetition and randomness, emergence and return. In a single word, we have come to see pattern as expressions or manifestations of *systems*: the nature of pattern is *systemic*.

We have begun to understand how much our patterning not only allows us to "see", but how much it actually *determines* what we see, simultaneously opening horizons while hiding others from view. The more we appreciate the ways in which pattern informs our conceptions of reality at all levels, the more we realize the freedoms and limitations that the underlying assumptions of subject/object, self/other, man/beast relationships so often bring to it.

There have always been two broad approaches to pattern. The first is reductive, seeking to account for pattern by looking minutely at the parts out of which it arises: know the parts and you will know the whole. The second is integrative, tending to avoid detailed study of parts in isolation and caring more for over-all function, movement, and inter-relationship. These two approaches are often seen as antagonists, especially in the field of medicine where traditional medicine with its extreme specialization is held in contrast to integrative medicine which tries to see sickness in terms of the whole being. Traditionally Western thought has been seen to favor the first route, largely assuming that the integrative part will follow from the details. Eastern cultures, notably those derived from Chinese thought, have favored the second, based on ideas of emergence from and return to an encompassing whole. There is still a tendency to see the two approaches as being exclusive to each other.

In actuality we need both. The parts can only be understood in terms of the patterning of the whole, and the whole cannot be understood without taking into account the patterning of the parts. The rise of Western science has been dominated by reductionism, and to incredible effect. Yet, over the course of our own lives, we have seen how modern science is increasingly aware of this two-way road. Disciplines that for ages have been separated, jealously guarding their own territory and even mutually antagonistic, are now seen to deeply inform each other. Consider climate change and ecology, or the vast neural hierarchical complexity of the brain and its profound intimacy with the physical world. The reductionist/integrative dichotomy is less and less helpful. The deeper aim of this book is to show just how integrative the idea of patterning is, and how intimately connected to planet Earth and the encompassing Cosmos we are. We are creatures of this cosmos, and patterning is the conceptual basis of our understanding of it and ourselves.

1.3 Why mathematics?

One of the first issues we have to deal with is that in any exploration of pattern or patterning we are inevitably going to use our own innate and learned systems of patterning in doing so.[1] To free ourselves as much as possible from dragging our preconceived ideas and prejudices into the process requires a large element of abstraction. Whenever there is a collection of things or events in which we seek or find a commonality, that commonality will be an abstraction, of which the individual things or events are then representations or examples. Man, dog, animal, transportation, house, verb, word, emotion, these and thousands of other words that we know, are abstractions, each of which can be instantiated in many ways. Inevitably though, these abstractions, though easy to exemplify through particular examples, are hard to define.

One of our teachers, the famous twentieth-century geometer H. S. M. Coxeter, used to talk about a simple game he called *vish*. This consists of looking up a word in a dictionary, and then following the trail: that word is defined in the dictionary by using words, and those words are defined by other words, and so it goes. Inevitably, of course, we end up in a *vicious circle* (hence the name "vish"), finding the very word we were looking up being ultimately used to define itself. We try to define words with words, and end up defining a word in terms of itself. When we look up the word "dog" we are told that it is a type of "wolf". Looking up "wolf" tells us that it is a "canid". Looking up "canid" gives us dogs and wolves (plus jackals and foxes). And that's as far as we are going to get with the word "dog". The point of this is, of course, that we cannot define everything. In the sciences, we have to start somewhere, and these are primitive concepts that we don't try to define.

One of the more suggestive definitions of mathematics is that *mathematics is the science of patterns* [159]. There is a good deal of merit in this definition, but it does rely upon us knowing what patterns are! This book puts things the other way around: how can we use mathematics to give meaning to the elusive word "pattern". We seem to be right in the middle of the game of vish. Following *The Princeton Companion to Mathematics* [67], we choose not to try and define mathematics, but are content to allow the book itself to answer that question. Ultimately we may well conclude that pattern is something that lies beyond mathematics, but we will certainly see that it gives us genuine ways to elucidate pattern and amazing tools to explore it.

Pattern as an idea is an abstraction, and it is a particularly vicious one from the point of view of definition. It quickly becomes clear that anything we can say, words or pictures, or anything involving the physical senses is ultimately itself an expression of pattern. Whatever the basic principles that underlie patterning may be, they have to be principles that can appear in the countless different forms that pattern can assume. This is where mathematics comes in, largely because it is so context free. Think for instance of the ordinary numbers $1, 2, 3, 4, 5, \ldots$. These are abstractions, free from any particular physical context, but capable of being exemplified in thousands of ways.

The fact that all of us use numbers in all sorts of ways, and so effortlessly, is a good indication of how good we are at abstractions. In fact language itself is immensely abstract, and the letters that you are reading right now are even more abstract. Mathematics is often considered as difficult, but in some ways it is no different from any other form of language. It too has to deal with definitions and cope with the problem of vicious circles. It does so by assuming a few "primitives"—concepts for which it makes no attempt at definition. The primitives are simply assumed to be understood or accepted as givens. Most often, and in particular in our case, they have strong metaphorical associations to things we know very well, so we feel comfortable with them. After that there follow the axioms, or we might say principles, that spell out assumed *relationships* between the primitives. The mathematical theory then arises out of this in the form of accumulative outcomes of these concepts and relations, and new definitions and concepts derived from them.

The most famous example of an axiomatic system in all its grandeur is Euclid's *Elements* (c. 300 BCE). This is one of the great achievements of the ancient Greeks, and by extension of the Western world. In the *Elements*, points and lines are primitives. Based on the primitives are the axioms: for instance, there is the axiom that for any two distinct points there is a line on which both of them lie. Beyond that there are definitions, for instance isosceles (literally equal legs) triangles, and after that theorems like " all isosceles triangles have two equal angles", and later the famous theorem that "the interior angles of all triangles add up to two right-angles" (180°).

The mathematics of Greek geometry, and much of mathematics in general, is based on metaphor. The points, lines, circles, of Euclid's *Elements*, and their mutual relationships defined by axioms, are based on what we already have deep experience with. But now they are made into abstractions, and it is the metaphor (for example, that we call the mathematical abstraction a "line" thereby bringing to it our long-standing familiarity with lines) that provides us with the meaning and intuition to understand it. What is sometimes difficult, and we will become increasingly aware of this as we go on, is that abstracting an idea and using a familiar name in a new abstract sense can bring about unexpected confusions. We will see many words, used in new senses—events, measures, partitions, information, randomness, maps, operators, and so on—that derive from common ideas but have more specific meanings in mathematical and scientific contexts. Using metaphor to extrapolate meaning out of these special words has to be done with awareness and care.

A famous example of this problem actually occurs with Euclid's axioms. As stated by Euclid it is not possible to prove that if a line cuts across one side of a triangle and enters into the triangle then it must exit by cutting across one of the other two sides. The idea of betweenness, as in the points *between* two points on a line, is so obvious from experience that it was never included in the axioms. This lacuna is filled in with Pasch's axiom, see Fig.1.2. David Hilbert, who appears several times in other situations in the book, was one of several people who presented complete lists of axioms that characterize our intuitive picture of Euclidean geometry.

What mathematics does is to produce conclusions out of its definitions and axioms by using the rules of inference (logical deduction). The results are often impressive, revealing structure implicit in the definitions and axioms that we might not otherwise have realized. Pythagoras' theorem is a famous example: in any right-angled triangle, the square on the hypothenuse (the longest side) is in area equal to the sum of the areas of the squares on the other two sides. This and other such theorems are truths about this system, consequences of logical reasoning from the assumed axioms, yet far from intuition.

Often these conclusions can be brought back into the physical world by means of the metaphor, now used in reverse. Returned to the physical world, Pythagoras theorem

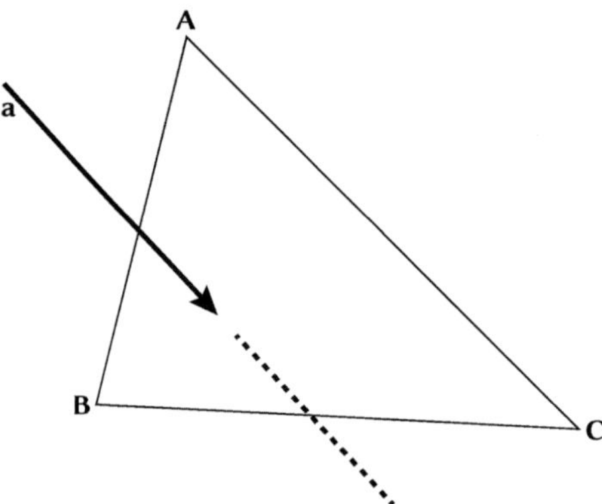

Figure 1.2: Pasch's axiom (1882): *let A, B, C be three points that do not lie on a line and let a be a line in the plane ABC which does not meet any of the points A, B, C. If the line a passes through a point of the segment AB, it also passes through a point of the segment AC, or through a point of segment BC*, Moritz Pasch (1843-1930).

becomes an important tool for determining distance. For if h, x, y are the lengths of the hypothenuse, and the two sides, then $h^2 = x^2 + y^2$ gives us a way to find h given x and y. Returning again to the mathematical world, this becomes a foundational tool in coordinate geometry for determining distance. It is used endlessly in the monitor displays of our computing devices, which are laid out in rectangular grids.

This back and forth transfer of concept from experience to abstraction is why mathematics is so extraordinarily successful in science. Rafael Nuñez, a scholar who with George Lakoff has thought a lot about this, writes:

> The effectiveness of mathematics in the world is a tribute to evolution and to culture. Evolution has shaped our bodies and brains so that we have inherited neural capacities for the basics of number and for primitive spatial relations. Culture made it possible for millions of astute observers of nature, through millennia of trial and error, to develop and pass on more and more sophisticated mathematical tools—tools shaped to describe what they have observed. There is no mystery about the effectiveness of mathematics for characterizing the world as we experience it: That effectiveness results from a combination of mathematical knowledge and connectedness to the world. The connection between mathematical ideas and the world as human beings experience it occurs within human minds [122].

Our development of pattern and patterning in this book works in exactly the same way. Our *undefined premises* are essentially just two:

- simple set theory (the idea of sets and elements that make up sets);
- the normal ideas that we have of two- and three-dimensional space.[2]

Admittedly, the latter is not especially primitive, but it is very intuitive, and going back any further would not enlighten us any further about the ideas of pattern that we wish to develop.

1.4 Process

We have already suggested that pattern is more about process than it is about things. Process already seems to imply some sort of *system* in which change takes place. It's not about things or even things changing in time, but about entities which are perceived and make sense only when seen in relation to other entities, and only within a dynamical context that unifies them all. The idea that seeing the world from a systems point of view might give us new insights into meaning (making sense of the world) seems natural enough, but it is hard to take it from this vague statement into something truly convincing. Often we tend to ascribe our thoughts and emotions to transcendental realms of mind or soul, beyond the physical world. Western philosophy has a long tradition of doing just this. But it does not necessarily have to be so. A suggestion of what a systems point of view might signify can be seen through a little vignette.

The Tang Dynasty (618–907) is widely considered to be the golden age of Chinese poetry. There was a long tradition of thought that the physical world and its changes are fundamental aspects of reality, and it was common for poets was to express their thoughts and emotions within the larger context of the physical world around them. Consider this example, "At the Mountain House of My Teacher, Waiting in Vain for Elder Ding" by the poet Meng Haoran (689/691–740):

> The sun angles below the western mountain range;
> valley after valley plunges abruptly into dark.
> The moon rises through pines, the night grows cold;
> I listen to the wind and the stream, full and pure.
> The woodsmen trudge homeward, wanting rest;
> birds search through cooking smoke for perches to settle.
> You promised to come and stay for a time:
> I'm alone with my qin at this vine-draped path.[3]

The first three lines give us scenes from nature. There is no mention of who is making the description. We are plunged into the landscape as if we are there ourselves. The inexorable passage of time is apparent. In the next two lines, the yearning of woodsmen and birds for a home within the vast wilderness is clear. Only in the fourth line does the poet begin to speak placing himself in the sounds of wind and the stream. Finally the human element enters: there's a promise, one that parallels the return of woodmen and birds. The word "alone" in the last line, is tied to a *qin* (a seven-stringed zither) and a vine-draped path. Since the qin is a solo instrument, it heightens the fact that his friend is not be there to hear the song. The poet's melancholy and isolation are immediate because we are already beside him. The world is seen as a physical system, imbued with meaning through the poet's relationship within it. We share his feeling at the exact moment that the poet realizes his own sadness.

The idea of systems and how they can deepen our insight into our relationships to the world of which we are a living part, is what this book is about. We treat it in terms of pattern because we think pattern is the basic element of cognition. But our approach to pattern is system based and it is this line of thought that leads to the principal mathematical concept lying behind our development of patterning : the *dynamical system*.

Dynamical systems arose initially in the study of the dynamics of a system of several planetary bodies moving around a sun under the mutual influences of gravity. In time this idea morphed into a mathematical theory that could frame and address any sort of question that involved temporal change. Later it morphed again to be a theory (or better, a collection of theories) that could address any form of change whatever, whether it be temporal, spatial, or otherwise; framed either in discrete or continuous forms; with or without aspects of chance and randomness. We will see that it is an idea that can express not only change, but also the nature of change in the forms of recurrence, equilibrium, evolution, emergence, chaos, and chance.

Although the idea of a system in which things are changing is quite natural and not in itself necessarily mathematical, the mathematical conceptualization of dynamical systems has proven to be important. As we will see, it offers extraordinary insights into what pattern is, as well as a rich vocabulary with which to express and develop those insights. This mathematics is not much about technical or computational skills, or the ability to slug through complicated looking equations and long chains of logical reasoning. It is really about concepts. And although these concepts can seem difficult at first, they are not difficult because they are complicated. They are difficult because they are unfamiliar.

Innovations in science are often inspired by questioning the conventional ways in which we have been conditioned to see things, and learning to see them in a different way. The framework of dynamical systems is like that. The forms in which we develop it here are the outcome of a great deal of very hard conceptual thinking by a large number of people over a period of several hundred years. Yet once we see them we wonder why they were so difficult to see. It is like the concept of heat. For a long time people thought that heat must be a material thing (it was called *caloric*.) This seemed like a reasonable idea, yet upon analysis the behavior of heat described in such terms was incomprehensible and there was no scientific progress in understanding it. Once it was realized that heat is nothing but the motion of atoms and molecules, the whole thing became clear.[4]

Fortunately the origins of dynamics, as Nuñez suggests, are deeply connected to our experience with the world. For sure the theory goes far beyond metaphor, but its roots are founded in something we know well from experience. The abstraction allows a certain universality, which is what we are seeking in our concept of patterning, but those abstractions arise in the first place from examples of pattern or patterning that we have already experienced.

Perhaps what is missing in the single paragraph of Nuñez that we have quoted is the fact that once the mathematical formulation has been made, we are often led to see things, find things, and understand things, in ways we would scarcely have imagined otherwise. Mathematics is full of its own insights and new ways of seeing things. It is full of its own internal forms of pattern, many of which are wonderful in their beauty and coherence. One of the secondary features of this book is to get a feeling for the beauty and insight that comes out of the coherent architecture of mathematics.

1.5 Founding ideas

Lest this begin to sound overly complicated, take a look at what we consider to be the absolute foundation of pattern or patterning, namely *distinguishing*. Distinguishing is the active side of *difference*. There is no sense of pattern if there is no distinguishing, and no distinguishing if there is no difference. And what is distinguishing really, except yes/no, on/off, yin/yang? There is nothing complicated about it.

The easiest mathematical abstraction of it is $0/1$, and the very simplest mathematical pattern system is based on the two-element set $\{0, 1\}$. Yes/no, on/off, yin/yang all come with their associated meanings: the mathematical $0/1$ is neutral, and not surprisingly, more amenable to other mathematical ideas.

Underlying this lies an often overlooked feature: *context*. Distinguishing and difference, these only make sense within some context. In our $0/1$ model, it is the two element set $\{0, 1\}$ that serves as the context. The 0 means something only because there is also the 1. In fact, the distinguishing already creates a context. Context, explicitly stated or not, is an essential part of any concept of pattern, the playing field on which the game is to be played. Think of the common frustration of having to deal with a robotic automated telephone system. If the conversation does not a fit one of its pre-defined contexts, distinguishing fails, there is no pattern recognition and information fails any longer to be information. Many jokes work on the principle of setting up a scenario that suggests one context, only to switch it in the punch line. The joke only works if the listeners are aware of both contexts.

Although the $0/1$ system may look overly simplistic, what lies at the heart of a computer are just strings of these yes/no distinctions. The brain, though of vast interconnected complexity, is essentially dependent on its neurons, which in themselves are basically capable of two relevant actions—firing or not firing, dependent on reaching a certain internal threshold potential. We know that each of our senses is ultimately linked to distinguishing—sensing difference—and in a situation where there is no detectable difference, they effectively cease to function.

Yet difference goes beyond just distinguishing, because there is also the flip side of this, which is *sameness.* We are not only good at distinguishing, we are good also at deciding two things are really to be considered as the same. Each dollar bill is different from every other one, but for most purposes we treat them as all the same. It is the same for buses, apples, chairs, dogs and cats, ... We could not function without distinction, but neither could we function if there were no attributes of sameness.

Lying within distinguishing there lies the fundamental principle of *change.* Somehow change cannot be described without difference, and difference cannot be detected save for change. Our senses are tuned to differences which they experience as change. However change is inherently dynamical in nature. The mathematics has to go beyond just listing differences and allow for differences to take place in temporal or spatial ways. It has to be able to express how situations progress or evolve under the influence of change. We will have to think about how that is conceptually realized in abstract terms. Is change in discrete steps, like the action of a computer, or is it continuous, like flowing water? Is change deterministic, with the future determined by the present, or is it subject to chance? If the latter, then what is the pattern of chance, and how do we deal with it?

As far as we know, the laws of physics, the principles which we can infer from living in the universe, are stable and pervasive both in time and space. This is the concept of invariance. There is constant change, but there is also invariance within that process. Like many things to do with pattern, we see a play of opposites that express an underlying integrity. It is not an antagonism, as Heraclitus would have it, but what Hindus might call the dance of Shiva. Order and disorder, creation and destruction, emergence and return, change and invariance revolve around each other, and all of them are aspects of the patterning process of which we are a part.

Whatever the answers, we can see that in pattern we are not dealing with things, but with process, and not only process caught in a moment of time, but process evolving through time. That's why we think in terms of *patterning.* Most patterns, perhaps we can say all, are the outcome of process. Think of the puffer fish making his circle in the sand. Nothing in this universe appears without process. That is why the book is called *The Pattern of Change.* The mathematical outcome of this will be what we call *pattern systems.* Not pattern as an isolated thing, but pattern as something that only exists within the context of a system in which it appears, or out of which it evolves.

1.6 A physical world

When we began this book we did not anticipate how much the sheer physicality of the world would impact our story. Perhaps the wholeness and completeness of things would be a better way to say it. As the book took shape, it was this insistent patterning of the world itself that kept arising. There is a leitmotif that runs through the book that revolves around atoms, these shy invisible units of reality that reveal themselves to us in such subtle patterns—Brownian motion, crystal structures with their rigorous symmetries and the simple ratios of their constituent elements, the discreteness of diffraction patterns, the puzzling discreteness of energies of photons emitted from substances, the exquisite sensitivities of our eyes to light, and ultimately in the amazing pattern of the periodic table.

At the biological end of our understanding we have the corresponding "atoms" of the mind, the neurons whose intricate networking and dynamical action is the ultimate source of all our mental activities, and all our knowledge of pattern. We may find it hard to define mind itself, but we do know that it is a manifestation of the very physical brain. Whatever else we are, we are creatures of this universe and all its physicality, and like every other life form, we are exquisitely adapted to its physical nature.

In the end we doubt that we will ever be able to know anything about the world except as pattern, even if we don't immediately recognize it as such. And it is a consistent theme that our sense of pattern is primarily founded in the physical world. We have come to the conclusion that it is by patterning that we, and all other creatures too, know and function in the world, and that this patterning can always be modeled into the context of pattern systems.

However, there is a cautionary note here. The mere fact that we make a mathematical model and use impeccable logic to draw conclusions from it, doesn't make it true! Saying that patterning is the way in which we see the world does not mean that any particular models of pattern are genuine reflections of reality. Science is riddled with theories that were useful in their time but have been overthrown later because their foundations were eventually seen to be incorrect. Ptolemy's famous book on astronomy, *The Almagest*, was the standard for all Western astronomy from the second to the sixteenth century. It is a beautiful mathematical work, internally consistent, and justly famous. But it is based on the false assumptions that the Earth is the center of the Universe, that the Sun and planets all revolve around it, confined to various concentric spheres, and that all their basic movements are circular around a fixed center or circular around a moving center that was itself in circular motion (epicycles). It was founded on a preconceived idea that the heavens must by nature be "perfect", and perfect in very particular ways. It was fundamentally flawed, but still it successfully served the needs of astronomers, astrologers, and navigators for centuries because it could make reasonably accurate predictions.

Many of our treasured models may be wrong. Even the best physics of today has large and mysterious gaps, and as yet there are no successful models that have unified all the known forces of the Universe. Some of the models we do have are stunningly accurate and predictive within their own domains, but may ultimately be replaced as

Figure 1.3: The famous kare-sansui (dry landscape) garden of Ryonan-ji in Kyoto. Bgabel, CC-SA 2.5 DEED, en.wikipedia.org/wiki/Ryoan-ji .

our understanding of the nature of things deepens. We depend on mental models to understand reality, but the fact is that we will never reduce reality to mental models, mathematical or otherwise.

There is an additional feature too, about our patterning of reality. No matter how we pattern the world, there will always be an incompleteness about it. There is always some conditional aspect since it always stands in a relationship with we who have framed it. A subtle lesson in this unavoidable relativeness is offered by the famous Ryoan-ji (Temple of the Dragon at Peace) Zen temple in Kyoto. This dry landscape garden is an arrangement of fifteen stones set in an extensive rectangle of meticulously raked gravel. The garden is viewed from a veranda. The stones are arranged so that the entire composition cannot be seen at once: only fourteen of the fifteen stones are visible from any single position. One's perspective changes as one moves, but it is impossible to see all the stones from a single vantage. Its designer (unknown) may or may not have been trying to convey this undeniable truth about knowledge, for no guide to its meaning was ever offered. Still, the garden is a wonderful abstraction, and it arose in a culture that was deeply sensitive to the relativeness of all perception.

As for mathematics itself, there have always been questions as to whether it is simply an invention of the human mind, or whether it has an essence that is independent of the human mind, and even of any particular forms of physical reality. These are questions that go back to the times of ancient Greece and they still generate heated discussion today. No one will deny the profound effect of our experiences of the world in mathematics, but mathematics is remarkable for the patterns that seem to be woven into its very fabric. Even within such an apparently simple idea as the whole numbers $1, 2, 3, 4, 5, \dots$ there are patterns of astounding depth and mystery, seemingly quite independent of the physical world. It is almost impossible to see these abstract patterns as something that are simply a peculiarity of limited human minds, see Appendix 16.7.

These and similar philosophical questions inevitably arise as we attempt to express and interpret our ceaseless patterning of the world. We bring them up as we go along because they seem to call for questioning and reflection. Our idea is not so much to draw philosophical conclusions as to ask the questions that naturally arise, and to stimulate discussion beyond the limits of mathematics or science themselves.

The nature of reality and existence is ultimately a mystery and will always remain so. There may be no firm answers, but through the book we can offer a broad context within which to think about them.

Chapter 2

The patterns of heaven

> Look up to contemplate the patterns of heaven. Bend down to examine the principles of earth ...
>
> Great Appendix, *Yijing* [43, §34]

2.1 Patterns of the stars

The epigraph above is taken from the *Great Appendix* to the famous *Yijing* (of which we will have more to say later). It dates back to Confucius (551–479 BCE) but no doubt has far earlier origins. It states a basic fact about human existence: our survival depends on looking at the world and seeing patterns that surround us. Before our age of almost universal light pollution, the heavens were ablaze for everyone to see—tantalizing star arrangements, their seasonal repetitions, wandering planets, and the beautiful waxing and waning of the Moon. How could all of these forms of pattern that were spread so dramatically over the heavens fail to be taken as a source of wonder, awe, and important knowledge about both the present and the future? Puzzling out the mysteries of the heavens has in some sense been, and no doubt will continue to be, the unfolding story of science. Today, as much as ever, the heavens continue to evoke awe and wonder, though now with centuries of theory and evidence that have shaped our understanding.

When we look at the night sky, our eyes are drawn to bright stars and the geometrical patterns—*asterisms*, as they are called—that they seem to form: the Big Dipper, Orion (the Hunter), the Twins, the Pleiades, Centaurus (the centaur), and the Southern Cross. These well-known constellations are striking enough to be common across many cultures, but we know there are scores of asterisms used to chart the skies which vary greatly according to culture. Essentially they are arbitrary—in most cases, there is no real relationship between the stars involved. They appear as they do only because of the position of our sun in its local stellar neighborhood. The choice of which star clusters to group together and the figurative names given them by different cultures are testaments to both our desire to reduce the unknown to familiar patterns and to our powers of abstraction. In themselves these are not the sorts of patterns that we are concerned with, but the patterning of their grand procession and their periodic repetitions in the sky are. It is the dynamic nature of their changing positions that made them so important.

2.1.1 Settling the seasons

One of the early accounts of how people used the patterns of the skies comes from China. The *Book of Documents* (書經 Shujing) (also known as the *Classic of History* (Shangshu)) tells the legendary story of Emperor Yao (c. 2356–2255 BCE).[1] This was the time of the beginnings of organized agriculture in China and timing the procession of the seasons had become essential.

> The emperor commanded the brothers Xi and He to observe the sky carefully and to calculate the motions of the sun, moon, and stars, along with the revolving divisions of the sky. He ordered Xi Zhong to move to Bright Valley in the east, to greet the rising sun and set the calendar. During that season, the days were of medium length and the star Niao (Bird) signaled mid-spring. The people were in the fields, and the birds and the beasts mated.
>
> *Book of Documents*, Canon of Yao, 2.

Emperor Yao aimed for a level of standardization, and that required a modeling of pattern with abstract principles. He sends Xi Zhong as an observer to set the calendar and to tally the days with the primal changes that occur on earth. As the story continues the Emperor sends Xi and He to make observations to the south, the west, and to the north in line with the seasons of summer, autumn, and winter, again taking note of the changes in the feathers and coats of birds and beasts.

> Emperor Yao said to the brothers Xi and He: "A year consists of 366 [*sic*] days. By means of a leap month, we can fix the four seasons and know a complete year. This is how we will grant and manage the numerous kinds of work to be done in the empire and how we will make all our accomplishments."

Every culture in history saw stories in the constellations. What the Chinese saw as a white tiger became central to another society's story in an altogether different time and place. Those storytellers were the Dineh (Navajo), indigenous American people of southwestern United States. What the Chinese called *White Tiger* and we call the *Pleiades* the Dineh named Dilyehe, the Planter.

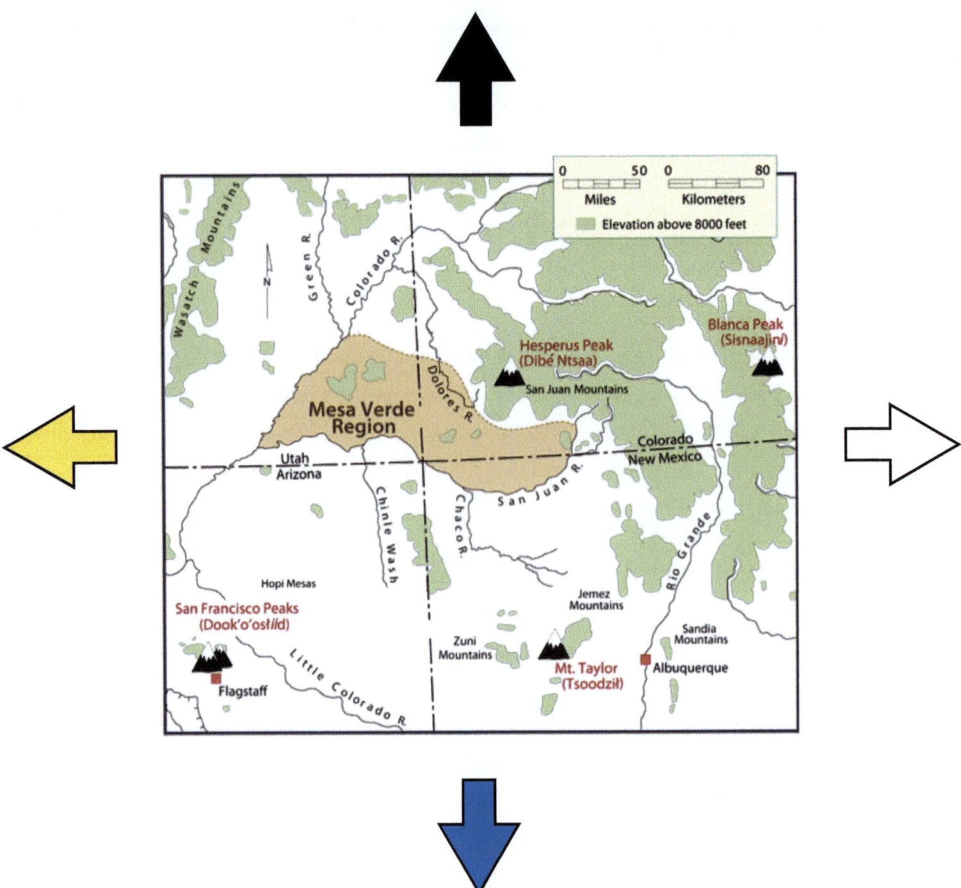

Figure 2.1: The four cardinal directions along with four associated colors are set by four mountains sacred to the Navajo nation. These are Blanca Peak in Colorado, east and white; Mount Taylor in New Mexico, south and blue; the San Francisco Peaks in Arizona, west and yellow; and Hesperus Mountain in Colorado, north and black. Map by Neal Morris, Courtesy of the Crow Canyon Archeological Center.

The *Dineh* (Diné) consider themselves descended from people, animals, and other beings who came from the First World to the current Fifth World.[2] Once they arrived, four sacred mountains were formed from soil carried from the Second World to set the boundaries of the Dineh homeland. Each one indicates a direction and a color, see Fig.2.1. The stars and constellations in the night sky are seen in relationship to these four sacred mountains with each direction associated with a cosmic process. Ha'a'aah (east) is "where the sun rises". Shadi'aah (south) is where "the sun travels with and for me". E'e'aah (west) is "where the sun sets". Nahookos (north) refers to the Nahookos Biko (North Star), Nahookos Bikaii (Big Dipper), and Nahookos Bi'aadii (Cassiopeia).

In the Navajo tradition the rising of the Pleiades constellation in combination with

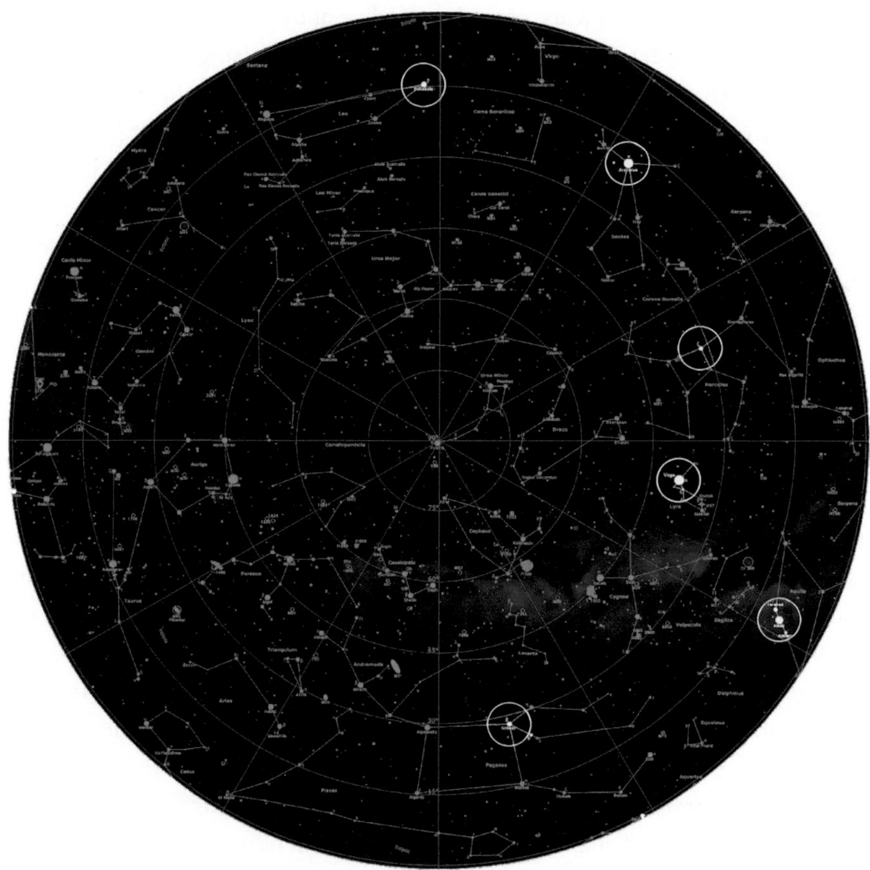

Figure 2.2: A star map of the northern hemisphere showing the six stars and their order of appearance as the head-feather of the Dineh (Navajo) constellation Ii'ne, Thunder. The helical risings of these stars mark off the seasons, starting at Denebola (at the top) in autumn and stretching across the winter season through Arcturus, Vega, and Altair to Beta Pegasus (at the bottom) in early spring.

the Moon is used to set the months. The Pleiades disappear in early May and reappear in late June or July. By correlating with other factors such as the lifecycle of plants and the runoff and waterfalls from the mountains, the Dineh use the constellation to find the preferred times for planting. Later, the first twilight appearance of the constellation in the east marks the first autumn frost. In the fall and winter, the constellation is used as a clock throughout the night.

Most remarkable is their constellation *Ii'ni*, or *Thunder*, which is a temporal sequence of stars spread out over a four-month period. In [110] it is described as stretching across time and space. Spatially the body of Ii'ni is comprised with stars from Pegasus, but the long feather rising from the head is made up of six stars including Denebola, Arcturus, Vega, Altair, and Beta Pegasus, spread across five of the Western constellations, see Fig.2.2. The temporal aspect is the helical rising (the predawn rising of stars that quickly fade from sight with the rising of the Sun) of the stars of the feather, which spreads out over some five months, October through to February when the first lightning and thunder signal the beginning of spring. Although Ii'ni can be represented figuratively as constellations are usually done—in this case stars represented as a figure with a tall tail feather—it evidently looks entirely different on a star map. It is truly a pattern that takes place in time. This sidereal sequence and its timing are clearly related both to the Earth-Sun cycle and to the actual geographic location of the Dineh people.[3]

2.1.2 Context and abstraction

The story of Emperor Yao and the Dineh constellations of Dilyehe and Ii'ne offer considerable insight into the nature of pattern. They both begin with the taking note of differences in the distribution of stars in the sky, and the conceiving of these differences as recognizable patterns. Though these asterisms are products of the mind and of our own making they allow us to orient ourselves both within the physical dimensions of our environments and within the cyclical succession of seasonal changes.

Mapping the skies was not about seeing static pictures, it was about seeing movement and change within a larger context. This is already a sophisticated thought process. These people were observing phenomena, observing regularities in spatial and temporal patterns, and then extrapolating those observations into predictions for the future. They not only saw movement and change, but also the potential to predict recurrence.

This is a point that we want to explore. Often in looking at pattern we may see a part that seems rather arbitrary, but set into a context that is more far-reaching. Wallpaper is a simple example, where a more or less arbitrary design is repeated over and over to cover a wall. It is the repetition that is the real form of patterning that underlies the production and application of wallpaper. The design itself is also pattern (in the sense that we can recognize it once we have seen it), but that is not what makes wallpaper wallpaper.

There is nothing uniquely human about our innate abilities to recognize and react to the recurrent changing conditions. These are a matter of survival for all forms of life. However, humans are universally drawn into expressing themselves in term of symbols—the naming of things, story-telling, counting, and ultimately writing—in short, abstract representations of the world. Both the ancient Chinese and the Navajo cultures named the stars, the constellations, and the cardinal directions. They told stories to explain them, they counted days and months, and drew pictures and signs. Numbers, language, and stories evolved into distinctive conceptual abstractions which could then be manipulated, examined in relation to one another, converted into other forms, and transmitted to other individuals. This was, and is, their power.

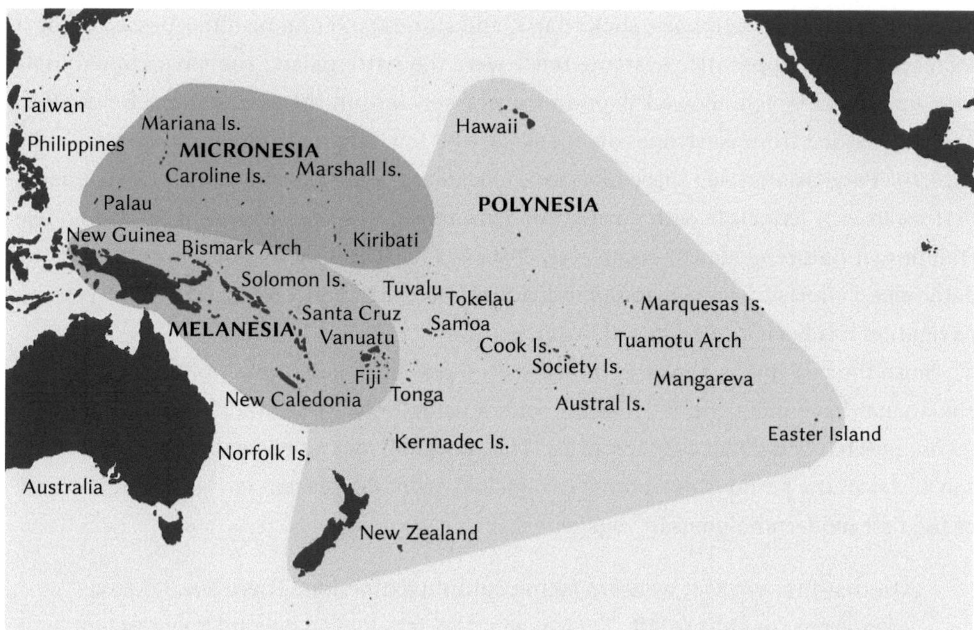

Figure 2.3: The Polynesian triangle—from Aotearoa (New Zealand) in the southwest, to the Hawai'ian Islands in the north, and to Rapa Nui (Easter Island) off the South American coast. Kahuroa, Public domain, via Wikimedia Commons, commons.wikimedia.org/wiki/File:Pacific_Culture_Areas.jpg.

What needs to be remembered in all this is that for all our natural ability and actual need to abstract the world, this is not easy. In truth much of it is incredibly hard. The history of civilizations is punctuated by great conceptual leaps that completely changed the ways in which people saw themselves and lived their lives. We will see over and over again that much of what we see as obvious was anything but obvious, and what in earlier times was obvious can now seem obscure. But we should not get fooled into thinking that cultural change means cultural advance and that what we see now is somehow the "correct" way of seeing. There is nothing unique or absolute about the patterns that we create and live by. So-called advances are often accompanied by cultural extinctions, some of them even precious and irreproducible, which we later come to regret. A striking example of this is Polynesian navigation.

2.1.3 Polynesian navigation

The incredible feats of Polynesian navigation are based on an exquisite awareness of the patterning of Nature that is intrinsically dynamical. Polynesia is a very small amount of land, in the form of about 1,000 islands, spread over a very large triangular patch of the Pacific Ocean, see Fig.2.3. Sun and stars, currents and wave patterns, the routines of birds, the virtual star compass, stick charts, and stories, and oral histories passed on by a secret master-to-apprentice system—these were the patterns and the navigational tools they developed which allowed them to travel freely within this vast region. Believed to have descended from Austronesian speakers who left Taiwan in the interval 3000–1000 BCE, the Polynesians used these tools for thousands of years, all without any instruments that we usually associate with navigation. Nowadays, this keen appreciation and understanding of nature is almost completely lost. In fact it is only by good fortune and the determined efforts of a few inspired individuals that the almost extinct art of Polynesian navigation has been saved from oblivion.

Since the mid-1970s, the twin-hulled sailing vessel, Hōkūle'a, and its descendant, the Hikianalia, have plied the Polynesian waters using only these traditional navigational techniques handed down over centuries. The Hawaiian master navigator Nainoa Thompson writes of the perilous position of the ancient art of Polynesian navigation at the time of the first modern Polynesian voyage:

> On that first voyage, we were facing cultural extinction. There was no navigator from our culture left. The Voyaging Society looked beyond Polynesia to find a traditional navigator to guide Hōkūle'a: Mau Piailug, a navigator from a small island called Satawal, in Micronesia. He agreed to come to Hawai'i and guide Hōkūle'a to Tahiti. Without him, our voyaging would never have taken place. Mau was the only traditional navigator who was willing and able to reach beyond his culture to ours.

The star-compass gives us some insight into the ways in which the stars were used as navigational aids. The Polynesian islands are in the equatorial regions, which means that most of the heavens, north and south, are visible at some part of the year. Any star that passed overhead (the zenith) is on the celestial equator. Each star, with its own specific declination, provides more references for the navigators as it rises and sets. Viewed from the perspective of a north-south axis, each star rises in the east, travels across the sky, and sets in a corresponding position in the west. The star-compass is a mental construction of the positions of the risings and settings of particular bright stars, Fig.2.4.

In effect, the Polynesian star compass is a circle divided into quadrants and a total of thirty-two segments, called *houses.* There are sixteen names assigned to the houses, one for each opposite pairs of intervals, and each star belongs to one house. Observing the serial rising of these stars and their settings allowed information about latitude and to some extent longitude. By keeping a memory log of speeds, directions, winds, and currents, from the outset of the voyage, location and navigation was possible. In more extensive versions, navigators had names for 150 stars, and they knew where each one appeared on the horizon or over specific islands. Once a navigators became oriented, they only had to watch which way the canoe was heading. As Thompson explains, in

Figure 2.4: A photograph of a recreation of the star compass of Mau Piailug depicted with shells on sand, showing the houses of the stars and the names of the stars and star clusters that occupy them. The arising and setting of a star occurs in opposite locations across the north-south axis. (As described by the Polynesian Voyaging Society, Newportm, CC BY-SA 3.0)

principle the ideas are not so hard, but in reality it takes years of training to be a successful navigator [128].

One of the most fascinating navigational devices was the *stick chart* (mattang) used by navigators from the Marshall Islands. The charts represented major ocean swell patterns and how islands disrupted those patterns. Most stick charts were made from the midribs of coconut fronds tied together to form an open framework. Island locations were represented by shells tied to the framework, or by the intersections of two or more sticks tied together. The threads represented wave-crests and directions as they approached islands and met other similar wave-crests. During the voyages, the navigators crouched down or lay prone in the canoes to feel the pitch and roll of the swells.

As they got closer to islands, the navigators looked for birds. Land-based birds that fly out from land in the mornings and back in the evening and, depending on the species,

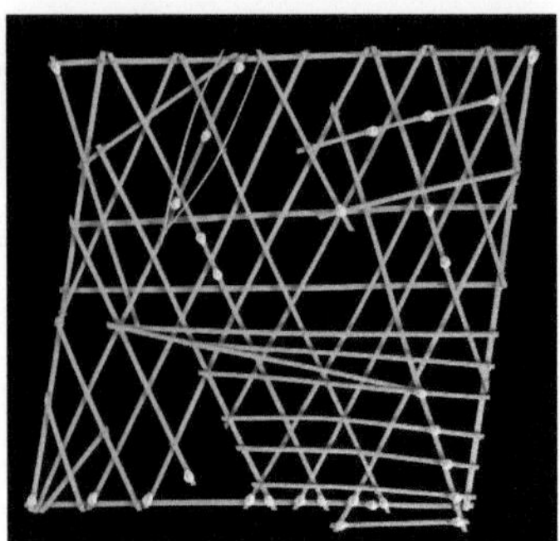

Figure 2.5: Navigational charts from the Marshall Islands made of wood, bamboo, sennit fiber, and cowrie shells. World History Commons, CC BY-NC 4.0, worldhistorycommons.org/marshall-islands-stick-charts, and Cullen328, CC BY-SA 3.0 Deed, commons.wikimedia.org/wiki/File:Micronesian_navigational_chart.jpg.

fly as far as a hundred and twenty miles out to sea, are visual clues to proximity and direction of not-yet visible islands.

There is a message in all of this. We live in a physical world, and the profundity of it is endless. We are so removed from the natural world in our cities, and our minds so often preoccupied with thoughts and emotions, that there is a tendency to forget the world from which we arise, that sustains us, and to which we return. The stories of our ancestors remind us of how intimate that connection really is.

2.2 The problem of the Moon

If we think in terms of the temporal periods of the Earth around its axis, the Moon around the Earth and the Earth around the Sun, each in itself is rather simple—simple periodic repetition. Put those three together, though, and you have a first-rate puzzle, and one that was to be a headache for calendar makers from very early on in the history of civilization. Thus we found Emperor Yáo saying: “A year consists of 366 days. By means of a leap month, we can fix the four seasons and know a complete year.” A leap month is necessary because whole numbers of lunar months match with neither whole numbers of solar years nor with whole numbers of earthly days.

All societies have needed some way in which to reckon the seasons of the year, and in this respect the most obvious candidates for recording time, the Sun and Moon, present a particularly confusing picture. The Moon is obviously a beautiful and important marker for counting the passage of days, with its easy to see phases of about seven days each. Unfortunately, though, the Moon’s 29.530589... day cycle relative to the Sun as seen from Earth is incompatible with our earthly days and year. Not every early civilization yielded to the temptations of the lunar cycle (the Egyptians had a solar calendar from ancient times), but most of them did, and thereby faced the problem of trying to rationalize the periods of the three basic cycles: the Earth's rotation around its own axis, its rotation around the Sun, and the rotation of the Moon around the Earth—days, years, and lunar months. These periods are incompatible, in the sense that none of them bears a simple numerical relationship (e.g. 3 : 2) to either of the others. Thus the difficult scheme of twelve-month years interspersed with thirteen-month years became a reality and the question of how many of each and how to intersperse them became a practical necessity.

So early on in our human attempts to comprehend the world, we were faced with the problem of understanding patterns that emerge when two or more incompatible — soon we shall say *incommensurate*—periodic phenomena must be reconciled. The idea was to look for a longer period of time that was made up of a whole number of each of the component cycles. It was out of such efforts that the famous *Metonic cycle* was discovered.

In reality, none of these periods is truly constant over the course of many millennia, so one would not expect any particular relationship between them to remain valid forever. But for the time scales involved and the accuracy of measurement of the times, such efforts were important. Meton of Athens was a mathematician, astronomer, geometer, and engineer who in 432 BCE introduced into the lunar-solar Attic calendar a system consisting of a cycle of 19 solar years.[4] The cycle consisted of 12 years with 12 lunar months in each, and 7 years with 13 lunar months in each, thereby dividing 19 solar years into 235 lunar months, or a total of 6940 days.

The Meton cycle is not perfect, but it is remarkably accurate—about one day in error every 219 years, or about seven minutes per year—and one can legitimately ask whether this kind of *almost periodicity* is an accident or something to be expected of co-occurring sets of periodic cycles, even if their periods are incommensurate. The answer is a bit of both. The Meton cycle is related to the phenomenon of almost periodicity, which, as we shall see, is guaranteed when a finite number of periodic cycles, no matter what their periods may be, are combined, §12.4.2. However, the rather low value of 19 years for such a good fit is probably an accident of good fortune, since there is no known relationship between the orbits of the Earth and Moon that would account for it.[5] Later we will look more at the problem of multiple cycles, but it is quite revealing to look at the problem that arises from trying to reconcile just two cycles with different periods.

The idea behind commensurability is a simple one. Two numerical quantities are *commensurable* (co-measurable) or *commensurate* if there is a common unit (or common measure), so that each is exactly a whole number of these units. Think of the unit as a measuring rod. In symbols, if a and b are the two numerical quantities, then they are commensurate if there is a number u, the length of the rod, so that $a = ku$ and $b = lu$ where k and l are whole numbers (integers), see Fig.2.6. If this happens, u is a unit of length with respect to which a and b are whole number of units and a and b are

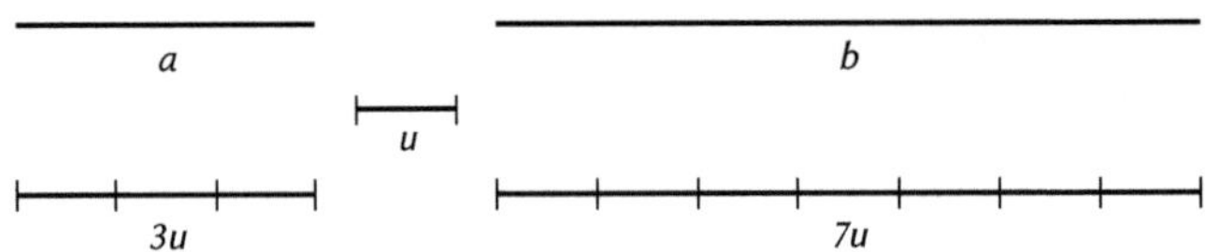

Figure 2.6: The line segments a and b are commensurable (or commensurate) because they both can be viewed as being made up of an integral (whole number) of the smaller line segment u. Thus u is a common measure for them.

commensurate. When it comes to reconciling the periods of two different cycles, what we are looking for is a common unit of time in terms of which both periods are whole numbers.

We can see that in the case of the Earth-Moon-Sun problem, Meton wanted to make the unit of time to be one day, and then to find some number of days that would be a whole number of years and a whole number of lunar months. His solution was

$$6940 \text{ days} = 235 \text{ months} = 19 \text{ years}.$$

It was not perfect, but it worked well enough to organize the calendar.

2.3 Commensurability

Commensurability plays a role in explaining the quantization that appears in pattern at the scale of individual atoms. The intuition behind this becomes evident if we reinterpret the periodicity of a wave in terms of a wave-form on a circle. If we consider some whole number of cycles of the wave, we can arrange them into a circle by joining the start to the end. Then we see a circle with some whole number of cycles of the wave that fit perfectly into it, as in Fig.2.7. Think of it as a *standing wave* on the circle. For a fixed circle, the only standing waves that it can support are those whose wave-forms will exactly fit it. Such a circle can exist precisely with wavelengths that are mutually commensurate. Simple though this idea is, it is one of the fundamental ways in which whole numbers show up in Nature. We will see exactly this phenomenon (and exactly the same figure) in the far more fundamental way in the periodic table (see §14.1).

Now, if we have two wave forms that we want to fit into the same circle there has to be a whole number of cycles of each. This is precisely the requirement that their wave-lengths should be commensurate. All of this leads to a very obvious question: are all pairs of waves mutually commensurate? Or, to put more directly, *is it true that any pair of numbers, whichever we choose, are always commensurate, that is, always have some common measure?*

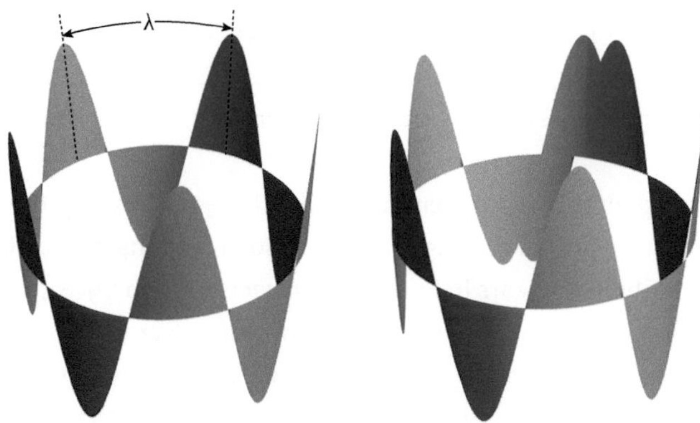

Figure 2.7: A wave moving around a circle will join up with itself perfectly if the circumference is equal to a whole number of wavelengths. It will then appear as a *stationary* or *standing* wave. Thinking of the circumference as representing one unit of time, the number of waves that fit into the circle is the frequency (number of waves per unit of time). Only those waves whose frequencies are whole numbers of this unit of time will do this. This is commensurability—the existence of a common unit of measure. On the left side there is a perfect match with 5 complete cycles and a corresponding wavelength λ which is $1/5$ of the circumference. On the right side there is a mismatch.

2.3.1 Pythagoras and incommensurability

A famous story from classical times tells of the discovery of Pythagoras (570–495 BCE) and his followers that the answer to this question is no: not all pairs of numbers are mutually commensurate. From the modern perspective this counts as one of the remarkable discoveries of the ancient Greeks. For the Pythagoreans of the time it was actually considered a serious setback. The Pythagoreans are credited with one of the earliest attempts to make a comprehensive connection between mathematics and the Cosmos, and *number* was a crucial part of their understanding. Their scientific discoveries are said to have included the Pythagorean theorem, Pythagorean tuning, the five regular solids, the theory of proportions, the sphericity of the Earth, the identity of the morning and evening stars (Venus), and the music of the spheres.[6]

For the Pythagoreans numbers meant *whole numbers.* In Greek mathematics of that time, there were no Hindu-Arabic numerals, no decimal representations of numbers, and no notation for fractions as we write them now. Nowadays we can conceptualize the entire range of numbers by identifying them with points on an infinite line, see Ch.16. However, in the days of the Pythagoreans there was no conception of the entirety of all numbers. Whole numbers were taken as the basic entities and were fundamental to their world view. Given our modern ideas of the quantization of physics in quantum mechanics (see Ch.13), their emphasis on the importance of whole numbers in the workings of Nature seems increasingly appropriate. Still, for the Pythagoreans it also suggested that any two numerical quantities, and notably any two lengths, ought to be related by whole numbers, and hence ought to be commensurate. This is not true, and the fifth century BCE Pythagorean, Hippasus of Metapont, is said to have been the person to have discovered it.

The most famous pair of incommensurable magnitudes appears as the length of the side of a square and the length of its diagonal (the ratio of diagonal to side is $\sqrt{2}$). A proof of their incommensurability appears in Euclid's *Elements* (see also ‡§16.5) and this is usually considered to be the first known example of incommensurability. This is certainly possible, but an intriguing suggestion, and one that is full of irony, is that it was the geometry of the pentagram, the very symbol associated with the Pythagorean school, that may have been the source of the discovery of incommensurability [49]. This suggestion is based on a Greek discovery that allowed one to determine the common measure of two lengths. It is called the *Euclidean algorithm*, or method of computing the greatest common divisor, though no doubt it was known long before Euclid's time.

It is interesting to go through this little piece of mathematics, because it leads to something that we might initially think is impossible to show: there are pairs of numbers that are incommensurate. The simplicity of the argument together with the surprising outcome are what makes it beautiful. It also introduces the notion of algorithms.

An *algorithm* is a prescription for performing some mathematical task. Much of arithmetic consists of algorithms, for example for addition, subtraction, and multiplication. Nowadays, in the age of computers, the word "algorithm" is commonplace and has a decidedly modern ring to it. Nonetheless, what we now call the Euclidean algorithm dates back well over two millennia to classical Greek times. More about the etymology of the word algorithm appears in §17.2.2.

Given any two lengths a and b, we wish to find a common measure u for them. Then we should have that

$$a = ku \quad \text{and} \quad b = lu \quad \text{for some } whole \text{ numbers } k \text{ and } l\,.$$

Let us suppose that b is the larger of the two lengths. Then

$$b - a = lu - ku = (l - k)u\,,$$

and since $l-k$ is also a whole number, u is also a measure for $b-a$. So u is also a common measure for a and $b - a$. We can also reverse this and see that if u is a common measure for a and $b - a$ then it is a common measure of a and $a + (b - a) = b$.

Given this, we can now switch our interest to the new pair a and $b - a$. We have gained something from this: in total we have reduced the size of our problem from $b + a$ to $a + (b - a) = b$. So we must be getting closer to the common unit u that we are trying to find.

We now repeat the process and try and find a common measure for the pair $a, b-a$, subtracting the smaller from the larger and thereby producing a still further reduced pair. So the idea is simply to keep repeating this. We stop when the two lengths become equal: this number is then is the common measure.

We wrote it out in terms of "lengths", but it works the same way with pairs of positive integers (whole numbers) a, b. The idea is illustrated by finding a common measure for the pair of integers 30, 12. The sequence goes:

$$(30,12) \to (12,30-12) = (12,18) \to (12,18-12) = (12,6) \to (12-6,6) = (6,6),$$

so the common measure is 6. This "common measure" is in fact the *greatest common divisor* of 30 and 18. Note that continuing the algorithm beyond this point would lead to $(6,6-6) = (6,0)$ and thereafter the algorithm would just continue to produce the same pair $(6,0)$ forever. The Euclidean algorithm determines the greatest common divisor of two integers.

This Euclidean algorithm is used millions of times every day in the algorithms of public-key encryption, though the numbers are much larger, with hundreds of digits.[7]

Now we can return to the question: is it true that for every pair of lengths, no matter what they are, there is a common measure? It would seem clear: just use the Euclidean algorithm for finding a common unit, and whatever comes out of it will be the common measure that we are looking for. There is a hidden trap in this. How do we know that the algorithm will actually stop? With whole numbers this is not an issue for the simple reason that whole numbers already have a common measure, namely the number 1. Each integer is a whole number of 1s. The Euclidean algorithm finds the *largest* common measure. But arbitrary lengths of geometric objects are not necessarily so nice.

The suggestion is that Pythagoreans discovered that it is not true that every pair of lengths has a common measure, and worse, this fact was staring them in the face in what is believed to be their very own emblem, the pentagram or the five-pointed star. This is the figure shown in red in Fig.2.8. Letting a be the length of the side of the surrounding pentagon and b the length of the edge of the pentagram (sometimes called the *diagonal* length of the pentagon), we want to find a common measure for these two lengths. Following the Euclidean algorithm, we replace the pair (b, a) with the pair $(a, b-a)$. But as the figure and its caption show, although the scale has been reduced, the geometry of the situation is exactly the same. In other words, we face the very same question as we began with. We can continue. The picture reduces in size, but the problem remains unchanged and we are no closer to a solution! If the diagonal and the side were actually commensurable the Euclidean algorithm would eventually find a common measure, at that point the pair of remaining lengths would be equal and the algorithm would stop. That simply isn't going to happen here. The conclusion is that there is no common measure for these two lengths: they are incommensurable. The ratio of the side length of the

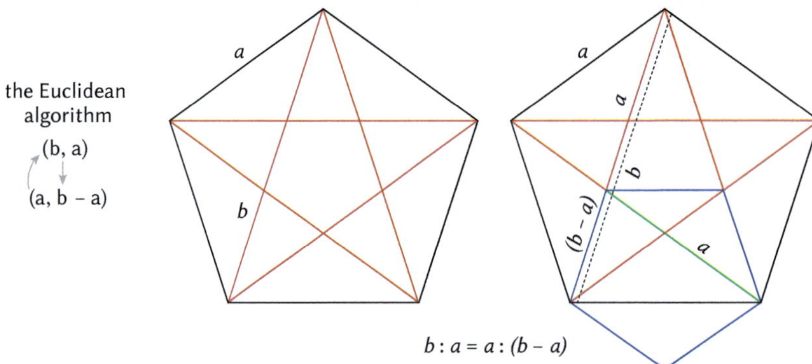

Figure 2.8: A pentagon in black and the corresponding inscribed pentagram in red. We are looking for a common measure for the two side-lengths a and b. We apply the Euclidean algorithm and pay attention to the ratio $b : a$ of the side b of the pentagram to the side a of the surrounding pentagon. Above is a reminder of how the Euclidean algorithm works: subtract the larger from the smaller (in this case $b - a$) and then repeat this subtraction process with the new pair $a, b - a$. Repetition of this process is supposed to continue until the two remaining pieces are equal to one another, or to put it another way, until their ratio is $1 : 1$. The figure shows that, geometrically, the effect of the Euclidean algorithm is to replace the original diagonal-side (b and a) of a pentagon by the diagonal-side (a and $b - a$) of a smaller pentagon (in blue). The scaling reduction is clear, but the geometry of the situation is the same: the ratio $b : a = a : (b - a)$ is unchanged, and we are left with the exactly the same problem. We can repeat, but only to get the same problem scaled down to yet a smaller pentagon/pentagram pair, with the ratio unchanged. In particular it will never become $1 : 1$. The geometrical fact used here is that the three sides that we have marked a in the right-hand figure are indeed equal, which we will leave as a little puzzle (the diagonals of the pentagon are parallel to the sides opposite to them).

pentagram to the side length of the pentagon is actually the famous *golden ratio*, often denoted by φ.[8]

2.3.2 The real line

We can put this into more modern terms. The principal connection between numbers and geometry appears in the form of putting coordinates on a line. This is explained in Fig.2.9. The line $\mathbb{R}$ coordinatized in this way is called the *coordinate line* and also the *real line*. True, the name "real line" makes no sense at this point, but we will see why it came about when we discuss the complex numbers, see §11.5. The coordinate line or real line appears many times in this book, so we have denoted it by the letter R, but written in the fancy font $\mathbb{R}$ (black-board bold), which is quite standard in the mathematical world.

Actually, the interpretation of the real line as a set of numbers raises a hard question. Do we really know what we mean by saying that all the points on the line match up with

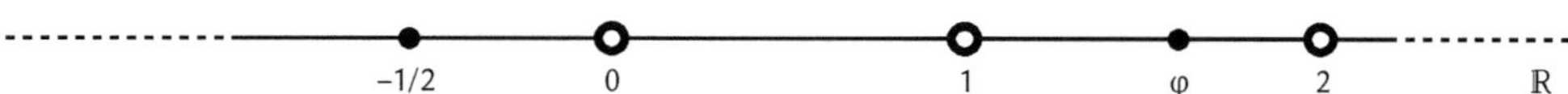

Figure 2.9: The line ℝ is coordinatized by choosing one point as the origin, labelled 0, and another (conventionally to the right) labelled 1. In terms of this unit length all other points correspond to numbers and all other numbers to points. Positive numbers are to the right of the origin with the positive integers appearing at equal steps. Negative numbers are to the left of 0. Every real number corresponds to exactly one point on the line, for instance the *golden ratio* $\varphi = 1.6180\ldots$ is also shown. Thought of in this way the line is called the *coordinate line* or the *real line*. The intuitive idea is that it is a link (effectively a metaphorical link) between the algebraic world of numbers and our geometric sense of a continuum. See also online ‡§16.5.

numbers and all numbers match up with points on the line? To see that there is an issue consider the situation of two commensurate numbers $a = ku$, $b = lu$ written as whole numbers of some unit u. Immediately we have

$$\frac{b}{a} = \frac{lu}{ku} = \frac{l}{k},$$

which is the ratio of two integers. A number which is the ratio of two integers is called a *rational number*. It is easy to see from the same equation that we can reverse this: if the ratio of two numbers is a rational number then they are commensurate. However, we know already that not all pairs of numbers are commensurate and hence there are numbers that are *irrational* (not rational). In supposing that all pairs of numbers are commensurate, the Pythagoreans had in effect presumed that all numbers are rational. But numbers like the golden ratio and $\sqrt{2}$ show that this is not true.

Where does this leave our line ℝ? What are all these irrational numbers, how can we name them, and how many of them are there? We can see some of this question arising in ancient Greek times. Euclid used the Greek word αλογος, alogos, for these numbers. The word logos had a variety of meanings in the Greek of the times—word, speech, reason, reasoning by utterance—so we can interpret alogos as meaning "unsayable". The Greek discovery was that there are unsayable numbers, or numbers for which they had no name. The translation into Latin of αλογος was *irrational*.

There is nothing irrational about irrational numbers, but the idea that these numbers don't have names (they are "unsayable") is an interesting one. Since we are so used to the expression of whole numbers using base 10 representation, we completely forget that even naming the whole numbers was a significant human invention. Think about it: the pattern of the whole numbers $1, 2, 3, 4, \ldots$ is obvious, ascending in steps of 1. However, how to give *names* to these successive integers which go on forever, is not at all obvious. Think of Roman numerals and ask how to write down $2, 649, 904$ for instance, and you see that there is a problem. In the Roman system each new power of 10 is given a different letter name, I, X, C, M. But going on this way is going to require an endless set of additional letters. Our own base 10 representation, which uses only ten symbols, is a creation of India, notably found in the seventh century in the work of the famous Indian mathematician Brahmagupta (c. 598–c. 668 CE) and his use of zero as a number.[9] The Hindu-Arabic numerals were adopted and standardized by Arabic mathematicians in Baghdad, including extending their use to fractions and introducing the decimal point. They were gradually introduced into Europe around the eleventh century, and most famously by Leonardo Pisano (more familiarly, Fibonacci (c. 1170–c. 1240–50), son of Bonacci) in his 1202 book *Liber Abaci* [54].[10]

Our system for writing down (naming) rational numbers is again so familiar that we hardly notice it. Really a fraction, say 5/17, is really nothing but saying "divide something into 17 pieces and take 5 of them". But how do we go on to name "unsayable" numbers?[11]

We do have a way to name all the points of the real line, and it is something we all learn in school and use every day, and again it is the decimal system. Thus, for example,

$$\begin{aligned} 3\frac{1}{4} &= 3.25 = 3.25000000000000000000000000000000000000000000000\ldots\ ; \\ \frac{124}{350} &= 0.354285714285714285714285714285714285714285714285714285714285714285 7\ldots\ ; \\ \varphi &= 1.6180339887498948482045868343656381177203091798057628 62\ldots\ . \end{aligned}$$

Every point on the coordinate line $\mathbb{R}$ can be expressed in decimal form, and conversely every decimal number represents a point on $\mathbb{R}$. Here we find that there is a link between periodicity and rationality: the decimal expansion of a rational number is eventually periodic, that is, it will start to repeat (note the repetition of the digits 428571 in 124/350). In fact this characterizes them: if the decimal expansion is eventually periodic then the number is rational, while the decimal expansions of *irrational* numbers *never* become periodic. The golden ratio φ, the square root of two, and π are famous irrational numbers: their decimal expansions never become just purely a matter of periodic repetition. When we think about it, these are probably just about the only three irrational numbers that most of us are familiar with, yet surely a "randomly chosen" real number is not likely to have the remarkable property that its decimal digits start to be perfectly repetitive after some point. The vast majority of real numbers must be irrational! Even so, there are infinitely many rational numbers and infinitely many irrational numbers in any finite length along the line $\mathbb{R}$, and indeed there are irrational numbers between any two rational numbers and rational numbers between any two real numbers. For more about the vastness of the real numbers in comparison to the rational numbers, see §7.2.

The point of bringing all this up is to show that there are numerous assumptions that we bring along when we try to match mathematics to reality (in this case making a perfect connection between a straight line and a number system). This is an important theme in the book: what implicit and perhaps unnoticed assumptions are we bringing to our conceptions of the world?

As a point of reference we introduce the standard notation for the sets of numbers that we have seen in this section and will see in the book :

- $\mathbb{N}$, the *natural numbers* $1, 2, 3, 4, 5, \dots$, also called the *positive integers*;
- $\mathbb{Z}_{\geq 0}$ or $\mathbb{N}_0$, the *non-negative integers* $0, 1, 2, 3, 4, 5, \dots$;
- $\mathbb{Z}$, the *integers* $\dots, -3, -2, -1, 0, 1, 2, 3, \dots$, also called *whole numbers*;;
- $\mathbb{Q}$, the *rational numbers*: all numbers that can be written in the form k/l where $k, l \in \mathbb{Z}$ (with $l \neq 0$);
- $\mathbb{R}$, the *real numbers*;;
- $\mathbb{C}$, the *complex numbers* (still to come).

The ascending inclusion relations (*set inclusion* are denoted by the symbol $\subset$)

$$\mathbb{N} \subset \mathbb{Z}_{\geq 0} = \mathbb{N}_0 \subset \mathbb{Z} \subset \mathbb{Q} \subset \mathbb{R} \subset \mathbb{C}.$$

2.4 Conclusion

From the earliest times, and in all cultures, people have searched for patterns. We are predisposed to do so, for it is the essence of survival. Sun, Moon, stars, the seasons, weather, seas, and lands, these are the dramatic forces that shape our world and that we have learned to respect and have tried to understand. In this chapter we have looked at how people have recognized and interpreted these patterns and used them as the bases of their unique civilizations. Distinction, recognition, repetitiveness, correlation, and predictability lie at the heart of this.

What is especially impressive in all of this is the way in which we see that pattern, though in an important sense omni-present and universal, is not something absolute. All cultures develop around recognition and adoption of pattern, but what is seen, and so taken as pattern, and how that information is interpreted is very varied—seasons, calendars, agriculture, navigation, language, writing, the stories, and how they are told.

One aim of this chapter has been to see the extensive nature of pattern and also how much it is inherently dynamical in nature. We tend to think of pattern in static terms, but once we think of recognition, repetition, or even the process of distinction, we realize that at its heart pattern is really dynamical. One of the main points of the book is to try and understand what that means and how it can be formalized into mathematics.

As we can see, the complexities of Nature lead rather naturally to questions that are inherently mathematical. We have taken the difficulties of predicting the motions of the Sun-Moon-Earth system as an entrance point. The issue of commensurability arises totally naturally both from the point of view of the lunar calendar and from the early attempts to understand the concept of number. We shall find repeatedly that understanding pattern or patterning inevitably leads to mathematical questions.

As often as not, the difficulties involved have little to do with complicated symbolic manipulations of equations, but rather lie in finding appropriate conceptual ideas. The transition from rational numbers, expressible as fractions, to the decimal system that enables us to express any number, whether it be an integer, a rational number, or an irrational number, was an immense conceptual step that took centuries to bring to conclusion. The fact that all of us learned this in school and it all seems quite natural is a clue to the great differences between *creating* new conceptual approaches, subsequently *learning* them once they have been established, and finally *using* them effortlessly once they are understood.

This book is full of concepts that really have taken centuries to evolve and have involved some of the greatest minds among us. Learning about these ideas and the struggles of their creators, and even then their difficulties in getting them understood and accepted by others, is a reminder of just how precious ideas can be and how hard it is to shake off our ingrained assumptions. We take so much for granted, often assuming that everything we now know is obvious and wondering how our ancestors could have been so blind. But we too are just as blind.

The road that we are on is not one that ends with some triumphant conclusion. The search for who we are and how we fit into the fabric of all things has no foreseeable end.

CHAPTER 3

The pendulum

3.1 Time and the pendulum

"All truths are easy to understand once they are discovered; the point is to discover them", so said Galileo. This might well serve as the subtitle to this chapter, for it is about discovering another way of expressing movement and change beyond what we see by experiencing it. The question is how are time and movement brought into some sort of formalism that matches our intuitive responses to them yet leads to deeper understanding?

The chapter begins with the whole issue around measurement of time and the great achievement of the mechanical pendulum clock. Using the simple motion of a pendulum as an example we show how the idea of a state space arises. It sees the back and forth swings of the pendulum now represented by its trajectories of motion through time. Later we will see how this develops into one of the foundational ideas of pattern, the idea of a dynamical system.

3.1.1 The sense of time

On November 14, 2017 the Tsukuba Express train scheduled to leave at 9:43:40 a.m. from Minami-Nagareyama left twenty seconds too early at 9:43:20 a.m. This occasioned a deep public apology from the railway company involved. Of course this seems like a rather amusing story that reflects more on the Japanese respect for the punctuality of their railways and our experiences with other railways than on the seriousness of the incident. However, it also reminds us of just how precise our conceptions of time have become, that we can fully expect to catch a train if we arrive at the platform ten seconds before its scheduled departure time and fully expect to miss it if we arrive ten seconds after. We have come to measure ordinary events in our lives to within seconds.

It was not always so. For the vast extent of human existence time has been framed in the cycle of days and nights, the varying phases of the moon, the slow turnings of the seasons, and the arrival of babies. It was only with the invention of sundials, slow burning candles, water clocks, and hour glasses that our perception of time started to become more acute.

Time is inextricably related to change, for that is how we come to measure it. Without change there would be no notion of time; yet change takes place in a "dimension" that we call time and it is hard to imagine change without some aspect of time. The pattern of change, the name of this book, is the way in which we personally have come to think about pattern, and not surprisingly some marker of successive changes is an important part of it.

If we think of the earliest time-keeping devices, those that we have just listed, we see that their appearance is continuous. Days, phases of the moon, sundials, candles, and water clocks, all involve continuous and smoothly flowing change, and hence suggest also smoothly flowing time. This is also the modern view of time—the clock may tick, but beneath those ticks and tocks runs continuous time. When it comes to memory, things look different. What we remember is events, and the sequential order of events, and it would seem that it is through them that we physically embody ideas of time. Our sense of it is more often in discrete rather than continuous terms. With future time it is the expectation of an expected significant event that embodies our conception, but the waiting can seem long and drawn out and more continuous. The nature of time, whether in some ultimate way it is continuous or discrete, or even what it is at all, is a profound mystery. We will have more to say about it later. However, in our study of pattern we will need both forms. This chapter derives from the history of how we keep track of continuous time, though ironically we will see that the great hurdle was how to discretize it.

With the rise of more complex societies, there was the need to divide the day and night into useful intervals, hours. Initially "hours" did not refer to equal intervals of time, as it does most often now. The etymology of the word "hour" goes back to the Greek "hora" which was used to indicate any limited time within a day, or even within a season, or year, and we still use expressions like "his hour has come" in this way in which the word is linked to an event (a change!). Already in the times of the Roman Empire, bells were rung in the forum to mark off the important periods of the day: prime, terce, sext, nones, ... , and these were taken into both Jewish and Christian religious practice. What is particularly interesting about the use of bells to mark time is that bells discretize time, they mark a sequence of discrete events.

This discretization became even more pronounced when the first mechanical clocks were created, for they did not have dials as we might expect, but rather chimed with bells at the appropriate hours. In fact the word "clock" is in the family that includes the French "cloche", the German "Glocke", the medieval Latin "cloche", all of which referred to bells. Ordinary people would not have been able to read clock dials, and in any case, in those times when clocks were rare, the town clock that sounded bells was far more useful than one that needed to be seen.

3.1.2 The challenge of making a mechanical clock

In his commentary in 1271 on a book called the *De Sphera Mundi* by Johannes Sacrobosco, the English astronomer Robertus Anglicus speaks of the great challenge in creating a satisfactory clock: in effect the challenge of making a wheel that would rotate exactly once with every revolution of the Earth. The way in which this problem came to be solved involved two fundamental inventions. The first was the design of a suitable escapement (the verge escapement) which could replace the continuous turning of the wheel with discrete incremental steps—in other words, a mechanism for changing the continuous flow of time into discrete steps, the step-wise rotation of a toothed wheel. The second, and harder, step was to find a really good way to drive the system. The available methods were to use the downward flow of water or the descending of weights, in other words, simple gravity. But neither of these is inherently based on a natural source with a regular beat or oscillation, and neither on its own could produce accurate time-keeping. That had to wait for another four centuries.

The earliest escapement seems to have been implemented by the Greek engineer Philo of Byzantium (third century BCE) in the form of an automatic filling device in a wash stand, and he mentions its use in water clocks. The Chinese astronomer Zhang Sixun (Chang Ssu-Hsün) in 976 CE built an escapement into a magnificent machine with the power being supplied by water pouring into buckets hung around a large wheel. When a bucket was full enough it forced the wheel to turn, with the escapement then halting the turning so that the next bucket now occupied the position of the former. In fact this, as with other Chinese devices described in literature of the time, was not primarily a clock, but rather a sophisticated astronomical instrument that modeled the motions of the Sun and Moon and the planets around the ecliptic, as well as dividing the day into intervals. Fig.3.1.2 shows an even more impressive astronomical clock tower from about a century later.

These devices solved half of the problem, but in effect they were water clocks, and no more accurate than any other water clocks. The first mechanical clocks in Europe used what is called a *foliot*, a sort of primitive balance wheel, consisting of a horizontal rod that swung back and forth with a period of several seconds, driven by the descending of a weight by gravity. By the late thirteenth century, clocks using a foliot started to appear in the town squares and churches of Europe. These continued to be refined, but the foliot was still not a true source of a regular beat. It was the discovery of a natural harmonic oscillator, the pendulum, by Galileo and its eventual implementation into clocks that was

Figure 3.1: An image of the huge astronomical clock constructed by the polymath scientist and statesman Su Song (1020–1101). His was not the earliest astronomical clock tower in China, but it was an early example having an escape mechanism and it did have the oldest known example of a power-transmitting chain drive. Wikipedia Public Domain.

to revolutionize time-keeping, increasing the accuracy from minutes a day to seconds a day.

It is said that Galileo Galilei discovered the regularity of the beat of a pendulum while attending mass in the cathedral of Pisa (around 1588) when, with the windows wide open, he observed the regularity of the swinging of the hanging lanterns, and measured their regularity by comparing them with his own pulse. In the early 1600s Galileo experimented with pendulums and arrived at three conclusions about the period of the swing of a pendulum:

- the period is independent of the weight of the bob at the end of the pendulum;
- the period is proportional to the square root of the length of the pendulum;
- the period is independent of the size (amplitude) of the swing.

It is amazing to think that it took until 1600 for anyone to take note of these things. At the same time, it shows just how difficult the idea was of thinking in this sort of way, isolating physical things from their environment, expressing relationships in precise mathematical terms, and deliberately testing hypotheses with experimentation. It is not just a question of observation, but a question of deliberate reduction and experimentation with the objective of testing hypotheses. This is how we have come to understand the meaning of science in our times, and Galileo is often regarded as the "first scientist" in this sense. Implicit in this there is also a question of faith—a faith in the ultimate rationality of the natural world that suggests it obeys certain rules, that these rules can be understood by the human mind, and that the appropriate language for them is mathematics.

Although science today is inexorably linked to consumerism, big business, and politics, these fundamental approaches to science remain today, just as they were in Galileo's time. But we shall also see that for all mankind's ability to recognize pattern, the true

Figure 3.2: The anchor escapement, used in pendulum clocks. Chetvorno, CC0, via Wikimedia Commons, en.wikipedia.org/wiki/Anchor_escapement.

advances in science do not come easily. It will be a recurrent theme that the great mathematical ideas and the great concepts of science are far from obvious and arise from exceptional minds. And we shall find that even when they do arise, new ideas are not always welcomed with enthusiasm. Objectivity is not among humanity's strongest traits. If new ideas seem contrary to the leading philosophical and conceptual ideas of the time they will be bitterly opposed. The story of Galileo and his persecution by the classical scholars and the Inquisition of the Roman Catholic Church is famous [61].

It is the first of Galileo's observations, *isochronism*, that was to be the critical one for clocks—even though, as we shall see, it is not exactly true. Strangely, though, it was not until much later in his life that Galileo actually drew up the plans for a pendulum driven clock, and tried to have one built. It was never completed. Instead it wasn't until 1656 that Christiaan Huygens, mathematician, astronomer, physicist, overcame the technical issues, including introducing the anchor escapement (Fig.3.2) and a clever suspension of the pendulum, and produced the first successful pendulum-driven clocks. The result was an amazing increase in time accuracy from around fifteen minutes a day to fifteen seconds a day.[1]

The importance of the pendulum from the perspective of clocks, and from our perspective of understanding basic dynamical principles, is that a pendulum is a natural oscillator. It swings back and forth in a perfectly regular way and will do so indefinitely as long as energy is supplied to restore energy lost by friction. Figure 3.3 is taken from Huygens' book describing his clock in detail.

Galileo is considered to be one of the fathers of modern science, perhaps the first scientist in the sense that we have come to view science, and the introduction of the pendulum was one of the great steps forward in time keeping. Galileo and his pendulum seems like an appropriate place to start looking more deeply at the pattern of change.

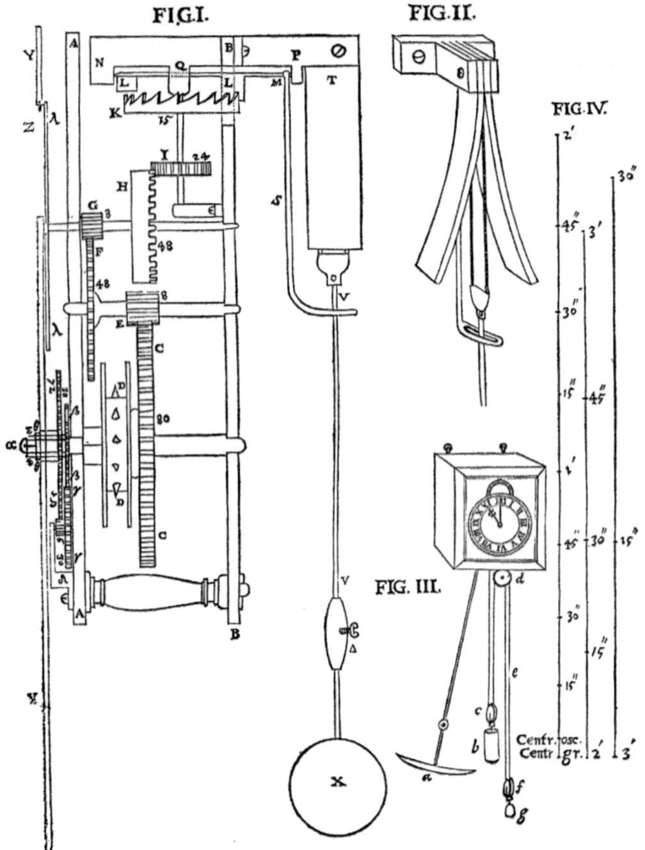

Figure 3.3: Huygens' pendulum-driven clock. This figure is taken from his book describing the clock in detail [87]. Huygens realized that a pendulum is not truly isochronous when the length of the arc of pendulum's swing varies. The figure shows (at FIG.II) the flexible mount and surrounding cycloid shape that corrected for this. The Print Collector/Alamy Stock Photo.

3.2 The pendulum and its state space

A pendulum consists of a stick or rod, usually assumed to have negligible mass, which is suspended by a pivot at one end and has a bob (or weight) at the other end. The pendulum is free to swing back and forth in a *plane* around its pivot. This is the familiar form of grandfather clocks.[2]

Allowed to swing freely, the pendulum automatically produces a regular repeating cycle. If we ignore friction and air resistance, there is nothing to stop the cycle once it has begun, and it will continue indefinitely. The underlying physics behind the motion of a pendulum is extremely simple: it is a continuous interchange between potential energy and kinetic energy. When an object is lifted against gravity, energy is required to do so.

The object is then said to have potential energy: it has the potential to do work, and work is the dispensation of energy. If the object is subsequently released, it will fall, converting that potential energy into movement, or kinetic energy.

As a pendulum swings downward it is converting potential energy into increasing angular velocity. At the bottom of the swing, the potential energy is used up but the bob has its maximum kinetic energy in the form of angular rotation and this carries the pendulum upwards, passing its natural rest point and converting this kinetic energy back into potential energy. Once the kinetic energy is all converted into potential energy, the process reverses and the pendulum begins its motion in the reverse direction. This conversion process is given by an exact mathematical expression that depends on the mass of the bob, the length of the pendulum stick, and gravity. As long as no energy is lost in friction at the pivot or from air resistance against the rod and bob, the principle of conservation of energy guarantees that the cycle of interchange of potential and kinetic energy will continue indefinitely and completely regularly. This then is a natural oscillator, and can be used to measure out equal intervals of time.

Simple though it is, the pendulum offers an excellent example of what will become one of the central ideas of this book, namely that of dynamical patterning. Informally and intuitively it certainly makes sense to say that a swinging pendulum is some sort of dynamical system. But to go deeper into the dynamics of pattern and see it in some sort of unified way requires a very different conceptual approach. What we are about to do now is to introduce a set of ideas that will put us on the path to understanding modern dynamical systems theory.

Returning to our pendulum, the first thing to note is that for it to do anything at all requires a context, namely the presence of gravity. Under the influence of gravity what is important in fully describing its motion is not a static visual image of the pendulum, but rather a description of its physical state at any moment. Its physical state is summed up in three things:

- the angle that the pendulum makes with respect to the vertical,
- the direction in which it is moving,
- the speed at which it is moving.

The angle zero is taken to be the position of the bob when it is straight down. The angles to the right, counterclockwise, are taken to be positive, those to the left (clockwise) are negative.

The speed is the rate at which the bob is traversing the circular arc on which it is constrained to travel, and since that depends only on how fast the angle is changing, we can reduce everything to the angle and how fast it is changing. We can incorporate the direction of motion by attaching a sign (+ or −) to the rate of change of the angle, and this is done in the same way: motion is taken to be positive if it is in the counter-clockwise

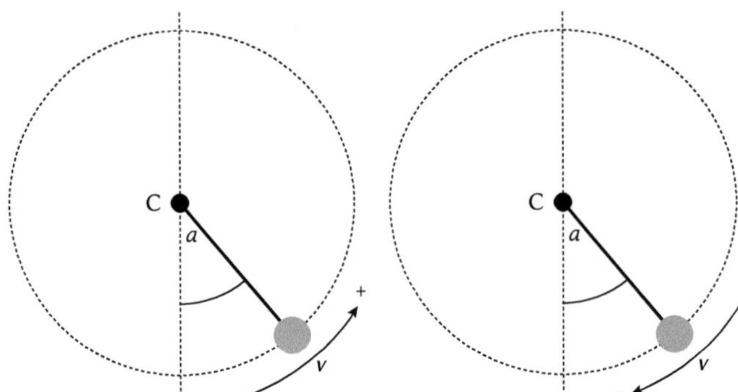

Figure 3.4: A pendulum is free to move around its pivot at the point C. Its position at any instant is described by the angle a that it makes to the descending dotted vertical line. Straight down corresponds to $a = 0$. The other variable involved is the rate at which the bob is moving along its circular path, its velocity v, with the counterclockwise direction being considered positive and clockwise negative. The pair of numbers (a, v) at any given moment determine the subsequent behavior of the pendulum for all subsequent time. The state is continually changing as the pendulum moves under the force of gravity.

direction, and negative if it is clockwise, see Fig.3.4. It is usual then to refer to this signed rate of change as *angular velocity*. Thus the physical state of the pendulum is reduced to two numbers: a, the angle of the pendulum's rod, and v, its angular velocity. In fact we will refer to this pair of numbers (a, v) as the *state of the pendulum*. Think of pairs like $(20, 35)$ referring to the bob being at angle 20° and the angular velocity as being 35 degrees per second counter-clockwise, $(0, 0)$ as being straight down and stationary, and $(20, -47)$ as angle 20° and the angular velocity as being 47 degrees per second clockwise.

We are familiar enough with pendulums (ideal ones with no friction) to know that with these two numbers in hand at *just one instant* in time, we will know what the behavior of the pendulum will be at all subsequent times. If we take the bob to a certain angle of the pendulum and impart some initial angular motion to it, then we can simply sit back and watch what it does. Of course the angle and the velocity are continually changing, and every so often the direction changes too. To understand the dynamics we need to understand how these two numbers, the angle and the angular velocity, are related and how they change in time. So the emphasis is switching from the simple picture of a pendulum swinging back and forth to a description of how the *state* of the pendulum (the pair of numbers that characterize its physical state at any particular moment) is changing in time.

3.2.1 Pendulum behaviors

We have no real need of the exact details—we all know what pendulums are like.[3] However, one thing that makes our pendulum unusual is that we allow it to swing over the top and come down the other side. This is certainly not the sort of thing that is desirable in clocks, but here we are interested in *all* the possible dynamics of our pendulum. Whether or not the pendulum swings in entire circles depends on how large the initial angular velocity is, but it is possible and we will allow for it to happen. With this in mind we see that the system can exhibit five very different behaviors:

(i) the usual type of pendulum motion, swinging back and forth (*librations*);

(ii) swinging in full circles counterclockwise;

(iii) swinging in full circles clockwise;

(iv) stationary, at rest, straight down (a stable situation);

(v) stationary, at rest, straight up (very unstable!).

This overall breakdown of what the pendulum can do is perfectly clear, but what do these scenarios look like when seen from the perspective of the states? As we have said, the state of the pendulum at any moment is described by the pair (a, v), and given that information the motion of the pendulum is (in principle) known for all future times. To understand better what all the possibilities are we look at the way the state (a, v) changes in time. To do this we set up a coordinate framework with the values of the angle a given in horizontal direction and the values of the velocity v in the vertical. With this coordinatization in hand we can graphically indicate every possible angle-velocity pair that the pendulum can possibly have, each such pair corresponding to one point in the plane. This plane itself is called the *state space* of the pendulum, and each point (a, v) in it refers to a particular possible a *state of the system*. If the pendulum is in some particular state, then as time proceeds it will continually change its state and so trace out a path or *trajectory* in the state space. There are many of these trajectories, depending on where we start. We sometimes speak of these trajectories as *evolutions of the states*. Fig.3.5 shows what some of these trajectories look like.

This new representation of the physical situation takes some getting used to. Fig.3.6 gives some visual clues as to how the correspondence between states and the actual physical motion of the pendulum works. Fig.3.7 shows more clearly how the state space separates into those states involved with librations and those involved with entire circular swings. So summing up, once we have created this state space we can trace out the path of the states that a pendulum will take, starting from any state. The idea is that no matter in what state the pendulum starts, its state will change in time and so trace out a trajectory in the state space.

3.2.2 Time and the state space

It doesn't take long to notice that time, though obviously implicit in the physics, does not itself appear explicitly in the state space. We are generally more used to graphs in

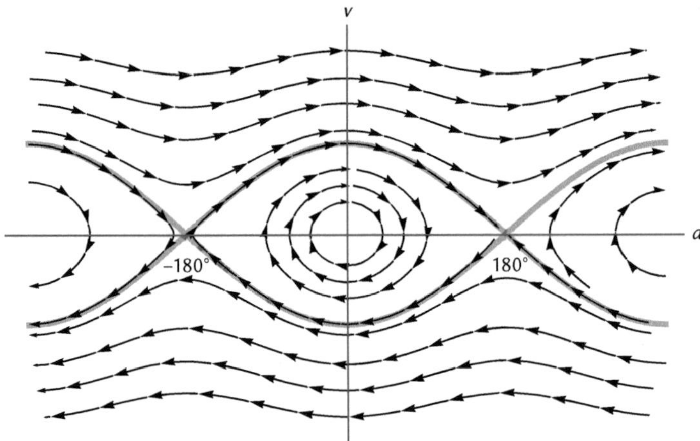

Figure 3.5: Some trajectories shown in the state space. The horizontal direction represents the angle of the pendulum, and hence its displacement from straight down. The vertical direction indicates the velocity of the pendulum—counterclockwise above the horizontal axis, and clockwise below. The arrows show the flow of angle and angular velocity in time. The inner elliptical trajectories are those of the familiar clock-like pendulum swinging back and forth. The outer ones (which correspond to higher energy levels) represent situations in which the pendulum swings in circles around its center. In the case of complete swings there arises the question of whether the angle, which is increasing or decreasing by 360° with each complete swing around the center, ought to be reset to the $[-180^\circ, +180^\circ]$ range after each rotation, or allowed to continue increasing or decreasing. Both versions have their benefits. Here we have allowed the angles to keep growing, and the picture repeats itself after every 360°. Fig.3.8 shows what happens if we reset instead. Computed on Mathematica, after a version of Housam Binous, Ahmed Bellagi, and Brian G. Higgins.

which time is on one axis and some physical attribute, like distance, on the other. In our state space we have all the potential states of the pendulum and the trajectories that show how states evolve as time goes by, but there is no indication of how quickly or slowly these trajectories of states are actually traversed, nor is there any indication of the periods of the various librations (back and forth swings). The situation as shown here is actually typical of the way in which dynamical systems are represented, and this is a good moment to describe how they are perceived.

No matter what state the pendulum system is in, that state determines all future states. That is to say, there is only one possible trajectory passing through that state. The reason is easy to see: if we know the velocity of the bob and also the angle that the pendulum makes to the vertical, then we know precisely what the pendulum will do in the next moment. We don't have to know how it got into that state—whether naturally from the prior swinging of the pendulum or artificially by being put into that state—once it is in that state, if it is left alone, the pendulum will swing by itself and trace out its unique trajectory of states. At each moment it will move at the speed and in the

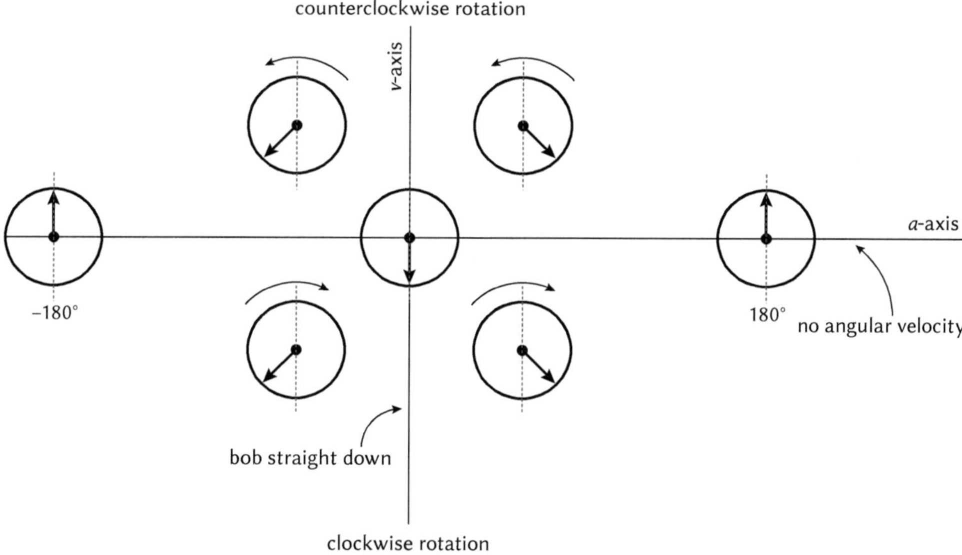

Figure 3.6: The figure shows the general relationship of the regions of the state space to the corresponding behavior of the pendulum. The horizontal axis, representing the angle of the bob, separates clockwise from counterclockwise motion. The vertical axis represents angular velocity and separates which side of the vertical the bob is in. Notice that the two points shown with $a = \pm 180^\circ$ actually represent the same state (namely vertical upwards with no angular velocity). We come back to this point later.

direction indicated by the velocity, and the angle will change accordingly.

Although we don't have to use any physics in looking at these, it is useful to realize that these trajectories must actually be equal energy contours. Since the system is assumed to be frictionless, the principle of conservation of the total energy requires that the states of the system evolve so that they have constant energy. So the trajectories must follow paths of equal energy.

Dynamical systems of this type, and in fact we will see that they are ubiquitous, are called *deterministic* since their future states are determined by their present states. It is important to understand that no particular state of the state space is any more important than any other. The state space portrays *all* the possible states in which the pendulum can be. Of course, we may be interested in our pendulum starting at some particular time in some particular state, and then studying its consequent evolution as time progresses. Since the evolution of any particular state of a deterministic system depends only on that state, not at what particular time it is in that state, we can see why time is not explicitly attached to the state space. So the correct way to use the state space is to say that this particular pendulum is in this particular state at this particular choice of time, and then let the state space trajectories tell us how the pendulum will behave from that point on.

Later, we will come to see how time or some agency of change is formally introduced into the conceptual framework of dynamical systems, and our whole conceptual approach to dynamics will acquire a comfortable familiarity. But the realization that time might be viewed as *implicit* in the *change* of the system, rather than being the direct cause of it, is notable.

What we see in Fig.3.5 is often called a *phase portrait* or what we will call a *state portrait.* The words "phase space" and "state space", and the accompanying phase or state portraits, need some clarification. First of all the word "space" should not be thought of in some technical sense of three-dimensional space or space-time. The state space is simply *the set of all possible states of the system* that we are studying. It is just a set of states, and in principle it has no other meaning than this. In particular cases, see for instance in §3.3.1, the set of all states may have a natural interpretation that gives it additional structure, but for the moment the state space is just a set whose elements are called states.

The words "state space" and "phase space" are used somewhat interchangeably in the scientific world. We have chosen to call them state spaces since the word "phase" has connotations that we will later use in the context of waves and phase angles. The word "state" is more in tune with our common ways of speaking of the state of the weather, the state of one's health, or the state of the union.

3.3 The state space in more detail

Let us look more deeply into what information lies in the state space and state portrait of the simple pendulum, now referring to Fig.3.7. Here are shown a number of trajectories in the form of sinuous lines with arrows indicating the direction of flow. No starting positions are indicated: wherever one starts on one of these trajectories, the state will follow the trajectory in the direction that the arrow indicates. In reality, there is one trajectory through every point (state). Although only a few trajectories are shown, one can clearly see their patterns. There are elliptical trajectories on the inside and long wavy ones on the outer sides.

At the center, we see the black dot indicating the state $(0, 0)$, a stationary state with the bob of the pendulum straight down. This is a stable equilibrium, and in terms of energy it can be considered as a state of zero energy. As we move outwards in the phase space (which amounts to increasing the energy) we enter the regime of librations. The red elliptical trajectory is typical. Increasing the energy increases the height to which the pendulum bob will swing. Eventually we hit the boundary—the place where the pendulum is about to change from oscillating back and forth to swinging around in circles. The green trajectory is this boundary. It looks like the red trajectory we just studied, but we notice now that at the state where angle a reaches $180°$ (straight up) the angu-

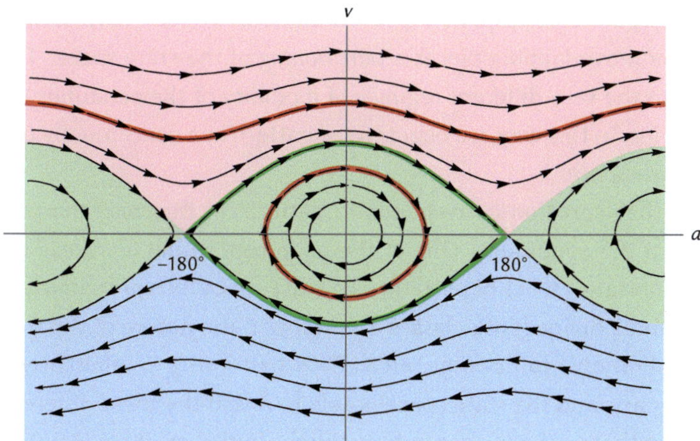

Figure 3.7: Several state space trajectories are shown. Starting anywhere on one of these and following the direction of the arrow shows how the pendulum's state will evolve. The trajectory shown in green marks the boundary between three totally different regimes. States inside the green boundary (the area in light green) belong to the modality of the pendulum in one of its oscillating states (librations) that we see in pendulum clocks. States outside the green boundary belong to the modality of the pendulum swinging in full circles, counterclockwise in the upper trajectories (the pink area), and clockwise in the lower (the blue area). This basic partitioning (pink-green-blue in the figure) corresponds to pattern, the three basic modalities of the pendulum. The overall periodicity of states that differ in angle by 360 degrees is also evident.

lar velocity is 0. In other words the pendulum just stops in the vertical position. This is another equilibrium state, but in practice it would be well-nigh impossible to realize this particular state since it is totally unstable (and, remarkably, though we can't see it, requires an infinite amount of time to achieve). But the boundary itself is significant.

A slight increase in energy and we are into different territory. Look at the next trajectory shown, also in red, just above the green one. It is a wave. As the angle a increases the angular velocity of the pendulum increases and decreases, but now it never reaches zero or goes negative. The pendulum is swinging in circles, always counterclockwise. Its angular velocity slows as the bob moves upwards and speeds up as it moves downwards, but it never changes direction. The lower wavy trajectories in the purple region represent swinging circles in the clockwise direction.

3.3.1 Pattern in the state space

The state space is the entire plane, and the trajectories that we have drawn show clearly that this state space/plane is divided into five pieces: the inner area corresponding to the usual motion of a pendulum (librations), the two outer pieces where the pendulum is swinging around in complete circles in the two possible directions, and the boundaries between these types where the pendulum moves to an unstable equilibrium. Finally there are the stationary points corresponding to no motion. We call such a division of the state space into non-overlapping pieces a *partitioning* of the state space. The partitioning corresponds to the very different regimes of modality of the pendulum—its different patterns of behavior. In this way, we begin to see pattern in a conceptually different way: a *partition of a state space.*

We see here that partition corresponds to pattern, in this case the pattern of the overall behavior of the pendulum. One of the main conceptual features of this book is that this is an important aspect of pattern. The state space interpretation makes it clear that the states can be divided into classes according to the nature of the trajectories on which they lie (librations, full swings, etc.). This patterning is inherently dynamic in nature, not stationary, and the state space is able to reveal the these differences.

The patterning that we have chosen to highlight in this example is not the only one possible. In fact there are innumerable ways in which we may decide to partition the state space, according to what we see as important. In our case we have seen that the dynamics itself points to a natural partitioning of the states into librations, full swings, etc. In a more physical presentation we might be interested in dividing the state space according to the frequency ranges of the librations, the lengths of their periods, or the physical stresses on the pivot that depend on the angular velocity. The same dynamics can be looked at in various ways, according to our interests and intentions. It has to do with what differences we care to distinguish, what differences we actually can distinguish, or importantly, even what differences we do not wish to distinguish.

The partitioning of the state space is a very important part of our approach to pattern, and it reflects a significant aspect of our cognition. The level of detail to which we care to, or can, make distinctions is crucial to how we operate in the world, and it is something that we change without conscious thought thousands of times a day. When we drive a car we do not internally pattern the street scenes in the same way as when we are window shopping. When you are looking for your keys, you are shaping your perceptions of your house and its contents very differently from when you are vacuuming the floor. Are we talking of forests or individual trees; are we looking at just automobiles, or individual brands, and different styles?

Not only do we endlessly partition the same phenomena in different ways, but in fact we have no other way of making sense of the multitudinous amount of data that our senses are offering us at every moment. Partitioning is a central aspect of patterning. Here we need only be aware of it. Later we will develop its role further.

At this point, we might take a look at the shape of the state space itself. If the pendulum swings in a complete circle, it comes back to the same *physical* state that it was in before. But in sweeping out an entire circle the angle a increases or decreases by 360°, depending on which way we go around, so the coordinates are not the same. For example, the states $(40, 10)$ and $(40 + 360, 10) = (400, 10)$ both indicate the pendulum at 40 degrees counterclockwise of downwards with an angular velocity of 10 degrees/sec in the counterclockwise direction. So too do $(40-360, 10) = (-320, 10)$, $(40+720, 10) = (760, 10)$ and so on. These are different states, at least as we have defined states, but they really do represent the same physical situation. Our state space is far larger than it need be. It contains many points that represent the very same physical state.

There is a simple way around this: keep the angles between -180° and 180°: simply discard the two parts of the plane where the a coordinates are outside this band of values, thus leaving a vertical band. This is okay, but it messes up our intuitive picture of the continuity of a pendulum swinging around in circles. If our pendulum is swinging counterclockwise in circles, then as it goes over the top, its angle will suddenly switch from $+180^\circ$ to -180° (and the other way around if it is going clockwise). This discontinuity misses out something important, and in fact there is an easy fix, see Fig.3.8. This new version of the state space is a *cylinder*, and illustrates the somewhat elusive idea of giving "shape" to the state space.

The idea of shape here is what we would call *topological* in nature and, just as we have seen it here, it can be important in revealing aspects of pattern. In our case it plainly manifests something important about the pattern modalities of a pendulum: they are cyclical (or periodic) in nature. Wherever there is periodic pattern we can expect to see a circle appear somewhere (note the word cyclical). We explore this in detail in Ch.11, where it stands as a foundational principle in the study of waves.

Which state space is correct? The cylindrical version presents the situation most accurately, but the flat version is often more convenient (for instance it is much easier to draw on paper). Cyclical phenomena are common and this same issue often arises. Since it is so easy to mentally switch from one picture to the other, both are used.

3.4 Galileo's conjectures and beyond

So what then is the fate of Galileo's three conjectures,§3.1.2? For a simple pendulum with no friction or forcing, the motion of the pendulum is genuinely periodic and furthermore the first conjecture is correct: the weight of the bob is immaterial.

The other two statements (about length and amplitude) are approximately true as long as the pendulum does not have too large an arc in its librations, that is, as long as the angle a is kept small. In learning the physics of a pendulum, most of us will have seen this assumption of small angles being made since it simplifies the mathematics immensely (the approximation is that $\sin(a) = a$, which is not especially good).[4] However, in the building of a satisfactory clock, Huygens did need to take account of the fact that the period of the pendulum depends on the length of its arc, see Fig.3.3.

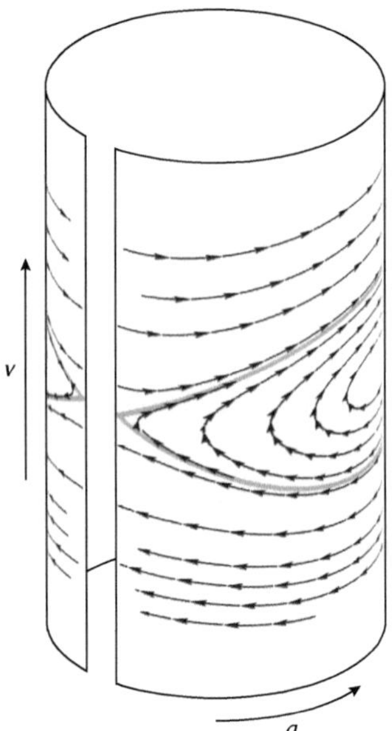

Figure 3.8: Wrapping the state space around a cylinder solves the problem of redundant states: states that are 360 degrees apart are really the same state. The cylinder is the appropriate topological shape for the state space. The pendulum's basic motion is periodic, and the cylindrical representation makes this emphatically clear. The vertical separation of trajectories from the center represents increasing energy.

3.4.1 Forcing and friction

So far our discussion comes from an ideal world with no friction. Real pendulums swing in the real world and friction, or damping, is the reality of the situation. The usual correction, which fits the general way that friction works, is to include a frictional force acting in the direction opposite to the direction of movement and proportional to the speed of that movement. The consequences of this are not too surprising. Energy is lost and the system falls into less and less energetic trajectories. Even if the pendulum starts off swinging in full circles, eventually energy loss will put an end to that modality and it will change to librations. These will get shallower and shallower, and after enough time the pendulum will come to rest, pointing downwards at its stable equilibrium state.

For real clock pendulums there has to be a compensating forcing mechanism, usually energized by weights, that nudges the pendulum very slightly at every step of the escapement and prevents it from descending into its stable equilibrium—in other words, from stopping. The general system is still just a pendulum, and we still have states and a state space just as before. The dynamics still proceeds on the basis of force and its effect on mass, except that now this is influenced not only by gravity acting on the pendulum bob, but also by a frictional force proportional to the velocity of the pendulum and a supplied rotational force that keeps it going. In addition, as we have mentioned, there is the serious technical issue of the varying arc length of the swing, something which is typical for a wind-up type forcing mechanism. With all this the general outcome is the regular tick-tock pendulum that we expect.

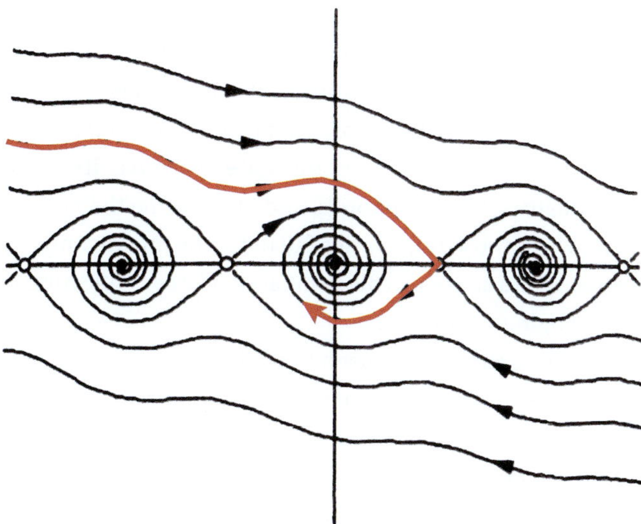

Figure 3.9: In the case of a damped pendulum, the trajectories in the state space are changed. Now energy is lost and trajectories decay towards total rest at the stable equilibrium. The figure traces one trajectory, and we can see where it switches from complete rotations to librations, and how these spiral in towards total rest.

3.5 Conclusion

The story of the pendulum marks a crucial moment in the long history of time-keeping. Although candles and water clocks can be used to mark time, they lack an essential feature: a natural source of steady oscillation or repetition. The rotation of the Earth on its axis, its rotation around the Sun, and the cycles of the Moon are all such sources, and of course they were our earliest forms of time-keeping. But they can only mark off longer intervals of time, not so easily minutes or seconds. Furthermore, they are vastly complicated by the fact that they can only be observed in terms of each other, and we have already seen some of the issues that arose out of that.

Galileo seems to have been the first person to realize the natural time-keeping possibilities of a pendulum. But he did more than that. The pendulum is abstracted out of a swinging lantern and reduced to a rod with a weight on the end. Friction and energy loss are thrown away at first, totally unrealistic though that is. What become essential are not the material constituents, but gravity and its effect on moving a mass in a constrained system. And that is the key: what Galileo was studying was basic relationships that can be interpreted in terms of a *system*; and what he was establishing was the nature of the patterning that arises within that system as it changes—the pattern of change.

Later we can refine it, adding back more realism to the underlying processes: friction and a driving force to overcome it. It is still a system and its importance still lies in change. Trajectories are still formed out of the passage of states as the pendulum swings. Patterns still appear, but they can now be more complicated. Dynamics guided by even a very simple set of rules can lead to a multitude of possibilities, and these interwoven possibilities are part of an intricate wholeness. That is the system idea: how parts become aspects of a unified whole. In our example, the state space has its distinctness in the uniqueness and individuality of its trajectories, and at the same time it has its wholeness as a unified system in which those trajectories fall naturally into distinguishable forms of pattern. Even the entire state space reveals itself as a cylinder, reflecting the underlying periodic nature of a swinging pendulum.

The main purpose of this chapter has been to see how we can put a particular process of change and its consequent evolution of pattern into a mathematical context. It is just a "simple" example, but it introduces new ideas that collectively form a new conceptual framework. As we go on we will see these same ideas arise in many more forms, for wherever there is a process of change this same conceptual framework can be found. We will argue that the idea of pattern is ultimately best seen in the context of change. This idea begins to emerge here, but it will take a good deal more effort to make a convincing argument and to bring out its full meaning.

The form of change that we have seen here is that of *continuous change*—the pendulum swings back and forth in smooth continuity. But there are many forms of change that occur in discrete steps. The entire digital world is based precisely discreteness, and it's enormous effect on society is an indication of its importance. In the next chapter we will look at change and pattern in the context of discreteness.

Chapter 4

Difference, change, and information

4.1 Distinction, difference, and change

We start with an important observation of the wide-ranging scientist and philosopher of science, Gregory Bateson (1904–1980):

> In lecturing, I commonly make a heavy dot against the board to achieve some thickness in the patch. I now have on the board something rather like a bump in the road. If I lower my fingertip—a touch-sensitive area—vertically onto the white spot, I shall not feel it. But if I move my finger across the spot, the difference in levels is very conspicuous. ...
>
> What happens is that a static, unchanging state of affairs, existing, supposedly, in the outside universe quite regardless of whether we sense it or not, becomes the cause of an event, a step function, a sharp change in the state of the relationship between my fingertip and the surface of the blackboard.
>
> This example, which is typical of all sense experience, shows how our sensory system—and surely the sensory systems of all other creatures (even plants?)—can only operate with events, which we call changes.
>
> The unchanging is imperceptible unless we are willing to move relative to it.
>
> Gregory Bateson [15]

In this short passage, Bateson puts his finger on the spot, so to speak. What alerts our tactile senses is change, and in the absence of change there is no signal and ultimately no awareness. As his finger passes over the chalk spot, it is the change at the boundary where the surface switches from smooth to rough that is signaled to the brain as difference.

Continuing, he says,

> We see what looks like the stationary, unmarked blackboard, not just the outlines of the spot. But the truth of the matter is that we continuously do with the eye what I was doing with my fingertip. The eyeball has a continual tremor called micronystagmus. The eyeball vibrates through a few seconds of arc and thereby causes the optical image on the retina to move relative to the rods and cones which are the sensitive end organs. The end organs are thus in continual receipt of events that correspond to outlines in the visible world. We draw distinctions; that is, we pull them out. Those distinctions that remain undrawn are not.

In the words of the neuroscientist A. K. Shakhnovich [147, Ch. 2]:

> Physiological *micronystagmus* is essential for visual perception, because it maintains constant displacement of the image on the retina, and thus ensures that different receptor elements are stimulated in turn. ...During the creation of a stabilized image on the retina, all visible differences in the visual field disappear after 1 to 3 seconds.

The examples that we have discussed thus far in the book suggest the ubiquity of pattern as a primary way in which we humans understand, interpret, predict, and interact with the world. They also suggest the intimate connection of pattern to dynamics, or more simply to change. Now, trying to deepen our understanding and uncover its underlying principles we find that nothing can be made of the world without our abilities to distinguish.

What Bateson and Shakhnovich are pointing out is that our abilities to distinguish begin at the level of neurons, and their ability to respond to difference. As such, we at once arrive at a connection that runs across the entire animal world, for all animals with sensory systems engage with the world through neurons. In fact it runs much deeper than this. Even single-celled life-forms distinguish between which molecules will pass through their bounding membranes, and which will not; even atoms behave differently according to the energies of photons with which they interact.

We can think also in terms of pattern recognition, for this immediately brings up ideas, first of distinguishing and then of memory. If there are no distinctions, there is nothing to recognize. If there is no memory, there can be no re-cognition. Memory is a perhaps unforeseen and hidden aspect of pattern, and we will have to return to it. But here we concentrate on the most basic feature of pattern, and that is distinction.

Pattern begins with distinction. This simple observation is the point from which we have to begin our deeper conception of what constitutes pattern.

Distinction is made on the basis of difference, and difference appears only in the context of change. Whether that difference is spatial, temporal, or neural, something has to change for there to be awareness of existence. Bateson moves his finger in order to sense the chalk spot on the blackboard; the eye has to be in continual movement to retain visual acuity.

This works both ways: for there to be change there has to be something that is different. Thus, at the beginning of our investigation into pattern we see two deeply related and foundational concepts: *difference and change.* And these lie at the origin of the concept of the pattern of change. Pattern inherently has a dynamical component to it.

4.1.1 The simplest pattern system

For pattern there must be difference, and for there to be difference there has to be distinction. Distinction means division into at least two things. So it is that the simplest of all patterns is the system consisting of just two states. This sort of yes-no, on-off, zero-one pattern is the fundamental type of dichotomy through which we divide the world into this and that. It exemplifies the word "dichotomy", literally to cut in two. As we have already noted, it is also the underlying feature in the action of a neuron, see §10.1.2, and so lies at the root of our entire sensory interaction with the world.

In framing this dichotomy we already see two of the important constituents of pattern: a *state space,*

$$\{0, 1\},$$

which provides the *context* and the two states, $0, 1$. Being mathematical, we have chosen to take the zero-one model as our primary example, and we will see that this fits in very nicely with the modern world, awash in binary digits. However the names that we give to the two states that make up a dichotomy are, in the end, just names and can be as variable as the distinctions that we care to make.

Here we introduce a standard way of describing a set, namely by listing its elements between braces $\{\dots\}$. Thus $\{0, 1\}$ is notation that stands for the set consisting of the two elements 0 and 1. Their numerical values are not particularly important, we could have use the letters y and n for yes/no or any other pair of symbols. Still, in the end the choice of 0 and 1 does prove useful, and it is a very standard choice.

Simple though the example seems to be, we should emphasize the important idea here. Distinction requires a dichotomy—one part cannot exist without the other, and it is only the two together that bring any sense to it. Thus the *context,* which is the pair taken together, is as much a part of the pattern as its individual elements.

Although the two possibilities for "the pattern" that is described by this system are simply 0 and 1, it is better to describe the pattern as the partition of $\{0, 1\}$ into the two subsets $\{0\}$ and $\{1\}$.[1]

This distinction between elements and subsets may seem overly fussy, but we make it because we shall see that the study of subsets of states is at least as important as the states themselves. As we saw in the case of the pendulum, it seems natural to collect together states that represent different types of motion of the pendulum: librations, full swings, and so on. In the natural world the underlying processes often operate at levels far finer than we care to distinguish or even can distinguish. Think of the local weather, for instance, that obviously derives from the movement of the vast number of atoms in the air and land or sea surrounding us. This distinction is the same point that Bateson makes in the epigraph that heads this chapter, where he is careful to mention that there is a threshold to be reached before sensory perception is possible. We naturally see the world at various levels of granularity (is it hot or cold outside, is it sunny or cloudy, the temperature is in the thirties, and so on).

This does not necessarily represent some lack of ability on our part. There are indeed physical limits to our ability to resolve granularity, but in the end that may be an essential constituent of how we frame reality. It is a point that we will see in more detail when we speak of events and entropy. Of course, in our present example of the simple idea of dichotomy where there are just two potential states, this issue of the degree of distinguishing doesn't exist. Still, it is often better to think in terms of subsets of the state space rather than in terms of individual states.

4.1.2 Yin-yang

> Light and darkness are two facets of the same reality ... and in every place where light is born shadows fall upon us.[2] *Natsume Soseki* [14]

The realization that dichotomies always arise within a context in which each part is an essential and logical complement of the other is often lost in daily life. But it is an ancient one. It occurs explicitly at the very beginning of the *Daodejing* (or *Tao Te Ching* in the older Wade–Giles romanization, see §15.3):

> From old, existence and nonexistence generate each other,
> difficult and easy complete one another,
> long and short contrast each other,
> high and low alternate with one another,
> pitch and tone harmonize with each other,
> before and after follow one another.[3] *Daodejing*, Ch.2 [42]

Although this quotation from Laozi (or Lao Tzu) may seem like a very simple observation, albeit put in poetical terms, it carries a deep idea that is easy to either miss or forget. In choosing to recognize a quality, or in the act of making a distinction, there is always a pair "A" and "not A". In as much as the one does not exist without the other, the two

Figure 4.1: The iconic taijitu (literally meaning the *diagram of taiji*) offers a beautiful and visually poetic presentation of the simple pattern of dichotomy. Notice that the white side always has a small black dot, and the black side a corresponding white dot, suggesting that neither can stand alone, but each depends on the other for its existence. Public Domain.

exist as a unity. Thus, though to say that "A *and* not A" are both true is a logical fallacy, if we don't take the "and" as a logical operator but simply as the statement that when one is invoked the pair is invoked, then the pair is a logical unity.[4] This interweaving of duality and unity is basically at the bottom of the (often contentious) reductionist-holistic discussion—another logical pair! There is no concept of unity without a notion of division. The very opening of the *Daodejing* makes this point, suggesting that it is in articulation that we frame our world of entities.[5] As Hegel wrote in 1816 (quoting from Ch. 1 of the *Daodejing*):

> We still have his [Laozi's] principal writings; they have been taken to Vienna, and I have seen them there myself. One special passage is frequently quoted from them:
> Nameless: the origin of Heaven and Earth,
> Naming: the mother of the ten thousand things [1].

Earlier than the *Daodejing*, the fundamental nature of dichotomy and its role in making sense out of the world had been formulated in the ancient Chinese theory of yin-yang.[6] Today, the yin-yang symbol in the form of the *taijitu* is familiar in tattoos, car decals, and jewelry, Fig.4.1.

The characters for yin and yang are 陰 and 陽, which are depictions of the shady and sunny sides of a hill and have built-in attributes of change, the one becoming the other as the day progresses. From early times the yin-yang pair was more simply represented by a broken (yin) stroke and an unbroken (yang) stroke:

▬▬ ▬▬ ▬▬▬▬▬

Of course the particular fascination of the taijitu lies in the way that it portrays the yin-yang coupling as a unity of dynamic interplay and balance. Somehow around this more subtle conception there arose the idea that the entire cosmos might be seen as an intricate dance of yin-yang polarities. This was not seen as some cosmic war of opposites

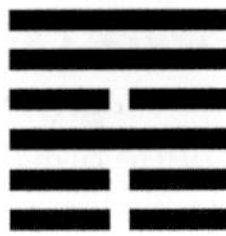

Figure 4.2: Hexagram 53: *Gradual Growth.* In terms of trigrams it reads wind over mountain.

as it has often been in other philosophies, but as a moving weave of balances in a sea of change, something far closer to the way we see the forces of Nature today.

The model that arose and that was to influence Chinese thought for millennia was formulated in a book, the famous *Yijing* 易經 the *Book of Changes* (often seen written in the older way *I Ching.*) Today the *Yijing* often evokes new-age images of divination and fortune telling, and it is true that prognostication has always been part of the *Yijing* culture. However, from at least the time of Confucius right up until today there have always been more nuanced and philosophical renderings of the significance of the *Yijing.* Within these perspectives there are insights that are relevant to our ideas of pattern. The *Yijing* is in many ways a beautiful example of a discrete pattern system.

The basic entities of the *Yijing* are the sixty-four hexagrams, the set of all possibilities for arranging six yin-yang lines. Each hexagram has evolved its own particular associations over time, and it is these associations that bring meaning to them. For instance, we have the hexagram of Fig.4.2, which seems particularly appropriate for the authors trying to write this book. Its association is *gradual growth.*

Fig.4.3 is a pictorial arrangement of all sixty-four hexagrams. Putting this into our language of pattern we consider the set consisting of these sixty-four hexagrams as the state space, and each hexagram as one state of the system.

The usual method of using the *Yijing* is to use a rather clever algorithm based on random divisions of a bundle of fifty sticks (ideally yarrow stalks) to choose each yin-yang line. Starting from the bottom and working upwards, repetition yields the six lines, one at a time. However, what makes the process so original, and this is crucial to the underlying philosophy of the *Yijing,* is that the algorithm has not two, but actually four outcomes, usually expressed through the numbers 6, 7, 8, 9. These correspond to *changing yin, yang, yin, changing yang.* The lines appear in two varieties, changing or *old* (6 and 9) and unchanging or *young* (7 and 8). The *Yijing,* as its name suggests (literally "Changes Classic") always sees things as process. The yin-yang lines are conceived as dynamic entities. Whenever a changing or old line is drawn, the corresponding opposite young line should also be considered. Thus with every hexagram that appears out of the algorithmic process, one is asked to consider also the hexagram(s) arising by changing one or more of the old lines to its opposite.

There are several features to this system that should catch our attention. The first is

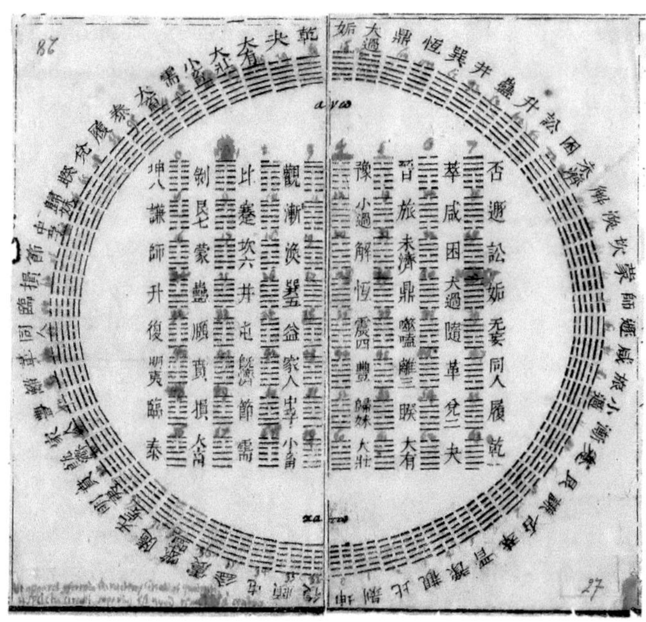

Figure 4.3: Here we see two arrangements of the 64 hexagrams, once in the great circle and once in the square array. In the array they read as a simple ordering of the binary forms of all the numbers between 0 and 63: 000000, 000001, 000010, ..., 111111. The circular array is similarly arranged. The actual linear orderings of the hexagrams found in all existing texts of the *Yijing*, including their modern versions, are very different from these. The origin of its present ordering is attributed to King Wen (1112–1050 BCE). The particular image here is of historical interest. In 1701, Joachim Bouvet (1656–1730), a French Jesuit working in China, sent the picture to the great philosopher and polymath Gottfried Wilhelm Leibniz (1646–1716). Having read the *Confucius Sinarum Philosophus* (1687) of Philippe Couplet (1623–1693), a book that was the result of more than a century of Jesuit study of Chinese thought, Leibniz saw similarities between Confucianism and his own views. He was particularly interested in whether Chinese ideographs supported his idea of a "universal characteristic" (*characteristica universalis*)—a language to be based on ideograms and a universal idea of knowledge. Leibniz added the Arabic numerals that we see here, see [102, 140]. Image from the Leibniz Archive, Niedersächsische Landesbibliothek. Public Domain.

that it is discrete. This is an attempt to see the state of things, seen from the perspective of the individual consulting the *Yijing*, as fitting into one of a discrete set of outcomes. It sees this outcome as a moment within a process of change and chance, and suggests potential directions of future change on the basis of the *uncertainties* around changing lines.

Perhaps the significant feature of the Yijing from our perspective is the way in which it treats change as fundamental to the way things are. The understanding that change and process are more relevant than objects and things came to dominate early Chinese language and thought. Things exist because of what they have become, not because they stand as prior entities with some independent existence. It is not only what they have become that is important, but what they will become. Western science has been slow to express these ideas so explicitly. But in his book *The Order of Time* [138], physicist Carlo Rovelli devotes a chapter entitled *The World is Made of Events, Not Things* to this very point. He writes

> We can think of the world as made up of things. Of substances. Of entities. Of something that is. Or we can think of it as made up of events. Of happenings. Of processes. .... Newton's mechanics, Maxwell's equations, quantum mechanics, and so on, tell us how events happen, not how things are. ... We understand the world as its becoming, not its being.

This is a crucial ingredient in our approach to pattern. We will come to think of it in terms of events.

The *Yijing* and the way it is used together form a pattern system. Ultimately, because of its dependence on human interaction and the actual circumstances of the individual using it, and because of the ambiguous language in which the potential outcomes are expressed (so typical of oracles!) the *Yijing* diverges from the sort of pattern system models that we will study here. But in some essential details it remains relevant. First of all, it is an example of dynamics within a discrete symbolic setting. Today, we live in a highly discretized world. All modern computers are discrete and as such any process modeled on them will be ultimately discretized. Of course by switching from kilobytes to gigabytes, the resulting processes can look very continuous—a digital photograph with a few megabytes of pixels can produce a gradation of tone that looks perfectly smooth to our eyes—but ultimately it is just the outcome of a large number of zeros and ones, or if we prefer, yins and yangs. Almost any dynamical system can be viewed as discrete if we so wish. It is even conceivable that at its deepest level the Universe has aspects of discreteness.

Second, chance is a crucial part of understanding change as we see it every day both in the natural and in our manufactured human worlds. Pattern is an outcome of change and within that chance can play a vital role. Not all chance leads to unpredictable outcomes, nor is all chance something to approach with trepidation. Boiling a saucepan of water on a stove means increasing the velocities of a vast number of water molecules which move quite randomly and independently. But the outcome is both predictable and useful, if you are making tea.

In what follows next, we shall see how both discrete dynamics and aspects of chance can enter into the creation of coherent pattern systems, which are quite predictive in their own ways and which also have proven to be extremely useful.

4.2 Strings of bits: the shift systems

In the digital world of today, the basic entity is the binary bit, the simple zero-one system that began this chapter. The Chinese combined their yin-yang lines six at a time into the hexagrams, and those hexagrams served as the basic symbols for their Yijing pattern system. The fathers of the computer age chose the ubiquitous eight-bit *byte* as their basic unit. Now we have kilobytes, megabytes, gigabytes, and terabytes as we find it necessary to store more and more information in digital form.[7]

Binary bits are ideal for machines because each binary bit is really just an on/off configuration. Computers operate as machines by processing long strings of binary bits. The simplest machines would process such strings *serially*, that is binary bit-by-binary bit. Speed and efficiency make *parallel* processing far more attractive, and modern computers at the time of writing now, tend to use a number of processors that can deal with 64-bits simultaneously. They take 64-bit "words" as their basic units.

In either case, the dynamics of digital processing involves reading and responding to a sequential flow of information, whether that be serially bit-by-bit, byte-by-byte, or in some larger units. We are going to look at various forms, but our basic object of study is simple serial bit-by-bit dynamics. Our aim is neither speed nor efficiency, but clarity. What's more there is a certain universality to the model that we are about to introduce. In some sense, it is the ancestor of all discrete processes.

With this in mind our attention shifts to strings of binary bits, that is, strings of zeros and ones. Think of these strings as a flow of information. A CD or a DVD is, underlying its lovely iridescent disk, basically a very long string of zeros and ones. The CD or DVD player reads these digits serially and interprets them in a temporal setting in the form of audial or visual information. This in a nutshell is the basic idea that underlies what we are going to call *dynamical shift systems* or more simply *shift spaces.*

The actual mathematical formalism is straightforward—strings of zeros and ones and the concept of reading strings serially or sequentially. However, there are three features that are special:

(i) Although there are a lot of binary bits on a CD (over 1.4 million per second of playing time), and even in a whole life time of reading out loud we could never get past even a minute part of it, it is still obviously finite. Mathematically it is far simpler to say that the string goes on without end. We assume that we are dealing with strings of *unbounded length.* Commonly we just speak of *endless strings*, or *infinite strings.* They have a start, *but no end.* We can express them symbolically in the form

$$\mathbf{x} = x_1\, x_2\, x_3\, x_4\, x_5 \ldots\,, \tag{4.1}$$

where each x_k is a 0 or a 1 and the dots mean that the string has no end.

(ii) We treat this entire unbounded string as a single *state* of the system. This is like making the CD itself one state of the system. Each possible unbounded, or infinite, zero-one string (of which there are of course infinitely many!) constitutes one state and the entire set of all possible unbounded zero-one strings is called the *state space* of the system. So the state space consists of all possible strings of the form (4.1). Below, we will see that each state amounts to an infinite upwards walk along the branches of a binary tree.

(iii) There is a simple formalism of the idea of "reading" a string sequentially. This is accomplished by simply shifting it to the left. In other words, reading the first digit of the string $\mathbf{x}$ of (4.1) is simply to replace it by

$$x_2\, x_3\, x_4\, x_5 \ldots .$$

This does not look like reading in the sense of comprehending, and it is not. It is just that after the first symbol is x_1 is "read", we dispense with it and go on to the next symbol. After this "left shift", we are still left with a string that goes on without end, so it still is a state. But it is a different one. The left shift changes the state $\mathbf{x}$ into another state.

Later on we will find it useful to interpret these zero-one strings as outcomes of a random process, like coin tossing, where the zeros and ones represent heads and tails.

4.3 Shift systems

Collectively what we have just done is to create a mathematical object that captures the essence of reading long strings of zeros and ones. We call them *states* and use boldface notation like $\mathbf{x}$ to distinguish states from the zeros and ones that make up the a state.

The states are unbounded zero-one strings, the *state space* is the collection of all such states, and the dynamics is supplied by the left shift that replaces one state (an infinite string) by another. We incorporate all this into a symbolism that we will use extensively throughout the book. Starting with our simple zero-one system $\{0, 1\}$ we are now looking at infinite strings made out of these two digits. We denote the set of all such strings by the suggestive notation[8]

$$\{0, 1\}^{\infty} .$$

Finally for each state $\mathbf{x} = x_1\, x_2\, x_3\, x_4\, x_5 \ldots$ we have its *left shift*, and for that we introduce the notation $s \triangleright \mathbf{x}$. In symbols,

$$s \triangleright \mathbf{x} = x_2\, x_3\, x_4\, x_5 \ldots \tag{4.2}$$

The s stands for *shift left*, and the way to think of it is as some sort of action, or operation or device that moves the string along by one step leftwards, dropping off the first digit

x_1. We need to keep in mind that the shift, being an operation or *operator*, is a different type of thing than the zero-one strings that it acts on. The shift operator s *acts*, on each state $\mathbf{x}$, taking it to the new state of (4.2). Notice the notation here: $s \triangleright \mathbf{x}$. We use the symbol "$\triangleright$" both to clearly separate the operator s from the state it is acting on $\mathbf{x}$ and, at the same time, to suggest that s is acting as an agent of change. This notation is used throughout the book.

The choice of unbounded (infinite) strings, as opposed to limiting them by some length, has the great advantage that shifting does not change the nature of the thing we are discussing. After the left shift, what remains is $x_2\, x_3\, x_4\, x_5\, \ldots$, which is still an unbounded string. If we were to use finite strings we continually run into the issues of just how long they are and issues around ending up with an empty string.

Thus,

$$\begin{array}{lcccccccccccccl}
\mathbf{x} & = & 0 & 0 & 0 & 1 & 1 & 0 & 1 & 0 & 1 & 0 & 0 & 1 & \ldots \\
s \triangleright \mathbf{x} & = & 0 & 0 & 1 & 1 & 0 & 1 & 0 & 1 & 0 & 0 & 1 & \mathbf{1} & \ldots \\
s^2 \triangleright \mathbf{x} & = & 0 & 1 & 1 & 0 & 1 & 0 & 1 & 0 & 0 & 1 & \mathbf{1} & \mathbf{1} & \ldots \\
s^3 \triangleright \mathbf{x} & = & 1 & 1 & 0 & 1 & 0 & 1 & 0 & 0 & 1 & \mathbf{1} & \mathbf{1} & \mathbf{1} & \ldots \\
s^4 \triangleright \mathbf{x} & = & 1 & 0 & 1 & 0 & 1 & 0 & 0 & 1 & \mathbf{1} & \mathbf{1} & \mathbf{1} & \mathbf{0} & \ldots,
\end{array}$$

showing the initial digits of an unbounded zero-one string and the following digits (shown in a different font) as they come into view through its successive left shifts. The exponents on s indicate the number of times that s has been applied.

There are several things about unbounded strings of zeros and ones that may seem peculiar. It might seem more natural to think of a state as referring to the leading symbol of some fixed string rather than an entire string. But in that case the state space would simply be the original state space $\{0, 1\}$ and we would not have advanced in taking account of any process of change and the intrinsic patterning that change may bring. In dealing with strings, however, as soon as we remove the leading symbol of a string, we are no longer looking at the same string. It is still an unbounded string, but it is a different string.

In this new picture there is the potential for change and seeing the effect of change. Still, there is a new and important detail. Unless there is some additional feature of memory, once we make a left shift of a state, thus dropping its first symbol, there is no record of what that symbol was. There is no way to reconstruct the original string: was it a zero or was it a one? The left shift is not invertible.[9] So the dynamics, in which *states change to other states*, is one that changes one string into another and it is the strings themselves that are taken as the states.

The fact that we cannot possibly "see" the entire unbounded or infinite string constituting a state $\mathbf{x}$ might also seem to be strange, but we will soon see how it is used. In the world around us we see sequences of events, for instance the sequence of full moons, and though such sequences are potentially unbounded into the future, we can observe and make inferences only from finite sequences of observations. Thus reading the initial digits of a state may lead to inferences about what is to follow, although not to complete knowledge of what the state actually is. We can understand a state as being an unfolding of process that reveals itself stepwise through repeated application of the shift. Comparing with the state space of the pendulum of Ch.3, it is also true there that the vast majority of states are not finitely expressible. A random angle and angular velocity of a pendulum bob is just a pair of real numbers, and real numbers typically involve infinitely many digits. But when we look at the way in which we expressed pattern, it was not specific states that defined the pattern, but whole families of states corresponding to the various different behaviors of the pendulum (librations versus whole circular swings, and so on) that were of interest. That is also how pattern is introduced into shift systems.

The set of infinite strings $\{0, 1\}^{\infty}$ together with the left shift s is called the *full shift space* on two letters.

Here the two "letters" are 0 and 1, but exactly which two symbols or "letters" we put there is irrelevant to what the shift space is. It is called the *full* shift because we have placed no restrictions on the strings: every possible string is included in the state space $\{0, 1\}^{\infty}$. Often, however, there are good reasons to impose definite restrictions on the strings we allow, in which case it is natural to deal with only a subset of the full shift space. For example, in the case of CDs there is an imposed restriction that amounts to saying that there must be at least two and at most ten 0s between every 1.[10] Shifting works as before and obviously strings obeying this rule still satisfy it after shifting. So the result is still called a *shift space*, but it is no longer a *full* shift space.

4.3.1 The ways of abstraction

It is interesting how mathematics tries to capture the essence of what is going on and abstract it. We have used unbounded zero-one strings as an idealization that capture the essence of finite, but very long, zero-one strings. Our mathematical object is a mental construction that cannot not exist in a finite world. A Platonic extrapolation might suggest that very long finite strings are just poor approximations of the more important ideal strings, which go on forever. That is not our understanding. Idealisms are important because they can abstract the essence of some quality—in this case the concept of *very long strings*, which is actually pretty vague! They are important because there are things that can be stated about these ideal entities that are only "in essence" true in the real world. An example is the law of large numbers that we discuss below in §5.1.3. If our strings of zeros and ones have a probabilistic origin (repeated tossing of a coin, for instance) then there are definite things one can say about the averages that will emerge in the ideal context of infinite strings. Back in the real world, these truths have strong implications

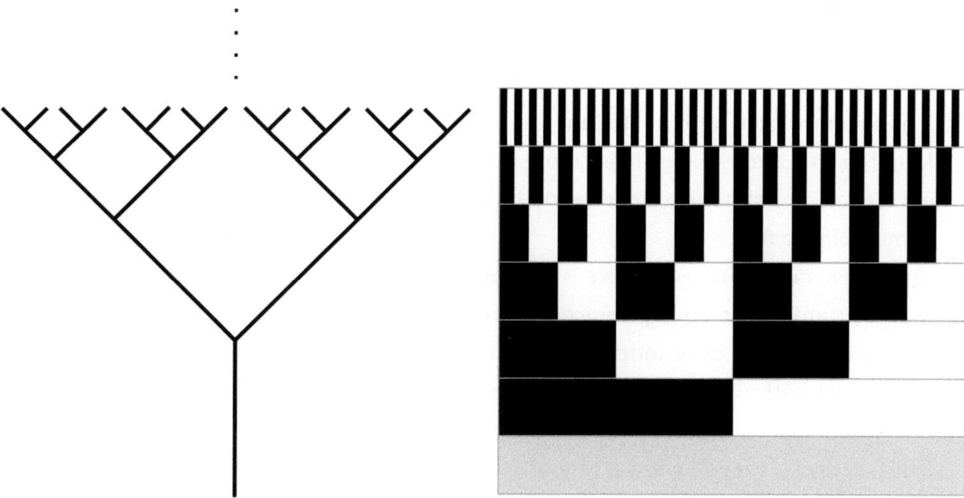

Figure 4.4: The infinite binary tree. Here we show the first few levels of a binary tree in a very plain version: a tree where each branch further branches into two. On the right we see the same logical structure in the form of successive binary divisions of a unitary origin at the bottom. This presentation has been called FuXi's keyboard by John Minford [90]. Each path amounts to a sequence of left-right choices starting at the first bifurcation up from the root. If we put a 0 in each left (black) and a 1 in each right (white) rectangle then an upwards path through successive rectangles produces at 0/1 sequence; indeed, every such sequence can be so attained.

about what will happen "in the long run". The "long run" may have no absolute sense, but knowing what we should expect is important. Casinos are very profitable.

Again, notice how the use of the word "reading" in the context of strings has been divorced from any conception of meaning, or interpretation. "Shifting" now simply becomes "drop the first symbol". The state space has become the set of all possible *infinite sequences*, not because we have any thought of dealing with all of them or even the entirety of one of them, but because in doing that shifting takes one infinite sequence to another infinite sequence. Because we have no way of going backwards or shifting right, there is no "memory" in this simple system. Abstract though it is, we see how very useful this idea becomes once we connect it metaphorically to the empirical world .

We explain and explore the dynamics more in the next section, but before we do that it is informative to look at a less abstract picture of our new state space $\{0,1\}^\infty$. In spite of its initial strangeness, $\{0,1\}^\infty$, is actually not so unfamiliar: it can be viewed in terms of a sort of "tree". Imagine a tree in which each branch opens out, or *bifurcates*, into two new branches. Fig.4.4 gives the picture. Superficially it is very much a tree as we normally think of as a tree. The differences are that there is no end to the branching—each branch gives forth exactly two more branches, and this goes on forever. This is called a *binary tree*. We can interpret each element of the state space $\{0,1\}^\infty$ as being a set of left-right instructions to walk an infinite path upwards along the branches of the tree, starting from its root. Think of walking upwards in the tree, starting at the root, with each 0 in the state understood to mean taking the left branch and each 1 meaning taking the right as we move from junction to junction. Each state of $\{0, 1\}$ produces a unique infinite walk up the tree and the tree itself represents the set of all such possible walks. In this interpretation, the shift space can be thought of as the set of all upward walks through an infinite binary tree.

Quite generally, whenever there is a situation in which there are sequences of yes/no, left/right, or any other binary choices, there is a binary tree involved, and if those sequences are very long or of indeterminate length, then $\{0, 1\}^\infty$ is an appropriate mathematical object with which to model it. Soon we will find the need to give $\{0, 1\}^\infty$ even more mathematical structure.

The fact that binary sequences amounted to successive bifurcations did not escape the attention of the Chinese thinkers and their hexagrams. One particularly appealing form of it is what has been dubbed FuXi's keyboard, Fig.4.4. Extrapolated into the full setting of an infinite keyboard, it would truly be a cosmic instrument, worthy of any theory of the music of the spheres! The diagram is known in Chinese as *Fuxi liushisigua cixu*, literally "FuXi's sixty-four hexagrams sequence", and whimsically referred to as FuXi's Keyboard. FuXi was a legendary emperor and culture hero. Any dating, indeed even if he ever existed, is uncertain, though traditional records claim that he ruled in the mid-2800s BCE. The diagram was actually created much later by Shao Yong (1011–1077), a philosopher, neo-Confucianist, cosmologist, poet, and historian.

4.4 Shannon

In 1948 a young mathematician/engineer from Bell Laboratories published a paper that was to revolutionize the theory of communication. One might expect to hear that his work involved a new technological breakthrough or development like the invention of transistors (which actually were invented at the Bell Laboratories). In fact it was purely theoretical. In his *A mathematical theory of communication* [148], Claude Shannon (1916–2001) laid down the principles of communication of information along binary channels in the presence of noise. "Binary channel" is just a convenient name for any of various ways of communicating information in the form of strings of binary digits. Electrical pulses down a wire, strings of binary bits on a CD or a DVD, signaling by flashes of light between ships at sea, transmission of bits of sound sent by a fax machine, or the radio transmission of information used in long distance communication via satellites: all of these and many others are examples of binary channels in action. Noise simply means randomness that confuses zeros and ones, something that is familiar from poor connections on telephones, and that is inevitable with any real-life communication system.

Figure 4.5: Ticker tape, used during the period 1870-1960 for transmitting stock prices. Each tape is in effect an "infinite" string of alphabetic symbols, though the transmission was binary. Public Domain, Women_in_Waldorf-Astoria.jpg .

4.4.1 The formalism and randomness

At first sight, this would hardly seem to present any special issues: send the binary bits as fast as possible and repeat things where they are confused by noise. But as we shall see, Shannon was interested in transfer of *information*, not just binary bits, and that raises the question of what information is. He also realized that there are far more subtle and efficient ways of correcting for noise than by simple repetition. However, one thing is readily apparent. The basic form here is strings of binary bits which are processed sequentially, and since such data strings can be indefinitely long, the appropriate place to start is with the shift system $\{0, 1\}^\infty$.

Shannon was an English speaking American, and so his language of communication was English. His idea of communication was that of sending English language texts along a binary channel. The ordinary alphabet of 26 letters extended by a "blank" character to act as a spacer can be used to form strings of English text. Let us denote this extended alphabet of 27 letters, which we will simply refer to as the *alphabet*, by $\mathscr{A}$. Strings of English text then appear as strings of symbols from the alphabet $\mathscr{A}$. It seems natural to denote the set of all possible alphabetic (infinite) strings by $Y = \mathscr{A}^\infty$. We are in the process of creating another dynamical system, this time based on infinite strings of letters from the alphabet $\mathscr{A}$. We can use the same idea of shifting left to "read" the letters of these strings.

As the 27 symbols of our alphabet can easily be encoded in the form of 5 binary bits $(2^5 = 32)$[11], we can rewrite any element of Y into a binary string, and so as an element of $\{0, 1\}^\infty$. Thus part of the process of communication is the step of encoding alphabetic

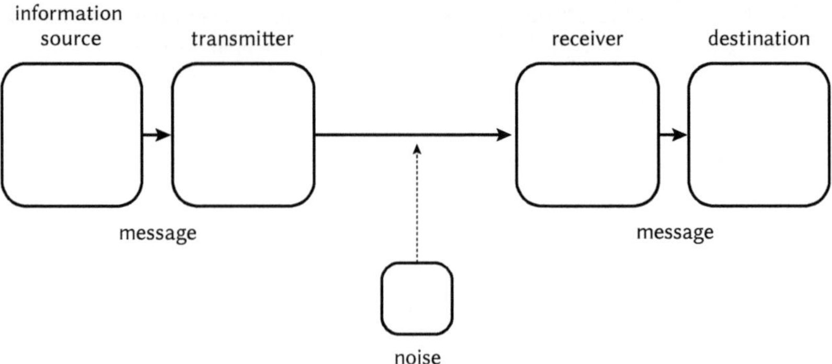

Figure 4.6: Shannon's schematic view of a communication system. A source of English text, say, is first changed (encoded) into a string of binary digits. The binary string is transmitted in the presence of noise or random interference. A receiver takes the resulting transmitted message of binary bits, errors and all, which is then decoded back into English text. In this section, we assume a noise-free environment. We will come back to the (very important) role of noise in § 8.4.9.

letters into binary bits, a process that we might indicate as

$$Y = \mathscr{A}^{\infty} \overset{\text{encoding}}{\longrightarrow} \{0,1\}^{\infty}\,.$$

This is the first step in the schematic view that Shannon had of a communication system. This has nothing to do with encryption. It just means that we need a way to convert the symbols of $\mathscr{A}$ into binary digits. Fig.4.5 recalls what was once a familiar example of this process. We will see a simple example of how encoding is accomplished in Tables 4.1 and 4.2.

Now English words are not just arbitrary collections of letters, and most random strings of letters don't look remotely like English:

XFOML RXKHRJFFJUJ ZLPWCFWKCYJ FFJEYVKCQSGHYD QPAAMK BZAACIBZLHJQD

is an example from Shannon's paper in which letters of $\mathscr{A}$ (including the blank space) are selected at random with each letter having the same chance of being selected. But letters appearing in English sentences (and that includes spaces between words) have very different frequencies of occurrence. For instance, the letter E is the most common letter whereas X or Z are far less frequent—look at the previous sentence for example. Shannon next looked at random strings of letters where the letters are chosen according to their usual frequencies:

OCRO HLI RGWR NMIELWIS EU LL NBNESEBYA TH EEI ALHENHTTPA OOBT TVA NAH BRL is an example. It's better, but things get even better if we still choose randomly but this time use frequencies based on letter pairs (also called *digrams.*) For instance, ST is far more common than SJ. Here is a sample from Shannon's paper:

ON IE ANTSOUTINYS ARE T INCTORE ST BE S DEAMY ACHIN D ILONASIVE TUCOOWE AT TEASONARE FUSO TIZIN ANDY TOBE SEACE CTISBE

Repeating this idea, moving to frequencies of three and then four letter groups (trigrams and quadrigrams) we have examples like

IN NO IST LAT WHEY CRATICT FROURE BIRS GROCID PONDENOME OF DEMONSTURES OF THE REPTAGIN IS REGOACTIONA OF CRE

and then

THE GENERATED JOB PROVIDUAL BETTER TRAND THE DISPLAYED CODE ABOVERY UPONDULTS WELL THE CODERST IN THESTICAL IT TO HOCK BOTHE

There is neither grammar nor semantic content in any of these, but amazingly they start to look like English! Of course there is nothing magical about English. The same analysis can be done with any language written in the form of sequential strings of alphabetic letters. In other languages, even though written in the standard roman letters, the frequencies are different, and of course the random strings of letters written using those frequencies will begin to look like those languages.

4.4.2 Probability and dictionary sets

There is a concept of pattern emerging here, and it is arising on the basis of the dynamics of chance shaped by some probability distribution. Shannon based his theory on efficient and optimal communication of messages precisely on just such ideas. We may question the value of considering "sentences" that have absolutely no meaning, but one of Shannon's insights was that for communication systems, at least from an engineering point of view, one should treat messages to be communicated as random strings of letters, ignoring completely the complicated issues of semantic content, but instead relying on randomness shaped by a probability. The engineer could hardly be expected to predict and design in the semantic requirements of all potential uses of a communication system, but could take advantage of the intrinsic nature of the principal language in which the communications were being made.

Incorporating this stochastic (i.e. based on randomness) structure into our formalism requires two steps. Imagine that we are receiving a long string of alphabetic text on a paper tape, but we do not know what the message is. In our mathematical language, this long string can be thought of as a single state $\mathbf{x}$ from the state space $\mathscr{A}^{\infty}$. We don't know what state, but we start to read the letters one by one. As we read letters we shift the tape one symbol, dropping off the first letter.

Suppose the first letter is the letter I. We now know something about the state $\mathbf{x}$ that we did not know before: it begins with I. The entire set of states (strings) in $\mathscr{A}^{\infty}$ that start with the letter I is a subset of $\mathscr{A}^{\infty}$, and if we denote this subset by [I], we can say that our string $\mathbf{x}$ lies in [I], or more mathematical looking $\mathbf{x} \in [\mathrm{I}]$. If the first letter in $\mathbf{x}$ had been V we would have said that $\mathbf{x}$ lies in [V], or $\mathbf{x} \in [\mathrm{V}]$, and if it were a blank we would say that $\mathbf{x} \in [\square]$. We will call sets like $[\mathrm{I}], [\mathrm{T}], [\square]$ *dictionary sets.* Just as in a dictionary we group together all the words that begin with a certain letter, here we do the same for our letter strings.

After reading the first letter, we move to the next state $s \triangleright \mathbf{x}$, which is $\mathbf{x}$ with the I dropped off and tape moved leftwards, so that we are looking at the next symbol. Suppose it is T. Then we know that the state $s \triangleright \mathbf{x}$ lies in [T], or equivalently $s \triangleright \mathbf{x} \in [\mathrm{T}]$. Shifting again, suppose that $s^2 \triangleright \mathbf{x} \in [\square]$. So far we have read "IT " (note the space). We are making no assumptions that our tape is starting at the beginning of a text or even at the beginning of a sentence. If we were to continue, we might get perhaps "IT WAS THE BEST OF TIMES IT WAS THE WORST OF TIMES ...", or perhaps a part of a word from a string of text that was reading "(THE DOG B)IT (THE MAN)".

What is relevant is that the shift space $\mathscr{A}^{\infty}$ has been broken up into 27 pieces (dictionary sets) $[\mathrm{A}], [\mathrm{B}], \ldots, [\mathrm{Z}], [\square]$, because every state (i.e. infinite string) of $\mathscr{A}^{\infty}$ must begin with some letter. As we read (shift) our state $\mathbf{x}$, transforming it into its subsequent states $s \triangleright \mathbf{x}, s^2 \triangleright \mathbf{x}, s^3 \triangleright \mathbf{x}, \ldots$, we see each of them move into these various pieces, according to what their first letter is.

What Shannon is suggesting is that we should look at the relative frequencies of letters. The frequency of the letter I is about 5%, that is it occurs 5 times per hundred letters of English text, on average. Since we are including blanks as letters here, we should also note that in English text the number of letters between blanks averages at about 5.1.[12]

It is better to work in *probabilities* than frequencies, so instead of saying that the frequency of the letter I is 5% we will say that the probability that a randomly chosen letter from a piece of English text is I is about 0.05. Probabilities always lie in the range 0 to 1, with 0 indicating the letter has no chance of being selected, and 1 indicating that the letter will be certain to be selected (and hence the others have no chance of being selected). Putting it another way we can say that the probability that the randomly chosen state $\mathbf{x} \in \mathscr{A}^{\infty}$ is actually in the set [I] is 0.05. For [T] we get a probability of 0.08. The probability for $[\square]$ is 0.17. The normal way to say all of this is that the probabilities associated with [I], [T], [□] are 0.05, 0.08, 0.17 and write

$$\mathbf{p}([\mathrm{I}]) = 0.05, \quad \mathbf{p}([\mathrm{T}]) = 0.08, \quad \mathbf{p}([\square]) = 0.17\,.$$

Since every state of the system, that is to say, every state, has to start with some letter from the alphabet $\mathscr{A}$, we must have

$$\mathbf{p}([\mathrm{A}]) + \mathbf{p}([\mathrm{B}]) + \cdots + \mathbf{p}([\mathrm{Z}]) + \mathbf{p}([\square]) = 1 \;\; :$$

with probability equal to 1 some letter will be the first letter of the string.[13]

If we think back to our discussion of the pendulum, we saw that our idea of pattern had focussed on dividing the state space into regions that collected together all states that have a certain common feature (librations, circling around clockwise, etc.). In our present situation we can see the same thing. We have divided our state space into 27 parts that represent a very coarse type of patterning, namely every state must begin with some letter of $\mathscr{A}$ (which also includes the "blank"). This has been refined by associating with each of the parts the probability that a randomly chosen alphabetic string of English text $\mathbf{x}$ will lie in that part (i.e. begin with that letter).

Recall that breaking a state space into a bunch of pieces is called *partitioning* the state space, and the result is called a *partition* of the state space (see also §4.4.3). In our present example, we have created a partition with twenty-seven *parts*, one part for each letter of our alphabet $\mathscr{A}$). What Shannon suggests next is that we will get an even better description of the patterning underlying English text if we look at letter pairs, also called *digrams*. Now we get refined sets like [IT], [T□], [□W], [WA], [WW], ..., where [IT] stands for the entire set of states $\mathbf{x}$ that begin with IT, and similarly for the others. Then we have, say,

$$\mathbf{x} \in [\mathrm{IT}], \quad s \triangleright \mathbf{x} \in [\mathrm{T}\square], \quad s^2 \triangleright \mathbf{x} \in [\square\mathrm{W}], \ldots\,.$$

Obviously letter pairs, or digrams, have much lower frequencies or probabilities. We have $\mathbf{p}([\mathrm{IT}]) = 0.014$, for instance, which is relatively high (the highest is [TH]), and the others are much lower. Now we have a more refined partition of $\mathscr{A}^\infty$ into $27 \times 27 = 729$ letter pairs

$$[\mathrm{AA}], [\mathrm{AB}], [\mathrm{AC}], \ldots, [\mathrm{ZY}], [\mathrm{ZZ}], [\mathrm{Z}\square], [\square\mathrm{A}] \ldots, [\square\mathrm{Z}], [\square, \square]\,, \tag{4.3}$$

each part with its own probability, which collectively total probability 1. Do not make the mistake of thinking that the probability of letter pairs is simply the product of their individual probabilities. If that were the case then there would be no change in the nature of random text created using single letter probabilities and that using letter pair probabilities. Consider, for instance, T and W. Both are common letters, but the pair [WT] is practically non-existent as a letter pair in ordinary English text (we can think of rarities like jawtwister, jowter, and a standard abbreviation for 'weight'). *Scrabble* players are well aware that not all pairs of letters are equally easy to use.

Continuing with our terminology, we will also refer to these digram sets as *dictionary sets*, just as in a dictionary all the words that start with IT will be grouped together. The same idea works for any number of letters, e.g. [IT □ W] is the dictionary set consisting of all states that start IT □ W ....[14]

What we have just done we can do again with *trigrams*, and *quadrigrams*, and as we have seen, we get an increasing semblance of actual English text even from totally random strings of text—as long as we follow their probabilities! It is noteworthy that the probability function $\mathbf{p}$ (usually called a probability distribution) can be formed at different levels of refinement: on single letters, on digrams, on trigrams, etc. As for its values, these are not uniquely predetermined things. Obviously, the probability distribution of interest depends hugely on the language. It also differs between different writers, and between different English speaking cultures or times. This is clearly about pattern, but it is pattern that emerges in a dynamic context of time, place, and the actual unfolding of English text in particular samples. These frequency analyses based on digrams, trigrams, and so on, fall into the class of statistical processes called *Markov processes*, which we look at more closely in Ch.5.

Recapping, studying long strings of English text leads to letter probabilities, digram probabilities, and so on. If we embed these probabilities into our model—the shift system on the alphabet $\mathscr{A}^{\infty}$—then even randomly chosen strings of letters respecting these probabilities will produce text that is recognizable as English-like.

That in itself was not the intention of Shannon. His problem was how to efficiently encode English text, or for that matter text produced from any language whether natural or artificial. Reducing genuine English, with real dictionary words and proper grammar and syntax, to the far simpler structure of partitions with attached probability distributions, was a huge step in abstraction. Even if it looks a little strange, we can see its value in simplification and objectivization. Compare what we have just written with trying to explain to a person all the rules that allow us to write grammatically correct English sentences!

4.4.3 Partitions

You might be noticing that in discussing pattern within the general framework of state spaces, our attention is not really on particular states but rather on *subsets* of states that show some common behavior under the dynamics. In discussing the frictionless pendulum it was not so much particular states that we were interested in—there are far too many of those—but rather those sets of states that produced librations, or those that produced clockwise or counterclockwise circular swings. We partitioned the state space into pieces that gathered together states that produced the various types of behavior which interested us.

In the foregoing section we have seen our state space $\{0,1\}^\infty$ partitioned into pieces, dictionary sets, and it is through these dictionary sets that we gain insight into the patterning. The dictionary sets gather together states with properties that interest us, e.g. all states that start with the letter I or all those that start with the pair of letters IT, and so on. The fact is that though the states themselves are crucial to understanding the exact dynamics, individual states are generally totally inaccessible (they are after all of infinite length) and not what is really important to us. Patterning that is relevant and that we can actually detect or observe, exists at a totally different scale. In our discussion of the frictionless pendulum the individual states are determined by two numbers (a, v) representing every possible angle and velocity that the pendulum can take. However we were not generally interested in the actual values of these two numbers (each of which could, in principle, be expressed to absurd degrees of accuracy, say a thousand or ten thousand decimal places), but rather in the general behavior of the pendulum. This again points to thresholds of distinguishability mentioned by Bateson above. This is not an issue of approximation or failure of knowledge, it is quite generally the way the world presents itself to us through our senses. It is a recognition of the way things are.

There may be some confusion around the word "partition". In normal use a partition can mean a wall that breaks a room in two. But it can also mean simply to break into parts, like victors partitioning land after a war. In mathematics it is this latter interpretation that carries the right intuition. A *partition* of a set is the breaking of it into a collection of disjoint (i.e. mutually non-intersecting) subsets. These subsets are the *parts* of the partition and the word partition refers to the entire collection of these parts. Thus for example $\{1,3,4\} \cup \{2,6\} \cup \{5\}$ is a partition of the set $\{1,2,3,4,5,6\}$ into three parts (the $\cup$ symbol being the mathematical symbol for union). In our present context, the display at (4.3) is a partition of the state space $\{0,1\}^\infty$ into 729 parts which are the digram dictionary sets. There are all sorts of ways to break a set into parts—there are 203 different ways to partition the set $\{1,2,3,4,5,6\}$, for instance; see also Fig.6.1 in §6.4. Which partitions are relevant depends on what we are trying to understand. Sometimes they are obvious and at other times deeply buried in the dynamics. We shall see both types.

4.4.4 Information

This same objectivization with which Shannon approaches semantics appears in his approach to the idea of *information.* Presumably the idea in a communication system is to transmit information as fast as possible. But this raises a very difficult question. Exactly what is it that constitutes information and what do we mean by transmitting it?

We are said to live in the "information age", and are said to be flooded with information, yet like our word pattern, we all know what it is, and yet somehow we don't. Some people, when pressed to define information, define it as pattern but, as is becoming rather clear, this just shuffles the problem somewhere else. Actually, we do think that information is very closely related to pattern, or more accurately to pattern systems, and it is a recurring theme in the book that we will eventually address directly. However, in the present context what we need is something rather specific to the problem at hand. Shannon's problem was not the question of transferring information in the form of meaning *per se.* He needed a notion of information that was devoid of semantic content but relevant to the issues of efficiency and accuracy of transmission of the symbols that made up the language in which communication was taking place.

His approach was to start with the idea of the binary bit. A binary bit can be considered as the most elementary unit of information. One binary bit codes for one of two possibilities, 0 and 1 which, as we have seen, is the elemental foundation of distinction. Two binary bits presumably carry twice as much information, and together they code for four possibilities, $00, 01, 10, 11$. Continuing, three binary bits can carry code for eight possibilities, and carry three times as much information as one binary bit. Quite generally n bits code for 2^n possibilities and n times as much information. It is for this reason that it is

> the *logarithm* of the number of potential possibilities or states that is the basic measure of the amount of information.

The logarithms of $2, 4, 8, 16, \ldots$ are $1, 2, 3, 4$. Detailed descriptions of information typically involve logarithms for just this reason.[15] More about this, including more information about logarithms and the formula for defining entropy, appears in Ch.8.

But it is not quite that simple. Consider the information obtained in receiving a binary bit, say a 0. Potentially this binary bit has one unit of information. But suppose that we *know* that the chance that the bit is 0 is not the same as the chance that it is 1—in fact, suppose that we happen to know the probability of the bit being 1 is virtually zero. Then we are almost certain to receive a 0 and so now when we receive this 0 it carries almost no information. In the case that we know for sure that that we will only get zeros, there is no information at all. So the information depends on the probability of the event occurring.

Ultimately we will make some claims that there really is more to information than just an average based on probability, but certainly information as Shannon discussed it, and as it is normally understood in information theory and physics, the value associated with events is primarily probability based. Fig.4.7 shows how numerical values of information are assigned to events of varying probability. We will get into a deeper discussion of these curious graphs later, but for now note that the information associated with an event depends on its probability of occurrence and can be read off the graph.[16]

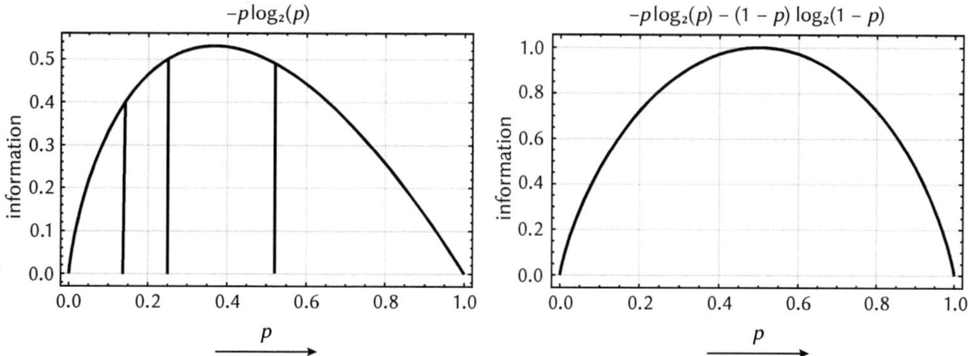

Figure 4.7: The plot on the left shows the information (vertical axis) associated with an event with probability p (horizontal axis). This is totally general and applies to the calculation of information wherever it appears. The vertical lines are to help in seeing the values of the information connected with events that have probabilities $1/2, 1/4, 1/8$ (namely $1/2, 1/2, 3/8$), which are used in the text. The plot on the right shows the total information carried by a binary bit for which the two possible outcomes, zero and one, have different probabilities of occurring. If the binary bit has probability p of being 0, then it has probability $1 - p$ of being 1. The total information carried by the binary bit then is the sum of the information carried by those two events. Note that the values on the information axis go from 0 to 1. The situation is symmetrical since the roles of 0 and 1 are completely interchangeable. The information carried by a binary bit is greatest when the probability is equally distributed between 0 and 1, and there it is indeed equal to 1, as it should be: the binary bit carries one unit of information and $\mathbf{p}(0) + \mathbf{p}(1) = 1/2 + 1/2 = 1$.

This may seem strange. It is. Notions of information (and its complementary sister entropy) take time to absorb; they have been—and to some extent remain—subject to much debate. They are relevant to our ideas about pattern, and will appear every so often. As they do so, they become more accessible and deepen our understanding, see §8.3.

In any case, let us return to Shannon. His question was about the efficient transmission of English text and hence of alphabetic symbols. In terms of the channel, the question is about how many bits/second can be transmitted. In terms of text it is about how many symbols/second. In terms of information it is about how much information per second. Information is about bits, and text is about symbols. The real issue comes down to how many bits/symbol are required. We are talking about averages here, of course—what is the average number of bits/symbol required to transmit English text?

Without further thought, one might be tempted to say that it is five bits per symbol since we need five bits to encode all of our 27 symbols. That would be an *information rate* of about 5, i.e. we need five bits to encode for each symbol. But this is far from the truth. For instance, since the letter E is so common in English text we can be much more efficient by using a very short code for the letter E and use longer ones for less frequently used letters. Morse code does this where the code for E is a single dot and the code for

Table 4.1: An obvious encoding

S	O	A	T
00	01	10	11

Table 4.2: A prefix encoding

S	O	A	T
0	10	110	111

T is a single dash, while the code for the infrequent letter is J is dot-dash-dash-dash. In fact the information rate per symbol in English is about 2. Put another way, we get about one bit of information for every two letters, on average.

Rather than deal with the full complications of the usual alphabet, here is a simpler example that illustrates how this comes about. Imagine that we have a "language" that is written with an alphabet of only four letters: $\mathscr{A} = \{S, O, A, T\}$. To keep things to a minimum, we have not included a space character here.[17] Suppose that these letters occur with probabilities $1/2, 1/4, 1/8, 1/8$. Here is a random string of letters from this alphabet:

AOSATTSOSOSSAOSSSTASSSSOAOSOASOOSASSASSO ....

What number would describe the rate of information that this language conveys per symbol? The information associated with each of these four events, which can be extracted from Fig.4.7, is $1/2, 1/2, 3/8, 3/8$, so the total information is their sum $7/4$. The information rate is 7/4 bits/symbol. Usually the letter H is used to denote the information, so we would write $H = 7/4$ in this case.[18]

It would seem to take two bits to encode the four different outcomes of this four-letter language, say, given by Table 4.1. But two bits can encode for total information of 2 (bits) and we need only $7/4$ bits, so to use this encoding seems wasteful.

A more clever way to improve the encoding is to use a *prefix encoding* which encodes the letters so as to use fewer digits for the more common letters, Table 4.2. It is called a prefix encoding because none of the codes is a prefix of any other, and hence the code can be unscrambled without having to waste bits for spacing. For instance the beginning of the random sequence above encodes to

$$1101001101111110100011 0\ldots$$

which can be uniquely read as

$$110\ 10\ 0\ 110\ 111\ 111\ 0\ 10\ 0\ 0\ 110\cdots \rightarrow \text{AOSATTSOSSA} \ldots.$$

Comparing the efficiency of sending text with these two encodings, we see that the first is clearly 2 bits per symbol, while the prefix encoding, taking into account the frequencies of the four symbols, gives

$$\frac{1}{2}(1) + \frac{1}{4}(2) + \frac{1}{8}(3) + \frac{1}{8}(3) = \frac{7}{4}$$

bits per symbol. The latter is more efficient, since it matches the information rate precisely.

One cannot always achieve such perfection, but in a limiting sense one can. This is Shannon's fundamental theorem [148] for a noiseless channel:

> *If the source has an information rate of H bits per symbol and the (noiseless) channel has a capacity of C bits per second then the maximum rate of transmission is C/H symbols per second, and furthermore there are encodings that come as close to this maximum rate as one pleases.*

In our simple example we were able to hit C/H on the nose. However, in general, one can only get as close to that ratio as is desired. What does that mean? The cost of more efficiency is to use codes that encode for finite strings of symbols (e.g. digrams or trigrams), taking into consideration their frequencies, rather than just individual symbols. In reality, theory has to be balanced with the practicality of more complex encodings. But the importance of Shannon's work is that it establishes the limits of efficiency in information transfer, and also shows that those limits can actually be approached in practice. As we have described it, this is the theory for a noiseless channel. We will come back to Shannon's approach to channels when noise (random errors) is included in the theory, §8.4.9.

We have modeled a communication system on a shift system, the shifts of strings of characters or letters from some alphabet. The notion of information is bound in the frequencies of occurrences in texts written in a language using these letters. Whether those probabilities are based on single letter frequencies, digram frequencies, or trigram frequencies, or even higher length groupings, the result is to introduce patterning into random strings that is based on those frequencies. Those letter frequencies then shape what these strings are likely to look like and at the same time enter into efficient ways to transmit information in that language. It really seems that pattern and information are related in deep ways. But we have to fully understand what we mean by pattern before we can engage that conversation in any useful way.

4.4.5 Entropy

Shannon's ideas on information were not entirely original with him. The idea that the basic unit of information was the binary bit had already been suggested by another researcher at Bell Labs, R. V. L. Hartley (1888-1970) in 1928, and Shannon was also aware of the work of the great Austrian physicist Ludwig Boltzmann (1844–1906) on entropy. Most of us are aware that there is this concept of entropy and it has something to do with disorganization and uncertainty. The famous law that in a closed system the entropy is always increasing (the second law of thermodynamics) is the discouraging nature of decay and disorder that we ever see around us. Yet, at the same time, it is what makes the assembly of order that living beings achieve so incredibly impressive. But, of course, a living creature is not a closed system! Order comes only in relation to a larger context in which it may be considered as balanced by entropy.

Entropy would seem conceptually to be far from information, but in fact Shannon saw the two as being the logical opposites of each other. Information is what removes uncertainty, entropy is what increases it. In fact the very same measure is used for both, indeed the very same formula. Some people use the terminology of neg-entropy for information, suggesting that it is the negative of entropy. Above, we expressed Shannon's theorem in terms of information, but he actually expressed it in terms of entropy. In recent times, the age of computers, information has taken on a much larger role in science, and we constantly think in those terms. But it is enlightening to realize that it is really a measure of order or certainty and is inseparable from its sister entropy. Another yin-yang pair.

Boltzmann was a pioneer in the theory of statistical mechanics. He was an atomist in an age when, at least in physics, atoms were not a popular idea in physics, and atomists were not suffered gladly. There is a tragedy to his life that hinges on this negative, and ultimately unfounded, reaction to his work. He hanged himself in 1906. His ideas of *statistical mechanics* arose from the realization that in a world in which even a liter of gas or liquid consists of an astronomical number of atoms, there is a need to understand the *statistical* outcomes of their random motions in order to understand the world at the macro-scale of human beings. We live in a world constituted of countless atoms, moving and interacting in random ways. Yet out of it emerges forms of order—forms that we can see at the human scale, and forms that we use and manipulate every moment of our lives. Today, statistical mechanics is a fundamental discipline in physics.

We will come back to entropy again, for as we have stated it it is not fully integrated into the dynamics. Entropy based on the frequencies of single letters can be replaced by the entropy based on the frequencies of digrams, then of trigrams, quadrigrams, and so on indefinitely. A full account of the pattern dynamical system ought to take into account the effects of the longer and longer patterns that lie buried in its symbol strings.[19]

4.5 Conclusion

In the dynamics of pattern there have to be two features: difference and what brings about difference, or to put it the other way around, change and that which changes. This chapter is about patterning that arises in the context of discrete changes within a finite set of symbols.

The simplest idea of distinguishing is the yes-no, on-off, zero-one, yin-yang duality. This is formalized into the primitive "alphabet" $\{0, 1\}$. Within this tiny context, change arises in the form of strings of these two symbols or, as we come to call them, binary bits. Simple though this idea is, the ability of long zero-one strings to carry pattern and information is undeniable in our computer age where we are surrounded by devices that are built entirely around this principle.

An important, but often overlooked, aspect to this is context. Zero and one lie here within the context of the set $\{0, 1\}$ which again, though very simple, provides the setting in which distinguishing can take place. Once dynamics (the process of change) is added, we are looking at long strings of zeros and ones, which we view as running from left to right, just as we see strings of letters writing English text. The set of all these possible strings is the larger context that includes the possibilities of change.

The dynamics operates by dropping off the first letter of the string and moving it leftwards by one step. Here a typical mathematical abstraction takes place. We would normally speak of this as "reading" the first symbol of the string, but mathematically no reader is assumed, nor is there any assumption of understanding or extraction of meaning. Reading has been replaced by the simple action of dropping off the first symbol. Within our minds we may think, if we wish, that we have read the first symbol, but that is something we add in to give meaning to the shifting operation.

We chose to make our strings to be unbounded, because doing so makes the dynamics easier to deal with: dropping a symbol and shifting left just replaces one infinitely long string by another. The cost of this abstraction is minimal—in most data sets, the number of binary digits looks pretty much infinite anyway. The new objects are infinite strings of binary bits and the new context is the set of all such strings. Each individual infinite string is a single state of this new system.

When we move to strings of English text, the primitive alphabet $\{0, 1\}$ is replaced by an alphabet $\mathscr{A}$ consisting of our familiar alphabet of letters together with an extra symbol for spacing to separate words. Again we have a shift system, $\mathscr{A}^\infty$, whose states now are the unbounded strings of letters from $\mathscr{A}$. Indeed we can form such shift systems with any finite set of symbols.

Genuine patterning develops when on top of this we introduce the idea of the frequency or probability of occurrence of symbols. The probabilities of the various letters A,B, ..., Y,Z,□ occurring are attached to the corresponding dictionary sets $[\mathrm{A}], [\mathrm{B}], \ldots,$ $[\mathrm{Y}], [\mathrm{Z}], [\square]$, the subsets of states (elements of $\mathscr{A}^\infty$) that begin with these various letters. Thus probability is attached to subsets of a partition of $\mathscr{A}^\infty$. This can be refined to probabilities of all the letter pairs (digrams), and hence to the subsets of a finer partition $[\mathrm{AA}], [\mathrm{AB}], \ldots, [\square\square]$. If one knows enough about the system, the probabilities or frequencies of trigrams, quadrigrams, and so on might be available. With each of these the structure of the patterning becomes clearer.

What we have created here appears in many forms in our interactions with the world. Of course writing in alphabetic languages is entirely based on strings of symbols from a fixed alphabet. But if we think about it, speaking is formed out of long strings of units of sound. People who read braille are dealing with long strings of tactile sensations, and signaling can be done with flashes of light, pulses of electrons down a wire, or bursts of photons. In fact the idea of strings taken from a finite set of symbols is *the* basic form of discrete dynamics, and it is fertile ground for understanding pattern. No doubt, however you are reading these words right now, at least one of these discrete dynamical systems is directly implicated.

The formation of strings of sounds from a finite set of sounds is as old as human language, and probably is co-existent with the origin of our species. Written language has been with us for millennia. We have seen how the Chinese of ancient times tried to encapsulate the changing Cosmos and the shifting place of the individual within it by means of a form of discrete dynamics.

The work of Claude Shannon was the first attempt to study the role of communication within the mathematical setting of discrete dynamics. His study brought in other ideas, especially the statistics of letter frequencies and how they shape the pattern of written text. Pattern is something emergent from the process of change guided by probability. Then there is the important idea of information and how the vague and seemingly indefinable term can be abstracted into a usable (and even computable) form.

Out of all of this comes a clearer picture of what a pattern system is about. The main constituents are:

- context: a set, the state space, which serves as the context. Its elements are the states of the system;
- change: a "mechanism" or "operator" that invokes change. Here it has been the left shift. In the pendulum model it was time;
- difference: partitioning the state space, the breaking of it into disjoint parts which individually collect together the states with some common property, hence establishing a level of 'sameness" and "difference". Later, we will expand this idea considerably and adopt the probabilist's convention of calling these parts "events");
- law, or probability: a measure of the frequency with which states will be found in these parts (the frequency with which the events occur).

This is the basis on which we will continue to develop our ideas of pattern: the *pattern system*. Here we have considered aspects of the dynamics of *discrete serial* or *iterative* change. By contrast the pendulum operates under continuous dynamics, and we will see the importance of both forms of dynamics as we go on.

CHAPTER 5

Chance

5.1 The question of probability

5.1.1 Introduction

Questions of chance, randomness, and fate are deeply rooted in human consciousness. No wonder—our lives seem so dependent, often precariously so, on chance. The balance of fortune and fate seems always to be standing in front of us as we wait to see which way life will go. Gamblers thrive on the excitement of risk, but generally we dislike it—even fear it—particularly as we get older. We tend to think in terms of misfortune—trees falling down, rogue waves at sea, the stock market crashing, just being in the wrong place at the wrong time. Alternatively we can think on the brighter side of things, good fortune—meeting the perfect partner, actually being in the right place at the right time, having the winning lottery ticket, or even just the good fortune of having been born at all.

Nowadays we know much more about chance, and we know that in spite of its negative connotations, chance is a fundamental aspect of the functioning of the Universe. Far from being bad, without chance there could be no life at all, nor any of the wonderful processes of change that produce everything from galaxies to grasshoppers. We see the laws of Nature as the guides that order events and steer them in predictable ways. But laws are binding, and indeed by the end of the nineteenth century physicists had come to the pessimistic conclusion that the world was essentially deterministic. "Give me the starting conditions and I can tell you the outcome of all events for all time", was the sort of idea.[1]

The twentieth century brought in very different ideas as it was realized that classical physics could not describe events taking place at the atomic level. The outcome was quantum mechanics, a new kind of physics in which chance is no longer an outside spoiler but a fundamental player with a vital role. But quantum mechanics is not the main consideration in the importance of chance in our understanding of the physical world. Already by the late nineteenth century physicists were coming to the understanding that we live in a physical world made up of vast numbers of atoms and molecules, all of which are jostling around and affecting their immediate neighbors. Even if it were the case that all of these atoms were moving deterministically, in other words, not really randomly, still the only approach to understanding of macroscopic effects (like temperature or pressure) that appear at the scale of living beings like ourselves would have to be based on the statistical behavior of this vast number of minute particles.

The ultimate nature of randomness and chance remains a thorny and often disputed question, but nonetheless, the observed effect is the same. Nowadays we can see the effects of chance operating at all scales—the millions of spermatozoa involved in the fertilization of eggs in an ovary, the sensory perception of smell which depends on the chance motions of molecules, the spread of disease, the evolution and stability of ecosystems, and genetic drift in biological species.

Far from being just something we simply have to contend with, chance, along with the laws of physics which direct all physical processes, is in the very fabric of things. As much as laws are binding, chance is liberating. The modern view is to see that laws, or *principles* as we shall often call them, tame the wantonness of randomness. But we can also read it the other way: randomness frees us from tyranny of those laws. We live in a dance of ever changing entanglements of the two great processes of stability and change.

In spite of its omnipresence in our lives and the deep impression that it makes on our minds, it is a strange fact that we human beings have very poor intuitions about chance. Our notions of randomness and chance are vague and it has proven remarkably difficult to develop mathematics that can deal with them effectively. Mathematics is ancient, and has been developed to considerable levels in many cultures, but a real theory of chance (probability theory, and its applications in statistical analysis) is remarkably recent, starting only in the mid-seventeenth century, and even then mostly in the context of gambling and financial risk in insurance and annuities. Much of its relatively short history can be seen as struggles to conceptualize the illusive concepts of chance, randomness and their quantification, and the accompanying disagreements about the nature of probability. It was only in the 1930s that the foundations of what we now call probability theory were finally laid down, see [72].

From the point of view of our story, chance is a vital component of pattern. The leading idea in our conception of pattern is that it arises as an outcome of process. Anything that affects the process affects the patterning that emerges: in particular, randomness or chance are important. Chance can make outcomes less certain, but surprisingly, as we shall soon see, chance can also produce almost deterministic-like outcomes. Boiling water involves activating vast numbers of water molecules into random motions, but the boiling temperature emerges as something that is for all practical purposes completely invariant, dependent only on the air pressure. At the level of individuals, human or otherwise, things may be essentially random, but collectively their behavior may be quite predictable. Insurance companies depend on this—and very successfully too. The arrival and departure of cars from a stretch of freeway is essentially a random process, but beyond a certain density of vehicles traffic jams are a virtual certainty, and that threshold, that critical state, is quite predictable. The genesis of the neural networks that integrate sensory organs and the brain in the development of a baby are struggles between competing neurons that search for targets in an environment of chance and hazard. Still the outcomes are almost always quite normal and completely functional.

The purpose of this chapter is to introduce some of the basic ideas of probability theory, to establish probability as an important part of our conception of pattern systems, and to show that it is just as much a creator of pattern as any of the more deterministic looking ingredients.

5.1.2 Randomness

Without a doubt the idea of randomness, though we feel that we know what it means, is extraordinarily hard to pin down. By the time we are done we will see some of the modern ideas that deal with this elusive idea, but to get there we have to work slowly through examples that guide the way.

We begin with the familiar situation of tossing a fair coin and looking for heads or tails. By "fair" we mean that the coin has equal chances of falling head side up or tail side up. Simple as it seems, what we have just said already raises difficult questions. What does it mean to say that the coin has equal chances of falling head-side up or tail-side up? We can rephrase it in terms of probability: the probability the outcome of a toss will be a head is $1/2$, and likewise the probability that it will be a tail is $1/2$. But this just begs the question. Evidently this is not something we can witness by tossing the coin just once. The outcome of a single toss tells us nothing.

The statement has something to do with what happens in repeated tosses. But even there it is not so clear what it means to be fair. If we toss the coin a hundred times, should we expect heads to appear exactly 50 times? In fact that is quite unlikely—the probability of exactly 50 heads in a hundred tosses of a fair coin is less than $1/10$. On the other hand, if we tossed a hundred times and got 100 heads, we would have every reason to doubt the fairness of the coin. It is possible, but we are correct in surmising that this idea of fairness, or equal probability, is something that ought to emerge in long-range behavior. That is what we need to explore.

Let us start again. We have a coin which we can toss and get a head or a tail. We are interested in long strings of tosses of a fair coin, with the idea that in the long run fairness should mean that we more or less have the same proportions of heads and tails occurring. Mathematically it is easier to deal with ones and zeros instead of heads and tails, so we will make that equivalence in what follows, with heads being ones and tails being zeros.

A long string of tosses means that we get a string

$$\mathbf{x} = x_1\, x_2\, x_3\, x_4\, \dots$$

of ones and zeros. This looks very familiar from our shift system, and it becomes exact if we allow the string of tosses to go on forever. Then $\mathbf{x}$ is just an element of the shift space $\{0,1\}^\infty$, which here we will denote by X. Given the context, we call such an infinite or endless string a *trial*. The terminology is suggestive: we are interested in testing the hypothesis that heads and tails appear, on average, equally often, so we make a long *trial* of tosses to see what happens. Mathematically the "long" part is made very long indeed—unending. We will see that it is out of this "unendingness" that exactness appears.

In making this connection of coin tossing to states in the shift space on two symbols we return to the pattern system ideas that match those that Shannon brought to alphabetic strings and the frequencies of individual letters.

5.1.3 Average values and expectation

In order to test the hypothesis of fairness of tossing a coin, we should consider the ratio of heads to the total number of tosses: that ratio ought to be $1/2$, at least in the long run. Since we are giving heads the value 1 and tails the value of 0, we see that $x_1 + x_2 + \cdots + x_n$ is actually the number of heads appearing in these n tosses, and so

$$\frac{1}{n}(x_1 + x_2 + \cdots + x_n) \tag{5.1}$$

is exactly the value of this ratio after n tosses, e.g. if in a particular trial x the values of the first five tosses are 1 1 0 1 0 then the average is $3/5$ which is the ratio of the number of 1s to the total number of tosses. Our intuition is that in the long run this *average* will converge on the expected value $1/2$. In fact this value is called the *expected value* of the process.

This intuition is essentially correct, but there is a subtlety to it. At this point we have just seen that a trial amounts to a single element or state from the shift space $X = \{0,1\}^\infty$. In fact any state from X could possibly be such a trial. After all, the states of X are just endless strings of zeros and ones, so each possible state amounts to an infinite string of heads and tails, and *any* such string is potentially possible. So here is the question: is it true that

$$\frac{1}{n}(x_1 + x_2 + \cdots + x_n) \overset{n\to\infty}{\longrightarrow} \frac{1}{2} \tag{5.2}$$

holds no matter which element $\mathbf{x} = x_1\, x_2\, x_3\, x_3\, \dots$ we take in X?

Well, the answer is clearly no. For example, $1\,1\,1\,1\,\ldots$ (all ones) is a perfectly good string in X and obviously its average is equal to 1, not $1/2$. And of course there are countless other strings that fail to converge to the expected $1/2$, e.g. for $1\,1\,0\,1\,1\,0\,1\,1\,0\,1\,1\,0\,\ldots$, where every third toss is a 0, the average converges to $2/3$. Indeed there are countless strings that fail to converge at all.[2] Yet, in spite of this setback, the answer is essentially yes! To see what is going on we need look at probability from the perspective of the shift space X.

If we toss a fair coin, the chances that we get a head or a tail are supposed to be equal. So it is natural to assign probability $1/2$ to each of the dictionary sets $[0]$ and $[1]$. Remember that $[0]$ and $[1]$ partition the set of all states (trials) into those that start with a 0 (a tail) and those that start with a 1, a head. We would expect that if we make series of trials, half the time the first toss would be a head, and half the time a tail. Using $\mathbf{p}$ to denote probability, we write

$$\mathbf{p}([0]) = \mathbf{p}([1]) = 1/2\,.$$

When we toss the coin a second time, the necessary assumption is that the outcome of the first toss does not affect the outcome of the second toss. The coin does not remember what it did last time. This property is called *independence*. Thus whether or not the first toss was a head or tail, the second toss has equal chances of being head or tail.

The potential outcomes of a pair of tosses are $00, 01, 10, 11$ so in terms of strings we are looking at the four dictionary sets $[00], [01], [10], [11]$, and each has equal probability $1/4$ of appearing:

$$\mathbf{p}([00]) = \mathbf{p}([01]) = \mathbf{p}([10]) = \mathbf{p}([11]) = 1/4\,.$$

Continuing this way, we see that this ought to work in the same way for dictionary sets of any length and so for each dictionary set $[a_1\,a_2\,\ldots\,a_k]$ we should have the probability

$$\mathbf{p}([a_1\,a_2\,\ldots\,a_k]) = \frac{1}{2^k}\,,$$

since there are 2^k possible outcomes, and they are all equally likely. This is how we expect a fair coin to behave.

The idea of independence that we have stated here is actually something that our human instincts are loath to really accept. Even the famous mathematician d'Alembert (1717–1783) made the statement, "Therefore the more head occurs successively, the more it is likely tail will occur the next time. If this is the case, as it seems to me one will not disagree, the rule of combination of possible events is thus still deficient in this respect." He doesn't think that the correct analysis of possible events is actually true. Don't all of us think that after three heads have appeared that there is a higher probability that the next toss will produce a tail rather than another head? But it is false. What does the coin know about the previous tosses made with it? If four identical coins were tossed at once, would the outcome of the fourth one be altered because we saw that the other three were heads?

The general rule of probabilities for the occurrence of two *independent events* is the product of their individual probabilities. Thus if we toss a fair coin and at the same time toss a fair six-sided die, the probability of the event **heads-4** is $1/2 \times 1/6 = 1/12$. The probability of two successive tosses of a coin both being heads is $1/4$ and the probability of four heads in a row is $1/2 \times 1/2 \times 1/2 \times 1/2 = 1/16$ (corresponding to $p([1111]) = 1/16$), as we saw before. *This multiplication rule is the signature of independent events.* In particular, the probability of a head appearing after heads have appeared in the previous three tosses is still $1/2$. In fact the probability of any particular arrangement of heads and tails (in a fixed order) in a string of four tosses of a fair coin is $1/16$.[3]

Now if we forget about order, then for sure in four tosses there is a higher probability that there will be two heads and two tails than four heads, but that is only because there are more ways to get that outcome. The probability of exactly two heads in four tosses is the probability that it lies in the union

$$[1100] \cup [1010] \cup [1001] \cup [0110] \cup [0101] \cup [0011],$$

namely,

$$\begin{aligned}\mathbf{p}([1100]) + \mathbf{p}([1010]) + \mathbf{p}([1001]) + \mathbf{p}([0110] + \mathbf{p}([0101]) + \mathbf{p}([0011])\\ = 6/16 = 3/8,\end{aligned}$$

whereas there is only one way to get four heads, and that is $\mathbf{p}([1111]) = 1/16$.

This distinction between the "shapes" of various outcomes versus the number of different ways that they can appear plays a crucial role in natural events, as we shall see in § 5.2.

As we see here, we are gradually extending the scope of our probability function $\mathbf{p}$. It is now defined on all dictionary sets of $\{0, 1\}^\infty$, and also on (disjoint) unions of such sets by addition. Later we will define the concept of *measures* and refer to probability measures rather than probability functions. To prepare for this we will include the word measure when speaking of probabilities, though at this point it is just a technicality.

We can now return to the question about the convergence in (5.2), where we expect the outcomes of our coin tossing trials to produce the ratio $1/2$. Some trials (i.e. some states from X) are "bad" and don't produce the averages that we would otherwise expect. But surely most of them do, and this is the key point: the vast number of trials do behave the way we expect: so much so that if we gather together *all* the bad trials (states), the ones which don't average out to be $1/2$, into a single subset F of all bad trials then we find that the set F has probability 0, i.e. $\mathbf{p}(F) = 0$.

Even to write this down is in itself a revelation. The same probability function, or probability *measure*, $\mathbf{p}$ that, so far, we have defined only on dictionary sets, extends naturally to allow us to talk about $\mathbf{p}(F)$. Now we are claiming that the probability of any of the entire set of bad trials appearing is zero. This is quite an extension, and we shall see what lies behind it in §6.6. Nonetheless, taking it to be meaningful, we have arrived at a statement of the famous *strong law of large numbers*. It can be expressed by saying that if $\mathbf{x} = x_1\, x_2\, x_3\, x_3\, \ldots$ is a coin tossing trial, that is to say, the string of outcomes (1 for heads, 0 for tails) of tossing a fair coin, then the averages, as we take into account more and more of the string, will converge to $1/2$ almost surely. Written in the language of mathematics this reads

$$\frac{1}{n}(x_1 + x_2 + \cdots + x_n) \stackrel{n\to\infty}{\longrightarrow} \frac{1}{2}, \quad \text{almost surely.} \tag{5.3}$$

Here the term *almost surely* is given a technical meaning:

"almost surely" means
"except on a set of trials of collective probability equal to zero"
(with respect to $\mathbf{p}$).

A more positive way to say the same thing is that the convergence will happen with probability one. If the probability that a certain thing doesn't happen is p, then the probability that it will happen is $1 - p$.

We will see that the strong law is far more general than this simple case might suggest.[4] It finally finds a link between our efforts to formalize the notion of probability and our ordinary intuitions about it—*long term behavior*. We have the notion that the probability of tossing a head is $1/2$, and similarly the probability of a tail, and so we ought to see an equal number of both, on average, *in the long run*. Now we can put meat onto the bones of this intuition: random trials will almost surely produce exactly these expected values. The meaning of the arrow $\stackrel{n\to\infty}{\longrightarrow} 1/2$ is not something vague. It means that, with probability 1, the limiting value of the average, *the expected value* of the distribution will be exactly $1/2$.[5] Without the notion of "almost surely"—the allowance of a set of exceptional trials which collectively have probability zero of happening—this theorem could not be stated.

Notice the often misunderstood fact that zero probability does not mean that things can't happen! If we consider coin tossing with 0 and 1 outcomes, then the probabilities for one head, two consecutive heads, three consecutive heads, etc. is $1/2, 1/4, 1/8, \ldots$. The probability of a thousand consecutive heads is less than .000000 ... 0001 where there are 300 zeros. What would the probability look like for a million consecutive heads, or a billion, a trillion? The probability of an entire infinite string entirely consisting of heads is clearly less than any particular positive number we can think of, no matter how small we choose it. That leaves only one possible value for the probability—zero. But still, the string consisting of all 1s (all heads) does exist as an allowable string.

In exactly the same way, the probability of any particular *pre-assigned infinite* string of heads and tails is zero. Each individual state of X has zero probability of appearing. This is one of those paradoxes of the infinite: every individual string has zero probability of appearing, but nonetheless, if we were capable of tossing a coin forever, some infinite string would appear. Perhaps another way to look at this is to think of shooting arrows at a target. If we examine things with infinite accuracy then certainly there are infinitely many points making up the target, and the probability that the archer will hit a pre-chosen point with infinite accuracy is zero. Nonetheless, his arrow will hit some point. In passing, it is worth noting that the word *stochastic*, which is often used as an alternative for *random* when referring to dynamical processes, derives from the Greek word στοχος *stokhos* "a guess, aim, fixed target, erected pillar for archers to shoot at".

The power of the strong law of large numbers

The strong law of large numbers is much more useful than it first seems. As the brief discussion on information in §4.4.4 in the case of $X = \{0,1\}^\infty$ suggests, it is not sufficient to consider only the case where the probabilities of 1 and 0 are equal. We might be in a biased situation where we have a probability p for heads (1) and $1-p$ for tails (0) where $0 \le p \le 1$, and $p \ne 1/2$. Everything still works, though now the *expected value* of the ratio (5.1) of 1s to the total number of tosses converges to $p1+(1-p)0 = p$, and again, this happens almost surely, i.e. with probability 1, *but now with respect to the* new probability function (measure) $\mathbf{p}$. Let us write $X(p)$ for $X = \{0,1\}^\infty$ in which 1 occurs with probability p.

The physical way in which the process of making a trial is imagined is just as before: we flip the p-weighted coin again and again, without end, and thereby create one state $\mathbf{x} = x_1\, x_2\, x_3 \ldots$ of $X(p)$. From our perspective of flipping the coin, the idea is that we are revealing more and more about some as yet unknown state $\mathbf{x}$: first that it is in $[x_1]$, then in $[x_1\, x_2]$, then in $[x_1\, x_2\, x_3]$, and so on. This is the same old shift space, but now with a different interpretation, in effect a different probability function (measure) $\mathbf{p}$. Now the probability of $[1,0,1,1]$ is $\mathbf{p}([1,0,1,1]) = p^3(1-p)$ while $\mathbf{p}([0,0,0,1]) = p(1-p)^3$. If $p = 3/4$ then $1-p = 1/4$ and these values are $27/256$ and $3/256$ respectively, very different from the equal values of $1/16$ that we would have had before, when $p = 1/2$. With this interpretation X is called a *Bernoulli shift* and trials (states) from $X(p)$ are called *Bernoulli trials*.[6] The strong law of large numbers still holds, but now it reads:

$$\frac{1}{n}(x_1 + x_2 + \cdots + x_n) \overset{n\to\infty}{\longrightarrow} p\,,$$

almost surely. The value p is then the expected value of the Bernoulli process. Again the "almost surely" is there, indicating that whatever "bad" strings arise , say ones which are simply long strings which are all heads, or ones which are all tails, or ones in which we put in two heads for each tail, collectively these exceptions amount to a set of states to which our new probability function (measure) $\mathbf{p}$ will assign the value 0. For all other trials the average will be as expected, namely p.

5.1.4 Random processes and dynamics

It is evident from the whole discussion that randomness is a feature that emerges from process, and hence has a dynamical origin. The process of coin flipping is a process of producing a sequence of head/tail or one/zero outcomes, and this has the nature of a dynamical process. In fact this process can be interpreted as the now familiar shift dynamics of the $\{0,1\}^\infty$ shift system. But as with many things in mathematics, interpretation is crucial in making the connection between mathematical systems and the "real world". This is an example.

To keep things simple, suppose we go back to the situation of tossing a fair coin, the case $p = 1/2$. We have interpreted coin tossing trials as infinite strings of tosses that have the outcomes 1 or 0. Each trial is one infinite string $\mathbf{x} = x_1\, x_2\, x_3\, \ldots$ and hence, a single element, or state, of the shift space $\{0,1\}^\infty$. The dynamics of the shift space is the left shift s that moves a string one step to the left, dropping off the first symbol, replacing $\mathbf{x}$ by $s \triangleright \mathbf{x} = x_2\, x_3\, \ldots$.

The obvious interpretation of this is that we have made a trial and the shift reads off the successive outcomes of the tossing. But this all looks a bit backwards: why would we be interested in the results in this order given that we have made the trial and seen all these outcomes already?

In fact the entire thing is in a sense backwards, because the way we have to deal with randomness is not by going out and deliberately creating random events (which is what we are doing in coin tossing), but rather observing them as they arise in some process and in watching their progression, trying to make probabilistic inferences about what will appear in the future. A natural way of coming to grips with chance is to take the position that there is a process that is creating an infinite (or at least indefinitely long) sequence $\mathbf{x} = x_1\, x_2\, x_3\, \ldots$ of events, and we are watching this sequence unfold as we see first the outcome x_1, then next the outcome x_2, and then x_3, and so on. The string of $0/1$ events appears to us one event at a time and it is in effect the string $\mathbf{x}$ shifting leftwards in front of our eyes. We read the events as they appear, learning more and more about the statistical nature of the process that we are observing. It is exactly the process of the standard shift space. Furthermore this dynamics is crucial to our attempts to predict future outcomes. The nature of the probability distribution that underlies the system will only appear as the process unfolds. As we see more and more, we know that the law of large numbers ought to lead us to valid conclusions about the nature of the probability law that we are observing.

So the reinterpretation is that there is implicitly some state $\mathbf{x}$ of the shift system that we are observing, but we don't know what it is. Our task is to make valid interpretations about the process that we are observing and so gain some insight into what to expect next. This is entirely in accord with the actual way in which we have to deal with chance. We observe, and on the basis of observation we predict. This interpretation is one we can make equally well for any Bernoulli shift $X(p)$, and in fact quite generally in thinking of stochastic processes.

We will eventually subsume the strong law of large numbers itself into the totally dynamical framework of the Birkhoff ergodic theorem, which rephrases these ideas into a powerful and more intuitive form.

5.1.5 Is this deterministic?

Although we have not brought it up so far, both the pendulum and the shift space are what are called *deterministic* systems. What that means is that the future of a state under the dynamics is completely predetermined by the state itself. The future behavior of a pendulum is set once its angle and angular velocity are known. The outcome of shifting an infinite zero-one string $\mathbf{x}$ is known once we know the string. These dynamical systems are deterministic, meaning that the dynamics proceeds from any given state without any choice or randomness involved. All dynamical systems have this property. That does not stop them from being used in probabilistic situations, as we have here.

Look more closely at coin tossing: a trial amounts to "choosing" some state $\mathbf{x}$ of the dynamical system. The probability aspect is embedded in the fact it is the ongoing making of the trial (coin tossing) that reveals the state. We can think of a trial as randomly choosing a state, but in reality a trial is the long chain of tosses that gradually makes up a state. So the way it is conceived is that an infinitely long trial will produce some state, and we can, if we wish, think of this state as somehow having been pre-chosen, with the long process of the successive tosses of that trial revealing what it is. It is on the basis of what we see being revealed that we make probabilistic guesses as to what is to come and how frequently various finite patterns of zero and ones will appear.

There is definitely a subtlety here. When we are studying a supposedly random process, we may not know whether or not it is truly random. In fact there is a major philosophical divide on this question: are there truly random events in our Universe or is everything deterministic, even if some of the deterministic features seem to be totally hidden from us? This is not a question that the mathematics takes any side on. Whatever the process, random or not, it will produce a sequence of events. If it is continued long enough it might be considered as infinite, in which case it has ultimately determined a single state of our state space. The sequence of events that we observe is interpreted as the outcome of the repetitive shifting of this state. As for some ultimate "cause" of the particular state that is revealed, it is left unsaid whether that be deterministic or truly an outcome of chance. This type of dynamical interpretation is not some special fabrication created just for this book. It is a standard way these situations are modeled. They are called *stochastic processes.*

5.1.6 Randomness itself

Although we will wait until Ch.6 to make it really precise, the general idea of this approach to random variables is not far away from the examples we have just seen. What is perhaps most surprising is that randomness is not specifically defined in the way we might hope for. Rather it relies on interpretation of the formalism of measures.

In the discussion that follows, it might be helpful to keep in mind the words of the great mathematician Andrey Kolmogorov (1903–1987), who (in the 1930s) first put forward the approach to randomness that we will follow here.

> Every axiomatic (abstract) theory admits, as is well known, of an unlimited number of concrete interpretations besides those from which it was derived.
>
> [99, A. Kolmogorov]

The key word here is "interpretations". The actual mathematical theory involved in probability theory is, as we have said, the formalism of measure theory, something that we will look at later. What Kolmogorov points out is that measure theory was never invented with probability in mind, but it can be *interpreted* so as to fit our intuitions of randomness. This idea that our mathematical ideas draw meaning from the way we interpret them is important to understand. Mathematical reasoning may produce mathematical conclusions with great certainty, but the validity of the interpretation will always be put to the test by how well it accords with our actual experiences of the world. In fact the mathematics of probability theory fits it extremely well!

First a bit of terminology. When we are tossing a coin we will often speak of this as being about a *random variable* which takes two possible values tails/heads, or more conveniently from the mathematical point of view, $0/1$. The outcome has two values, hence it is a variable, and this outcome is supposed to be "random", hence it is called a random variable. In the case of a six-sided die we have a random variable with six possible outcomes, namely $1, 2, 3, 4, 5, 6$. We are not necessarily supposing that there is any fairness or equiprobability here: the die could be loaded. The only requirement is that the probabilities of the various outcomes $1, 2, 3, 4, 5, 6$ add up to one.

Quite generally, then, we start with a finite set of values that are the possible outcomes of the process we have in mind:

$$V = \{v_1, \ldots, v_k\}.$$

These are the values that the "random variable" can take on. Thus for instance in the coin tossing situation $V = \{0, 1\}$ and for a six-sided die $V = \{1, 2, 3, 4, 5, 6\}$.

With this in hand we can construct the shift space V^{∞} of all infinite sequences made up from the elements of V. This formalizes the idea of an extended process: each of these sequences can be viewed as a *trial* in which we watch the unfolding of the random process as a sequence of values from V.

The probability part arises by assuming the existence of a probability function (measure) $\mathbf{p}$ that assigns values to all the various dictionary sets that we can make with the set V. Here there is a major shift in point of view. In the coin tossing or the die tossing scenarios, successive tosses were considered to be independent, in the sense that the present throw does not affect the outcome of the next throw. This is not something that we wish to assume! In looking at Shannon's work we saw that he definitely was thinking in terms of random strings of text, but he took into account that consecutive letter pairs (digrams) do not simply occur with the product of the probabilities of appearance of the two letters of which they are composed. In many stochastic processes the present situation very much affects the future.

Still, the idea is that $\mathbf{p}$ assigns values to all of these dictionary sets. This function is subject to the properties of a probability function (measure) that we expect: for instance that it only takes values that lie between 0 and 1, that the $\mathbf{p}$ assigns the value 1 to the entire space V^{∞}, and that the probabilities for disjoint dictionary sets add, just as we have seen before. For instance, in the coin tossing scenario we have

$$\begin{aligned} \mathbf{p}([00]) + \mathbf{p}([01]) + \mathbf{p}([10]) + \mathbf{p}([11]) = 1, \text{ and} \\ \mathbf{p}([000]) + \mathbf{p}([001]) + \mathbf{p}([010]) + \mathbf{p}([011]) \\ +\mathbf{p}([100]) + \mathbf{p}([101]) + \mathbf{p}([110]) + \mathbf{p}([111]) = 1, \end{aligned}$$

and so on. But even if we had some strange coin for which successive coin tosses were not independent, still these equations must hold, since in both cases the dictionary sets account for the entire space.

But that's it. Just three things go into the definition of a random variable:

- a set of potential values of the random variable,
- the construction of the corresponding shift space,
- the assumption of a function defined on the dictionary sets that behaves as we would expect a probability function (measure) to behave.

We have been deliberately vague here. In Ch.6 we will formulate this idea with complete precision.[7] But here we see already the basic ideas behind the definition of a random variable.

The most straightforward situation, like that of coin or dice tossing, is where successive outcomes of the process are independent of everything that has occurred before.

Recall that we call a random process *independent* if what happens next as we carry out a trial is totally uninfluenced by what has happened up to that point. In this case the situation of probabilities is straightforward: in effect it reduces to the probabilities for the simple dictionary sets $[v_i]$. The probability of v_i appearing is some value p_i, and this is the probability of v_i appearing wherever we are in a trial. The probability of the sequential pair $v_i\,v_j$ is $p_i\,p_j$ wherever in the sequence we are, and so on for all other dictionary sets, just multiplying the probabilities for the sequence of events. Given any set of non-negative numbers

$$P = \{p_1, \dots, p_k\} \text{ subject only to } p_1 + \cdots + p_k = 1\,,$$

we can produce a random variable in this way. Such a variable is usually referred to as an *independent identically distributed random variable* or simply i.i.d.

In the fair coin tossing situation we have $V = \{0, 1\}$ and the probabilities are $P = \{1/2, 1/2\}$. For a Bernoulli shift $V = \{0, 1\}$ and $P = \{p, 1 - p\}$. For a six-sided fair die $V = \{1, 2, 3, 4, 5, 6\}$ and the probabilities are $P = \{1/6, 1/6, 1/6, 1/6, 1/6, 1/6\}$.[8]

The concept of a random variable would not be useful if it did not go beyond simple independence. As we have seen, many random processes *do* depend on what the previous outcome was, or even what the previous outcomes were. The concept applies to any situation, just as we have described it, no matter what the complexity of the probability measure is, as long as it obeys the simple rules for dictionary sets that we have suggested. This will look a lot more natural when we get to Ch.6.

What is perhaps surprising about all of this is that we seem to be actually creating a random variable here, according to our wishes. We are not assuming a pre-given situation. We only need to give ourselves a set of numbers (the potential values of our desired "random" variable), form the space of infinite strings of these numbers, and then attach probability numbers to the resulting dictionary sets, subject only to the natural conditions required to make sense. Then, voilà, we have a random variable which behaves in the way we designed it to do.

What is fascinating about it is that it so neatly skirts the problem of trying to define what randomness and chance are. There is actually no sign of randomness or chance in the definition. That is reserved for our conceptualization and interpretation. In doing so it avoids any of the metaphysical assumptions that tend to surround chance. Ultimately what makes this way of defining random variables viable is that it can be shown that the law of large numbers applies. In other words, the key intuition that we have about probability and randomness, namely that the underlying statistics will emerge if we allow the process to go on long enough, is not just a hope but is an actual theorem.

The process produces the outcomes that we would expect of a random process. All we have done is to lay a probability distribution (in effect, numbers for all the various dictionary sets) onto the shift space formed from the set V. The mathematics of it (which actually takes some work to do) then shows that our intuitions about how randomness ought to be manifested are upheld. The virtue of it is that it is a plain mathematical setting that can be explored like any other area of mathematics.

In the case of i.i.d. variables, the result is that for any trial the average of the string of values from V that emerge will tend towards the expected value

$$p_1 v_1 + p_2 v_2 + \cdots + p_k v_k ,$$

almost surely, i.e. each variable contributes to the expected (or average) value with its associated probability.

Experiments in randomness

The deeper one looks into randomness, the more strange the idea seems to be. The foundations of probability theory manage to evade the actual production of randomness, but people who run trials and test theories need sources of randomness and have no choice but to try and find or to try and create such sources. We have been using the standard idea of tossing coins or dice to produce random sequences. We could have talked about shuffling a deck of cards or shaking a jar full of colored stones and drawing "randomly" from that, but that simply begs the question. Is flipping a coin guaranteed to produce what we would be willing to call randomness? Who does the flipping? How high should it go, how many spins should it make to be called a good flip? Some people are actually skillful enough to very much affect the outcome.

In a study [45], the authors (who are very respected probabilists and statisticians), observing that a coin tossing machine could be set to produce heads every time, began looking more deeply at the physics of a rotating coin. They found that the way in which the coin is flipped, the precession that accompanies the spinning, can introduce a bias into the expectedness of head and tails. In other words it was possible to introduce bias, unexpectedly or intentionally, into coin flipping. So, if these humble but time honored ways of producing random sequences are not in fact random, what other ways are there?

As a matter of fact, the actual creation of random strings of say zeros and ones is a serious business, and it is difficult (to put it mildly).[9] It is hardly surprising that strings of random zeros and ones are crucial for computer simulations or for any testing of statistical processes, but exactly how does one go about producing such things in a computer setting? How can a deterministic machine like a computer, that simply follows the steps prescribed by the algorithms with which it is presented, produce something that is truly random or unexpected? How can we even test a random number generator to see how good it is at really producing random strings? Certainly the mere fact that there tends to be an equal number of zeros and ones is no proof of randomness. The string 0101010101 ... does that, but it doesn't look random. In fact one can probably fabricate strings of digits that will pass any particular test for randomness, while failing others.

There are suites of tests for randomness and there are computer algorithms, pseudo-random number generators, that do produce sequences that pass such frequency tests, though not of indefinite length. Often they are evoked by initiation with a "random seed", say the present state of some register in the computer that varies frequently in time and is unassociated with the random number generator itself. But still, since the total number of states in a computer is finite, repetition will inevitably occur, and given the same seed it will produce the very same string of zeros and ones as it did the last time.

Chance underlies the outcomes of many quantum processes, and these outcomes are believed to be entirely random, see Ch.13. Relying on a physical facility that can supply random sequences of numbers via quantum processes is inconvenient, but such facilities are available. However, even they are not without issues. An interesting study [27] looked at five sources of "random zero-one" sequences including two that derive directly from quantum mechanical processes. For each of the five sources, its authors took ten samples, each consisting of 2^{32} (over 4 billion) binary bits (zeros and ones). Two were generated by quantum mechanical devices, two from the standard mathematical computer platforms, *Maple* and *Mathematica*, and one from strings of digits taken from the binary expansion of the ever fascinating number π. The computers used the pseudo-random number generators of the two mathematical systems, and π is just π. The two quantum mechanical devices (the commercially available QUANTIS and the source provided by the Vienna Institute for Quantum Optics and Quantum Information) both use a source of photons that are separated into two paths by a beam separator. The physics behind the two is slightly different, but both involve the issue of the exact adjustment of the beam separator. The study passed the data through a whole suite of tests, including random walks (see §5.3) and the obvious one called the *Borel normality test*, which simply says that dictionary sets should have the frequencies we expect, e.g. every dictionary set of length three should occur with frequency $1/8$.

Surprisingly the results showed statistically relevant differences between computer sources and quantum sources, and even between the two quantum sources. What does this mean? The authors conclude, "We close with a cautious remark about the impossibility to formally or experimentally 'prove absolute randomness'. Any claim of randomness can only be secured relative to, and with respect to, a more or less large class of laws or behaviors, as it is impossible to inspect the hypothesis against an infinity of—and even less so all—conceivable laws."

Thisbrings us back to our opening thoughts about randomness—how difficult it is to pin down this deeply intuitive notion. Mathematically speaking, we have a seen that in principle we can formalize the notion of a random variable in a consistent and logical way. However, this in itself does not amount to a definition of randomness. As a case in point, certainly π can hardly be called a random number, and its digits can be computed as far out as time, patience, and computing power will allow. However, the binary digits of π (or its digits to any other base) are very successful in passing all the standard tests for randomness, see for instance [188], though there is no known explanation of why this might be so. Whether or not in the physical world the output of such and such a process is really random is not answerable. Nonetheless, we can model random variables, and we do know that such models can be powerful predictors.

It is in *process* that our intuitions about chance really arise. Typically, human societies find themselves faced with processes that have a stochastic nature. We may not know whether their origin is purely random or whether they arise from hidden deterministic causes, but on the basis of observing a process in action we are made aware of, or can deduce, its possible outcomes and then form estimates of their frequencies of occurrence. Using those potential outcomes and their relative frequencies (interpreted as probabilities) we are creating a random variable. The mathematics justifies our intuitive sense of chance by placing the process into a simple model built around shift system. Laws of large numbers or their equivalents are mathematical consequences that then fulfill our expectations of what chance is about.[10] The question of hidden variables in the apparent randomness of quantum processes is an example of the continuing interest in these types of questions. A discussion of this is taken up in Ch.14.

We can, and should, ask whether or not this sort of two-way process—that is going backwards from observed frequencies to a genuine mathematical model and then using that model to predict the statistics of future events—is really justified. Is it really the case that if we are witnessing some sort of random sequence of events that this can be explained in such a mathematical way? There are many issues around this that will arise as we go on. But there is one obvious one which we shall discuss in the form of ergodic theorems that deal with just this type of question, §8.5.

5.2 A simple example of probability and law

Here is a very simple random process that takes place on a checkerboard. This is just a board marked off into 8×8 squares. We pay no attention to any coloring of the board itself. For us, each square is just a blank space. There are checkers of two colors, black and white. In each square of the board there is either a black or a white checker. Initially all sixty-four squares are filled randomly. The dynamics proceeds in the following way. One of the squares of the board is chosen at random, with equal probability for each square, i.e., $1/64$. Whatever checker is on that square is replaced by one of the opposite color. This process is repeated over and over. That's it. The set-up is illustrated in Fig.5.1.

This simple-minded random (or stochastic) process is actually a dynamical system, and in fact we shall model it as a pattern system. Before doing that, we take a quick look at what is going on. We have our checkerboard filled with black and white checkers,

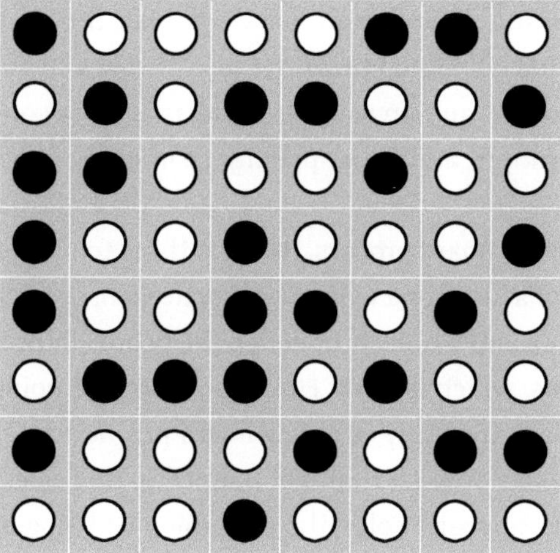

Figure 5.1: A checkerboard filled with black and white checkers. The dynamics proceeds by choosing one checker at random and changing its color.

and as time goes by the arrangements of checkers on the board change due to random checker replacements.

What should we expect? What sort of patterning arises in the long run? For instance, how often can we expect all the checkers to end up being black? How often can we expect an equal number of black and white checkers? The process proceeds entirely by chance, but it soon becomes apparent that pattern emerges.

The first thing to notice is that there is a sort of pull to keep towards equal numbers of black and white checkers. We can see this rather dramatically if all the board is covered with one color, say black checkers. Then the next replacement, wherever it occurs, will be with a white checker. In fact, whenever there are more black checkers than white, there is a greater chance that the next randomly chosen square will be black and so get replaced by a white checker. Similarly, if there are more white checkers than black it favors change to an additional black checker. So there seems to be a self-stabilizing feature emerging.

Although there is a random selection process working here, and in fact at each step the process of choosing the next square is totally independent of any previous choices, nonetheless the outcome of choosing a square, which is to alter the color of the checker on that square, *does depend* on the state of the board at that moment. The process is no longer independent of the past. The outcome of the next step depends on the white and black checkers on the board as a result of the process up to that point. On the other hand, it does not depend on how the board got to that configuration. It depends only on what that configuration is.

A process in which the next step depends only on the present configuration but not on how that configuration came about is called a *Markov process.* One can compare this with the die throwing Bernoulli processes, where neither the past nor the present state of the die enters into what the outcome of the next throw will be. The situation is now no longer one of *independent* random variables, and so is more complicated. Still Markov processes are relatively simple, and their appearance in many processes involving chance makes them important.

Returning to our checkerboard process, the next thing to realize is that since there are 64 squares and a choice of a black or a white checker in each, there are in total 2^{64} different arrangements of the black and white checkers on the board. In decimal terms, this is well over 10^{19} board positions, which is vast. However, from the point of view of the actual step-by-step process, there are really very few board positions it can be at the next step. In fact there are only 64, because all that will happen is that one of the 64 checkers will be replaced by one of the opposite color, the remaining 63 being left untouched. What's more, the questions we wish to ask can be put into a far simpler setting. Above, we asked about board configurations with no white checkers, or with equal numbers of white and black checkers. As long as we are interested only in questions about the numbers of white or black checkers, with no regard to their actual positions, we can drastically simplify matters by grouping all possible board positions into 65 sets $P_0, P_1, \dots, P_{64}$, where P_k is the set of all board positions with exactly k white checkers. Then we can write the set $\mathscr{B}$ of all possible board positions as

$$\mathscr{B} = P_0 \cup P_1 \cup \cdots \cup P_{64}\,.$$

This leads to a remarkable simplification. Suppose that we are in a particular state of the system, that is, a particular board position—with say 21 white checkers. Then there are 43 black checkers. Our next board position depends on which square we choose on the board. On the one hand, with probability 21/64 it will be occupied by a white checker. In that case that white checker will be replaced by a black one, and our next board position will have only 20 white checkers and 44 black checkers. On the other hand, with probability 43/64 our choice will be black and then the next board position will have 22 white checkers and 42 black. This little observation applies to every single one of the board positions in P_{21}. So we can summarize it by saying that if there are 21 white checkers, then at the next step there will be 20 with probability 21/64 and 22 with probability 43/64.

With this we can reduce our thinking to a much simpler system which we can think of as taking place with the "alphabet" consisting of the numbers

$$\mathscr{N} = \{0, 1, 2, \dots, 64\}\,.$$

If the present situation is that there are k white checkers, then the transition probabilities at the next step are

$$k \to \begin{cases} k-1 \text{ with probability } k/64 \text{ if } k > 0\,, \\ k+1 \text{ with probability } (64-k)/64 \text{ if } k < 64\ . \end{cases} \tag{5.4}$$

We can now think of the game as producing a long sequence of values in $\mathcal{N}$. Thus, if the initial board setting has 43 white checkers, the ensuing game will produce a string, something like:

$$43, 42, 43, 44, 43, 42, 41, 40, 39, 38, 39, 38, 37, 36, 37, 38, 37, 36, 35, 34, 33, 34, 33, \ldots\ .$$

What we have done is to rework our checker game into the setting of a shift space on the alphabet $\mathcal{N}$. We can grasp this in the simple picture presented in Fig.5.2. This is a typical way of displaying a Markov process, see §5.4.

To recap, our methodology is to start from some particular state or board position, which can be taken at will, and then follow the trail of subsequent states that arise as it is made subject to the random process of checker replacement that we have discussed. In doing so we construct a 'trial' or state of the state space $\mathcal{N}^\infty$.

We might guess that the strong law of large numbers might work here to tell us that no matter from what state we begin, in the long run a fixed pattern will arise. This is true, but it is slightly different now. From the point of view of the numbers of white or black checkers, with no attention paid to the actual positioning of the checkers, there are 65 different outcomes, and every single one of them has chances of recurring. Over time they will all recur over and over. However, their rates of recurrence are quite different. There is just one board position with no white checkers, there are 64 with exactly one white checker, 2016 with two white checkers, all the way up to $1,832,624,140,942,590,534$ with exactly 32 white checkers. Thereafter they decline again, with the same numbers now referring to black checkers. What happens in the long run is not a single number but a probability distribution, showing how the probability is distributed over the various possibilities from zero to sixty-four white checkers. This is what we show in Fig.5.3. Going from the various numbers of possibilities to probabilities is just a matter of dividing by the total number of possibilities, 2^{64}. The outcome of the law of large numbers is that, unaffected by where the game starts or what actual random choices we make, in time this

Figure 5.2: The 65 states of the checkerboard game shown as a graph along with the transition probabilities. The top numbers are the transition probabilities from right to left, and the bottom numbers the probabilities from left to right.

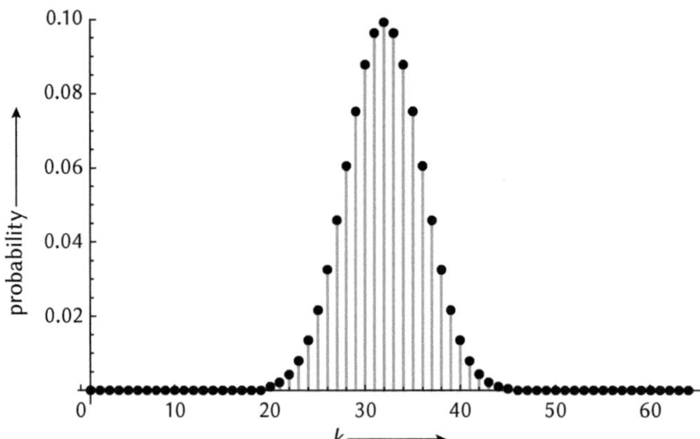

Figure 5.3: The distribution for the probability of k white checkers and $64 - k$ black checkers in the random checker-switching process. This is an example of a binomial distribution, see the Endnote 11. The same graph can be viewed as describing the relative numbers of the various board positions with $0, 1, 2, \ldots 64$ white checkers. The height of the graph at k is the number of board positions in $P_k/2^{64}$.

probability distribution will appear, though again there is an "almost surely" exclusion clause.

As expected, the distribution is steeply concentrated towards middle values, around $k = 32$, in other words steeply concentrated towards equal numbers of white and black checkers. In fact the probability that the number of white checkers will find itself outside of the range 20 to 44 is 0.0016 and the probability of it ever becoming 0 or (64) is 5.42101×10^{-20}. Running the dynamics endlessly we would expect to see all the tiles to be black on average about once in every 1.8×10^{19} steps. To put this in perspective, if there were a billion transitions per second, then we would expect to see a completely black board position about once in every six hundred years. On the other hand a board position with 32 checkers of each color occurs about one in every ten steps.[11]

The lesson is clear. In individual events, chance seems like just chance. But large numbers of repetitions in space or time reveal underlying pattern. The probability distribution arising from a random process is called the *law of that process*, and in the end it can totally shape what happens. Pattern arises out of dynamical process in the form of a probability distribution.

There is another lesson to be learned here. It has to do with the power of grouping. Given what we have just said, it may come as somewhat of a surprise that in the long run every possible board position is just as likely to show up as any other. The probability is distributed over all of its 2^{64} board positions completely equally, each state having probability $1/2^{64}$. This is a very small number—in fact the same number 5.42101×10^{-20} that we saw above. However, as we have seen, when we don't pay attention to position but only to the *number* of white checkers in the board positions then things change radically. It is not only distinguishing that matters, but the various ways in which we *choose* to distinguish that are crucial. This is again something that points to the important role that partitioning makes to pattern. We only see what we care to (or can) distinguish.

5.2.1 A simple change and an extinction

The checkerboard process that we have just described leads to a sort of stability, but it is not a situation of equilibrium. All board positions are possible and in a true trial (infinitely many plays on the board) we can expect all of them will show up endlessly, all with their particular frequencies reflecting their varying probabilities. The dynamics is essentially self-correcting with an over-abundance of one color leading to a strong bias towards equalizing the numbers of white and black checkers. But it does not settle down to one single state.

A very simple rule change leads to a very different picture. Suppose that now when we randomly choose a square whose checker we plan to replace, we don't choose the opposite color but instead make another random selection from the other sixty-three squares on the board and use the color of the checker we see there.

What happens now? We can get an idea by looking again at the extreme situation in which every checker is black. We choose any square whose checker we plan to replace. It is a black checker, of course. Now we choose any other square, at random, and use the color of the checker we see there. Since there are nothing but black checkers, it is also black, so we replace the black checker with a black checker. Nothing has changed. The board is in equilibrium. No further plays will ever change it. The same goes if all the checkers are white. There are two equilibria, one with all black checkers and one with all white.

What about other situations? Any imbalance simply biases the system to more of that imbalance. If there are more black checkers on the board, then in the second choice we are more likely to choose a black checker. The more the bias the more likely it is to get more biased. The board will always end up all white or all black. Which one? Even if the board is originally set up with an equal number of white and black tiles, no matter how they are placed on the board, in the long run a bias shows up, and it will be sufficient to drive the board to one of its two equilibrium positions. Once it is there, it can never escape. There is an equal chance that the equilibrium that it goes to will be all white or all black, but almost surely the system will end up stuck forever in one of them. The resulting probability distribution for the equilibrium states is shown in Fig.5.4.

What we are seeing here is how even in situations that are operating entirely on chance, the dynamics may lead to the certain emergence of very pronounced effects. A more mathematical way to say this is that an underlying *probability law* will emerge out

Figure 5.4: The expected distribution for black and white checkers in the extinction game. The process is random, but now biases are not self-correcting, but actually self-enhancing. The long-term outcome is just one of two states, both equally likely, namely all black or all white. Thus we see probability $1/2$ for each of the states with zero or sixty-four white checkers, and zero for all others.

of the process. The law of large numbers is a statement to the effect that this hidden probability law, whatever it may be, will make itself known as we observe the process over longer and longer periods of time.

5.2.2 The Ehrenfest model

The checkerboard game described in Fig.5.1 that we have just described in some detail appears as an example in the fascinating book *The laws of the game* [50], and it was designed to illustrate a common real-world phenomenon, the *Ehrenfest model* of the equalization of pressure when two vessels containing gas are brought together.

Fig.5.5 shows illustrates the idea. Two containers of equal size and at the same temperature contain atoms (or molecules) of the same gas, but not necessarily the same number, and therefore not necessarily at the same pressure. The containers are joined together and their atoms are free to pass from one side to the other. Suppose that the total number of atoms is N. To keep in mind the size of the numbers that we are dealing with here, remember that for a liter of gas at standard temperature and pressure, N is of the order of 6×10^{23}. So let's suppose that in total we have a liter of this gas, and let's call the two containers the left and the right containers. We will label an atom black if it is in the left chamber and white if it is in the right chamber, so what was changing colors in the checkerboard process is now equivalent to atoms changing sides. The atoms of a gas move at random in all directions (their average speed is the temperature of the gas). In spite of all the complexity, the overall effect is that each atom is free to move from one side to the other. Of course any imbalance in numbers raises the chance of some atoms from the denser side moving to the less dense side. The primary difference between this

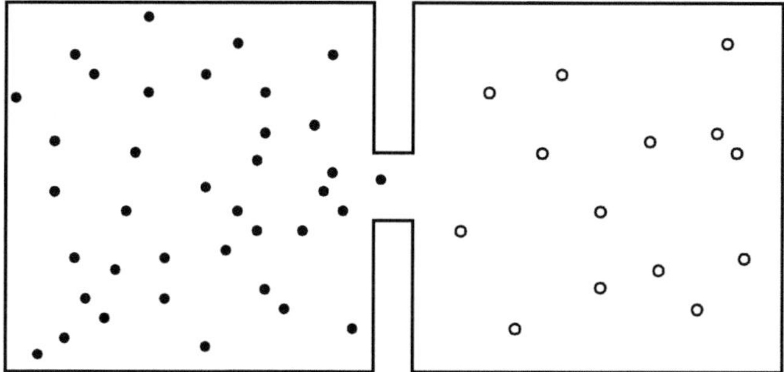

Figure 5.5: Two containers containing freely moving atoms of the same gas are joined together. Although the distribution of the atoms remains in the hands of chance as far as movement of the atoms is concerned, still, with almost certainty, it quickly becomes indistinguishable from being equally divided between the two containers. To match with the checkerboard example, atoms on the left side are colored black, those on the right white, so they are imagined to change color as they pass from one side to the other.

atomic picture and the checkerboard game is that the number of "checkers" has gone from 64 to the astronomical number N. The distribution is again a binomial distribution, but the shape is now incredibly sharply spiked at the center, that is at equal numbers of atoms in each container. Just how spiked this is is indicated by looking at the probability that the ratio of the number of white atoms to the number of black is greater than 1 by a hundredth of a million (10^{-8}). This turns out to be of the order of $10^{-500000}$—in other words, even this apparently minor deviation won't happen [62]!

This then gives a strong intuitive understanding of how and why the equalization of pressure occurs. It is a random process, the outcome of which, though random, is nonetheless for all intents and purposes completely deterministic. This is a sample of the idea lying behind statistical mechanics, which uses statistical methods to understand the bulk effects of vast numbers of randomly moving atomic particles. At any moment there is a vast number of potential micro-states (the actual locations and velocities of the individual particles), but viewed at the macro-state level (the pressure) the outcomes appear as equilibrium states, and they do so with some sort of statistical certainty.

The Ehrenfest model is due to the physicists Paul and Tatyana Ehrenfest (1907) who were interested in how equilibria are established. Often this model goes under the more amusing name of the *dog-flea model*—how the distribution of fleas tends to equality when two dogs spend time together, the fleas jumping randomly from one to the other.

5.3 Random walks

The year 1905 is called Einstein's *Annus Mirabilis*, the miraculous year in which he wrote four papers which were to change the direction of physics forever. The most famous of all is of course the paper in which he announced his theory of special relativity. Out of that came another paper announcing the theory of mass-energy equivalence, which we all know as $E = mc^2$. A third paper was on the photo-electric effect, in which he took up Max Planck's ideas on the quantum nature of light and went on to explain the photo-electric effect. In modern terms it says the emission of electrons from a cathode by the stimulation of light comes about by the transfer of the energy of single photons into the kinetic energy of single electrons. This is the paper for which he got his Nobel Prize in Physics. We will discuss this more in Ch. 13.

The fourth paper, and the one that we turn to now, is the paper on Brownian motion. *Brownian motion* is the visible motion of very small particles when they are suspended in a solution. They are said to have been first mentioned by the Scottish biologist Robert Brown (1773–1858) in 1828 after he observed, under the microscope, minute pollen grains jiggling around in a solution of water. Brown quickly found that this phenomenon had nothing to do with pollen grains, or even with living botanical materials, but that it appeared with any material particles as long as they were sufficiently small.

In fact the effect had been noted nearly two thousand years earlier by Lucretius and used by him to support the Epicurean atomic hypothesis, suggesting that the visible motion of very small particles is caused by atomic motion:

> Watch carefully whenever shafts of streaming light are allowed to penetrate a darkened room. You will observe many minute particles mingling in many ways in every part of the space, illuminated by the rays and, as though engaged in ceaseless combat, warring and fighting by squadrons with never a pause, agitated by frequent unions and disunions. You can obtain from this spectacle a conception of the perpetual restless movement of the primary elements in the vast void, insofar as a trivial thing can exemplify important matters and put us on the track of knowledge [108, pp. 2.120 –2.144].[12]

Nowadays we are familiar with the idea that Brownian motion is a result of the random motions of the atoms of the surrounding solution. But in 1905 there was not much support in the physics community for the reality of atoms. They were not "visible" with any technology of the time, so the value of the atomic idea was seen more as a theoretical device than a true feature of reality. It is true that Brownian motion was observed to increase with increasing temperature, and that fitted into a kinetic view of heat (that it was nothing other than the energy of moving atoms), but there were equally compelling reasons based on the assumed overall homogeneity of liquids and gases that impossibly small atoms could not produce random motions on particles the size of pollen grains. Einstein showed mathematically that it was possible.

Of course there is a whole story around the accumulation of evidence that ultimately made the atomic hypothesis unavoidable, but Einstein's work was a convincing step in that direction. The key principle now goes under the name of *random walks.* As it turns out, many random processes can be reformulated as being random walks. They are all Markov processes, random processes in which the present state of the system is relevant to what happens next, but not at all on whatever random incidents brought it to its present state.

The random walk is often imagined as the haphazard path of a drunkard as he lurches in random directions. Questions revolve around how far he is likely to stray from his starting point, and what the probabilities are that he will (randomly) find his way back to that point. In the case of Brownian motion, these lurches in a particle's path are due to collisions of atoms of the solution with the particles embedded in it.

The simplest random walk is in one dimension (1D) where the drunkard steps one step right or one step left, the choice being equally likely. In principle it does not matter whether or not all the steps are equal in length, but it is easiest to assume that they are. In this case the model is a line marked off with the integers. If he is presently located at the position marked by the integer n, then at the next step he is at either position $n-1$ or $n+1$, each equally likely. Evidently this is a Markov process: the next step depends only on his present position, but not all the steps he took before to get there. Fig.5.6 shows several runs of a (1D) random walk.

The two-dimensional (2D) random walk takes place on the grid in the plane made from all the points (x, y) where x, y are integers, but works in much the same way. Each step is in one of the four available directions, left, right, forward, backward, and each direction is equally likely, see Fig.5.7.

It is an easy step to move to 3D where the steps are now in six possible directions, left, right, forward, backward, up, down. In fact, though we don't have names to distinguish all the directions , the walk can be imagined to take place in any number of dimensions. All are Markov processes.

Bacteria are known to engage in 3D random walks. Certain species of bacteria (e.g. Escherichia coli) have flagella which they use to move themselves around. These flagella operate in two distinct modes—moving straight ahead or tumbling around—and they use them alternately, first taking a straight run for about a second and then tumbling for a moment before moving off again in a randomly selected direction. In effect it is a random walk. "An E. coli cell is about 2 micrometers in length and swims an impressive speed of about 30 micrometers per second, tumbling off in a new direction every second or so" [131, Ch. 11].

There is one constant feature that arises no matter what the dimension:

> The average (root-mean-square) distance between the final and initial positions of an n-step walk is proportional to $\sqrt{n}$.

Further details and a sketch of a proof of this are given in the Endnotes.[13]

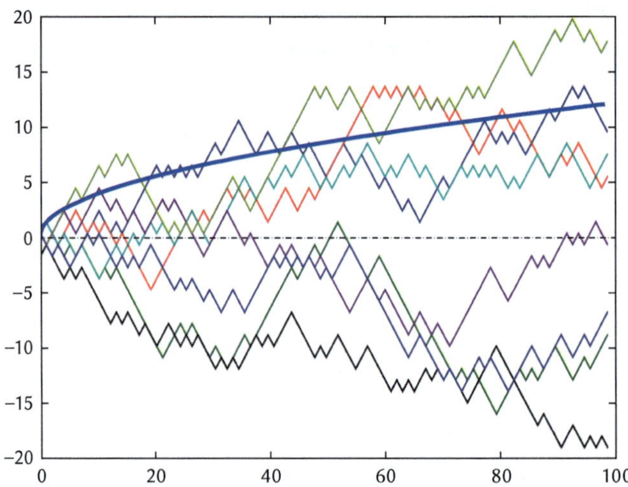

Figure 5.6: Here we see several 1D random walks, each with one hundred equal size steps. The steps are numbered along the horizontal axis and their accumulated effect along the line is shown on the vertical axis. Note that the graphs here are not pictures of the walk (which is on a line), but show how far the walk has strayed from the origin. A walk returns to its starting point, zero, whenever its graph crosses the horizontal axis. Some return to the start, even several times, some seem to move away. Not surprisingly, none gets anywhere near a hundred steps from the origin. Theory says that the average (actually root-mean-square) distance away from the starting point after n steps of the walking is proportional to $\sqrt{n}$, shown here with the blue curve. With probabilistic certainty, every 1D walk will actually return to the starting point (and indeed it will do so infinitely often since the Markov property implies that at any return, it is just like starting a brand new walk, and so it too will almost surely return, and so on). But there is no guarantee of how long one will have to walk before the next return.

Random walks are often thought of in terms of diffusion. If a point in space is considered as a source of particles that each undergo independent random walks, there will be a diffusion of those particles in space (1D, 2D, or 3D according to the dimension of the space). If we think of n steps taking n units of time then on average the diffusion takes place at a rate that is a constant times the *square root of the length of time* involved.

Einstein deduced this result, but was also able to work out the rate of diffusion on the basis of the physics of the situation and found that it was inversely proportional to the radius of the assumed spherical particles being pushed around. Putting these two facts together along with known physical constants he deduced the average step length. Measuring in micrometers ($1\ \mu m = 10^{-6}$ m $= 1/1000$ mm) and assuming particles of diameter $1\ \mu m$, Einstein concluded that the actual average rate of diffusion is about $0.8\ \mu m$/sec., or something close to $6\ \mu m$/minute. The limit of observable resolution of a conventional optical microscope is about $0.2\ \mu m$. In short, the motion of micro particles would indeed be observable in a microscope.

The central impact of Einstein's theory was its evidence for the truth of the atomic

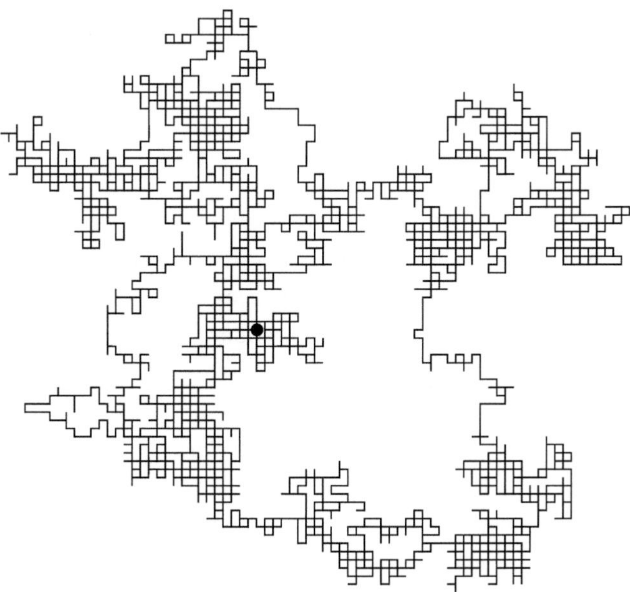

Figure 5.7: Here we see a single 2D random walk with 5000 steps. The actual order of the sequence of steps is lost in the figure, but we can see that the walk returns to the starting point (the black dot) several times. As in the 1D situation, it is a fact that almost surely every walk will return to the origin. Also, as with the 1D situation, the average distance from the starting position of a walk after n steps is proportional to $\sqrt{n}$. In fact this square root phenomenon is a universal property independent of dimension or the artificial unit steps in just horizontal or vertical directions. Importantly the almost sure return to the origin is *no longer true* in 3D.

hypothesis, namely that atoms actually existed. Key to the success of his theory were the lengthy and careful experiments of Jean-Baptiste Perrin (1870–1942), in which he experimentally confirmed most of Einstein's predictions for the stochastic behavior of suspended particles [93]. He received the Nobel Prize for Physics in 1926 for this and other work on discontinuous processes. In 1912 there also appeared the first indications of point-like diffraction in crystals, and that was another huge boost for the atomic theory. We will come to this in §12.3.

Random walks in 1D and 2D will almost surely (again with that probabilistic meaning) return to their starting points. Not so for three dimensions and beyond: something new happens. Although the average distance of an n-step walk is still proportional to $\sqrt{n}$, the chances of a return to the origin in a 3D walk is only about one-third. There's just too much room to get lost in 3D, and not surprisingly the chances of return continue to decrease in higher dimensions.

The difference between random walks in two and three dimensions is something that emerges in living systems. The efficiency of various enzymes is enhanced when they are situated on a membrane (2D). Substances utilized by enzymes have an affinity for the membrane surfaces where their random movements are more likely to bring them into the appropriate points of reaction [50, p.90] .

5.4 Probability in Markov processes

The checkerboard game and random walks are Markov processes. The potential states are the possible board positions and the location of the drunkard on the line. What makes them different from a process like coin tossing is that the next arrangement of checkers on the checkerboard depends on its present arrangement, and the next position of the drunkard depends on where he is now. Successive values of these new states are *not* independent: the present situation *does* affect the next. How the process got to its present state does not matter, but the present itself does. This is the hallmark of any Markov process. They are named after Andrey Andreyevich Markov (1856–1922).

This means that we are no longer in the scenario of independent events, and the probability of sequences of events can no longer be obtained just by multiplying the probabilities of those events. What is relevant are the *transitions* between states and the probabilities of these transitions occurring.

Fig.5.2, which we have already seen, is an example of a suggestive and very general way of creating simple Markov processes. In essence it pictures a *graph*, a set of nodes joined to each other by edges, along with arrows that indicate possible transitions. We should add in a word of caution. There are two separate uses of the word "graph" in mathematics. The first is the "graph of a function", see for example Fig.5.10. We will usually call them *plots*. The second is the one we are using here, a set of nodes or vertices and lines or edges joining them. The playing of the checkerboard game amounts to a "walk" on this graph, making choices at each node of which way to go. These choices are random, but are made on the basis of the probabilities indicated with the arrows.

This idea is easily generalized. Consider Fig.5.8. Here we have a three-letter alphabet $V = \{a, b, c\}$ viewed as nodes of a graph where there are arrows between the nodes. This is often referred to as a *directed graph*. Starting at any node and following arrows we trace out sequences of letters, say *abaaccabbbcccba....* Ignoring probabilities, which are indicated by the numbers over the arrows, we can transition from node to node at will, and each such (infinite) walk on the graph amounts to describing a single state of the shift space V^{∞}. In fact V^{∞} is nothing else than the set of all possible infinite walks on the graph.

However, we wish now to use the numbers beside the arrows as probabilities, so that the choices that we make at each step of the walk are determined randomly, but according to the probabilities on the arrows. It is immaterial how we make the choices as long as they are at random with the probabilities shown. Here we suggest the tossing of a regular six-sided die and a coin to make the choices, as suggested by Table 5.1.

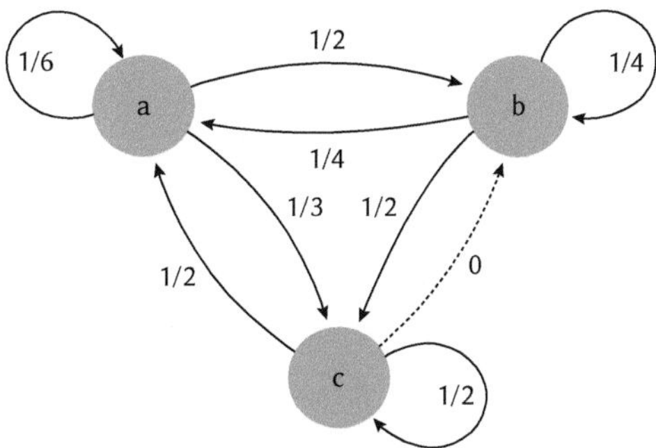

Figure 5.8: The figure shows a directed (meaning that it has arrows) graph with three nodes marked a, b, c. The idea is that one can walk around this graph from node to node following the arrows. At each node one has three choices of the next arrow to follow. The fractions on the arrows indicate the probabilities for choosing that arrow. We call these steps *transitions*, the corresponding probabilities *transition probabilities*, and the graph itself a *transition graph*. We have indicated the transition $c \to b$, which happens with probability zero (effectively it doesn't happen), by a dotted arrow.

Table 5.1: This table suggests a method for making the choices of direction at each step of the way, using a six-sided die and a coin. The table is laid out with the transition row letter → column letter, so if one is at node a, the transition $a \to b$ is made if the die rolls to one of $2, 3, 4$. A coin flip is used in addition to get the transitions from node b and a single coin flip at c. The transition $c \to b$ never occurs.

row → col	a	b	c
a	1	$2, 3, 4$	$5, 6$
b	$1, 2, 3, H$	$1, 2, 3, T$	$4, 5, 6$
c	H	—	T

Fig.5.8 enumerates the transition probabilities in what is called the *transition matrix*:

$$P = \begin{pmatrix} 1/6 & 1/2 & 1/3 \\ 1/4 & 1/4 & 1/2 \\ 1/2 & 0 & 1/2 \end{pmatrix} = \begin{pmatrix} \text{prob}(a \to a) & \text{prob}(a \to b) & \text{prob}(a \to c) \\ \text{prob}(b \to a) & \text{prob}(b \to b) & \text{prob}(b \to c) \\ \text{prob}(c \to a) & \text{prob}(c \to b) & \text{prob}(c \to c) \end{pmatrix}.$$

With this information we can compute the probabilities of walking any finite sequence of the nodes by simple multiplication. For instance, if we are standing at node a, then the probability that we will walk the transitions

$$a \to a \to c \to a \to b$$

on the path is

$$\left(\frac{1}{6}\right)\left(\frac{1}{3}\right)\left(\frac{1}{2}\right)\left(\frac{1}{2}\right) = \frac{1}{72}.$$

Since the transition probability $c \to b$ is zero, the probability of any part of a path that includes the transition $c \to b$ is always zero.

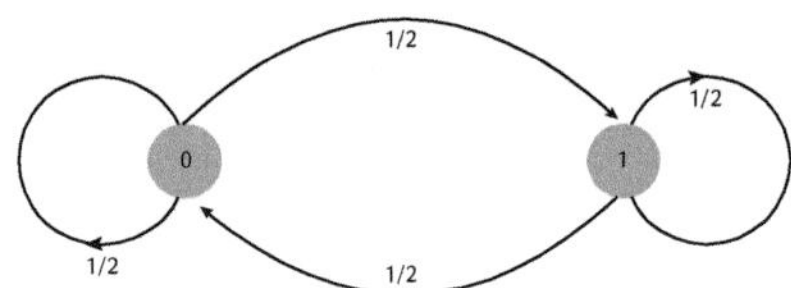

Figure 5.9: The full shift $\{0, 1\}^\infty$ can be modeled as a simple transition graph. Here, not only do all the transitions occurr with probability $1/2$, but also (not surprisingly given the symmetry of the situation) the probabilities of appearance of 0 and 1 themselves are $1/2$.

It looks like straight multiplication of probabilities, just as before. However, there is a difference. Strangely enough, although we are given all the transition probabilities of the two-letter dictionary sets, we don't seem to know the probabilities of the individual letters themselves! These must surely appear just by looking at the letter frequencies in long strings that arise from the process, in other words, they ought to emerge on their own from the process, and in fact they are directly computable[14]: we find that the probabilities of occurrence of a, b, and c are $1/3$, $2/9$, and $4/9$. Note that the transition probability for $a \to c$ is $1/3$, which is *not* simply the product of the individual probabilities for a and c: $(1/3)(4/9) = 4/27$. This is not a process of *independent* random variables.

In the example above we gave the probability for the sequence of transitions $a \to a \to c \to a \to b$, *given that we were already standing at a*. Since the probability that we will find ourselves standing at a is $1/3$, we see that the probability that if we jump into the midst of an infinite walk we will find ourselves looking at the string *aacab* is

$$\left(\frac{1}{3}\right)\left(\frac{1}{6}\right)\left(\frac{1}{3}\right)\left(\frac{1}{2}\right)\left(\frac{1}{2}\right) = \frac{1}{216}\,.$$

Quite generally the transition graph method produces true Markov processes, processes where the next outcome of the process is dependent entirely on the present state of the system. In special cases it can actually model the earlier model of identical independent distributions too. The graph in Fig.5.9 produces the full shift on two letters.

Rose-colored glasses

What we have just shown is that transition graphs give us new ways to assign probabilities to states in a shift space V^∞. The trials are still viewed as symbol strings of the form $\mathbf{x} = x_1\, x_2\, x_3 \ldots$ where all the x_j come from the set V. That is, they are all the states of the shift system V^∞. What is different is that we have put an entirely different type of probability measure on it by imposing transition probabilities for consecutive letters in the string. Some transitions may even be probabilistically zero.

All that is required is a directed graph and a matrix of transition probabilities. As for this matrix, the only requirements are that the entries be between zero and one (they are supposed to be probabilities!) and that the sum of the entries in each (horizontal) row of the matrix add up to 1. There is a simple reason for this requirement. At each vertex we must go somewhere on the next step, so the sum of the probabilities of the outgoing arrows must add up to 1.

The outcome is a shift system, in just the same way as before. What is really relevant is the probability measure that we have created to go along with it. This can be elaborated in the formalism of random variables that Kolmogorov proposed. Details can be found in the Endnotes.[15]

There is a somewhat subtle but important point about how a probability measure affects the way in which we look at the shift system V^∞. Although some potential paths are rare, or even impossible, we are still looking at the full entirety of all possible strings. We are not thinning out or removing states. Instead we have a probability measure and

we interpret the value of this probability measure on any particular subset as being the probability that a randomly chosen element from V^∞ will be found in that set.

Thus in our example the dictionary set $[bc]$ gets probability $1/2$ while the dictionary set $[cb]$ gets probability zero. Every dictionary set appears in V^∞, but with our probabilistic glasses on, some of them become invisible. Probability measures drastically affect pattern. In effect they are filtering glasses that change how things look from the viewpoint of the probability. It is just like putting on green, blue, or orange filtering glasses—the world looks different—whence the expression "wearing rose-colored glasses", in which our optimism about the world is made to increase. In this way we see the probability measure as really being the "law" of a process. The underlying dynamics is unchanged, but the measure changes everything by interpreting what is likely and what is not, even to the extreme of saying some events will, probabilistically speaking, never occur. On this view the probability measure is the ultimate shaper of the outcome of the dynamics.

Often, as in this example, we use a probability measure to effectively "throw away" a part of a state space. We could have banned the transition $c \to b$ by changing the state space $\{a, b, c\}^\infty$ to a smaller state space in which all states with the pair $\ldots cb \ldots$ are thrown away. Instead we have the simpler option of allowing all possible states, but letting probability do the job of throwing away. The probability measure has the ability to significantly shape the dynamics of a pattern system.

5.5 The pattern of errors

It seems strange that the process of making mistakes can also manifest itself in the form of well-defined pattern, but it does. The origins of this story begin, as so much of science does, in the study of the heavens, and in particular in the need for accurate measurements. Astronomical measurements were expressed in terms of numbers even as far back as the Babylonian times over three thousand years ago. However, the trouble with multiple measurements of the same thing is that one gets multiple readings, and paradoxically, the more the accuracy employed the more likely it is that these readings will not all be exactly the same. So naturally the question arises as to which reading is the best, or in a more sophisticated approach, given these readings what is the best guess for the correct value.

According to [158], the famous Greek astronomer and geographer Hipparchus (c.190 BCE–120 BCE), who was known to have used Babylonian sources for his work, favored choosing the mid-range values. The great Alexandrian, Claudius Ptolemy (c.100–c.170 CE), whose Earth-centric astronomy book the *Almagest* was to dominate all European cosmology until Copernicus, preferred to use the value that best suited his theory. So much like all of us! Tycho Brahe and his famous collaborator Johannes Kepler, as well as their contemporary Galileo, all realized the importance of repeated measurements and raised the question of how best to use their outcomes.

Given a set of numerical measurements, say $x_1, x_2, \dots x_n$ of the same thing, the question is how to choose one number x that best matches them. The two most obvious are the *mean* (or average) and the median. The *average* is

$$(x_1 + x_2 + \cdots + x_n)/n\,.$$

The *median* is the middle of the range, the value x_i which divides the range into half below and half above. Thus if n is odd, say $n = 7$, and the measurements are assumed to be in increasing order, so $x_1 \leq x_2 \leq x_3 \leq x_4 \leq x_5 \leq x_6 \leq x_7$, the median is x_4. If n is even, say $n = 6$, then it is customary to go half-way between the two mid-values: $(x_3 + x_4)/2$. Medians are often used in giving a useful guide for the cost of housing in an area: a few very expensive houses can completely skew the average cost, but not the median cost.

As it turns out, the median and the mean answer two different optimization problems. Let's go back to our n measurements $x_1, x_2, \dots, x_n$. Given any number x, the numerical size of the difference between x and x_1 is $|x - x_1|$. The absolute value sign $|\ \ |$ makes sure that we are looking at the numerical size of the difference, not the difference itself, which might be negative: e.g. $|12 - 17| = |-5| = 5$. We can do this with the remaining numbers $x_2, \dots, x_n$ and then look at the sum

$$|x - x_1| + |x - x_2| + \cdots + |x - x_n|\,. \tag{5.5}$$

One idea is that the best choice we can make for x, given that our measurements were $x_1, x_2, \dots, x_n$, is the value of x that makes this sum a minimum. We find the optimum value in the sense that it is the one that will minimize the sum of the numerical differences between x and all the measurements. It turns out that this is exactly what the median does.[16]

A second idea is not to use the absolute value, but instead to look at the difference of squares, $(x - x_1)^2, (x - x_2)^2$, etc. Now we don't need to worry about the signs since the squares make everything greater than or equal to zero. This time we look at the sum

$$(x - x_1)^2 + (x - x_2)^2 + \cdots + (x - x_n)^2 , \tag{5.6}$$

and look for the value of x that will minimize it (hence the terminology *least squares*). This turns out to be exactly the mean.[17]

Needless to say there was plenty of argument about which was the more appropriate of these, over the course of a hundred years, but gradually the mean or average value won out. This history is described in the highly readable and informative paper of Saul Stahl [158].

5.5.1 The normal distribution

Along with this arose the interesting question of how the distribution of the measurements would fall around the mean. The question runs like this. A quantity is measured and various values for it are obtained. Presumably it has an actual true value. If we assume that their mean is the best guess that we can make then we can examine how all these measured values lie around this mean, some above, some below, maybe some even spot on. There is clearly a randomness implicit here. But how do we expect these values to fall around their mean? Is there something invariant about this, some pattern to the errors, or is it entirely something without form?

It turns out that there is indeed an emergent pattern in the way in which the errors distribute themselves around the mean. It is actually a form that we have all learned to recognize because we often see the results of statistical trials and see how the results distribute themselves. This is the *normal* or *Gaussian* distribution, the familiar *bell curve.* The bell curve is so common now, and we are so used to data being clustered around its mean or average in the form of a bell curve that it is hard to imagine that this did not jump out as something obvious to the mathematicians and scientists who first thought about it. However, there is quite a difference between seeing the shape of a curve and actually knowing what its precise definition is, and from there going on to show that it is essentially the assured outcome of certain types of statistical trials.

Conceptually the normal or Gaussian distribution actually refers to a family of probability distributions, $\mathcal{N}[\mu, \sigma^2]$ that depend on two parameters, which determine the location of the peak and the spread of the bell—the *mean* μ and the *standard deviation* σ. We will use only the simplest form $\mathcal{N}[0, 1]$.[18]

Fig.5.10 shows the normal distribution.[19] As a model to keep in mind, think of arrows being shot at a one-dimensional target with 0 being the center. The curve now indicates how the shots will theoretically distribute themselves left and right around the center. The potential hitting points of the arrow form a continuum, and it is important to note an important distinction between how discrete probability tables and continuous probability

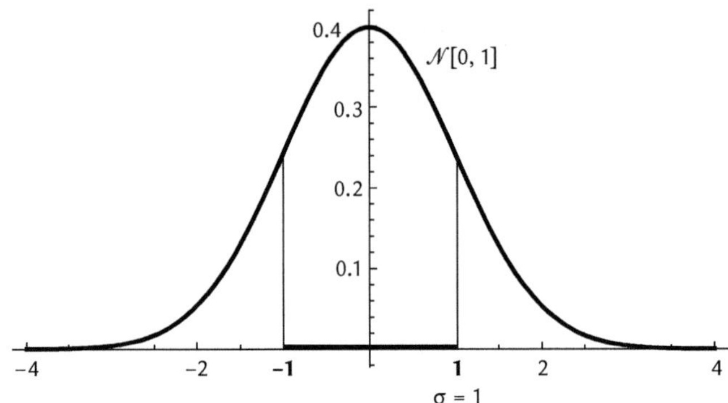

Figure 5.10: A plot of the normal distribution, $\mathcal{N}[0, 1]$, with mean 0 and standard deviation 1 (see Endnote 18). It is the common standard bearer of the family. It is the *area under the curve* that denotes probability. The area under the entire curve is equal to 1, as it should for a probability distribution. We have marked the interval which represents all points within one standard deviation of the mean. The area bounded by the curve between $-\sigma$ and σ (in this example $\sigma = 1$) is just over 0.68, indicating the probability that an event will fall within one standard deviation of the mean. The rapid decrease in the probability of outliers (events far from the mean) is rather apparent. The probability of an event lying over two standard deviations 2σ from the mean is less than 5%, and for three standard deviations it has dropped to well under half a percent. These statements apply without change to the entire family of normal distributions.

graphs are shown. We can see that the value of the normal distribution at 0 is around 0.4, but this does not mean that the probability of hitting the center is 0.4! In fact the probability of hitting the center dead on, or indeed any prescribed point on the target, is zero. What we can talk about is the probability that the arrow will fall in a certain range, and the graph indicates this in the form of area. Thus, for instance, the probability that the arrow will fall between -1 and 1 is the area under the part of the curve that lies over the interval $[-1, 1]$ (approximately 0.68). The area under the entire distribution is equal to 1—the arrow is bound to hit somewhere.

Normal distributions are common, but they are not the only ones that appear in Nature. They have one special property that is certainly not universal: in normal distributions, outliers (events distant from the mean) are very unlikely. Normal curves drop off so fast that for an event to appear that is even a few standard deviations (see Endnote 18) away from the mean is very unlikely (see the figure caption). The normal distribution is important, but many random processes in Nature are not normally distributed. Hurricanes do happen!

Returning to our story, Gauss was drawn to the question of the distribution of errors by the astronomical phenomenon of the asteroid *Ceres*. For various reasons, astronomers hoped to find a planet in the asteroid belt, the large spatial gap between the orbits of Mars and Jupiter, and sometime in 1800 an organized attempt was made to find it. In the meantime, on January 1, 1801, the Catholic priest Giuseppe Piazzi found Ceres by

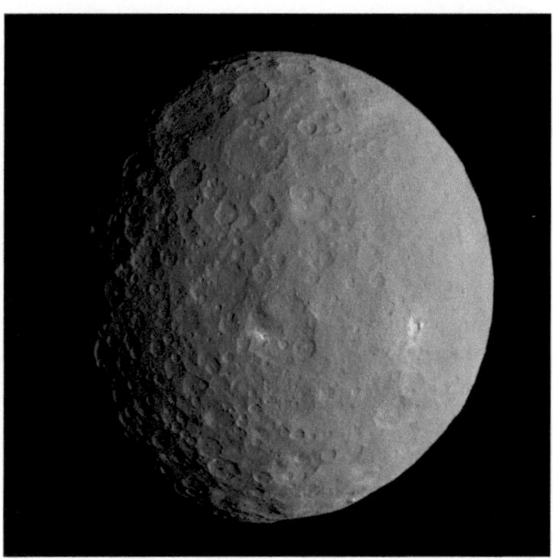

Figure 5.11: A photograph of the asteroid Ceres taken by NASA's *Dawn* spacecraft in May 2015. Its diameter is 945 kilometers. By comparison our Moon has diameter 3,474 kilometers. Justin Cowart, Ceres-1, Justin Cowart, Public Domain.

accident, while actually looking for a particular star. In the subsequent six weeks he observed Ceres some twenty-four times until sickness stopped him in February of that year. His observations were published in full in September of 1801. By this time the position of Ceres as seen from Earth was too close to the Sun to make it possible to see it, so the question became how to find it again when it reappeared.

It is at this point that Gauss, still young at the age of 24, and destined to become one of the greatest mathematicians of all time, decided to think about the problem. It was in solving this puzzle that he developed his method of least squares from which he went on to derive the explicit form of the normal distribution. More relevant to the story of the discovery of Ceres, he predicted the area in the sky where the search ought to be made; correctly as it turned out, and quite different from the areas suggested by other astronomers [158].

5.5.2 The central limit theorem

The most convincing argument that the normal distribution is a "pattern of errors" is the famous central limit theorem. This comes in various forms and with various degrees of generality, but in its most approachable form, which dates back to de Moivre (1667–1754) and Laplace, it says that the normal distribution arises all by itself when looking at how the outcomes of a random variable disperse themselves around its mean. Fig.5.12 gives an instant illustration of what this means.

Suppose that V is a real-valued random variable. Suppose that its probability measure, or law, is $\mathbf{p}$. Suppose that we repeatedly choose real numbers using the same random variable V, where each choice is totally independent of any previous ones. As before, we can treat this repeated selection process as being the outcome of a sequence of random variables $V_1, V_2, V_3, \ldots$ that are identical and independent (i.i.d.).

Remember that these random variables are not numbers. They are processes that *produce* numbers. Each time we make a sample, V_k will produce a real number x_k. The values that come out of it are distributed according to the probability law $\mathbf{p}$ that governs the randomness.

To see the cumulative effects of making samples with V, we look at the sums

$$S_n = V_1 + V_2 + \cdots + V_n ,$$

which themselves are random variables. After making n selections we end up with n numbers $x_1, x_2, \ldots x_n$, and we have their sum $x_1 + x_2 + \cdots + x_n$. This sum is one outcome of the random variable S_n. If we do it over and over again, with n being fixed, we get many random outcomes $x_1 + x_2 + \cdots + x_n$, but they are governed (indirectly) by the probability measure $\mathbf{p}$. The strong law of large numbers applies and it says that the averages S_n/n will converge to the mean value μ of V almost surely:

$$\frac{S_n}{n} \longrightarrow \mu . \tag{5.7}$$

We are interested in how the outcomes of repeated trials of V will distribute themselves around its mean μ. In other words, we are interested in $V - \mu$, or in terms of repeated trials we are interested in

$$(V_1 - \mu) + (V_2 - \mu) + \cdots + (V_n - \mu) = S_n - n\mu .$$

If we keep adding, things get larger and larger (think of this like a random walk) so we need to scale down by the appropriate factor to compensate for the expected growth. This factor is $\sqrt{n}$, which can be seen by the same sort of idea as described in random walks, see Endnote 13. The standard deviation σ of V relates to its spread, but for simplicity let us suppose that it is equal to 1 and also that the mean μ is equal to 0. We are led to look at $S_n/\sqrt{n}$.

The central limit theorem says that the probability distribution
$$S_n/\sqrt{n}$$
looks more and more like the standard normal distribution as n tends to infinity.

Just what this means is illustrated by Fig.5.12. We randomly chose one hundred real numbers in the range $[0, 100]$. The probability distribution is flat, completely uniform across the entire range from 0 to 100. We would expect the mean to be something around 50. We repeated this 400 times and made a histogram showing where these means came out. Comparing to the bell shape of the normal distribution, the resemblance is clear.

Of these requirements, the *finiteness of the variance* is particularly significant. The variance of V is the expected value of $(V - \mu)^2$. Events that are far from the mean, so-called *tail events* or *outliers*, have to be rare to keep the standard deviation finite. Random

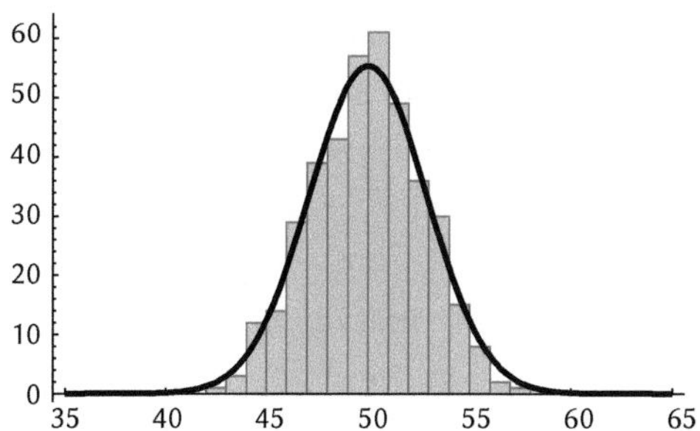

Figure 5.12: An example of the distribution of means of random process of 400 trials (see text) compared with the normal distribution that the central limit theorem offers as the expected distribution over all trials.

processes in which outliers are not sufficiently rare do not fall within the compass of the central limit theorem. Examples occur in the food-searching patterns of various animals. Often, after searching a local area for some time an animal will travel some distance before resuming a local search. Thus sequences of short steps are interrupted by a far longer step (an outlier, so to speak) so as to move into a new searching area.

The central limit theorem is an example of how pattern arises in the context of process —a pattern of errors.

5.6 Conclusion

This chapter may well have been called the *patterns of chance*. We don't normally think of chance as a generator of pattern. To our usual way of thinking it is a wild card, a reference to randomness and the unpredictability of things, a bringer of fortune and disaster. Chance is seen more as a disruptor of the smooth purposeful progression of things than a natural shaper of that progression. This is not to say that a falling asteroid won't upset the present order of things, but the more we have come to understand chance, the more we have come to appreciate its central importance to the way things are, and in particular to life itself.

This revelation has been slow in the making. Chance has always had a certain aspect of mystery to it, and success in placing the concept into a mathematical and scientific setting has really only been achieved in the past two centuries, and not without much misunderstanding and disagreement along the way. Our intuitions about chance are not very reliable. We still go to casinos with the hopes of beating the odds, even though we know that casinos make lots of money, and they are set up, based on the laws of probability, to do so. Far from being places where we can expect to be the lucky exception, casinos make money in an entirely predictable way and we should expect to lose. The laws of chance are relentless, shaping a totally predictable outcome—*in the long run*.

It is this "in the long run" that is the important point, and this is the way we have come to see the outcome of random processes. In the preceding chapter, we introduced the shift system on two letters. This is the simplest example of discrete dynamical patterning, laying out all the varying sequences of changes between two possibilities. It is also the most fundamental. In the present chapter, we have seen how this same model can be reinterpreted to serve as a basic model for probability. We interpret the strings as the potential outcomes of long trials of heads/tails (one/zero) tossing. Each infinite string represents one such trial.

The difference now is that we have imposed on this system a measure $\mathbf{p}$ of probability. Initially we have taken $\mathbf{p}$ to give equal probability to the appearance of head or tail. Since subsequent tosses are independent of any and all previous tosses, it is easy to obtain probability values for all the various finite sequences of tosses. It is out of this that the strong law of large numbers appears: namely that in the long run the average number of zeros and ones will tend towards equality. This is no longer a *feeling* about what should happen, but a *theorem* that follows out of the mathematics. It agrees with our intuition, but it has moved into something that is formalized and established in a precise way. This is the way that chance and randomness shift from gut feelings into formalizable processes.

If we think about this in terms of shaping or generating pattern, we see that it is the probability measure $\mathbf{p}$ that is responsible. If we change the probabilities so that the coin is biased with probabilities $2/3$ for heads and $1/3$ for tails, the shift system itself is just the same, but the resulting probability distribution on the finite sequences of tosses is different, and the law of large numbers produces the appropriate $2/3$, $1/3$ outcome. With this we appreciate the way in which probability becomes law. In fact $\mathbf{p}$ is called the "law of the process".

Underlying all of this is the mantra "in the long run". The outcome is not determined by single events, but by the outcome of millions of them. We may win at the casino today, but the laws of probability will win in the end, whether it be with us (if we play enough times) or on average over all of its customers.

Going beyond the gambling image, we can look into the ordinary processes that occur around and inside us all the time. The checkerboard example points in a new direction. The way the game is played, the moves are completely random and every board position is equally likely. There is no bias towards any particular arrangement of the black and white checkers. However, if it is not the particular arrangement of the checkers that is important, but just the *number* of white and black checkers, then the situation is immediately changed. Now the probability is steeply inclined towards board positions with more or less an equal number of checkers of each type. This phenomenon has significant physical implications. Boiling a kettle of water amounts to agitating into motion trillions upon trillions of atoms and molecules. The directions of movements, all their collisions and reversals, are random and beyond all reckoning. But their overall effect is almost deterministic. The law of large numbers applies and the kettle boils, just as we expect, at 100° C, every time.

A slight change in the rules of the checkerboard game, and the entire pattern of dynamics changes, and now, with equal probability, the system will reach a full board of black or a full board of white checkers from which there is no escape. The law changes and the pattern is totally different.

From atoms moving at random it is only a short step to the whole subject of random walks. If a particle moves in equal steps but in random directions, what can we say about its path? Here it turns out that dimension suddenly becomes important. If we start on a random walk what are the chances that we will arrive close to or at some preselected point? This is an important question in living structures: molecules (proteins, enzymes, and so on) move randomly, yet to work they must be able to find their targets. In one and two dimensions, chances are high. In three they are much lower. The laws of chance are vital to life. Molecules don't know where they are supposed to go. They are subject to endless disorganized collisions and recoils implicit in the randomness of crowds of particles jostling together. To function they need to arrive at places where there are appropriate pattern matches. Were it not for chance and the definite patterns that arise from chance, this would not have the reliability or consistency to sustain life. Our very existence depends on the patterns of chance.

One of the great steps in the advancement of the theory of atoms was Einstein's use of the mathematics of random walks to show that the famous chaotic movement of minute particles in fluids, Brownian motion, could be explained from the random movements of much smaller particles at the tiny nano-scale. The estimates he made from this were carefully established empirically, and this work became one of the arguments that finally convinced physicists of the reality of the sub-visible atomic world.

Perhaps one of the strangest ideas to emerge from the study of randomness is the pattern of errors. Even making random mistakes produces its own form of pattern, and this is the famous normal curve. This is formalized as the pattern that arises in looking at how data distributes itself around its mean or average value. This pattern is a standard in statistical analysis, the normal distribution, and the extent that it is fully expected is another testament to the pervasive outcomes of chance. In probability theory it goes by the name of the central limit theorem and it is widely applicable (and, we might caution, even more widely used). Its importance and usefulness is unquestioned.

All this being said, it remains the case that even within science randomness still carries considerable mystery. Their amazing successes in uncovering natural laws led nineteenth-century physicists to conclude that the world was deterministic—that if we really knew the exact positions and momenta of all the components of the universe then its entire future would be known. Of course we cannot answer the question of whether or not it will be raining in downtown San Francisco exactly one year from the moment that you are reading this sentence, but on this principle we could: it is all there, but so stunningly complex we could never figure it out. In this view randomness is understood more as an effect that covers for ignorance than an intrinsic quality of Nature itself. This really is a clockwork universe.

There is good reason behind such thinking. It is entirely intuitive to us that effects must have causes. Of course there are phenomena like the weather where the causes are so manifold that we cannot predict their combined effect, but that does not affect the general idea. Things just don't happen of their own accord, out of nothingness.

Nonetheless, by the later parts of the nineteenth century, and increasingly in the twentieth century, randomness was found to be essential in understanding physical processes and predicting their outcomes. Probability and statistics became an essential part of science (and also of commerce, manufacturing, economics, and the social sciences) and gradually whole disciplines arose to describe and quantify statistical or stochastic processes. Complex deterministic systems that could not be solved explicitly nonetheless led to outcomes that could be described statistically: weather forecasts, political polls, population studies, recovery rates for different diseases, insurance policies, quality control protocols, and thousands of others were models that involve chance in this sense. And all of them were instances of pattern systems.

None of this would have had much influence on changing our thoughts about the fundamental nature of chance were it not for the quite unexpected discoveries of the early part of the twentieth century. Physicists found that the laws of classical physics, which had proved so reliable for the tangible world, could not account for events being observed in the newly revealed atomic world. In quantum mechanics, the theoretical model that was developed to account for the new reality, chance was no longer extraneous interference or accident. It was now a fundamental player that had exact quantifiable outcomes in a world that seemed increasingly insubstantial, evanescent, and dynamical. Molecules and atoms were not only jostling with one another, they were themselves composed of smaller, constantly changing entities, whose properties defied any description in classical terms. Famously, even such givens as the speed and position of a particle could no longer be taken to be independently observable features. Suddenly determinism became a huge question mark.

The strangeness of chance remains to this day. Many attempts have been made to show that determinism—that the present determines the future, that all effects are the determinable outcomes of causes—can be incorporated into quantum mechanics, but none have been successful. The evidence continues to mount that *acausal* randomness and chance are truly embedded in the heart of things. Chapter 14 is devoted to the patterning of quantum mechanics and touches upon this great debate.

The fact is that chance remains mysterious. Even different methods of producing "random" zero-one sequences seem to be mutually distinguishable, although they pass all the conventional tests for randomness. Given all this illusiveness, it is not surprising that chance has proven so hard to accommodate to the precise ways of mathematics and the sciences. The present (quite standard) approach to randomness and random variables, which we have seen here as it was proposed by Kolmogorov, manages to create a logically consistent and rigorous basis to probability theory, but manages to do so without giving us any deeper answers to these philosophical questions. In itself it is an object lesson in how much of mathematics works, bypassing metaphysics but still producing formal systems that conform to our intuitions.

Whether chance is an outcome of our inability to take into account a vast multitude of deterministic data, whether there are hidden causal variables that we have yet to discover, or whether it is an intrinsic property of Nature itself, it is certain that chance is a vital ingredient in the formation of pattern. We can see the world as an intricate nesting of scales of magnitude that range from the smallest of the subatomic, through the nano-scale of molecular interactions, on to biological processes supporting life, up through the human scale that we encounter on a daily basis, further through planetary processes, onwards to stellar and gargantuan galactic formations, and finally to the entirety of an expanding universe itself. Chance enters into every one of them.

How do vast numbers of chance movements, encounters, combinations, and contradictions result in pattern and a coherent world? How can chance lead to consistency, and from constancy to the knowable? These are the questions that we have begun to address in this chapter. We may not know all the answers, but we can understand that chance and randomness do lead to pattern, and even decisively to pattern. Chance is a crucial ingredient in pattern, and our theory of pattern has to include it.

Chapter 6

Pattern systems defined

6.1 Introduction

The course of the book so far has to been to explore meanings of the ubiquitous word "pattern" in terms of basic ideas like states and state spaces, difference, partition, change, dynamical process, and chance. We begin to see how mathematical ideas blend with intuitive ones. In this more expansive view of what pattern means both its inherent dynamical nature and its universal appearance become apparent. It is a way of thinking about the world. We begin to see it everywhere.

We wish to incorporate all the basic ideas that we have seen, but now put them into a unified mathematical framework. This model ought to encompass what we have already seen and provide a foundation from which we can deepen our explorations. Thus the chapter stands Janus-like, both consolidating the past and serving as mathematical support for what is to follow.

We need to keep in mind that the ways in which we frame the world and its multitude of phenomena are ultimately inseparable from the (human) mind. We see pattern and patterning everywhere, and we gradually learn new ways of conceptualizing it. Pattern is not inherently mathematical, but framing it in mathematical terms lays forth its underlying principles and provides an over-arching context for studying it.

Since our purpose is to be able to frame a huge variety of dynamical phenomena using a small collection of basic ideas, the concept of a pattern system is necessarily abstract. All of these abstract ideas are ones that we have seen in more concrete forms in the examples we have discussed so far. A key to understanding mathematical abstractions is to link them to ideas and processes with which we are already familiar. In doing so we actually take part in one of the pattern processes that we will come to discuss later: we can articulate abstract ideas and attach meaning to them by associating them with other ideas that exemplify them and to which we have already extended meaning. For example, the infinite set of integers $\ldots, -4, -3, -2, -1, 0, 1, 2, 3, 4, \ldots$ is a considerable abstraction: the introduction of zero and negative numbers was a hugely important but seemingly unintuitive step in the development of mathematics. Yet, visualizing the integers as a line of fence posts going off endlessly left and right into the distance offers immediate insight. Fixing one fence post as a reference point, we can number the posts to the right by $1, 2, 3, 4 \ldots$, those to the left by $-1, -2, -3, -4, \ldots$, and the reference post by 0. The integers themselves can now be thought of as indicating the number of steps to the right or to the left of the reference post. The operations of addition and subtraction simply amount to shifting right and left. In ‡Ch. 16 we discuss the natural numbers in some detail, using this model as a natural starting point.

Ultimately the modeling process works both ways. Although concrete examples provide meaning, it is also true that in abstraction we find ways to extend our sense of meaning, revealing connections between diverse pattern systems that we had not realized before. Pythagoras is famous for his views that the whole numbers underlie the internal order of Nature. In quantum mechanics, which we will come to later, we will see that his intuitions have significant resonance. Abstractions can be seen to make the ways in which we frame the world more accessible.

Our formalism of pattern is about recognizing the key features that we believe underlie pattern, distinction and change, giving precise mathematical definitions of them and their logical inter-relationships, and exploring their inherent properties within a coherent abstract framework. This formalism not only makes it easy to apply the same ideas to many different situations, revealing unities that otherwise would be unrecognized, but also allows us to see how pattern systems can be compared with one another, how transference of pattern can be formalized, and how pattern systems can be combined into larger pattern systems. It is through this sort of synthesis that we will be able to better understand how wholes arise from parts. Beyond all this, or better said, behind all this, is an amazing range of mathematical tools and insights that have arisen out of the formalism of pattern, or what we call pattern systems.

We should offer a few words of guidance about reading this chapter. It contains a lot of material: the mathematical essence of the core ideas of sameness and difference, events and event spaces, and the concept of measure along with its connections with invariance. These assemble into a formal statement of what a pattern system is. But beyond the formalism there lie principles which are expressible without appealing to mathematics. This is the aim of the next section §6.2. At the end of the chapter in § 6.8 we encapsulate all of it into a single idea: pattern can be construed quite simply as *the physical or mental manifestations of the processes of change in a system.* The mathematical formalism offers an explicit interpretation of this and a means to discuss and explore pattern within the realm of science.

Much of what follows in the subsequent chapters can be read quite effectively without first reading it in its entirety. We have chosen its position in the book because we have reached a point where we have seen the principal ideas on which pattern systems are built and it makes sense to make them explicit in a formal sense. The subsequent examples then show the great variety of situations in which these same ideas reappear and how the theory deepens as new concepts arise.

In all this we might keep in mind the important words of the physicist Carlo Rovelli:

> We can think of the world as made up of *things*. Of substances. Of entities. Of something that is. Or we can think of it as made up of events, of happenings. Of processes. Of something that occurs. Something that does not last, and undergoes continual transformation, that is not permanent in time. The destruction of time in fundamental physics is the crumbling of the first of these two perspectives, not the second. It is the realization of the ubiquity of impermanence, not of stasis in a motionless time [138].

Pattern lies deep in our ways of interacting with and participating in the natural world. The Chinese character Li, 理, which we show in the frontispiece of the book, stands for a word whose variety of meanings covers much of what we understand pattern and the principles behind it to mean [182, 183]. If in the end we find that *everything* that we can know about the world comes about though pattern and the recognition of pattern, then the thought and care we put into the basic ideas are paramount.

6.2 Pattern systems in words

The basic ingredients of a pattern system are context, change, difference and sameness, and measure or frequency.

6.2.1 Context

Whatever aspect of pattern we look at, the starting point is a context. Nothing can be understood without knowing the contextual field in which one is speaking. How many times have we seen people taking other people's words out of context in order to make them mean something else? What do right and wrong mean without the setting in which we interpret them? Everyone who has had to interact with a robotic telephone answering system knows what happens when the conversation goes outside its contextual field. The first step to any form of information or knowledge is to know the field or universe of possibilities from which it comes. Context is the setting within which we are working, and in our case it is the *state space* itself. The state space is made up of all possible states in which the particular system of interest to us can occur. It serves as a "universe" in which discussion is possible. So at the outset we have a state space and that forms the context for the pattern system.

Mathematically the state space is just a set, its elements being the states. Neither the state nor its elements have any intrinsic meaning. For us they are simply denoted by symbols that in themselves have no meaning. The state space, this set of elements, can be finite or infinite. Of course we may, and usually do, assign meaning to the states. We have seen this in the states indicating angle and speed for a pendulum (see Fig.3.5), or zero-one strings in a shift system. It is like the weather. To us the state of the weather is something about experience. To the weatherman each may be just numbers (temperature, pressure, wind velocity, percentage of cloud cover, UV index, and so on). The state of the weather is just a collection of numbers. At the systems level each potential state is just one element of the state space, and that state space is in turn the collection of all possible weather states imaginable.

Again it might be useful to think of the coordinate plane. Literally we just have a set, each of whose elements is an ordered pair (x, y). The fact that we think of this set of number pairs as constituting a plane is an example of assigning meaning. We have also thought of the various states of a pendulum in terms of ordered pairs of numbers.

Because the mathematics does not preassign meaning, it is free to take on new meanings. What is important is that the built-in relations that are defined between the mathematical entities that comprise a pattern system match corresponding relations in the systems that we attempt to model.

How does one know at the outset what the right context for understanding a phenomenon is? In general we don't. Often part of understanding a pattern entails enlarging or contracting an established context, or even abandoning it altogether. Mathematical models, however internally consistent, cannot assign absolute truth to our theories about the world. Many useful models developed in the past served well enough in their time but since have been superseded.

However, whatever the context is, from our perspective it is a just a set of elements, which we call states.

6.2.2 Change

All life forms dwell in a sea of change. As human beings our awareness of change has given rise to numerous attempts to formalize it, to give it meaning, and often, to somehow avoid it. Greeks, Chinese, Hindus, Navajos, and countless others, all have thought about it. Yet, though the mathematical accomplishments of ancient cultures are striking and have served as the basis of what mathematics is today, learning how to incorporate change into mathematics is relatively recent—only the past 450 years or so. The problem is this: although a state space is about *things* (states are elements of a set), change, as we have so often said, is not a thing. It refers to action, process, movement, and flow. Yet both in written language and in mathematics we need to represent ideas in symbolic form. How does a static symbol represent something whose very nature is not static?

One solution, which one finds in the etymology of ancient Chinese characters, is to appeal metaphorically to things with which we associate action or change. For instance the character 行 represents a crossroads, from which is derived its meaning "to go". Similarly 易, change (the same yi of the ancient Yijing, *The Classic of Changes*) is commonly explained as being derived from a picture of a lizard or chameleon with its ability to change color [197].

In mathematics the way that has gradually evolved is to create a new class of "things", which commonly go under the names of *transformations, operators, or mappings.* In this chapter we will call them *operators.* In the context of a state space, what operators do is to indicate changes of states to other states in the same state space. They do this by *pairing* each state to some other state which represents the outcome of change. When change takes place in a state space, in our minds we think of states "changing to" other states. However, simply by pairing before-and-after states we can indicate change without actually moving the states or elaborating details of how that change happened, how much energy was consumed doing it, how fast it took place, what the path of the process of change followed, and so on. It's like your travel itinerary that may have items like Taipei → San Francisco. That's not to say we aren't interested or are unable to determine such details. If the physical situation is sufficiently described we can, and we may even have a formula or rule that prescribes how the change works—expressions like $x \to x^2$. But the concept of change itself is, in effect, reduced to a list of pairs of states, showing how change affects each state. And since states are just elements of a set, in the end a transformation or change or operator, is nothing but a pairing of each element x of the set with some element y of the set which is its "outcome" under change.

This starkly reduced idea of transformation is the outcome of several centuries of thought. The formative ideas incorporating change into mathematics (and physics) had become current in the seventeenth century after the work of such people as Galileo, Fermat, and Kepler. The full flowering of this is what we now call the calculus—the creation of Leibniz and Newton—and it revolutionized mathematics.

The calculus is about continuous change, but over the subsequent years, and especially since the start of the twentieth century, the importance of change that takes place in discrete steps has assumed great importance. The continuous-discrete distinction lies at the heart of the pros and cons of the analogue versus digital debate. Some people love the beautiful old phonograph records where sound is laid down in grooves in the form of smooth waves. Others prefer the more modern convenience of digitalized sound that is printed onto CDs or downloadable as binary digits from the cloud. The same might be said of film versus digital sensors in cameras. The debate will no doubt continue, but when it comes to pattern we really need both—the pendulum and the shift system. In this book we tend to stress the discrete forms of change, since the mathematics tends to be more intuitive and less technical, but ultimately the same formalism is used for both types of dynamics. We will begin with the discrete, and later in §7.3 see what lies behind the continuous.

The reality of the situation is that both discrete and continuous dynamics live side by side in the scientific world. The evolution of a continuous process is often too hard to determine without the use of a computer, and that most often means discretizing the process into minute temporal or spatial steps which approximate the continuity. In the other direction, the dynamics of some discrete processes are often far more easily described by replacing them by continuous ones. For instance, in making models of neural processes it is common to replace the complicated dynamics of a hundred discretely firing neurons that comprise a complex neural unit by a continuous process that approximates their combined effect. As we will see, quantum theory combines both. It is an unresolved question whether, at its very foundations, space-time should be considered as more discrete-like or more continuum-like, or whether indeed either of these descriptions is appropriate.

With a state space and one or more operators of change, we have the first outlines of a system. Within the state space there is a sense of relationship through change.

6.2.3 Difference

With the two basic concepts of a context and change within that context, we come to the question of what we might be able to say about what is happening in such a system. Our senses, the way in which we get to experience the world, are driven by difference. They react to difference, and it is these differences that play out the process of change. Since it seems impossible to talk about change unless there is difference, or difference without change, the two seem to depend on each other. This seems pretty straightforward: the states of the context are elements of a set and different elements are different. But it is more subtle than that.

Animals of all kinds know the world through their senses. Just as importantly, and perhaps most often overlooked, is that they can know nothing about the world that lies beyond those senses. The experiences gained through vision (with or without color), sound, smell, taste, touch, magnetic fields, electric fields, pressure, temperature—all of these along with their limiting thresholds—serve to make their worlds. All these senses work through receptors that respond to differences. We ourselves are limited in our direct experiences by the senses with which we are endowed and their various thresholds of activation. Because we have created tools that greatly extend what our own senses are capable of distinguishing, we have enlarged our worlds of experience. Even so we have no idea about what a bat actually experiences as it flies with its sonar vision, or a red diamond rattle snake might experience as it senses us through its heat-sensing pits. Even with all our tools, we cannot experience the reality of the world as experienced by others [198]. The lesson from this is simple. We don't necessarily function at the level of states. Animals of all kinds seemingly have access to the vast array of states around us, but in reality they are able to differentiate only a fraction of it.

Even in the cases where we can differentiate, we can see that we may choose to do so in varying ways. Traversing a city street is a complex process in which our brain decides what differences are important and what are not. As a driver it is one set of differences, as a pedestrian another. Even as a pedestrian our filtering of the world by distinctions is not the same if we are window shopping as it would be if we were hurrying to get to our dentist appointment. Again, each person is an individual whose uniqueness is vital to the lives of the people around them. With respect to the law, each is (at least in principle) the same. The biological classifications of species, genus, family, order are markers of sameness at different levels of detail. There are endless other examples: animal/plant, four legged/two legged, hooves cloven or not, male/female.

To a great extent ambiguity in our levels of distinction is unavoidable. The material world is made of atoms, but that is not a scale at which we function. It took millennia of speculation before we could mount enough evidence even to be convinced of the existence of the atomic world, yet it obviously affects everything. Difference is really a relative term. Without imposing metaphysical components to reality or turning to the mental world of mathematics we can question whether or not there is "difference" in any absolute sense. The world exposed to our senses at every moment is both vast and limited: too vast to be fully experienced and limited by the paucity of our senses and the tools we have to extend them. Understanding this is important, for in this way we can make it quite explicit that we approach reality with varying levels of resolution according to our needs and capabilities.

These simple observations alert us to the porous nature of any particular context that we might wish to fix. There may be an underlying context, but what we choose to, or can, distinguish is not necessarily the same thing. Often it is particular subsets of states that are important to us rather than the individual states themselves, while in a dynamical setting it is the changes in the courses of these subsets that are relevant. In the language of probability theory these subsets of states are called *events*, and that is the way we shall speak of them. We can think of them as providing a certain level of articulation, a particular level at which we are able or wish to say things about our pattern system.

It may seem strange to call certain subsets of the state space "events", but the process of interaction with a pattern system is one that involves articulation and change, in other words the sequencing of events. The conventional sense of events as things that happen is now put into a mathematical setting with precisely the same understanding.

At its heart then, what we are aiming for is a formalization of articulation: what we can say about a system is what we can articulate, and what we can articulate is based on what we can distinguish. This is what *partitioning* of the state space is about. The parts of the partition are the basic *events*, the foundational items of articulation. The expansion of this articulation to include logical operations and the effects of change are the basis of what we call the *event space*. The event space is a coarser version of the state space, but not one that is uniquely determined or dictated by it.

Mathematically articulation begins by imposing a concept of difference/sameness directly on some prior established state space, lumping states into various subsets according to whether they are to be considered as the same or different. In mathematics we have the freedom to choose whether or not we are willing to go beyond the physical and allow for perfect resolution.

All this discussion can return us to the question: how do we establish a state space in the first place? The revelation of a previously unconsidered class of distinctions may lead us to refine or to redefine an earlier established state space. In the other direction, we might better understand a pattern system by reducing its state space and its complexity, reducing until only the essential variables are left.

Thus the state space and the event space, just like the general and the particular can be seen to influence each other. Just as we doubt any idea of absolute difference, it is doubtful that we can establish an absolute state space. We always will be limited by what we don't know. The great teaching of the classical Asian philosophies and of the contemporary philosophy of science is that we should be wary of pronouncements of bedrock certainties or absolute articulations of reality.

6.2.4 Frequency and measure

The most basic outcome that we are likely to become aware of in an ongoing context of change and distinction is the frequency of events, a measure of how often they reoccur and in what order. The recurring phenomena of the Sun's daily rising and setting, the great cycle of the seasons, and the Moon's recurring waxing and waning are ancient and obvious examples. But we shall discover that the concept of measure that we use to express frequencies can play a very different role than simply being a necessary outcome of a changing system. It can become a key component that makes a pattern system what it is. We have already seen this with shift systems involving strings of letters: letter frequencies are characteristics that in themselves can distinguish entire languages.

The mathematical entities that carry this type of numerical information are actually called *measures*. We might think of a mathematical measure as being a bit like a tape measure, which can assign numerical values to lengths. We will see them as mathematical entities that assign non-negative numerical values to events. Then we can talk about the *measure of an event*, meaning the numerical value given to that event. This might be the probability of that event arising, or its frequency of occurrence in the dynamical flow of a system.

The only real requirement of a measure is that the measure of the union of two disjoint events (events that have no states in common) is the sum of the measures of those two events. As such there may be countless measures that we can define on a system. But often there is only one that truly reflects its dynamics (even though it may be hard to find!). In fact, one can look at it the other way around and see a measure as being like a *law* of the system—a law that shapes what the dynamics will look like and so to some extent defines it. Instead of simple quantification we can see measures as guiding the system—not so much in the sense of forcing, for chance is an important feature in many systems, but by assigning more or less likeliness to the sequencing of successive events.

In an important way we can see an appropriate measure as being a source of unity of a system—that brings relationship to its dynamical unfolding. As we introduce them and see them in ever more detail over the course of the book, their enormous importance will become evident.

6.2.5 Summarizing the four components of pattern

If we summarize this discussion, we can state that the formalization of pattern involves four concepts:

context—change—-difference/events—frequencies/law

These concepts are not necessarily mathematical in nature—we are familiar with all of them in the normal course of life. Together, though, they can be thought of as encompassing what we call a *systems way of thinking* about the processes. They can be applied to weather systems, traffic flows, the functioning of neurons, bio-chemical transformations within the body, and thousands of others, without any strongly mathematical flavor. However, they can be formalized precisely in mathematical terms, and doing so shows how beautifully they fit together in abstract terms. The final outcome is our formal definition of pattern systems in §6.7, where it takes the form of four requirements or axioms.

6.3 The dynamics of change in a shift system

With this in mind, we go back to the simple shift system $\{0, 1\}^\infty$. This is the set of all infinite strings of zeros and ones. The dynamics is defined by the very simple idea of shifting strings one step to the left, dropping off the first letter (0 or 1) as we do so. In this way one string becomes another.

In spite of its simplicity, this shift system can be considered as the universal model of dynamics which proceeds in distinct steps. We have already seen some of this in its interpretation in Shannon's theory of communication and in its extensive use in thinking about trials in processes of chance.

Although stating what the dynamics is about is easy, we need to find a more mathematical way to express it. In effect what the dynamics is about is states changing to other states. Each state

$$\mathbf{x} = x_1\, x_2\, x_3\, x_4\, x_5 \ldots ,$$

whatever it is, changes to the corresponding

$$x_2\, x_3\, x_4\, x_5 \ldots ,$$

where the first letter has been dropped off.

This fits the idea suggested in §6.1 that change is encoded mathematically in the form of pairs that we think of as "before and after" states. For the shift operator s these are the pairs

$$(x_1\, x_2\, x_3\, x_4\, x_5 \ldots \quad , \quad x_2\, x_3\, x_4\, x_5 \ldots),$$

for which there is the more suggestive notation

$$\mathbf{x} = x_1\, x_2\, x_3\, x_4\, x_5 \cdots \longrightarrow x_2\, x_3\, x_4\, x_5 \cdots = s \triangleright \mathbf{x}\,.$$

Seen in this way s can be thought of as an operator (or mapping or transformation) that takes states to states by means of this pairing. As such, s is a new type of mathematical object. In the language of dynamical systems it is usually called an *operator*, and in the particular case here it is called the *shift operator*. The operator (shifting) makes the change. The arrow here is commonly used notation to visually convey that change. Later on we will look at all sorts of operators that bring about change. In every case they amount to pairings that indicate all the various before and after states of the operator.

Still, it must seem that we have done nothing! We have introduced the idea that dynamics can be expressed in terms of operators that transform or change states to other states, but we don't see a machine moving a long tape of symbols along, no guillotine cutting off the first letter of every string. The mathematics here is not about *how* one state changes to another, only about the state to which it changes. How it happens is not the point; it is what happens that is important. Although phrased as dynamical, there is nothing that is actually moving. The states don't literally move. All that happens is that each state gets *associated* with another and *we* are the ones that are saying if you are looking at the state $\mathbf{x}$ right now, then the next state to look at is the left shift of $\mathbf{x}$, namely $s \triangleright \mathbf{x}$. The shift is telling us what to do no matter what state we happen to be interested in. It is an operator that applies to every possible state. This is symbolically represented in the notation

$$s : \{0, 1\}^{\infty} \longrightarrow \{0, 1\}^{\infty},$$

which we can read as saying that s is an operator that associates to *every state* of the state space $\{0,1\}^\infty$ another state of the same state space.

It is not necessary that $s \triangleright \mathbf{x}$ is actually a different state from $\mathbf{x}$ itself: if $\mathbf{x}$ is an endless string of 1s then

$$s \triangleright \mathbf{x} = s \triangleright (1\,1\,1\,1\ \ldots) = 1\,1\,1\,1 \cdots = \mathbf{x}\,.$$

It is often the case that some states stay the same under a particular operator. When we rotate a disk around its center, the center stays fixed.

There are countless ways in which we may associate states with other states, and so countless possible operators. The vast majority of them would have no conceivable interest to us. When it comes to strings of symbols, our operator s is one of the most natural we can think of, and it is the one that is used to define the dynamics of the shift system. Actually we can use s to create other operators, which are all quite natural. The operator s may be applied repeatedly, creating a sequence

$$\mathbf{x}, \quad s \triangleright \mathbf{x}, \quad s \triangleright (s \triangleright \mathbf{x}), \quad s \cdot (s \triangleright (s \triangleright \mathbf{x})), \ \ldots\,.$$

This is much more conveniently written as

$$x, \quad s \triangleright \mathbf{x}, \quad s^2 \triangleright \mathbf{x}, \quad s^3 \triangleright \mathbf{x}, \ \ldots\,,$$

and immediately suggests that along with s we have the operator s^2 which shifts strings left by two steps. And if we have that, then similarly we have s^3 that shifts things along by three steps, and so on.

This shows that if we have an operator then we can produce new operators by applying it repeatedly. In fact if we have *any* two operators on the same state space then we can apply them one after the other to form a new operator. We say that we *compose* the two operators and the result is called the *composition* of the two operators. This suggests that composition is algebraic in nature, a sort of multiplication that multiplies (composes) two operators to get a third.

This is the beginning of a type of *algebra* for operators. Thinking in these terms, it is convenient to add in one more operator, which we can think of as playing a similar role to that of 0 in the arithmetic of addition and to 1 in the arithmetic of multiplication. This is the operator that does nothing, that associates each state with itself. It is called the *identity operator* and in this book we will usually denote it by the symbol ϵ. In keeping with the notation above we can also denote it by s^0, so we have

$$\epsilon \triangleright \mathbf{x} = s^0 \triangleright \mathbf{x} = \mathbf{x} \quad \text{for all } \mathbf{x} \text{ in } \{0,1\}^\infty\,.$$

Whatever system we have, we will always have the operator that does nothing. Our convention will be to denote this operator by ϵ.

This "algebra" of operators, where multiplication of numbers has now morphed into an idea of multiplication of operators, marks the beginning of a whole new area of mathematics that is usually called "modern algebra". We will see how this becomes the basis of the patterning of symmetry in Ch.9.

6.3.1 The rule of composition

As a short digression into the *composition* of operators, consider s^3, the operator that shifts through three steps. If we follow it by s^2, the total effect is to shift through five steps, which is s^5. We write this as

$$s^2 \circ s^3 = s^5 \,.$$

The $\circ$ is used to denote the fact that we are composing the operators—it is the symbol for the "multiplication" which is composition. Note that the order of the two operators is to be read *right to left*[1]: first apply the operator s^3 and then the operator s^2, and the total effect on an arbitrary element x of the shift space is

$$(s^2 \circ s^3)(\mathbf{x}) = s^2 \triangleright (s^3 \triangleright \mathbf{x}) = s^5 \triangleright \mathbf{x} \,.$$

The effect is to apply the operator s^5 in one go.

This is of course quite general. For any positive integers k and l,

$$(s^l \circ s^k)(\mathbf{x}) = s^l \triangleright (s^k \triangleright \mathbf{x}) = s^{l+k} \triangleright \mathbf{x} \,.$$

This even works if $k = 0$ or $l = 0$ since s^0 does nothing.

As it stands, the order of application of the operators does not matter: $s^2 \circ s^3 = s^5 = s^3 \circ s^2$. However, this does not always happen, and in general the order of operators *is* important. The way we were taught it, you should put on your socks before you put on your shoes. It's not the same if you try it the other way around. When we study symmetry in Ch.9, we will see how important order is in the process of composition of operators.

We might note, as we have noted earlier in Ch.4, that the shift operator is not reversible. Given just $s \triangleright \mathbf{x}$, we cannot reconstruct $\mathbf{x}$. The first letter is gone and we have no way of knowing whether it was a 0 or a 1 unless we have in addition some sort of memory. As we have set it up, our shift system has no memory. So with the shift operator we do not have a corresponding operator s^{-1} (a so-called inverse) which reverses it. However, in §6.5.3 we will find an important interpretation for the notation s^{-1}, even for the non-reversible one-sided shift, and it will have a considerable role to play.

The take away from this is that we have a state space, in this case $\{0, 1\}^\infty$, and an operator which simply associates each state of the state space to some other state. It is this association that is interpreted as the action of change. If we are interested in some particular state, this tells us what the next state will be, and that is how we interpret change. We can compose this operator with itself repeatedly to form new operators.

6.4 Difference and Sameness

> Little Understanding said to Great Impartial Accord, "What is meant by the term 'community words'?" Great Impartial Accord said "'Community words' refers to the combining of ten surnames and a hundred given names into a single social unit." Differences are combined into a sameness; samenesses are broken up into differences. Now we may point to the hundred parts of a horse's body and never come up with a 'horse'—yet here is the horse, tethered right before our eyes."
>
> Zhuangzi [185, Ch.25]

At the core of pattern recognition, and thus of our concept of pattern itself, is the notion of difference.

In this section we begin the process of formalizing difference. As we have said in §6.2, a useful way to think about it is to think in terms of *articulation*. What we can say about a system depends on the level of distinction that we can make. Once that level has been established, we can articulate not only basic differences, but also combinations made by combining them with the basic *logical operators and, or, not*. States which conform to some level of articulation in this sense, are what we will call *events*. The level of articulation is vastly increased once we incorporate change, for now we can also express more complex events that correspond to sequences of events emerging as the system evolves. Thus difference extended to the systems level leads to a larger notion of events—in fact to an algebraic structure called the *event space*.

Like all qualities, difference comes with its yin-yang complement, *sameness*. What is interesting is that our ability to take things as the same is equally important as our ability to distinguish. As we have noted, what we choose to treat as difference and what as sameness is something that we impose upon the world. The primary instincts of all living forms have to do with survival, and survival requires devoting resources to paying attention and reacting to local environments and what is happening in them. Wasting resources on details that are irrelevant does nothing to promote survival. All life forms are adapted to be able to sense and react to what is important to their survival, and to either ignore, or simply be unable to perceive, distinctions that don't matter.

How then do we express this mathematically? There is no simple answer to this question. However, we can begin with thinking about some properties we may associate with sameness.

6.4.1 Equivalence relations and partitions

We can start with a reasonable set of properties that we can associate with sameness:

(i) a thing is the same as itself: x is the same as x;

(ii) if x is the same as y, then y is the same as x;

(iii) if x is the same as y and y is the same as z, then x is the same as z.

Each of these involves a pair, and if we were to replace "is the same as" by the symbol $\sim$ this would read:

$$\begin{aligned} &\text{(i)} \quad x \sim x \quad \text{(reflexive);} \\ &\text{(ii)} \quad \text{if } x \sim y \text{ then } y \sim x \quad \text{(symmetric);} \\ &\text{(iii)} \quad \text{if } x \sim y \text{ and } y \sim z \text{ then } x \sim z \quad \text{(transitive)}\,. \end{aligned} \tag{6.1}$$

This is exactly the mathematical definition of an *equivalence relation* on a set. Equivalence relations are relations involving pairs that are reflexive, symmetric, and transitive. Of course equality (=) is the most obvious example, but there are all sorts of other ways to define sameness on a set. A pair of familiar mathematical examples comes from the world of plane geometry. Two triangles are congruent if the lengths of their sides are the same and their corresponding angles are equal or, in simpler terms, one can be put on top of the other to get a perfect match. Congruence is an equivalence relation. Two triangles are called *similar* (here taken as a specialized term) if they have the same angles. This means they have the same "shape" but not necessarily the same size. This is another equivalence relation on the set of triangles. In congruence size matters, in similarity it does not. In less mathematical contexts, we find that names of the days of the week effectively put an equivalence relation on the endless succession of days. It's always interesting to see how dogs never fail to recognize other dogs as dogs, although the variation in sizes and shapes of dogs is enormous. Yet a dog will rarely pay any interest in a stuffed toy dog, no matter how visually faithful it is to the real thing. Their sense of sameness here is based on other factors, notably their significant olfactory abilities.

As soon as we have an equivalence or sameness relation on a set, the set naturally falls into "sameness subsets". Just group together all the elements into their sameness classes. This immediately *partitions* the set into parts. We see this on calendars where the days of the year are broken into classes according to the days of the week.

Looked at a bit more formally, suppose that we have a sameness relation (6.1) on some set X. Take one of the elements x and then look at all the elements that are the same as x. We will write $\tilde{x}$ for this set, indicating that we are looking at all the elements y for which $y \sim x$. This is called the *equivalence class* of x. If y is the "same" as x ($y \sim x$), then x is the "same" as y ($x \sim y$), and then transitivity shows that $\tilde{y} = \tilde{x}$: the two equivalence classes are identical.

Equivalence classes have to be either equal or disjoint. This means that the equivalence classes form a *partition* of the set. An equivalence relation divides up a set into sameness classes according to the particular sameness relation that we have in mind. Equality is an equivalence relation: the one in which each element is equivalent only to itself, and the partition of the set is into singleton sets $\{x\}$ as x runs over X. At the other extreme is the equivalence relation in which all the elements are equivalent to each other, in which case there is just one equivalence class, namely X itself.

Let's look at this the other way around. Formally, a *partition* of a state space X is a collection $\mathscr{P}$ of subsets $P_1, P_2, P_3, \ldots$ of subsets of X satisfying the single condition: every element of X is in one, and only one, part P_i. We can define an equivalence relation out of this by declaring that two elements are equivalent if they are in the same part of the partition. The parts P_i become the sameness classes.

Thus the important realization that partitions and equivalence relations are two aspects of the same thing! Put in this way, sameness seems to be almost arbitrary. Is that really the case? Yes. For instance, there are 52 different ways to partition a set of five elements and so there are 52 different equivalence relations on that set, and so, in turn, 52 different ways of interpreting sameness on that set, see Fig.6.1. The mathematics is impartial as to which of these relations may or may not be relevant. That is our choice. And, as we have seen, making such choices is something that we do endlessly, consciously or not.

Let us consider again the example of the familiar shift system $X = \{0, 1\}^\infty$. The state space X consists of all possible infinite strings made up from the two "letters" 0 and 1. In using this example, both in the way that Shannon used it and in the way we have used it in thinking about probability in the context of tossing a coin, we have seen that it is not the individual states that are so important, but rather the dictionary sets. Individual states are necessary to articulate what the state space is and how the shifting dynamics works. But our use of the system is at a different level. We began with the simple partition $X = [0] \cup [1]$. In words it divides the infinite strings into those that start with 0 and those that start with 1. Since it is a partition it can be viewed as a sameness relation. In this case two strings are taken as the "same" if they start with the same letter. With sameness comes difference, and so the same relation takes on the most basic distinctions we can make here: whether the string starts with a 0 or a 1.

To get more detail we consider the two-letter dictionary sets, $[00], [01], [10], [11]$. These form another partition and now sameness (and difference) is more refined, using the first two letters as the basis of distinction. It is the same way that we partition the dictionary, putting together words according to how they match in their successive letters. Beyond this we can go to three-letter dictionary sets, four-letter dictionary sets, and so on.

In any pattern system, we will have a state space, which is the underlying context

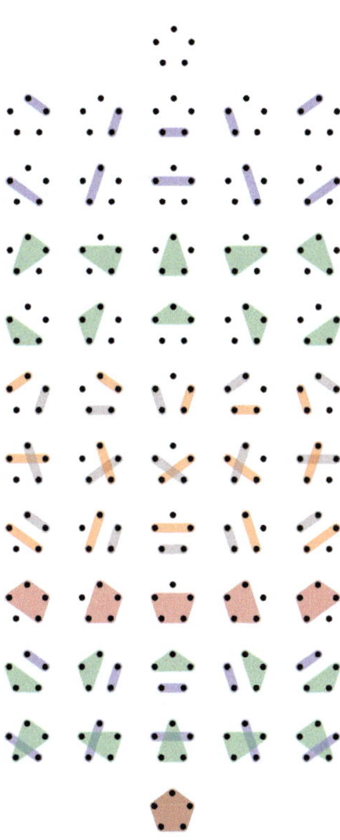

Figure 6.1: A graphical illustration of all the partitions of a set of five elements. At the top is the partition into five distinct parts (corresponding to the equivalence relation of *equality*), and at the bottom the partition into just one part (the equivalence relation in which every element is equivalent to every other element). In between we see points grouped using color to distinguish them. For instance the second and third horizontal rows deal with partitions into four parts where one part has two elements and the others just one element each. Tilman Piesk, CC-BY 3.0, en.wikipedia.org/wiki/Partition_of_a_set.

of the system, and on top of that a partitioning which establishes what sameness and difference are. As far as the idea of a pattern system in general is concerned, any partition is as good as any other. The particular partitioning that we impose depends on our interest or what seems relevant, especially in terms of the dynamics. In the study of the pendulum, we divided the state space according to the gross behavior of the pendulum: librations, full swings clockwise, full swings counterclockwise, and the two equilibrium positions (which are actually individual states).

So part of pattern is context, part is dynamics, and part of it is determining what it is that we care to be distinguishing: deciding what is "the same" and what is not.[2] Mathematically, in the definition of pattern sameness is something we are free to decide upon, but once chosen it is exact.

6.4.2 Other forms of sameness?

Looking over the definition of equivalence that we have introduced here, and thinking more generally of what sameness might mean, it is the third item "if x is the same as y and y is the same as z, then x is the same as z" that might be open to question. If we were to replace "same" by "love", it is certainly questionable. Suppose that we agree that two metal rods are the same if their lengths differ by at most 1 mm, and suppose that we have three rods a, b, c. Then the fact that a and b are the same, and similarly b and c are the same, does not imply that a and c are the same. They could differ in length by as much as 2 mm. This is the problem that we face in the almost periodic order of simultaneous cycles with incommensurate periods: the idea that 1 year is the same as 365 days rapidly falls apart if we chain the notion together over periods of centuries. Later, in Ch.12 we will discuss the notion of almost periodicity in more detail and look at other ways of approaching this notion of sameness.

As far as our definition of pattern goes, the idea of sameness that we have just introduced is a crucial component, and it appears in the definition in the form of partitioning of the phase space.

6.5 Events and the event space

6.5.1 The algebra of logic

At this point we have abstracted the pattern system idea into a contextual state space X, a sameness-difference partition $\mathscr{P}$ indicating the level of sameness and difference, and a dynamics produced by some set S of operators on X. The sameness subsets that make up the partition $\mathscr{P}$ are called *events* in this system. They represent the basic form of our capacity to express or note difference. In this section we will see how we can expand this concept of events. Ultimately what we are able to express about a system, *as seen from the perspective of the initial sameness-difference partition from which we begin*, is how the dynamics alters events and the basic logical statements that we can make from them using the logical operators "and", "or", and, "not". Articulation is expressed in the form of *events*, which are subsets of the state space, and the entirety of these events is what we will call the *event space*.

In the case of the simple shift space $\{0, 1\}^\infty$ we might ask, for example, about the collection of all those states (infinite zero-one strings) that contain three consecutive 1s in the first ten letters of the string. This might be the sort of question that could arise in a probabilistic coin tossing trial. It is a statement that involves both the basic $[0], [1]$ partition of the state space and the outcome of shifting the string ten times. We would call the *entire collection* of states that constitute the appearance of three consecutive 1s in the first ten letters an *event*. The *event space*, which we denote by the symbol $\mathscr{E}$, is the set of all possible outcomes of such questions. An important feature of this is that the event space $\mathscr{E}$ has its own algebraic structure.

We can break the discussion into four steps:

(i) the ways of representing logical operations in terms of sets and algebras of sets;

(ii) including the dynamics;

(iii) "completing" the algebra of sets to allow countable numbers of logical operations ;

(iv) defining the event space of a pattern system.

6.5.2 Event space (i): set algebra

The idea of *Venn diagrams* is probably familiar. Venn diagrams offer a visual equivalent of the basic logical operations of *conjunction* (and), *disjunction* (or), and *negation* (not) that we can apply to sets.

Figure 6.2 illustrates how this works. In each of the six parts to this figure we see a square box representing a set X and one or two boxes A and B representing subsets of X. To the left in the top line we show the *union* $A \cup B$ of the two sets A and B, first in the case when they have elements in common and second when they are *disjoint* (no elements in common). This corresponds to the logical operation of "or", expressing the set of all the elements that are in either A *or* B (possibly both). To the left in the bottom line we show their *intersection* $A \cap B$ expressing "and" — the set of all elements of X that are in both A *and* B. This intersection may be the empty set, as seen in the second of the two cases.

The *negation* of the set A— "not A"— is the set of elements of X that are not in A, shown at the top right. It is also called the *complement* of A and is denoted by $X \backslash A$. With negation it is important to know the full set X (the context) in which we are working. At the bottom right we have included the *exclusive or, xor*, which is often denoted using the symbol $\triangle$. These are the elements of X that are in A or B, but not both, that is, $A \triangle B = (A \cup B) \backslash (A \cap B)$.

This suggests something of fundamental importance:

the normal processes of logical argument can be interpreted in terms of the simple manipulation of sets.

If we think of the brain, its power to perform the basic logical functions of "and", "or", "not" can be realized by activating or suppressing groups of neurons: if two groups are simultaneously activated it is an "and"; if either of two groups is activated it is an "or"; if one group is suppressed it is a "not". What appears to be our elusive mental facility of logical reasoning may actually be based on very simple physical ideas.

In the case of a pattern system, we need our event space to allow all these logical operations to make sense. The outcomes of the operations of intersection, union, and complementation on events are also important distinctions, and are also be considered to be events and included in the event space $\mathscr{E}$ that we are creating. In effect we are creating a language for articulating logical reasoning. This language, made up of subsets of the state space, has an *algebraic structure*, which arises by insisting that unions, intersections, complementations of events are themselves events.

Our on-going mantra is that whatever subsets we include as events, the resulting collection of subsets should be *closed* under the operations *and-or-not* or, put in set theo-

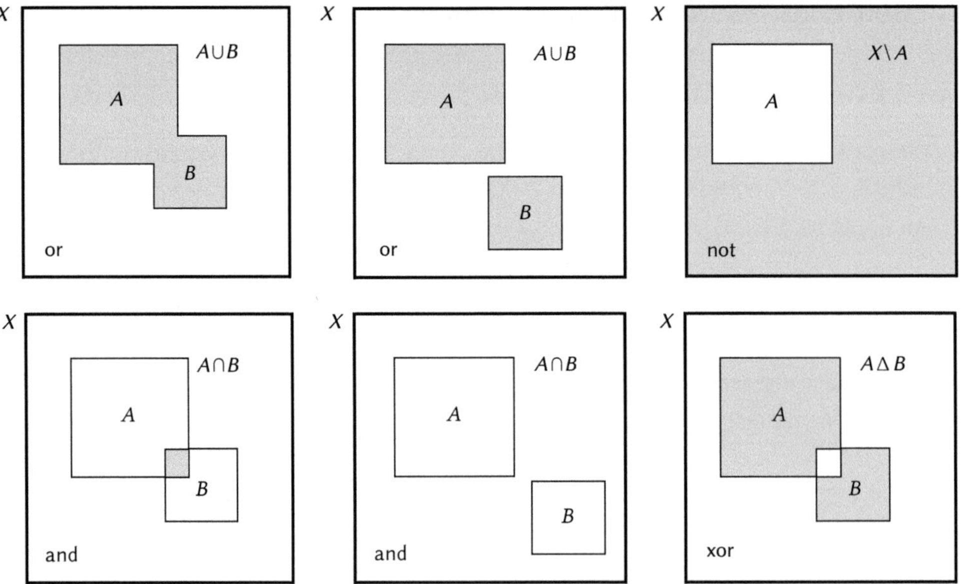

Figure 6.2: A visual representation of the set-theoretic operations of union, intersection, complementation, which correspond to the logical operations of disjunction (or), conjunction (and), and complementation (negation). We have also included the "exclusive or", see text.

retical terms, all the possible subsets resulting out of them by using operations of union, intersection, and complementation should also be events.

> Events are the subsets that we can articulate within the state and an event space is to be the totality of all these and further events obtained from the operations of and-or-not.

A collection of subsets that is closed under these three operations is often called an *algebra of subsets* or more precisely a *Boolean algebra* of sets.[3]

The Venn diagram idea shows how logic can be translated into something that is more tangible, namely simple operations on subsets of a specific set.

6.5.3 Event space (ii): dynamics

In pattern systems we need to extend this logic of sets to include change. That means incorporating the way in which the operators of change, the operators that make up our set S, influence the evolution of states.

A good place to start is the simple zero-one shift space, $\{0, 1\}^{\infty}$, with its iterative dynamics arising out of the shift s and its sameness-difference relation given by the simple partition with its two dictionary sets $[0]$ and $[1]$.

Dynamical considerations enter when we ask a very simple question about the future: which states of $\{0,1\}^\infty$ will move into the dictionary set [0] *after* one shift? Evidently these are the elements of the two dictionary sets [00] and [10], i.e., $[00] \cup [10]$. We can quite easily read this in the language of logic: a state will move into the part [0] of the partition $[0] \cup [1]$ after one shift if and only if it is in [00] *or* [10]. Earlier we have written this more succinctly as $[*0]$, where we treat the $*$ as a "wild card", meaning that both possibilities —either 0 or 1—are included: $[*0] = [00] \cup [10]$. We continue to use the notation here.

To ask another question, which states of $\{0,1\}^\infty$ will be in [1] after two shifts and [0] after five shifts? This time the question has the logical answer

$$[**1] \cap [*****0],$$

which we could expand out into a union of dictionary sets by putting in all the possible zero-one choices for the $*$ wild cards ($2^7 = 128$ dictionary sets in all, since there are seven wild cards $*$, each with two possibilities).

Because future states are so important, we introduce a notational device to deal with them. The answer to the question of which states move into [0] can be re-expressed as

$$\{\mathbf{x} \in \{0,1\}^\infty \ : \ s \triangleright \mathbf{x} \in [0]\}.$$

In words, it is the set of states $\mathbf{x}$ for which $s \triangleright \mathbf{x}$ starts with 0. This is forward-looking to the state $\mathbf{x}$ being shifted into the dictionary set [0]. But also think of it in terms of going backwards: in asking for all the states that will go into [0] after shifting by s we are asking about backing up from [0]. A suggestive way of writing this is $s^{-1} \triangleright [0]$:

$$s^{-1} \triangleright [0] = \{\mathbf{x} \in \{0,1\}^\infty \ : \ s \triangleright \mathbf{x} \in [0]\}.$$

The notation $s^{-1} \triangleright [0]$ can be literally read as *the set of all states that move into* [0] *after applying the shift* s. This way of thinking suggests the operator s undoes what the s^{-1} does.

Although the inverse notation suggests that we are going backwards with our shift, it is actually about the future. We know we can't actually recover the past once a shift has been made, but, mentally shifting the time frame, we can ask about all the present states that will lead to some future outcome, and that is what this does. So, although we have been at pains to note that there is no operator s^{-1} that acts on states, for that would be asking us to assign a definite state where there is no unique answer, it is legitimate to work at the level of *subsets* where we ask for the *subset* of all states that could give $\mathbf{x}$ after one shift by s. So s^{-1} makes sense *at the level of sets*, and that is the way we are using it here.

We can apply this quite generally.

For any pattern system with state space X and for any subset $Y \subset X$ and for any operator $r \in S$ we define

$$r^{-1} \triangleright Y := \{\mathbf{x} \in X \ : \ r \triangleright \mathbf{x} \in Y\}, \tag{6.2}$$

or in words, $r^{-1} \triangleright Y$ is the set of all states that *will* go into Y after application of the operator r.

Note here the use of the symbol $:=$. In mathematics there are two ways in which equality is used. In the usual way it means that two mathematical values or mathematical objects are "equal", $=$. The second way it is used is as a definition where we are saying that some symbolic expression is being *defined* to be the same as something else. That is what is happening here. To keep this distinction clear it is quite common to use $:=$ for this definition-sense of equality. So in the above equation we are *defining* $r^{-1} \triangleright Y$ to be $\{\mathbf{x} \in X \ : \ r \triangleright \mathbf{x} \in Y\}$.

Using this notation, we find that in our examples

$$[{*}0] = s^{-1} \triangleright [0] \quad \text{and} \quad [{*}{*}1] \cap [{*}{*}{*}{*}{*}0] = (s^{-2} \triangleright [1]) \cap (s^{-5} \triangleright [0]). \tag{6.3}$$

Going back to the question we raised earlier, about the collection of all those states (infinite zero-one strings) that contain three consecutive 1s in the first ten letters of the string, that set can be written in the form

$$[1\,1\,1] \cup s^{-1} \triangleright [1\,1\,1] \cup s^{-2} \triangleright [1\,1\,1] \cup \cdots \cup s^{-7} \triangleright [1\,1\,1].$$

It is an important fact that we can give a very natural meaning to the inverse of an operator (or mapping, function, transformation) at the level of subsets, even though that operator is not invertible in the sense of there being any operator that undoes it.

After this discussion, we can state the third requirement of our event space: if X is a state space and S a family of operators on it, our event space $\mathscr{E}$ should be an algebra of subsets of X with the property that

$$r^{-1} : \mathscr{E} \longrightarrow \mathscr{E}$$

for every $r \in S$. In other words,

if Y is an event and $r \in S$ is one of our dynamical operators then $r^{-1} \triangleright Y$ is also an event.

6.5.4 Event space (iii): completion/sigma algebras

For most practical situations the requirements that we impose on the event space could stop here: the event space is an algebra of sets that takes account of the dynamics. However, as we have seen with shift systems, though in the real world all things are finite, in the mental world we are not so constrained, and in some ways allowing infinite strings makes things easier. Infinite strings may lie in the world of ideal forms, but even so, working in these ideal forms can be useful.

Consider, for instance, question of states versus events in the simple zero-one shift space. The dictionary sets, which are the foundational entities of the event space, are expressed in finite terms, $[x_1\, x_2\, x_2 \dots x_k]$. A single state,

$$\mathbf{x} = x_1 x_2 x_3 \dots$$

can be written as

$$\mathbf{x} = [x_1] \cap [x_1\, x_2] \cap [x_1\, x_2\, x_3] \;\cap \cdots ,$$

but this entails an infinite number of intersections of dictionary sets. If we allow infinite unions and intersections of events to be events, then states will become events. If we don't, they won't.

We will say that the event space is *complete* if it is closed under finite and countable (or listable) infinite unions and intersections (see §7.2 for more on the definition of countability).[4] The technical terminology is that the event space is a *sigma-algebra.*

If we are given a Boolean algebra then we can always form its completion, simply by including all the subsets that arise out of countable infinite unions and intersections of the subsets that are already in it. Yet this already puts us into an ideal (as opposed to physical) world. Finite Boolean algebras are already complete since there are only finitely many subsets to deal with. But the completion of an infinite Boolean algebra, even though it is easy to say what we mean by it, is essentially beyond conception. In the case of the completion of the event space of the shift system, there is usually no easy way to say whether some arbitrary subset of states of the system is actually an event or not. Fortunately, the sets that actually show up in the normal course of things are always events.

6.5.5 Forms, natural and ideal

There is a significant point that arises here. To what extent are we able to articulate the states of a system? We do not suppose that the level of our capacity to express difference necessarily goes as far as distinguishing the individual states of the system. In the real world that we live in physical states presumably exist in subatomic regions about which we still know virtually nothing. What we distinguish is limited by either what we need (what degree of resolution do we need to drive the car down the road?), by natural thresholds to our sensory capabilities, or by our technological abilities to perceive. It is pretty safe to say that the resolution at which we perceive the world, no matter what technological advances we may make, will never be an ultimate resolution. This

Figure 6.3: Plato (c. 428–c. 348 BCE) on the left, pointing to the heavens, and Aristotle (384–322 BCE) on the right, pointing to the physical world. Detail from Rafael's famous painting *School of Athens* (1510–11). Public Domain, photo commons.wikimedia.org/wiki/File:Raphael_School_of_Athens.jpg .

limitation, which in itself might be construed as an absolute, can be seen as the source of entropy and the source of our uni-directional sense of time.

This difference in perception is, at its heart, an ancient point of contention in the Western mind: the differences between Aristotle's and Plato's ideas about forms. Aristotle points to the world, and sees things in themselves as forms. It is the physical world, accessible only to physical observation (events). Plato points to ideal forms existing in a heavenly world, or at least the mathematical world, that is not limited in this way and in which physical forms are only approximants. In Rafael's famous *School of Athens*, Fig.6.3 we see this displayed quite literally.

In mathematics we want both worlds. We want a mathematics that can model the world as we can experience it. At the same time we are stunned by the beauty of the ideal world that it allows us to uncover. Perhaps most wonderful of all is how together they can fit to give us genuine insight. The law of large numbers is a statement about the ideal world, yet it gives us insight into the real world. We toss the die repeatedly, and what come out are random outcomes, a product of the real world. But we know what will happen in the long run—and casinos thrive!

Later in §6.6, where we introduce the mathematical idea of measures, we will see that measures are defined on sigma-algebras and we will want to use the sigma-algebra version of event spaces. There is a good reason for it. Our notion of area, for instance, is based on the simple ideas we have about the area of a rectangle: length times width. But we also want to talk about the area enclosed by a circle, and clearly a circle cannot be decomposed into a finite number of rectangles. But if we allow ourselves decompositions into vast numbers of rectangles that we can take as small as we please, then we can reach, in the limit, a situation in which there is no further area to account for, and so find the exact area enclosed by the circle. The circles in their perfection are ideal forms—not to be found in this Earthly world. But they do exist in the world of ideas, and that is a domain that mathematics is happy to explore. Theorems like the law of large numbers or the famous Birkhoff theorem that we will come to in §8.5 are statements from the ideal world, but they are immensely practical.

So in the mathematics we have the choice between these two worlds: is the event space to be just the Boolean algebra arising from a finite partition, or do we wish to extend our ideas and mathematical powers by completing it to a sigma-algebra? In this book, we only occasionally need to be careful about this, but we have the freedom to choose, and it is important to be aware of this.

6.5.6 Event space (iv): the event space of a pattern system

With all this preparation we can now specify what we mean by the *event space* of any pattern system.

The *event space* of a pattern system, with state space X and dynamics given by a set of operators S, is a collection $\mathscr{E}$ of subsets of X that arise from the partition $\mathscr{P}$ by assuming that each set P of the partition is in $\mathscr{E}$ and:

E(i) **negation**: $\mathscr{E}$ contains $X \backslash Y$ whenever it contains Y;

E(ii) **and**: $\mathscr{E}$ contains $Y_1 \cap Y_2 \cap Y_3 \cap \cdots$ whenever it contains $Y_1, Y_2, Y_3, \dots$;

E(iii) **or**: $\mathscr{E}$ contains $Y_1 \cup Y_2 \cup Y_3 \cup \cdots$ whenever it contains $Y_1, Y_2, Y_3, \dots$;

E(iv) **dynamics**: $\mathscr{E}$ contains $s^{-1} \triangleright Y$ for all $Y \in \mathscr{E}$ and for all $s \in S$.

The subsets of the state space that arise in this way can be viewed as prescribing the possibilities of articulation within the context of the system and its evolution through dynamics. Notice that E(ii) and E(iii) do not specify whether the list of sets involved is finite or goes on forever (countable). *We will allow ourselves to make either assumption.* If we allow countable unions and intersections then we are talking about complete event spaces; otherwise $\mathscr{E}$ is simply a Boolean algebra.

6.5.7 The importance of event spaces

We have seen how an event space arises out of an initial sameness-difference relation in the form of a partition $\mathscr{P}$ and then the subsequent dynamics. The existence of an event space arising out of a pattern system is already assured once we have established a state space, the dynamics, and a partition for sameness and difference. In the end, however, what is important is the event space, not the partition from which we derived it. The event space is a formal description of the level of articulation. It consists of subsets of the state space and these subsets constitute the level to which we are able to make statements about subsets of the state space.

Sometimes we will find that it is more natural to put aside the idea of the event space beginning with some initial partition, and simply assume that there is an event space in the sense we have defined through E(i)–E(iv) in the box above. This is the way that we shall frame it in §7.5.

6.5.8 The evolution of a partition

Suppose that $\mathscr{X} = (X, s, \mathscr{P})$ is a discrete iterative pattern system. Here we have cut the system down to the minimum: a state space, a sameness-difference relation given by a partition $\mathscr{P} = \{P_1, P_2, \dots, P_N\}$, and an operator s that produces the dynamics.

The dynamics enters at the level of the partition as soon as we ask about those states of X that will fall into a particular part after one iteration of s. Thus $s^{-1}P_j$ is the set of states of $x \in X$ for which $s \triangleright x \in P_j$. These sets form another partition of X. Combining this partition with the original partition we end up with the collection of subsets $P_i \cap s^{-1}P_j$. Now these sets $P_i \cap s^{-1}P_j$, as i and j go through all possibilities, form a new partition of X (though some of these sets may be empty): every state x must be uniquely in some part P_i and $s \triangleright x$ must be uniquely in some part P_j (possibly the same part). Collectively,

$$\mathscr{P}^{(2)} = \text{the partition of } X \text{ whose parts are all the sets } P_i \cap s^{-1}P_j.$$

Since i and j run independently from 1 to N, there are potentially N^2 of these sets (less if some of them are empty, in which case we can ignore them), and we have a new and refined partition of X that takes into account the possibilities for what happens after one step of the iterative process. In the case of the zero-one shift, with the starting partition $\mathscr{P} = \mathscr{P}^{(1)} = \{[0], [1]\}$,

$$\mathscr{P}^{(2)} := \{[00], [01], [10], [11]\}.$$

According to our ideas about events and event spaces in §6.5.6, notably E(iv) and E(ii), these new sets are events and need to be in the event space. Having seen this, it is natural to continue to

$$\mathscr{P}^{(3)} := \text{the partition of } X \text{ whose parts are all the sets } P_i \cap s^{-1}P_j \cap s^{-2}P_k,$$

this time considering all the possibilities for how sameness-difference evolves over the first two iterations. In principle there are N^3 sets here, though again it is quite possible that some of them are empty. It may be, for instance, that there are no sets that go through the sequence P_2, P_1, P_1, i.e.

$$P_2 \cap s^{-1}P_1 \cap s^{-2}P_1 = \emptyset .$$

We do not include empty sets when we talk of partitions: it is the remaining non-empty parts that form the new partition. The evolution of pattern, as it transitions under the dynamics, continues to refine the initial partitioning of X producing $\mathscr{P}^{(4)}$, $\mathscr{P}^{(5)}$, $\mathscr{P}^{(6)}$, ... (in this notation $\mathscr{P}$ itself is $\mathscr{P}^{(1)}$). In the case of the simple zero-one shift, these sets are the familiar dictionary sets.

According to our idea of event spaces, all of these sets will be events. What's more, from the perspective of the sameness and difference, if the original partition establishes what we distinguish and what we do not, then these new events are the limit of what we will be able to articulate as the system evolves.

In the sequel we will call the sets $\mathscr{P}^{(n)}$ the *evolution* of $\mathscr{P}$.

6.5.9 The event space itself is a dynamical system

The definition of pattern systems allows us to see that the event space of a dynamical system is in itself a dynamical system. It is a set whose elements are events, that is to say, certain subsets of states. It has a dynamics in the sense that if $s \in S$ is an operator then there is a corresponding operator s^{-1} that acts on the event space. It is quite typical, and of considerable importance, that when structures are built out of the framework of pattern systems, these too become pattern systems. This points again to what we take to be the case: all systems combining sameness/difference and change are in essence pattern systems.

In the normal treatment of dynamical systems event spaces are usually simply understood implicitly rather than being explicitly mentioned. However, for us they underlie the basic ways in which we can talk about pattern. Event spaces arise as a natural consequence of what can be articulated given the starting premise of the partition $\mathscr{P}$.

The final way in which we have described them is in their "complete" form, that assumes closure of the event space under infinite unions and intersections. This is the way it is defined in mathematics and in probability theory. There is an overall gain in simplicity in going to the limit, and it is the way dynamics is typically understood. In reality, however, as finite beings we will never be able to articulate it to some infinite limit. The inexactness that is implied by some version of sameness-difference is an *a priori* given, and that has important consequences in how we can know the world. This is the point made so eloquently by Carlos Rovelli, which we quote in the epigraph heading this chapter, from his book *The order of time* [138]. We will bring this question up again in Ch.8 when we discuss the important role of entropy in pattern theory.

6.6 Measure and measures

The final feature in the composition of a pattern system is the introduction of a way of assigning values to events. Typically we want to know the frequencies of events and their probabilities of occurrence, or to assign some values like area, temperature, cost, or population size that depend on a particular aspect of a pattern system that we are looking at. The mathematical entities that do this are called *measures*.

6.6.1 The need to measure

One of the immediate problems that arose as early civilizations turned to agriculture for stable and reliable production of food was the problem of measuring out land in equitable ways. We read in Herodotus (*The Histories*) that the Pharaoh Sesostris divided the land of Egypt into plots of equal area and distributed them amongst the people. Not surprisingly there was a catch: they had to pay an annual tax on their land.

> Anyone with a plot eroded by the river (Nile) would go to the king and say what had happened. Inspectors would be sent out to measure the precise extent of the man's loss, so that his future tax-rate could be set at a level appropriate to the reduced size of his holding. It is my theory that this is what lay behind the discovery of geometry.[78, Herodotus Book 2].

Hence the origins of land measurement, and our word geometry.

The basic concept was, of course, area. For rectangles the formula for area came without much trouble (length times width) but for parcels of land with irregular or curved boundaries, things were far harder. The accomplishments of ancient civilizations in finding ways to compute areas and volumes of simple geometric figures can be found in [92]. Some were right, and some were just approximations, but it wasn't until the creation of calculus that a general method for computing areas was made possible. However, our concerns here are not technical issues of actually computing area, but rather the conceptual ideas that lie behind the problem of assigning measure, that is, numerical values, to otherwise numberless entities.

In order to deal with pattern systems in any quantitative way we need to introduce some concept of measuring the size of sets. The basic subsets of interest are those of the event space—the space of subsets of the state space that arise out of the processes of sameness and difference and the dynamics of the system—so what we are looking for are ways in which to associate measure to these sets. We have already seen some of this quantification in our section on chance, where the measure of a set represents the probability that a randomly taken state will actually lie in that set.

Here, then, we introduce the mathematical idea of measure—indeed the mathematical entities involved are called *measures*. We shall use them for the discussion of arbitrary pattern, for analyzing the frequency components of wave-phenomena, as a basis for describing probability or frequency distributions in pattern systems, for the definitions of entropy and information, and ultimately for measuring relative degrees of synthesis of composite pattern systems. The power of measures lies in their applicability to continuous or to discrete systems, or even to systems that seem to lie somewhere in between.

The important thing to understand at this point is that there are in general all sorts of measures that can be defined on a set X. We might be measuring land for area, for the total annual rainfall on it each year, for the amount of it that is arable, or for the number of burrowing owls inhabiting it. Each is a measurable attribute of land and parcels of land, but each measures a different type of thing. Length, area, volume are familiar examples of measures. To see that there are many other possibilities, look at Fig.6.4. What we need now is a general conception of what measures should do, rather than particular examples.

Suppose that we have a set X. It could be the set of all points in the plane, the set of all points inside a ball in three dimensional space, the set of numbers on the sides of a die, $\{1, 2, 3, 4, 5, 6\}$, or, importantly for us, the points of the state space of some dynamical system. The question is this: how can we assign a meaningful notion of numerical measure to subsets of X? Think, for instance, of the Nile delta and assigning numerical values to parcels of land being distributed to people. Each parcel P of land has some area, call it area(P). From what we informally know about area we can expect three things to happen:

(i) area$(P) \geq 0$, no matter what parcel of land we look at;

(ii) area$(\emptyset) = 0$ (the area of "nothing" is zero);

(iii) if P consists of two separate pieces of land, P_1 and P_2, then
area$(P) =$ area$(P_1) +$ area(P_2).

This simple idea is the basic one that underlies the definition of a measure. If X is a set then a measure on X is an assignment of numbers to subsets of X. We will use letters like $\mathbf{m}$, $\mathbf{n}$, $\mathbf{p}$ for measures, so if $\mathbf{m}$ is a measure on a set X, then $\mathbf{m}$ assigns values $\mathbf{m}(Y)$ to subsets Y of X. To be a *measure* it has to satisfy three properties:

(i) $\mathbf{m}(Y) \geq 0$, no matter what subset Y we look at;

(ii) $\mathbf{m}(\emptyset) = 0$;

(iii) if Y_1 and Y_2 are *disjoint* subsets of X, i.e. $Y_1 \cap Y_2 = \emptyset$, then

$$\mathbf{m}(Y_1 \cup Y_2) = \mathbf{m}(Y_1) + \mathbf{m}(Y_2).$$

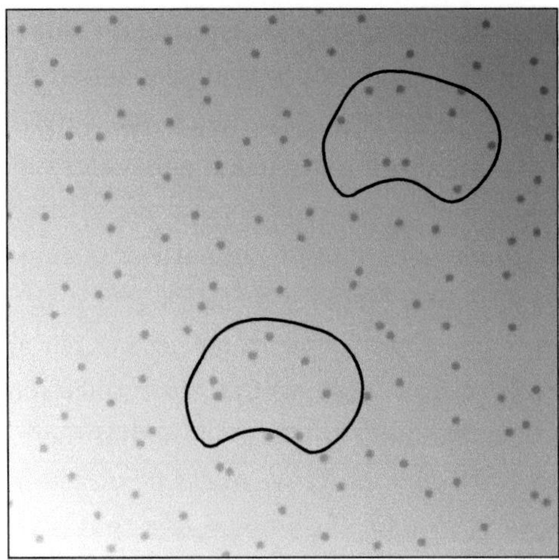

Figure 6.4: We see here a square region of the plane that we shall call X. Let's say it is a 1×1 square so its area is 1. It has a finite collection of points inside it and also a gradient shading. We can easily imagine three different measures on this set. Each assigns numerical value to any subregion (subset) of the square. The first is the usual one of area, so it assigns to each subregion of X its area. The second assigns numbers by counting the number of points inside each subregion; the third measures the density of black ink in each subregion. To illustrate, we have taken what is supposed to represent an arbitrary region in the plane (the region enclosed within the curved boundary) and a second copy of that region moved to a different part of X. The two subregions indicated have equal area, but have different numbers of points in them and also different densities of ink. These three types of measures, one based on area, one on counting points, and one on ink density, are completely different from one another, but each is a measure, and each has its own uses. It becomes rather obvious that there must be a vast number of measures we can define in this sort of way depending on what it is that we care to measure.

6.6.2 Measure defined

As is so often the case in mathematics, a common English word, here the word "measure", is used both in its conventional ways and its new mathematical way. As we have just defined it, "measure" is now also a technical term for a certain type of *mathematical object* that assigns numbers to subsets of a set. If $\mathbf{m}$ is a measure on a set X, then it does, in effect, "measure" subsets of X, and if Y is a subset then we speak of $\mathbf{m}(Y)$ as the *measure of Y* (with respect to the measure $\mathbf{m}$). In that sense its meaning is closest to the use of measure when we refer to a tape measure as a measure. Think of it as a mathematical tape measure, though mathematical measures can do far more varied things than tape measures—things like area, volume, quantity, density, or probability. The subtlety and usefulness of this concept become apparent as we go along. In any case, we try to avoid any confusion between its common and mathematical meanings when we use the word "measure".

There are three subtleties that need to be added to our tentative definition of a measure in order to arrive at the full mathematical definition of a measure, and these ultimately underlie its importance and power. The first is rather plain. It is useful to allow the measure of some sets to be infinite. For instance, in thinking about area in the familiar setting of the usual coordinate plane, the area of the entire plane is obviously not finite. It really doesn't have a numerical value since the total area exceeds any finite number. To allow for this kind of thing we just use the symbol ∞ and write $\mathbf{m}(X) = \infty$. That does not mean that infinity is now a number. It just is a very useful way of saying that the area of the plane exceeds any finite value. The rules for handling ∞ are what we might expect:

$$\begin{aligned}
&\infty + x = x + \infty = \infty \text{ for all real numbers } x\,;\\
&\infty + \infty = \infty\,;\\
&\infty\, x = x\, \infty = \infty \text{ for all real numbers } x \geq 0\,;\\
&\infty - \infty \text{ is not defined}\,;\\
&0\,\infty \text{ and } \; 0\,\infty \text{ are not defined}\,.
\end{aligned}$$

Note the last two are about what we *cannot* do, because there is no consistent sense that can be made of them.

The second refinement also involves the infinite and harks back to the same issue that we brought up in the case of events. Item (iii) above refers to the additive nature of measure when it is applied to the union of *two* disjoint subsets. It is quite easy to infer logically from this that the same idea works for unions of three, four, five, ...disjoint sets, just as we would expect. However, just as in our discussion of event spaces, we cannot assume from this that it will automatically work if we take the measure of the union of an *infinite* number of disjoint subsets. The genuine power of measure theory comes from *assuming* that this additive law *does hold* for unions of a countable (listable) infinite number of disjoint parts. We will explain why this is so important in §6.6.3.

The third subtlety is that although a measure defined on a set X gives values to subsets of X, it rarely gives value to *all* subsets of X. In defining a measure $\mathbf{m}$ on a set X one must specify a specific algebra of subsets to which the measure applies. In our case this algebra of subsets will be the event space.

With these three cautions in mind, we can give a full definition of what we mean by a measure (see box on the next page).[5] Notice that M(iii) is stated in terms of a list $Y_1, Y_2, Y_3, \ldots$ of disjoint subsets of X. This list may be finite or may go on without end (countable).

The measure is a *finite measure* if the measure of the whole state space, that is $\mathbf{m}(X)$, is finite. It is called a *probability measure* if the measure of the entire state space is 1, that is $\mathbf{m}(X) = 1$. It is in the context of probability that we will most often use measures.

A *measure* on a pattern system X with an event space $\mathscr{E}$ is an assignment (a mapping)

$$\mathbf{m} : \mathscr{E} \longrightarrow \text{the real numbers} \cup \{\infty\}$$

(remember that $\mathscr{E}$ is a set of *subsets* of X) with the following properties:

- M(i): for all events Y of X, $\mathbf{m}(Y) \geq 0$;
- M(ii): $\mathbf{m}(\emptyset) = 0$;
- M(iii): if $Y_1, Y_2, Y_3, \ldots$ are any *disjoint* events of X (they have no points in common) then

$$\mathbf{m}(Y_1 \cup Y_2 \cup Y_3 \cup \cdots) = \mathbf{m}(Y_1) + \mathbf{m}(Y_2) + \mathbf{m}(Y_3) + \cdots .$$

We have deliberately phrased our definition of measure in terms of the *event space* of a pattern system since that is the way we will see them. In defining a measure $\mathbf{m}$ on a set X we must specify a specific algebra of subsets to which the measure applies. The measure will be defined for the subsets of X that are in this algebra, but not necessarily for any others. The typical situation is that we have the framework of a pattern system and we want to define a measure on the event space. So it's the event space that we take as the algebra of subsets. Most often the event space itself is derived from an initial partition of X, but this is not necessary and is not part of the definition of measures.

The issue is one of existence. Even for ordinary area or volume on the usual plane or in usual three space, the requirement of translational invariance makes the existence of sets for which area or volume *cannot* be assigned inevitable! In practice these "non-measurable sets" never show up—they are denizens of the ideal world and their complexity is such that they are both hard to find and hard to describe. Although it is our duty to define the mathematics properly, non-measurable sets simply don't arise in practical problems.[6] In principle the idea of a measure is very straightforward, and that is all we need. It is subtleties like this that actually make a fully rigorous development of measure theory difficult.

Since for any subset $Y \subset X$ in $\mathscr{E}$, its complement $X \backslash Y$ is also in $\mathscr{E}$, and since $X = Y \cup (X \backslash Y)$ is a disjoint union and the measure of any set is non-negative, the axioms tell us that we have

$$\mathbf{m}(X) = \mathbf{m}(Y) + \mathbf{m}(X \backslash Y) \geq \mathbf{m}(Y),$$

showing that no subset of X has measure exceeding $\mathbf{m}(X)$. This is just what we expect.

Beware! M(iii) contains the explicit assumption of *disjointness*. The sum rule only works if the subsets involved have no overlap. What is true in general is that for the union Y of any two sets Y_1 and Y_2, we have

$$\mathbf{m}(Y) = \mathbf{m}(Y_1) + \mathbf{m}(Y_2) - \mathbf{m}(Y_1 \cap Y_2),$$

the reasoning being that the set $Y_1 \cap Y_2$ lies in both Y_1 and Y_2 and so has been counted twice: once in $\mathbf{m}(Y_1)$ and once in $\mathbf{m}(Y_2)$. If Y_1 and Y_2 are disjoint from one another, $\mathbf{m}(Y_1 \cap Y_2) = 0$, and we are back to M(iii).

The definition of a measure is based on the three assumptions M(i), M(ii), M(iii). Any assignment of sets to numbers that satisfies these conditions is a measure. It is possible to define a measure on any set; in fact, on any non-empty set there are infinitely many different measures. In the rest of this section we discuss some of these, all of which we will find familiar, even if we are not accustomed to speaking of them in the language of measures.

6.6.3 The need for countable unions

As we have noted, in M(iii) we are not limited to taking just a finite number of sets. We may also take an infinite (countable) number of sets (see also §7.2). In §6.5.5 we noted that although the usual idea of areas seems straightforward for rectangular and rectilinear figures, when we try to work out the area of something with curved sides, even something as simple as a circular disk we are in trouble.

Still, even with no special mathematics we would have some idea how to get an estimate of the area of a circle: we can approximate it by filling its inside with finer and finer polygonal figures whose area we know how to express, and thereby find its area as the limiting value of the sum of the areas of such figures. We can also do this from the outside too, covering the whole of the disk with finer and finer collections of polygonal figures that better and better approximate it, Fig.6.5. Essentially this was the method in pre-calculus times. There was an understanding that the area of a circular disk of radius r ought to be πr^2, where π is some number just over 3. The question was what is the value of π? By inscribing polygons and circumscribing polygons with large numbers of sides the area of the circle could be hemmed in between inner and outer estimates that closely approximate it.

Archimedes (c. 287– c. 212 BCE) calculated π to be in the range $22/7$ to $223/71$ using polygons with 96 sides, and Liu Hui (third century CE) got a value of π as approximately $3927/1250 = 3.1416$ (which is correct up to this number of decimal places) in two ways, using polygons of 1536 sides and some clever interpolations from polygons with 96 sides, Fig.6.5. So from the very earliest times, our assumption M(iii) was implicitly used, if not explicitly stated. Incidentally, the Chinese had even better approximations of π by the fifth century, but by the time the Jesuits arrived in China in the seventeenth century much of their knowledge of mathematics and astronomy, including these accurate approximations of π, had been forgotten [166]. Of course the simple estimate of π as $22/7$ is the common one that we all learned in school. As for π itself, it is an irrational number whose decimal expansion goes on forever with no identifiable regularities.[7] In fact the

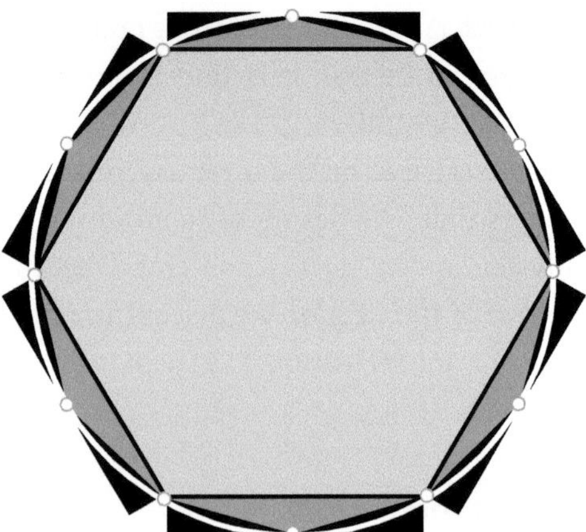

Figure 6.5: The figure shows the geometry that underlies Liu Hui's methods for computing π by inner and outer polygons. The figure shows a circle with an inscribed hexagon (light grey), and a corresponding set of rectangles (in black) that together with the hexagon completely cover the circle. The principle was to keep doubling the number of sides, 6, 12, 24, ..., deriving better and better approximations. The darker grey dodecagon (12 sides) shows the beginning of the next step. See Wikipedia article on Liu Hui's π algorithm.

digits of its decimal expansion, though computable, are so irregular that they pass all the usual tests for randomness, see §5.1.6.

Since the invention of calculus and the theory of infinite series, we have fast numerical ways to work out the value of π to any number of decimal places we please. Although these methods do not rely on inner and outer approximations, the process of limiting approximation by rectangles still lies at the theoretical heart of integration itself, both in its older forms and in the new ones based on measures.

We need our measures to be able to deal with both finite and infinite collections of subsets. We should realize, however, that the step from finite to countably infinite in the summation rule is entirely non-trivial. It can be interpreted as the step from Boolean algebras to sigma algebras, from the finitary world of our existence into the Platonic world of ideal forms that we discussed in §6.5. It is easy to appreciate why it is necessary, but it does take us into a world beyond which we no longer have such clear intuitions. This is not to cast doubt upon the process, which is done formally with a great deal of care, but to realize that dealing with infinite collections of things we are stepping into unusual territory.

6.6.4 Lebesgue measure and invariance

The ordinary concept of *area* goes by the name of *Lebesgue measure* in mathematics, thus honoring Henri Lebesgue (1875–1941) who created our modern theory of measure that we have been discussing. Lebesgue measure is the measure that assigns area to subsets of the plane. Given that there must be countless measures on the plane, we can reasonably ask why it is that our usual concept of area is the one that arose naturally and universally in human culture.

The key lies in the what we might refer to as "dynamics". In defining area, the real key, the thing that makes it useful, is that if we move a region around in the plane, as long as we don't deform it, its area will remain the same. If that were not true, the concept would be pretty useless. The most natural dynamics on the plane is that of translation, that is just moving things somewhere else, without any rotation, or stretching. It is what we do with a window on our computer monitors when we click on a graphical object and drag it somewhere else. Now, it is a theorem that our usual concept of area is, in fact, the *only* measure of subsets of the plane that has this translational invariance. Amongst all the translation invariant measures the only choice is that of scale—are we measuring in acres, hectares, square miles, square centimeters, ...? Once the scale is established there is only one measure possible, and it is the universal one. As it turns out, area is also invariant under rotation. That comes for free.

The same situation applies to volumes in 3D. It would be sad if our bottle of wine should shrink in volume as we brought it home. Translational invariance is indispensable to our concepts of length, area, and volume, and in each case there is, up to scaling, only one possible measure. These are the Lebesgue measures, one for each dimension (including four, five, and higher dimensions). Lebesgue *proved* that these familiar concepts are indeed expressible as measures as we have defined the word. He also proved that these measures exist!

Although we might not first think of it as such, from our point of view translation is dynamical in effect. If we think of the plane as a state space with the individual points as the states, then any translation, say a combination shift right by one unit and up by two units, is an *operator*. We would normally think of it as actual motion, though, as we have emphasized it amounts to associating to each state (x, y) the new state $(x + 1, y + 2)$. Replacing 1 and 2 by arbitrary real numbers, there are clearly infinitely many such translations. This is a form of dynamics (change) that we will call *spatial dynamics* rather than the more usual *temporal dynamics* that explicitly involves time. Requiring the measure to be invariant under the dynamics of translation, to be *translation invariant*, is a strong assumption—so strong that there is effectively only one measure available.

In the case of 1D, we are dealing with the real line (§2.3.2). We start with the so-called *closed intervals* on the real line, that is to say all the intervals

$$[a, b] := \{x \in \mathbb{R} \ : \ a \leq x \leq b\},$$

where a, b are allowed to run over all real numbers for which $a \leq b$. Not surprisingly we define the length of this interval by

$$\ell([a, b]) = b - a\,.$$

Next we create an event space out of these sets by taking the smallest collection of subsets that contain all these intervals and whatever we can get from them by using finite and countable unions and intersections (and-or constructions) and complementations (negations) of the event space axioms E(i) – E(iii). The (complete) event space that arises out of this is called the *Borel algebra*, denoted $\mathscr{B}$, and its sets are called *Borel sets*. The Borel algebra is the standard event space for the real numbers.

We could equally well have started with the *open intervals* on the real line,

$$(a, b) := \{x \in \mathbb{R} \; : \; a < x < b\},$$

where a, b are allowed to run over all real numbers for which $a < b$, and defining

$$\ell((a, b)) = b - a,$$

come to the same conclusion.[8]

It is easy to write down this definition, but it is impossible to imagine the complexity of subsets of the line that are contained within the Borel sets. Even so, it is not so hard to see how the standard measure of length that we applied to the intervals can be extended through unions, intersections, and complementations to a measure that is defined on the entire Borel algebra. The invariance under translation is built into the way the Borel sets are defined. In symbolic terms translational invariance means if we translate the interval by any chosen distance c, so that we are now looking at the interval $[c + a, c + b]$, we should find it to have the same length $b - a$. Indeed it does:

$$\ell([c + a, c + b] = c + b - (c + a) = b - a = \ell[a, b].$$

This then is length, now viewed as a measure—*Lebesgue measure*—and we will consistently designate it by ℓ.[9]

What we have done here can be done for the plane (2D), this time using rectangles, and for space (3D), using boxes. In fact it can be done in 4D, 5D,... (we just increase the number of coordinates, but reuse the same ideas), even if we can no longer visualize them. The set operations of countable unions and intersections, and complementation produce an event space, which is again called a *Borel algebra.*

Lebesgue measure: All of these measures (length, area, volume, and higher dimensional equivalents) are called *Lebesgue measures.* In this book, all are denoted by the same symbol ℓ, the underlying dimension being understood from the context.

This process of translation, which is simply based on addition and subtraction, underlies the basic symmetry of the real line. The real line looks the same everywhere. The same goes for the plane and for 3D space. We can think of its translation invariance as expressing a fundamental compatibility of Lebesgue measure with this symmetry. The process of translation, in the sense of sliding sets along on the real line, is its most basic expression of change, and the Lebesgue measure is invariant with respect to it. The same holds for the Lebesgue measures in higher dimensions too.[10]

6.6.5 The importance of invariance

Translational invariance is a necessity for making length, area, and volume useful. What we have just been saying is that, turning it around, this invariance is so powerful a restriction that the standard definitions for length, area, and volume are the only possibilities we have. There are infinitely many different measures that can be put on any of these spaces (we will see some of them below), but if we want a translationally invariant measure then the only freedom at each dimension is that of changing the scale of the corresponding Lebesgue measure. Our definitions of length, area, volume, arising over millennia of human culture, are not just human conventions—they are the unique solutions to our need of invariance.

We can phrase the idea of invariant measures quite generally. A measure $\mathbf{m}$ on a pattern system $\mathscr{X} = (X, S, \mathscr{P}, \mathscr{E})$ is called *invariant* or *stationary* if $\mathbf{m}$ is defined on all the sets of the event space $\mathscr{E}$ and

$$\text{for all } A \in \mathscr{E} \text{ and for all } s \in S, \text{ we have } \mathbf{m}(s^{-1} \triangleright A) = \mathbf{m}(A). \qquad (6.4)$$

Here again we see s^{-1} appearing. One might expect that it would be s, not s^{-1}, that is the right thing. However, if we think of $s^{-1} \triangleright A$ as talking about the future, the set of points that come from A after applying the operator s, then it makes sense. A stationary or invariant measure, whether that be length, area, volume, or something more abstract like probability, has the property that *it takes the same value on any set as it had on the set from which it was derived by change.*

Consider for example the usual heads-tails, zero-one shift with the probability measure $\mathbf{p}$ that assigns equal values of $1/2$ to zeros and ones. This really is a measure: it gives value $1/2$ to the basic dictionary sets $[0]$ and $[1]$, and quite generally the value $1/2^n$ to any dictionary set $[a_1\, a_2\, \dots\, a_n]$. Now, the probability of the appearance of heads and tails, ones and zeros, is unchanged by the fact that we may have already tossed the coin lots of times already, so we would expect that this probability measure ought to be invariant under shifting. Indeed we do find, for example, that

$$\begin{aligned} \mathbf{p}([1]) &= 1/2, \\ \mathbf{p}(s^{-1} \triangleright [1]) &= \mathbf{p}([*1]) = \mathbf{p}([01]) + \mathbf{p}([11]) = 1/4 + 1/4 = 1/2. \end{aligned}$$

But note that the perhaps expected definition is not correct:

$$\mathbf{p}(s \triangleright [01]) = \mathbf{p}([1]) = 1/2 \neq 1/4.$$

The invariance of a measure is revealed by s^{-1}, not s itself. We can repeat this where we are now tossing a biased coin, say with $\mathbf{p}([1]) = p$ and $\mathbf{p}([0]) = 1 - p$, and again the tosses are taken to be totally independent of each other. This time we have

$$\mathbf{p}([1]) = p,$$
$$\mathbf{p}(s^{-1} \triangleright [1]) = \mathbf{p}([*1]) = \mathbf{p}([01]) + \mathbf{p}([11]) = (1-p)p + p^2 = p\,.$$

For the measure $\mathbf{m}$ on a pattern system $\mathscr{X}$ to be invariant means that the invariance equation $\mathbf{m}(s^{-1} \triangleright A) = \mathbf{m}(A)$ must hold for all operators $s \in S$ that produce the dynamics and for all subsets A of the event space. Note that axiom E(iv) for event spaces in §6.5.6 is exactly the condition that we need here. For deeper insight into this, see §11.10.3.

It would be hard to overstate the importance of invariance within the context of change, for it underlies both modern physics and our whole approach to pattern. We know that we live in a cosmos of constant change, a cosmos in which all "things" are impermanent, more processual than material. We look for permanence, solid foundation on which we can rest, but it is not to be found in the material world: it is to be found in the laws or principles through which change occurs. Invariance is about what does *not* change when there is change. What does not change under change is the origin of order. If it were not so, how could forms arise and persevere?

Whether we think in terms of laws, principles, or invariants, that which does not change within a context of change is the dynamical counterpoint to sameness and difference. We will come back to this in a far more detailed way when we come to Ch.9, where we look at the extraordinary range of ideas around symmetry and how we understand the world.

6.7 Pattern systems defined

With all this preparation behind us, we can now define precisely what we mean by a pattern system.

> A *pattern system* $\mathscr{X}$ consists of four mathematical components
> PS(i) **State space**: a non-empty set X.
> PS(ii) **Dynamics**: a non-empty set S of operators on the state space X.
> PS(iii) **Difference**: an algebra (either a Boolean algebra or a sigma-algebra) $\mathscr{E}$ of subsets of X which is compatible with the dynamics S (that is $s^{-1} \triangleright E \in \mathscr{E}$ for all $s \in S$ and for all $E \in \mathscr{E}$).
> PS(iv) **Measure**: a (invariant) measure $\mathbf{m}$ defined on the event space $\mathscr{E}$.
> We typically express a pattern system in the form $\mathscr{X} = (X, S, \mathscr{E}, \mathbf{m})$.

As we mentioned before, often we use a partition $\mathscr{P}$ to initiate the event space. In that case the event space is the algebra of sets that arises out of the logical operations of "and", "or", "not", and the dynamics of the system. If that is the case, we can explicitly include this in the description of a pattern system.

The definition puts pattern into the framework of a *system*. It is no longer a static entity but a dynamical unfolding of process based on change and difference. Difference refers to distinguishability, and the event space represents the full extent of that, taking into account the effects of change.

Although it is not required, usually we are interested in finding *invariant* measures, for they fully integrate the dynamics of the situation. They are often described as being the *law of the process*. Often it is not obvious what the law of a process may be, and it is revealed only when we understand the dynamics sufficiently well. Still, the law of a process profoundly shapes the evolution of the system, as we shall see.

On this point it is interesting to observe that the Chinese character for *law* is 法. Its original meaning, however, which is implicit in the two parts of the character ("water" and "go" or "leave"), is that of a river ford. As Buddhism was introduced into China it was used as a translation of the Sanskrit word *dharma*, in the sense of crossing over. Hinton [79, Ch.Dharma] writes of this that the meaning of dharma, at least from a Chan point of view, is "the fundamental laws or patterns that govern the unfolding of Tao as it unfurls into the ten thousand things." This is the metaphorical idea that is implied by our systems idea of patterning.

6.8 What is pattern?

> The lord whose oracle is in Delphi neither indicates clearly nor conceals but gives a sign.
>
> [136, Heraclitus, Fragment 93]

Arriving at the mathematical definition of a pattern system has involved a lot of ideas, and as such it might seem to make the idea of pattern and patterning something more remote and difficult than it really is. Its immediate value is its formal nature and the explicitness of its definitions. As we go on we will see how deeply it stands as the foundation of many diverse wonderful mathematical and scientific developments. This is really what the rest of the book is about.

Even so, having arrived at the formal definition of a pattern system, we still have not defined pattern. We have a system—now, what is pattern? Here we make the obvious suggestion: pattern systems are dynamical in conception, and pattern is what a pattern system produces as distinguishable events. A pattern system, as we have defined it, can be construed as the definition of a system in general: that is,

> a *system* is an abstraction
> that models the concepts of difference and change within some context.

Pattern consists of the ways in which the system manifests itself, and that manifestation takes the forms of events. What we can perceive and articulate about any system is its events. Conversely, when we claim to understand some phenomenon, we are interpreting perceived events as the manifestations of some system. In this way the concepts of system and pattern mutually support each other.

There is an important, and often unnoticed aspect of this. We come to see the world through events, those processes of likeness and distinction that we use at every moment. But what we make as likenesses and distinctions are a product of our experiences of life. Putting it in the terms we are using here, the event space is a lens which shapes what it is that we "see" and what we don't. Speaking of middle life, but in fact applicable to entire lifetimes, James Hollis writes:

> Perhaps the first step in making the Middle Passage meaningful is to acknowledge the partiality of the lens we were given by family and culture, and through which we have made our choices and suffered their consequences. If we had been born of another time and place, to different parents who held different values, we would have had an entirely different lens. The lens we received generated a conditional life, which represents not who we are but how we were conditioned to see life and make choices. All generations are seduced into anthropocentrism, tending to defend their vision of the world as superior to others. So, too, we succumb to the belief that the way we have grown to see the world is the only way to see it, the right way to see it, and we seldom suspect the conditioned nature of our perception [85].

Writ large, on the cosmic scale, in both Daoist thought and in the philosophical tradition of Mahayana Buddhism we find the mutual 'system/manifestation' duality explicitly stated. As the *Daodejing* Ch.1 puts it:

> Oftentimes without intention I see the wonder of Dao.
> Oftentimes with intention I see its manifestations.
> Both of these are the same in origin; They are distinguished by names after their emergence. Their identification is called mystery [32, p. 122].

Then there is the famously cryptic *The Awakening of Faith Sutra* whose central thesis can be summarized as:

> [T]he principle of One Mind has two aspects. One is the aspect of Mind in terms of the absolute (*thathāta*; suchness), and the other is the aspect of Mind in terms of phenomena (*samsara*; birth and death). Each of these two aspects embraces all states of existence. Why? Because these two aspects are mutually inclusive [7, Part 3].

"Mind" here seems to serve as the same encompassing metaphysical principle as Dao, beyond words, but ever manifested to our senses.[11]

More simply the Buddhist nun, Aoyama, says

> The wind goes its own way and is without form. We know it is there when we hear it in the grass or trees, see the clouds scudding overhead, or feel it blow against us [5, Hearing the wind in the pines].

We experience events, and it is out of these experiences that we make the world, formulating mental conceptions of systems which "explain" them. The events are then patterns, and they are patterns because we experience them as manifestations of some system.

Put in one line, we might define *pattern* in a simple sentence of ten words:

Patterns are the (physical) manifestations of the processes of systems.

CHAPTER 7

Exploring the definition of pattern

7.1 Introduction

At the end of Chapter 6 we arrived at a formal mathematical definition of what we mean by a pattern system. The overall ideas of context, change, difference, and measure are intuitively familiar, but their translation into the language of mathematics and emergence as a new mathematical entity is something else. The whole enterprise may seem to have limited value. The purpose of this chapter is to put a bit more flesh on this skeleton, to see how the concept of a pattern system is a mathematical entity in its own right, and to show that it can shed light on how pattern systems may be alike or different. Later, in Ch.8, we will extend this further into the ways that pattern systems can interrelate and integrate to form larger pattern systems.

We start by filling in two gaps. The first is a discussion of how the infinite is spoken of in mathematics. Inevitably we find ourselves dealing with infinite sets, for even the set of natural numbers is infinite in the sense that there is no last or largest positive integer. Not surprisingly our intuition about just what the "infinite" is when speaking of infinite sets is very poor. One of the surprising outcomes of trying to understand what "cardinality" means—that quality that captures something of the size of a set—is that there are really very different versions of the cardinality of an infinite set. The set of natural numbers and the set of all real numbers between 0 and 1 are both infinite sets, but they are not the same cardinality. This difference, the difference between *countable* and *uncountable* sets as they are called, actually shows up many times in this book and can even be seen to be a point of confusion in the history of science. It is important to recognize it.

The second gap is that in arriving at the definition of pattern systems we used only discrete dynamics. We need to see that continuous dynamics fits the definition as well. To do that requires a bit of an excursion into how continuous dynamics actually arises in science and how the mathematical idea of a dynamical system arose in the first place.

Beyond this we want to see how the new definition actually helps us to understand pattern better. The overall story, one that we will begin thinking about here and will continue in greater depth in the next chapter, is that in the reality of things, pattern systems do not live in isolation. We may separate a pattern system, say the organization of a neuron in the brain, and study it in depth as a pattern system that can be modeled mathematically, but in reality neurons exist and function in vast networks of neurons and other biological/biochemical constituents of the body, so become parts of larger pattern systems which they both affect and are affected by. Comparing pattern systems and thinking about how pattern systems can be interrelated and combined into more complex pattern systems is really crucial. No matter at what scale we look at things, from the subatomic to the galactic, the same principles of context, change, distinction, and quantification are always there, and they are always intimately connected.

In this chapter we enter the beginnings of this. What does it mean for two pattern systems to be the same, and what does it mean for one pattern system to be "like" another, even if it is not really the same? We base the discussion around three examples. The first is the familiar shift system, the second a new system called the doubling system, and the third is an apparently wild and unpredictable system taken from a family of pattern systems called logistic systems. In the end these three systems will be seen to be very much related through the concept of pattern systems, the first two being essentially identical in spite of appearances, and the third being a sort of dynamical shadow of the other two.

One virtue of finding a connection between two pattern systems is the possibility of being able to transfer structural ideas that we know about one to the other. A famous example that we take up later is that of diffraction of X-rays. Ordinary visible light can diffract and this is accomplished using diffraction gratings. Once X-rays were discovered and assumed to be based on the same principles as visible light it was only a matter of time before someone asked about diffraction of X-rays. This then turned into an incredibly important tool in crystallography. In this chapter we see how to use the relationship of the doubling system to the logistic system to get an understanding into the natural invariant measure that can be associated with the latter.

Along the way, we also develop a new way of looking at a pattern system. Normally pattern systems are based on the dynamics of their state spaces. As we have noted, however, the level of distinction that we may have, which is encapsulated in the event space, may be far less detailed. The idea of tracking dynamics on the basis of changing events rather than at the level of changing states leads to a way of bringing the dynamics to a level at which we can actually articulate it. This brings us to event-space patterning. We can make the argument that the reality of our experience with the world is based on the changes amongst events. We basically operate at the level of event space patterning. Finally we find that the simple shift space is really the universal model of iterative dynamics at the level of events.

This all leads to a further discussion of what we think of as *paradigm systems*, systems that somehow grasp the essential characteristics of some family of pattern systems. The basic pattern system of a point moving around a circle is, as the language suggests, the paradigm for all cyclical processes.

The chapter ends with the statement of one of the famous theorems of dynamical systems, the Poincaré recursion theorem, which here is put into the setting of pattern systems. This theorem establishes, in a beautiful way, one of our deepest intuitions about pattern—that somehow it is about recurrence or return. It has to do with recognizing, *re-cognizing*, what in one way or another has already been experienced. This profound idea is one that we keep coming back to in the book. The Poincaré theorem shows that the idea is already implicit in the very definition of pattern systems. We do not go so far as to prove the theorem in the text proper, but it is short enough and readable enough that it makes sense to include the proof in the Endnotes (see Endnote 19) for those who would like to see it. On the one hand it uses nothing that we have not discussed, and on the other hand it uses every part of the definition of a pattern system to the fullest.

This chapter can be read selectively. The three examples consisting of the shift, doubling, and the logistic system can help to solidify basic ideas around pattern systems, and the creation of event-based pattern systems is used to a considerable extent in the next chapter on entropy.

7.2 Mathematics of the infinite

When we think of what we know about our universe we seem always to be stretching in two directions: towards the vastness of space and towards the minuteness of the subatomic world. Is the universe infinite or is it finite and expanding, and if so does it continue to expand forever? Can we continue to divide space indefinitely into smaller parts until we arrive at the infinitely small? Is time indefinitely divisible or is it ultimately granular in nature? These questions have been with us from ancient times, and when we look at the two extremes of cosmology and particle physics today, we see that we are still asking them today.

The infinite and infinity—these words are common enough, and we bandy them about as if we knew what they meant. But we cannot directly experience them—they are mental constructs, and it is hard to imagine how we actually can define them. How do we work with precision with concepts that don't really have approximate counterparts in reality?

We can see these issues arising in the writings of Aristotle. In his *Physics* IV.10, where he is trying to understand the nature of time, he notes that the "now" separates the past from the future, but neither the past nor the future exist in any physical sense. So what then is it that nows separate? He goes on to say that though nows are all that we can say exist, "time does not seem to consist of nows". What he means by this is explained later in VI.1 where he writes:

> Also, a point [on a line] cannot be successive to a point, nor can a now be successive to a now, in such a way that they form a length or a stretch of time. I mean, things are successive if there is nothing of the same kind as themselves between them, but there is always a line between points and always a stretch of time between two nows.

It might seem that this sows more confusion rather than clarifying anything, but Aristotle really does hit on an important dichotomy. When he uses the word "successive" he means that one thing follows another in some sort of temporal sequence. This is discrete in nature, like the shift systems that we have been talking about. When he thinks of time as a continuous process along a continuum, like a line, it is more like the swinging pendulum: there it would seem that the basic constituents are points—moments of time. But points on a line don't have successors in the sense that for each point there is a next point. Pattern systems can be discrete or continuous, they can be placed into the same overall formalism, but they are truly different from one another. This difference is important and it was made manifest in a remarkable way by Cantor in his investigations into set theory.

Before we launch into this in more detail, we should make it clear that we are not out to define some *thing* called infinity. What we do talk about is the *property* of being infinite. This is a quality, not a thing. In fact we try to avoid the word "infinity", though some mathematical notation explicitly uses the symbol ∞ and we call this symbol "infinity".

Quite literally, the quality of being infinite, is being "not finite". The positive integers $1, 2, 3, 4, 5, \ldots$ collectively form an infinite set because there is no end to them. There is no last positive integer. The real line, or the line of Euclidean geometry, is infinite because it is unbounded (in both of its directions). The standard Euclidean plane is infinite in the sense that it is unbounded in all directions. Our discussion starts from this understanding.

Set theory, as a theory, was first studied in depth by Georg Cantor (1845–1918), and though it generated a lot of criticism from some of the notable mathematicians of the time, sets and Cantor's ideas have become the principal foundational concept of mathematics. We have already seen this—with our state spaces being sets whose elements are called states, with sets of operators, and measures that take values on sets, and so on. Everything is based on sets.

Cantor's great discovery was to realize that not all "infinities" are the same. In the domain of infinite sets there is a dichotomy, this time between what are called *countable sets* (or what we have also called *listable sets*) and *uncountable sets*. This distinction might seem to be far too esoteric for our understanding of pattern, but it is not. In fact it occurs at a very familiar level: the set of all integers is countable, but the set of all real numbers, even the set of all real numbers between 0 and 1, is not. That is pretty much Aristotle's puzzle in a nutshell.

This distinction is what we discuss here. It is one of the few places in the book where we put down a fairly complete argument—partly because it is rather fun to follow, and more importantly because it is so revealing. It is fascinating to ponder the famous "diagonal" argument of Cantor that leads to the dichotomy. Perhaps it is something everyone should know.

7.2.1 Counting

The set of all positive integers, also called the natural numbers, is an infinite set. What makes it infinite is that it is unending. The natural numbers $\mathbb{N} = \{1, 2, 3, \ldots\}$ form an infinite set because no matter how many natural numbers we write down, we can always think of the next one. Every number has a successor. $\mathbb{N}$ does not have a last element. See §16.2 for the development of the natural numbers based on the "successor" idea.

7.2.2 Cardinality

We start by going back to the idea of how many elements there are in a set of things. From the earliest times people have needed to keep track of collections of things, whether they be sheep in a flock or bushels of grain. The simplest idea of counting is simply to pair off things with tokens—say stones, or fingers, or lines marked into clay. A flock of sheep goes out grazing in the morning, and for each sheep out of the gate a pebble is placed in a container. At the end of the day, returning sheep are paired off with pebbles and there should be a perfect match. If not, a sheep is missing (or with extraordinary luck, there is an extra one). This is the idea of *one-to-one correspondence.* Two sets are the same size, or to put it in fancier words, have the same *cardinality*, if they can be placed in one-to-one correspondence with each other.

With the advance of civilization we have created the abstract system of numbers $1, 2, 3, \ldots$, which has become a standard and universal set with which to pair things off. If you want to keep track of how many things there are in a set, the standard way is to count them, in effect pairing the items with the numbers $1, 2, 3, \ldots$ and recording the last number reached. We are so used to this set of symbols (and linguistic names for them) that we have made it an absolute by which we do most of our counting. But in the end, it is still putting things in one-to-one correspondence.

When we say that two sets have the same *cardinality* we mean that they can be put into a one-to-one correspondence with each other. This is another example of a sameness/difference relation. The two sets may look and be very different in other ways, but if they have the same cardinality, that is a kind of sameness. Look at Fig.7.1. All of

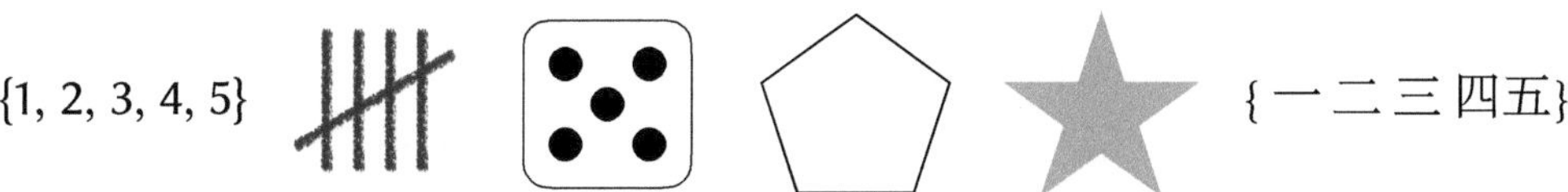

Figure 7.1: Will the real five come forward, please?

these have something in common, namely the number 5. We would say that they have cardinality five. But that is a totally abstract idea. If we were pinned down to really define what cardinality five means, we might raise five fingers, point to the pattern of the five on a die, or exhibit the list of five Hindu-Arabic numerals (our usual ones), and say "five", or "cinq", or "fünf", or "go" (Japanese). A Chinese person might point to the "same" numerals written as a Chinese character.[1]

The fact is that the cardinality five, which is a concept that five-year-old children understand, is an extraordinarily abstract idea. It not really a thing. More technically, the word "five" is non-referential. This type of situation is completely commonplace, but is rarely thought about.

We are putting a sameness condition on sets, and it is just the very sort of notion of sameness that we talked about before, see § 6.4.1. Cardinality is an *equivalence relation.* Two sets X and Y have the same *cardinality* if their elements can be put into one-to-one correspondence. See Fig.7.2. We could write $X \sim Y$ to indicate that the sets X and Y have the same cardinality.

Coming back to the cardinal 5 we could define it as all those sets X for which $X \sim \{1, 2, 3, 4, 5\}$. This is still language and symbol dependent, but it is set at a different level, as an equivalence class of sets. It doesn't matter if a Chinese person writes the same definition as we see it in Fig.7.1 or whether a follower of Pythagoras shows us the points of a five-pointed star, because in all of these cases the essence lies in sets that can be put into one-to-one correspondence with each other. In this way we have defined the cardinal 5.

So though we tend to think of "five" as an independent entity, in fact it is a concept that can only be instantiated in concrete sets, any of which can be put in one-to-one correspondence with whatever our favorite five element set is.

The virtue of cardinality is that it can be applied to sets that are not finite. So let us consider the set of natural numbers $\mathbb{N}$ from the point of view of one-to-one correspondences. The first thing that is a bit surprising is that $\mathbb{N}$ can be put in one-to-one correspondence with a *proper* subset of itself, i.e. with a subset that is not the entire set: if we leave out every second number then we have all the odd numbers, and we can certainly put the odd numbers in one-one-correspondence with the whole of $\mathbb{N}$:

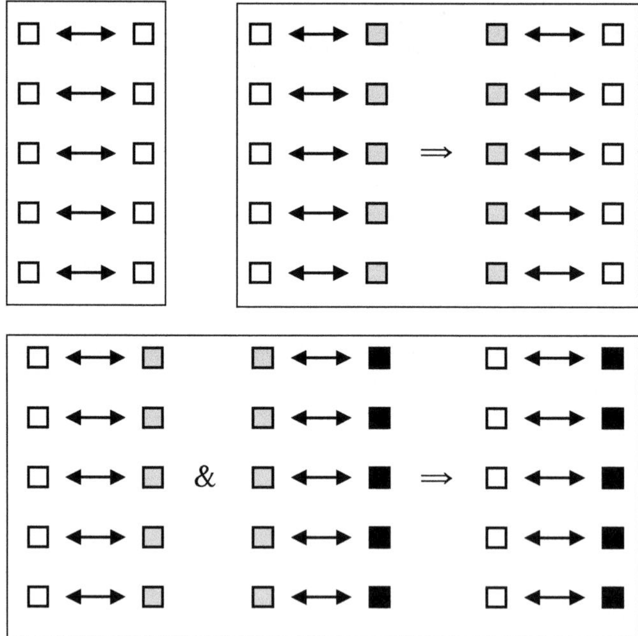

Figure 7.2: Cardinality is an equivalence relation. The figure shows the meaning of the three parts of the definition of an equivalence relation: reflexive, symmetric, transitive. The idea of one-to-one correspondence is a relationship between pairs that is reflexive, symmetric, and transitive.

$$\begin{array}{cccccccc} 1 & 3 & 5 & 7 & 9 & 11 & 13 & \ldots \\ 1 & 2 & 3 & 4 & 5 & 6 & 7 & \ldots . \end{array}$$

This sort of thing obviously doesn't happen with finite sets. In fact this basic distinction is one that is used to define what it is to be infinite:

> a set is *infinite* if it can be put in
> one-to-one correspondence with a proper subset of itself.

A set is called *countable* or *listable* if its elements can be put in one-to-one correspondence with the set $\mathbb{N}$. The terminology makes eminent sense—we can *count* off or *list* all the elements of the set.

Note that the way we have phrased it, the word countable excludes finite sets. It is not uncommon to allow the word countable to apply to finite sets as well, in which case they would call them finite countable sets, but in this book we will use it exclusively for infinite sets.

> In this book, the word *countable* always refers to infinite sets.

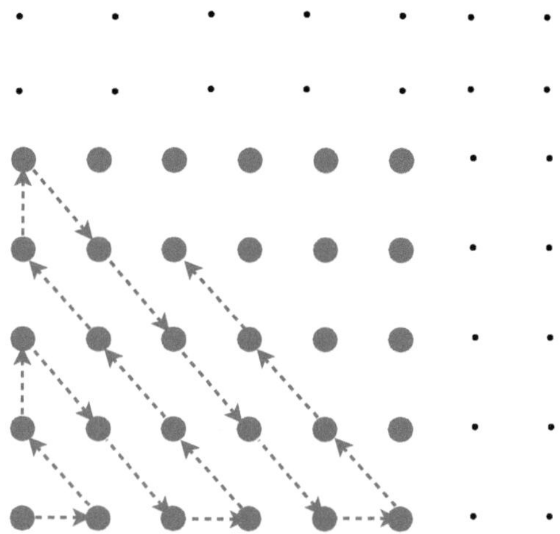

Figure 7.3: How to count a countable set of countable rows of points. Here they are laid out as an infinite set of rows each of an infinite number of points, like an infinite orchard of trees. The dotted path shows how the counting can be carried out.

It is not surprising that the union of two disjoint sets, one of which is countable and the other finite, is still countable. Just start by counting off the finite set and then continue to the countable set. This does point out, however, that we may need to be careful in the way we count off the elements in order to show that a one-to-one correspondence exists. If we were to start by counting the countable set first, we would never get to the end of it (because it has no end), so we would never get to start listing off the elements of the finite set.

More surprising is the fact that the union of two disjoint countable sets is still countable. This also is easy to see. Just start counting off the elements of the two sets, alternating between the two, so the odd numbers are used for counting one set and the even numbers for the other.

After this, it is clear that the union of a finite number of countable sets is still countable. But quite amazing is the fact that even the union of a countable number of countable sets is still countable. The easiest way to see this is to imagine an infinite orchard consisting of a countably infinite number of rows of trees, stacked one row over the next. Fig.7.3 gives the picture. There is some cleverness here in seeing how to do the counting. Trying to list the trees by going along one row completely and then going to the next is hopeless since one will never even get to the second row. But starting in the corner and scanning diagonally produces a way to label the trees that eventually will get to every one of them, see Fig.7.3.

After this it is no surprise that the entire set of integers $\mathbb{Z}$ and even the entire set of rational numbers $\mathbb{Q}$ are also countable. For instance, the positive rational numbers are *by definition* the numbers expressible as the ratio of two integers: every positive rational number is expressible in the form a/b where $a \in \mathbb{N}$ and $b \in \mathbb{N}\backslash\{0\}$. In the infinite orchard of Fig.7.3 we are, in effect, counting the set of all ordered pairs (a, b), where a and b are positive integers. In fact we count every rational number multitudes of times since, for instance, $1/2 = 2/4 = 3/6 = \ldots$. Tto get all the rational numbers we need only add in 0 and all the negative rational numbers—a finite and another countable set. It's still countable.

Next we find the surprising fact that the number of *finite subsets* of $\mathbb{N}$ is countable. So now we are not just looking at the whole numbers $\mathbb{N}$, but the entire set of all possible finite subsets of numbers. Yet, we can see that it is true. First, the number of one-element subsets is clearly countable, for there is just one for each element of $\mathbb{N}$. The number of two element subsets is countable by the tree-counting argument above, see the caption of Fig.7.3. Following this idea we can continue to three element sets, and see that they too constitute a countable set, and so on. Since we already saw that the union of a countable number of countable sets is still countable, the whole works is countable.

The set of all *finite* subsets of $\mathbb{N}$ is countable.

Of course this means that the set of all finite subsets of any countable set is countable.

7.2.3 Cantor's diagonal argument

At this point we are tempted to think that all infinite sets are countable. But it is Cantor's famous *diagonal argument* (nothing to do with the diagonals of Fig.7.3) that shows that the set of all real numbers, $\mathbb{R}$, is *uncountable* (i.e. neither finite nor countable). In fact, even a finite interval of real numbers like $[0, 1]$ is uncountable. This basic dichotomy between countable and uncountable infinite is important and will be used many times in the book. If we think of countable as meaning listable, in the sense that we can write them in a list $L(1), L(2), L(3) \ldots$ indexed by the positive integers, then we are saying that it is not possible to so list all the real numbers between 0 and 1.

At first sight it seems hard to think of any way we could show such a thing, especially since we have seen that just because two sets *can* be put into one-to-one correspondence does not mean that it is obvious how to do it. That is what makes the diagonal argument so interesting.

Since it is so famous and so easy to understand once one sees it, we will present a proof that $[0, 1]$ is uncountable here. Rather than dealing with the interval $[0, 1]$ directly, it is actually easier to show that the state space of the full shift $\{0, 1\}^\infty$ is uncountable: the set of all infinite strings of 0s and 1s. This looks totally different from the interval $[0, 1]$, but we will see the connection at the end.

Each individual state $\mathbf{x}$ is a countably infinite string of zeros and ones

$$\mathbf{x} = x_1\, x_2\, x_{,}\, x_4\, \ldots,$$

and we note that the subscripting itself uses the positive integers to produce this one-to-one correspondence. But the entire set of all these strings is not countable.

The diagonal argument is an example of a *proof by contradiction.* Such an argument runs by assuming that what one is trying to prove is actually false and then using this assumption to show that it leads to a logical contradiction. The argument runs like this. Suppose someone comes to us and says that the state space $\{0,1\}^\infty$ is countable. That means they claim that they can put all the states of the full shift into one-to-one correspondence with $\mathbb{N}$. So what they present to us is a list of states, that is a list of infinite $0/1$ strings

$$L = \mathbf{x}^{(1)}\,\mathbf{x}^{(2)}\,\mathbf{x}^{(3)}\,\mathbf{x}^{(4)}\,\ldots,$$

which they purport to be the entire set of elements of $\{0,1\}^\infty$, i.e., every possible infinite $0/1$ string is in the list. Here each $\mathbf{x}^{(k)}$ stands for an *entire infinite string* and we are using superscripts for the first, second, third, ...infinite string.

We don't actually know what is what in this list, but even so we can find a string that is not in it. It is useful to refer to Fig.7.4 in reading this.

The argument goes like this. Each string $\mathbf{x}^{(k)}$ in this list is a countably infinite $0/1$ string which we can write out. It will have the form

$$\mathbf{x}^{(k)} = x_1^{(k)}\,x_2^{(k)}\,x_3^{(k)}\,x_4^{(k)}\,\ldots,$$

where each term $x_j^{(k)}$ is a 0 or a 1. We look at the kth term of the kth string, that is $x_k^{(k)}$. This is either a 0 or a 1, we don't know which. Whichever it is, we define y_k to be the opposite, so $y_k = 1$ if $x_k^{(k)} = 0$, and $y_k = 0$ if $x_k^{(k)} = 1$. What we are doing here is looking at the kth term of the kth element of the list, and writing down its opposite.

Now we use these y_k to make a new string in $\{0,1\}^\infty$ by defining

$$\mathbf{y} = y_1\,y_2\,y_3\,y_4\,\ldots\,.$$

This is an infinite string of 0 s and 1 s. As such it is supposed to be in the list L. But the way that we have constructed $\mathbf{y}$ makes it impossible to lie in the list! Since the kth term of $\mathbf{y}$ is different from the kth term of the kth element in the list, i.e. $y_k \neq x_k^{(k)}$, it must be that $\mathbf{y} \neq \mathbf{x}^{(k)}$. This being true for every k, $\mathbf{y}$ is not in the list L. We have a contradiction: the list L is not all of $\{0,1\}^\infty$. Fig.7.4 illustrates what is happening.

It might look like we can correct the problem here. The infinite string $\mathbf{y}$ got left out of L, so we should put it in. But putting $\mathbf{y}$ into the list, say putting it first and pushing all the other strings one step further down the list, to produce a new list L' that contains $\mathbf{y}$ doesn't solve the problem. We again use the diagonal argument on this new list L' instead, which will produce another string $\mathbf{z}$. And if you notice, the first symbol in $\mathbf{z}$ is the opposite to the first symbol in the first item in our list, which is $\mathbf{y}$, and so right away our $\mathbf{z}$ is different from $\mathbf{y}$, and indeed from every other element in the list L' (which

y	**0**	**1**	**0**	**0**	**1**	**0**
$x^{(1)}$	**1**	0	0	0	1	1
$x^{(2)}$	1	**0**	0	1	1	0
$x^{(3)}$	0	0	**1**	1	1	1
$x^{(4)}$	0	1	0	**1**	1	1
$x^{(5)}$	0	1	0	0	**0**	1
$x^{(6)}$	1	1	1	0	1	**1**

Figure 7.4: Shown are the first six numbers in the first six elements of the state space $\{0, 1\}^\infty$ of some hypothetical enumeration of all the elements of this space. The table should be imagined as going on indefinitely, both to the right and downwards. The numbers written in boldface are the elements $x_k^{(k)}$, which lie on the diagonal. Across the top is the element $\mathbf{y}$ which we construct by simply taking y_k to be whatever $x_k^{(k)}$ is not. In constructing $\mathbf{y}$ this way, we guarantee that it is different from every string in the list.

happens to contain all of L). There is no way out of this conundrum. It is not possible to produce a list (a countable list) that contains all of the elements of the zero-one shift. And this is the conclusion that we are seeking. The state space of the zero-one shift is not countable, i.e. it is *uncountable.*

> The state space of the zero-one shift space (and in fact the state space of any shift system with an alphabet with more than one symbol) is uncountable.

We will use this important fact a number of times in the book. A very simple application of it is to show:

> The set of all subsets of the natural numbers $\mathbb{N}$ is uncountable.

Notice that we have already shown that the set of all *finite* subsets of $\mathbb{N}$ is countable. But if we allow infinite subsets as well then the situation is totally different. We can see this directly from the zero-one shift space. Any state in this shift space is an infinite string $\mathbf{x} = x_1\, x_2\, x_2 \ldots$ where each x_k is 0 or 1. The set of k for which $x_k = 1$ is a subset of $\mathbb{N}$, so we can think of each $\mathbf{x}$ in the state space as corresponding to one particular subset of $\mathbb{N}$, namely the one which is described by the positions in $\mathbf{x}$ where its 1s occur. Think of the 1s as being flags that say which positions are in the set and which are not. Every state of the shift space corresponds to exactly one subset of $\mathbb{N}$, and vice versa. So in proving that the state space is uncountable, we also proved that the set of *all* subsets of $\mathbb{N}$ is uncountable!

Initially we had set out to show that the interval $[0, 1]$ of all real numbers between 0 and 1 is uncountable. Instead, so far we have proved that the state space of the shift space $\{0, 1\}^\infty$ is uncountable. The connection is that every state $\mathbf{x} = x_1\, x_2\, x_2 \ldots$ can also be thought of as the binary (like decimal, but using base 2 instead of base 10) expansion

$$0.x_1\, x_2\, x_2 \ldots$$

of some number in $[0, 1]$, thus showing there must be an uncountable number of them. We leave the details until §7.5.3, where the connection arises in a different way.

The distinction between the countability of $\mathbb{Z}$ and the uncountability of $\mathbb{R}$ was first brought to light in a paper of Cantor in 1874. In 1891 he published a refinement of his uncountability proof—the diagonal argument that we have just seen.[2]

If we look back at the puzzle that Aristotle raised (§7.2), we can see that, at least in part, he was confronting the issue of countability and non-countability: there is no way that the set of points of $[0, 1]$ can be listed, *no matter what order we put them in.* We will come back to the elusive question of time in §8.4.11, but it does shed light on the paradoxical nature of trying to make sense of the infinite.

As a simple example of the implications of countability, consider the simple shift system $\{0, 1\}^\infty$ as a Bernoulli heads/tails system with the probability of heads being p, where $0 < p < 1$. Let $\mathbf{p}$ be the corresponding probability measure on the system. A typical trajectory of this system starts from a single state $\mathbf{x} = x_1\, x_2\, x_3, \ldots$ and then shifts it endlessly through new states as the leading terms are dropped one after another. This produces a countable trajectory of states. According to the law of large numbers, the sequence of resulting binary digits, x_1, x_2, x_3, ... will, with almost certainty, average out to having pn heads per every n consecutive symbols. Yet our trajectory is countable and constitutes a minuscule part of the entire state space $\{0, 1\}^\infty$ which is uncountable. In fact, the entire trajectory is a set of $\mathbf{p}$-measure zero in the state space, and so has zero probability of being chosen in a random selection of a state. Nonetheless, it has probability one of delivering the correct statistics of the system!

7.3 Continuous change

In formalizing the idea of a pattern system and arriving at the four underlying features listed in §6.7, we have relied on examples from discrete dynamics. This allows for a certain efficiency in the presentation without a distracting diversion into pattern systems based on forms of continuous change, but we need to see that those four underlying features apply equally well there too.

The mathematics of dynamical systems arose in the study of systems which change continuously in time, and to this day this remains a prolific area of research. The extension to discrete dynamical systems came later and there is a substantial difference between the two.

In discrete iterative dynamics as soon as we have the generating operator, for example the shift in shift spaces, the dynamics is completely defined. This is very different from the continuous situation where if, for example, we specify the operator which indicates how the states evolve in one second of time then although this certainly enables us to know how the states will evolve in two, three, four, ... seconds of time, it won't tell us how the states evolve in a tenth of a second or a thousandth of a second. No matter how small a length of time we take, just knowing how states evolve in increments of that length of time won't help to say how it evolves in yet smaller lengths of time. Of course we might hope that the dynamics taken in small enough time steps really is a good approximation to what is really happening in a continuous way, and indeed that is the basis of a huge amount of computational work in dynamics. But the fact is that continuous dynamics depends on what happens *instantaneously* at each moment. This is the popular "nowness" idea in its mathematical formulation.

Historically, developing mathematics that could express this nowness (infinitesimal time steps and infinitesimal changes of state) and how it stretches out to evolve across time is what calculus is all about. It took a few centuries before the entire process could be given mathematical rigor, but from the times of Newton and Leibniz (late seventeenth century) mathematicians and physicists happily embraced its consequences, even over the loud objections of some philosophers who quite rightly pointed out the logical gaps. After all, it worked!

A typical dynamical system from the world of physics involves the positions and velocities of one or more bodies with mass. In the example of the pendulum the positions were points in the plane, which we were able to express simply as the angle the pendulum makes with the vertical since it is restricted to motion along a circle. The velocities could then be expressed as the rate of change of the angle. The dynamics arose out of the physics of force. The force of gravity (later augmented to include the force of friction and a periodic driving force) acts directly downwards on the pendulum bob according to Newton's equation that equates force to mass times acceleration. Thus one begins with an equation that describes the acceleration and hopes to deduce from it the positions and velocities of the relevant bodies. This is the origin of the dynamics. The real question is how does that equation, which describes the moment-by-moment forces of motion, actually translate into the actual values of position and angular velocity as time progresses?

Since velocity is the instantaneous rate of change of angle (or position) and acceleration is the instantaneous rate of change of velocity, we have to somehow back up from an equation about acceleration and draw out of it the position and velocity that arise in time. The acceleration equation that arises out of the physics is a *differential equation* and the process of revealing the position and velocity over the course of time is in effect *integration*.[3] This is a case where technical terminology happens to fit quite well with our intuitive notions: differences in change seen at the momentary level lead to extended evolved patterns over time. Newton's equation is at the level of "now", and the pendulum's patterns are expressions of how "nows" become integrated into wholes.

More generally a typical situation involves several bodies, say in three-dimensional space, which all mutually influence each other through some force (say, gravitational, electro-magnetic, or biological) and from these mutual influences a dynamics of change arises. Newton created his theory in the context of planetary motions around the Sun and effectively solved the two-body problem of the motion and dynamics of two masses under the mutual influence of their individual gravitational forces. Since then those same sorts of methods have been applied in diverse areas like chemical processes, population dynamics, and the activity of neurons in a neural network. Although the actual notion of "force" may vary (e.g. birth and death rates in predator-prey dynamics or temperature in chemical processes), the mathematical objects are much the same.

The process of integration is a huge field in itself and is centuries rich in ideas and techniques. Many types of equations are very hard to solve, and many remain beyond any presently known techniques. In spite of this, there are very general theorems that prove that solutions to these problems do exist and are unique, even if we cannot actually explicitly write them down. This is important. It is possible to prove that solutions exist (that the integration is actually logically possible) without having to know *how* to explicitly write them down. Knowing that a solution actually does exist is different from actually producing it in some closed form.

Painted in broad strokes, the basic existence theorem for differential equations says:

> Given a finite set of objects whose mutual interactions through the forces connecting them are known (and hence also, thanks to Newton, their accelerations in space, moment by moment) then, given their actual positions and velocities at one particular moment of time (these are called *initial conditions*) their future positions and velocities are uniquely determined for all subsequent times.[4]

There are two statements buried in here: given the initial conditions: first there is a solution and second this solution is the only solution. Whether or not we are clever enough to be able to write down an explicit formula for the evolved process, the theorem states that a unique solution exists. Indeed in many (perhaps it would be more accurate to say, most) cases, we do not know how to write down an explicit formula for the solution. This does not mean that the solution cannot be explored computationally. It can, and very successfully, using computers to numerically churn out the development of the solution by using tiny discrete approximations to the actual dynamics—in other words by converting the problem from a continuous to a discrete one.

The famous problem of planetary motion is the *three-body problem* for which, in general, no useful closed solution has been given.[5] This was the very problem that led Poincaré to his new concept of a dynamical system. His idea was that we need not necessarily know the exact details (which we may term the quantitative details) in order to get a qualitative picture of the dynamics, and in the end the qualitative picture might be just as interesting and important as knowing the exact solution. That the planets move in elliptical orbits around the Sun is both interesting and important! Since then the concept of dynamical systems has evolved in all sorts of directions and has been generalized and applied to almost every mathematical model of change. This is of course why it is so important to us, and it is the *qualitative* aspects of change that we will continue to emphasize.

It was Poincaré who stressed the particular mathematical formulation of a state space, which captures the idea of the entire set of states of the system of interest and, so to speak, detached time so as to make it appear as a set of operators that trace out the evolution of states. It is revealing to see how this comes about.

7.3.1 Time as operators

To understand just why this way of looking at things arises naturally out of the mathematics, we go back to the basic theorem stated in the box above. Our system consists of some entities and their rates of change, let's say some finite number of material particles and their velocities. Quite typically a full description of a state will have many components to it. For instance, a particle which is moving in three-dimensional space will have three space coordinates and three velocity coordinates (its speeds in the three directions of the coordinate system), and so needs six numbers to specify its state. If there are three particles (bodies) and each state takes account of the positions and velocities of each of them then there are $18 = 3 \times 6$ numbers required to describe each state of the system at any given moment. To help keep this in mind but not clutter the mathematics with unnecessary symbols it is common, just as we have done with shift systems, to use a bold-face font to indicate states of the system. We will do this. Thus a single symbol $\mathbf{x}$ is supposed to carry all this detailed information. Each state $\mathbf{x}$ of the state space represents one of the infinite possibilities for the positions and velocities of these particles. The positions and velocities are bound by the differential equation that formalizes the moment-by-moment rate of change of these states.

Starting at some moment in time, let's call it time 0, and any arbitrarily chosen state $\mathbf{x}$, the fundamental theorem says that there is a *unique* solution that describes exactly how that state changes in time. There is an evolution of states, which we denote by $\mathbf{x}(t)$, as time t progresses. In this notation $\mathbf{x}$ itself is $\mathbf{x}(0)$, and $\mathbf{x}(t)$ is the state that has evolved out of our starting state $\mathbf{x}(0)$ after time t.

There is a simple and important logical outcome of the uniqueness here. Suppose we were to evolve through time t in two time intervals, t_1 followed by t_2 (not necessarily equal) and $t = t_2 + t_1$. Then after time t_1 our state $\mathbf{x}$ will have arrived at the new state $\mathbf{y} = \mathbf{x}(t_1)$ and after further time t_2 it will have evolved to $\mathbf{x}(t)$, while $\mathbf{y} = \mathbf{y}(0)$ will have evolved to $\mathbf{y}(t_2)$. Both the evolutions of $\mathbf{x}$ through time $t = t_1 + t_2$ and the evolution of $\mathbf{y}$ through time t_2 are uniquely determined by the dynamics, but their outcome must be the same since the evolution of $\mathbf{x}$ really did pass through $\mathbf{y}$, that is $\mathbf{x}(t_1) = \mathbf{y}$, so they carried exactly the same information. In short,

$$\mathbf{x}(t) = \mathbf{x}(t_2 + t_1) = \mathbf{y}(t_2) = (\mathbf{x}(t_1))(t_2),$$

see Fig.7.5.

This looks far more suggestive if we change notation a bit and write $t \triangleright \mathbf{x}$ for $\mathbf{x}(t)$. We are now thinking of time t as an operator that acts by taking each state $\mathbf{x}$ of the system into the state that it will be in time t later. Since we can think of t as linking before-and-after states separated by that length of time, it can be seen as an operator.

This operator works on the entire state space—every state will evolve somewhere after time t, even if it evolves to the same state or evolves stationary in the same state. There is an operator for time t_1 that produces the evolution of states after the time t_1, and similarly another one for time t_2. Indeed, for every non-negative value of time there will be an operator, and of course for time 0 the operator that leaves every state untouched:

$$0 \triangleright \mathbf{x} = \mathbf{x}(0) = \mathbf{x}.$$

In this language we can write what we just said as

$$\text{if } t = t_2 + t_1 \text{ then } t \triangleright \mathbf{x} = (t_2 + t_1) \triangleright \mathbf{x} = t_2 \triangleright \mathbf{y} = t_2 \triangleright (t_1 \triangleright \mathbf{x}). \qquad (7.1)$$

This is really a statement of algebraic structure:

$$(t_2 \circ t_1) \triangleright \mathbf{x} = t_2 \triangleright (t_1 \triangleright \mathbf{x}) = (t_2 + t_1) \triangleright \mathbf{x}.$$

It shows how the composition of time operators (following one by another) matches the usual addition of real numbers: *composition of these time operators is structurally related to addition in the real numbers.*

Thus we have transformed the consequences of the uniqueness theorem into a picture that we have already seen: a state space with operators on it. In principle we have an infinite number of operators since t can be any positive value we please. However, they evidently form a continuous family tightly linked to the ordinary real numbers by the fact that composition of these operators corresponds to addition.

This was Poincaré's insight.

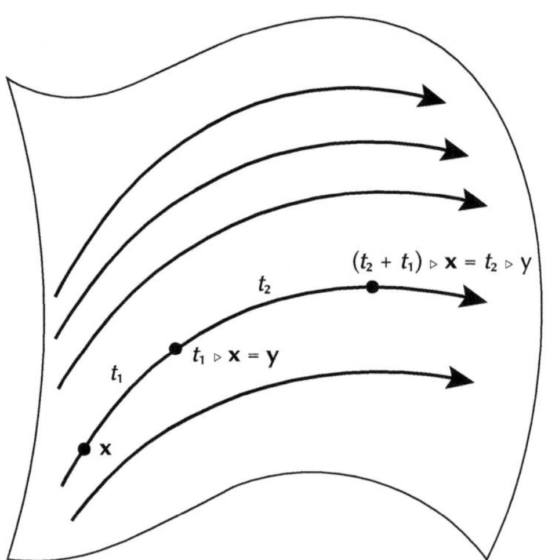

Figure 7.5: Starting from any state $\mathbf{x}$ the system changes in time and traces out a trajectory of states as it evolves. Whether we do this in one step or two successive steps, or indeed any number of steps, the result is identical.

7.4 Measure and probability

Kolmogorov's famous book of 1933 [99], that lays down the modern mathematical approach to probability, begins in the following way:

> The purpose of this monograph is to give an axiomatic foundation for the theory of probability. The author set himself the task of putting in their natural place, among the general notions of modern mathematics, the basic concepts of probability theory—concepts which until recently were considered to be quite peculiar.
>
> This task would have been a rather hopeless one before the introduction of Lebesgue's theories of measure and integration. However, after Lebesgue's publication of his investigations, the analogies between measure of a set and probability of an event, and between integral of a function and mathematical expectation of a random variable, became apparent. These analogies allowed further extensions; thus for example various properties of independent random variables were seen to be in complete analogy with the corresponding properties of orthogonal functions. But if probability was to be based on the above analogies, it still was necessary to make the theories of measure and integration independent of the geometric elements which were in the foreground with Lebesgue.

The interpretation of probability in terms of measure theory is one of the reasons (though not the only one) that we have laid so much stress on it. This probabilistic interpretation is already familiar, but it bears repeating. The setting is a collection of states that a system may take. This collection is the state space. As observers we observe certain of these states displaying an attribute of the system that we care to distinguish. Many states of the system may exhibit this same attribute. These various subsets of the state space that are of interest are designated as "events". Out of these arises the event space, see §6.1 and §6.5. Each event has a certain probability of occurrence and we can assign that probability as a measure (in the common sense of the word) of that event. In this way we create a measure (now in the technical sense) on the event space. The measure of the entire state space (which itself is always considered to be an event) is equal to 1.

Reversing the idea, given any measure on a set whose total value is 1, we can interpret it in the sense of probability. By definition a measure comes with its own event space (often only implied, but nonetheless there) and the measure of each event is then interpreted as its probability of occurrence. Much of the task and difficulty of probability theory is to actually determine the measure of some event (i.e. the probability of such and such happening), which, although arising quite naturally in some physical context, is nonetheless a rather complicated combination of the simpler events (e.g. dictionary sets) whose measures we actually know.

Expectation—an average arising out of the dynamical evolution of states—is a property that we have seen in the law of large numbers, and it appears even more explicitly in §8.5. It is properly expressed in terms of integration, which itself is a direct, though somewhat detailed, consequence of the definition of measure, §11.10. So, in fact, all the main features of probability theory are interpreted measure-theoretically.

Invariant measures are measures that respect the dynamics of the system—the ones that satisfy the condition given in (6.4). Of course invariant measures are valuable, *if we can find them*, because they point to something deep about the patterning of the system. Such a measure is often referred to as the *law* of the pattern system. There is a good reason for this. Assuming that we have all the ingredients of a pattern system except for some measure, then we know the potential trajectories of states in the system and we know the potential events, but we will have no idea of the expectation of those events. If we can find that its dynamics actually implies an underlying invariant measure, we will have revealed a lot about how the system actually behaves.

What is more interesting is when the dynamics does not imply any specific invariant measure but there are many available choices. Think of the Bernoulli systems that are all based on the zero-one shift space but then depend on the probability p that a 1 will occur. In effect this choice determines an invariant probability measure $\mathbf{p}$. Simply by imposing this measure on the system (i.e. assuming that it is the invariant measure of the pattern system) we immediately shape the patterning of the system by assigning expectations to the events.

We will see this used to great effect in examples of pattern systems in §12.5 where the outcomes of the pattern system are radically changed by using different invariant measures and there are countless ways of imposing such measures. As we mentioned before, this gives rise to the idea of the measure as a *law*. Though not necessarily deterministic such a law shapes the dynamics of the system—the likelihood of events and their sequencing. In some pattern systems there is but one relevant law; in others there are many choices.

7.5 Itinerary pattern systems

7.5.1 Sequences of events

A fundamental question we can ask in observing a system which is changing in time is whether we are seeing it at the level of states or really at the level of events. A system may be changing in some continuous way, but our senses are based on discrete change. If we understand events as our basis of articulation or description of what is going on in a system, then what we are really observing is sequences of events.

The formalization of this idea, which dates back to Jacques Hadamard (1865–1963), leads to the idea of itinerary pattern systems. We start with a discrete iterative pattern system $\mathcal{X} = (X, u, \mathscr{P})$ with a single operator u and a partition $\mathscr{P}$. We have usually used the letter s or t to denote an operator. Here, in order to avoid confusion with previous uses, we denote the operator by u. The situation here is supposed to be perfectly general and we make no assumptions on how u actually operates on the state space X.

A natural way to simplify the dynamics and make it more comprehensible is to replace each trajectory by the sequence of parts of $\mathscr{P}$ which it successively visits. If

$$\mathscr{P} = \{P_1, P_2, P_3, \dots, P_n\}$$

is the partition, then starting from any state x, we would look at the sequence of events rather than the sequence of states themselves. So rather than looking at trajectories

$$x,\ u \triangleright x,\ u^2 \triangleright x,\ u^3 \triangleright x,\ \dots\ ,$$

we might look rather at the sequence of parts to which they belong. This may be something like

$$P_3,\ P_1,\ P_5,\ P_2,\ P_2,\ P_2,\ P_4,\ \dots\ .$$

We call this type of sequence an *itinerary*. Each state x produces an itinerary. Maybe some distinct states have the very same itinerary. In any case it is potentially a coarser way of looking at the dynamical system. Trajectories follow successions of states while itineraries follow successions of the parts or the events that states occupy. They function in the domain of sameness and difference that the partition defines. If we adopt a different partition of the state space, the itinerary system will also be different.

It does not take long to realize that instead of writing down the sequence of parts as we have done, we might just as well write

$$3, 1, 5, 2, 2, 2, 4, \dots$$

and this gives us the idea that we are really looking at our pattern system $\mathscr{X}$ in terms of a new system which is just composed of strings of numbers $1, 2, 3, \dots n$, where n is the number of parts to the partition $\mathscr{P}$.

With this in mind we introduce a new set $I = \{1, 2, 3, \dots, n\}$ which will be our new alphabet, and then form the familiar

$$I^\infty = \text{the set of all strings of elements of } I.$$

Now for each state $\mathbf{x} \in X$ we have a corresponding state in $\mathbf{y} \in I^\infty$ which corresponds to the itinerary of $\mathbf{x}$. In our example,

$$x \mapsto 3\,1\,5\,2\,2\,2\,4\,\dots\,.$$

Thus there is a correspondence or mapping

$$X \xrightarrow{\varphi} I^\infty\,. \tag{7.2}$$

We call this mapping φ (phi). It takes states of X to strings of letters from I, our new alphabet.[6]

Now I^∞ is the state space for the familiar shift space, this time on the alphabet $\{1, 2, 3, \dots n\}$. Let us denote this shift system by $\mathscr{I}$. If we let the left shift on this space be denoted by s, we see that the dynamics of the shift space really match the dynamics on X. In our example, the trajectory starting from the point $u \triangleright \mathbf{x}$ is clearly $1\,5\,2\,2\,2\,4\,\dots$, which is the same as the left shift of $3\,1\,5\,2\,2\,2\,4\,\dots$. So

$$\begin{array}{rll} \mathbf{x} & \xrightarrow{\varphi} 3\,1\,5\,2\,2\,2\,4\,\dots & \\ u \triangleright \mathbf{x} & \xrightarrow{\varphi} s \triangleright 3\,1\,5\,2\,2\,2\,4\,\dots & = 1\,5\,2\,2\,2\,4\,\dots\,. \end{array}$$

This actually says something quite important that we should stress: it does not matter whether we first map over from X to I^∞ and then apply the operator s, or instead first apply the operator u and then map over from X to I^∞. The mapping φ is not just a mapping between two sets: it also respects the dynamics. We say that φ *intertwines* the two operators. This type of property is often expressed in a more visual way by a diagram:

$$\begin{array}{ccc} \mathbf{x} & \xrightarrow{\varphi} & \mathbf{y} \\ {\scriptstyle u}\downarrow & & \downarrow{\scriptstyle s} \\ u \triangleright \mathbf{x} & \xrightarrow{\varphi} & s \triangleright \mathbf{y}\,, \end{array}$$

where the idea is that starting from the left top corner ($\mathbf{x}$) one can proceed to the right bottom corner ($s \triangleright \mathbf{y}$) by following either set of arrows, and they both give the same answer. These are called *commutative diagrams* in mathematics.

This idea is of importance to us because it suggests how we are able to relate two different pattern systems. For us, pattern is about change, and any genuine relation between two patterns of change must take into account the process of change. That's what we see here. It is not just a mapping of X to I^∞, it is a mapping of pattern systems. If we want to emphasize this fact we can write

$$\mathscr{X} \xrightarrow{\varphi} \mathscr{I}\,. \tag{7.3}$$

We call $\mathscr{I}$ a *factor* of $\mathscr{X}$.[7]

A factor is like a *shadow*, perhaps less detailed than the system that casts it, but nonetheless following its dynamics. What we see here is that the itinerary system I^∞ is like a shadow of the pattern system X that creates it.

The new system, which is based on taking some initial partition of the initial state space, is a shift space, and though quite likely it is coarser than the original system, it still follows its dynamics. When we think about it, this process is entirely similar to the ways in which we observe the world. We often try to understand a complex system by measuring some quantity or quantities associated with its states and then observing the strings of resulting measurements that arise out of its dynamics. In effect we model itineraries. Our ancestors studying the pattern of changes of the Moon replaced a very complex dynamical system involving three massive bodies interacting through the forces of gravity by sequences of symbols representing the various phases, and then studied the patterns that they could discern in the resulting symbolic strings. We observe, make observations, take notes in the form of sequences of symbols, and so model the world. The *Yijing* §4.1.2 might be seen as based on a similar idea.

We should note two things about representing a pattern system in terms of itineraries, as we are doing here. First of all it is quite possible that two different states produce exactly the same itinerary. The itinerary system is in general coarser than the original when it comes to the level of states. Second, although each state of the original system gets represented as a symbol string in I^∞, that does not mean that every possible string in I^∞ will appear as the itinerary of some state. We will see in §7.7 that when we enlarge the scope to include measures there is a very simple way to take account of this.

This process of creating a shift system out of a discrete iterative pattern system reminds us of the famous image that goes far back to the *Book of Songs* (Shijing, ca. 400 BCE). It appears much later in Ezra Pound's translation of the *Doctrine of the Mean* [130, The Unwobbling Pivot], and it was picked up from there in the poetry of Gary Snyder:

> There I begin to shape the old handle
> With the hatchet, and the phrase
> First learned from Ezra Pound
> Rings in my ears!
> "When making an axe handle
> the pattern is not far off" [155, Axe Handles].

7.5.2 Things and events

> What encourages us with a Western metaphysical tradition to separate time and space is our inclination, inherited from the Greeks, to see things in the world as fixed in their formal aspect, and thus bounded and limited. If instead of giving ontological privilege to the formal aspect of phenomena, we were to regard them as having parity in their formal and changing aspects, we might be more like classical China in temporalizing them in light of their ceaseless transformation, and conceive of them more as "events" than as "things." In this processual worldview, each phenomenon is some unique current or impulse with a temporal flow. In fact, it is the pervasive and collective capacity of the events of the world to transform continuously that is the actual meaning of time [3, Ames and Hall, p.15].

Although this quotation refers to continuous time, whereas we might think of event-based experience of time as discrete, the point is that the "thingness" of things is their persistence in the some sense of sameness through sequences of events. They are always changing, but at some level of perception or distinction they do not. That is what *persistence* means.

7.5.3 The doubling system

The story continues with yet another pattern system, also based on the interval $[0,1]$, the set of all real numbers between 0 and 1 (including 0 and 1), as the state space. The dynamics is iteration of a new operator d, which is the simple process of doubling: $d : x \mapsto 2x$. Since doubling will spill outside the interval $[0,1]$ if $x > 1/2$, we subtract 1 when that happens. This is much the same sort of wrap around as we have in clocks, where seven hours after seven o'clock we are at fourteen o'clock, which on a clock reads as $14 - 12 = 2$ o'clock. Thus in our case we have, for example, $d : 0.7 \mapsto 2(0.7) - 1 = 1.4 - 1 = 0.4$. Explicitly the operator is defined by

$$d \triangleright x = d(x) = \begin{cases} 2x, & \text{if } x < 1/2\,, \\ 2x - 1 & \text{if } x \geq 1/2\ , \end{cases}$$

where, again, we have used the alternate operator notation $d \triangleright x$ for $d(x)$. We call this dynamical system the *doubling pattern system* and denote it by $\mathscr{D} = ([0,1], d)$. There is a revealing way to visualize this by using FuXi's keyboard again, see Fig.7.6.

Earlier on we used the FuXi keyboard, Fig.4.4, to visualize the zero-one shift system. Now we are using it to visualize the doubling system. There must be a direct connection, even though the two systems seem to look quite different. The zero-one shift system $\mathscr{S} = (\{0,1\}^{\infty}, s)$ has infinite zero-one strings as its states and the shift mapping as its

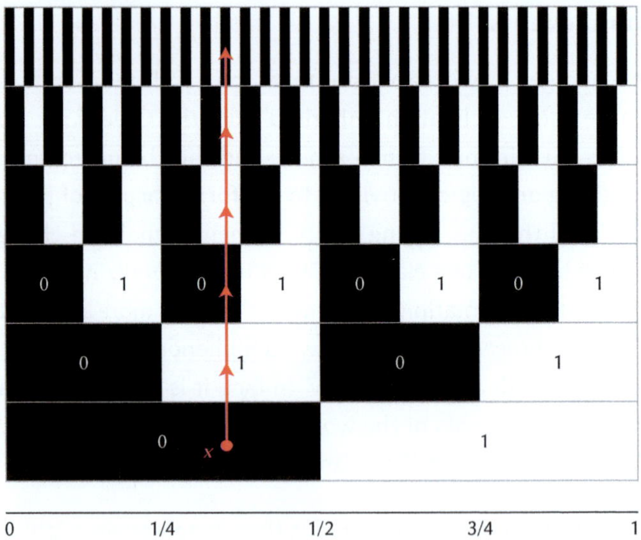

Figure 7.6: Although doubling sounds like things expanding, in fact our process can be thought of as contracting, for if we double the number of things we want to pack into fixed space (in our case $[0, 1]$) we have to halve the size of each of them. Looking at FuXi's keyboard from bottom to top, we first see a representation of the state space $[0, 1]$ figuratively divided into two halves by the black and white rectangles. In the second row this process is repeated, with the same state space represented twice over, each likewise divided into black and white rectangles, but scaled down by a factor of two so as to fit exactly. In the third row, we double again, so we now have our state space represented four times, and so on. Altogether there are five doublings shown here—we can imagine the process going on forever. The red line shows how our operator d works. An initial point x, shown here in the left half (black) of the state space (it is less than $1/2$), is taken to the point representing $2x$ in the next row, which we see in the right half (white) of the state space. The next doubling takes us to $4x$ which is greater than 1 so there is a wrap around and the point ends up in the first half (black) of the state space. And so on. The vertical red line traces the iterations $x \mapsto d \triangleright x \mapsto d^2 \triangleright x \mapsto \cdots \mapsto d^5 \triangleright x$, part of the trajectory of x. The zeros and ones are used to connect this picture to the zero-one shift space.

dynamics. The doubling system has the points of the interval $[0, 1]$ of the real line as its states and the doubling operator as its dynamics.

Superficially they look quite different. They are seen to be closely related by using an itinerary interpretation of $\mathscr{D}$. Divide the state space $[0, 1]$ of the doubling system into two parts, $\{P_0, P_1\}$, with P_0 being all the points of $[0, 1]$ from 0 up to (but not including) $1/2$, and P_1 all points of $[0, 1]$ from $1/2$ and up. Using this partition we can look at the corresponding itinerary system of the doubling system. If we just use the alphabet $\{0, 1\}$ to write down whether a state is in P_0 or P_1 then the itinerary system is simply our most basic shift space $\{0, 1\}^\infty$. Each state of the doubling system is a number in the interval $[0, 1]$ and lies in either P_0 or P_1, and so is recorded as a 0 or a 1. As we apply the doubling operator repeatedly, the state changes and it produces a trajectory, which is a corresponding string of zeros and ones. Each trajectory determines one state of $\mathcal{S}$.

FuXi's keyboard shows the itinerary mapping perfectly. It also makes something clear: each state in the doubling system has its own itinerary and, in reverse, every conceivable itinerary corresponds to some point x of $[0, 1]$.

There is a simple, but in its own way profound, meaning to this. Just as we can write each number x of the $[0, 1]$ in its decimal expansion using the digits $0, 1, \dots, 9$, so also we can write each in its binary expansion using the digits $0, 1$. Looking at Fig.7.6 and the particular point x we have chosen, we see that in following the red line vertically upwards from x we are actually looking at its binary expansion:

$$x = .010110\dots .$$

This goes on indefinitely and presents us with x in the form of an infinite zero-one string. That is the simple connection.

Actually it is better than this. Applying the doubling map d to the state x takes us to another state $d \triangleright x$ which is also in $[0, 1]$, and so, using the same example, we can read its binary expansion of the same red line: $d \triangleright x = .10110\dots$. Doing the same for $d^2 \triangleright x$, and so on, we see that it is just the left shift on strings. The reason is simple. Doubling moves the binary point one point to the right (just as multiplication by 10 moves the decimal point one step to the right). Since we "wrap around" if the result exceeds 1, we discard anything to the left of the binary point. This is just what the left shift does.

This is the intertwining that we could have already anticipated from §7.5.1: doubling corresponds to shifting in the corresponding itinerary system.

7.5.4 The binary and decimal systems as itinerary mappings

It is worthwhile to consider this more carefully, because the same idea lies at the foundation of our binary and decimal systems. Each $x \in [0, 1]$ can initially be thought of as a point on a line segment—this is pure geometry. Dividing this line segment in two in the middle so that it falls into two parts P_0 and P_1, x must lie in one of them, and this gives rise to the first binary digit, a 0 or a 1. Doubling x and wrapping back into the interval $[0, 1]$ if necessary, we observe again which part it belongs to, P_0 or P_1. This results in the next second binary digit. Endless repetition of this converts any point x of $[0, 1]$ into a string of 0s and 1—the binary expansion of x. Seen in this way the binary expansion is the outcome of a dynamical process.

What applies for the binary system applies equally well for our decimal system, where the line interval $[0, 1]$ is divided into ten equal parts $P_0, P_1, \dots P_9$ and the process this time is multiplying by 10, wrapping whenever the outcome is beyond 1. Now the left shift corresponds to multiplication by 10, the familiar rule that moving the decimal point to the right by one is multiplication by 10. The only difference is that in the shift system that we use here we keep only the part that lies after the decimal point. That is what the wrapping does.

The doubling system $\mathscr{D}$ on $[0,1]$, the binary system representation of the numbers of $[0,1]$, and the shift system $\mathscr{S}=\{0,1\}^{\infty}$ are essentially the same thing.

There are two take aways from this. The first is that we have natural mappings

$$[0,1] \xrightarrow{\varphi} \{0,1\}^{\infty}$$

and

$$\{0,1\}^{\infty} \xrightarrow{\psi} [0,1],$$

the first being the itinerary map that effectively gives binary expansions of real numbers, and the second that reverses this and takes strings of binary digits into real numbers. In both directions the mappings *intertwine* the dynamics of the two systems so that they match perfectly. Each of the two systems is a factor of the other. We call this type of relation a *conjugacy*, and the two systems in this relation are said to be *conjugate*. The two systems look different but from the point of view of pattern systems *they may be considered* as equivalent. However, note the subtlety to this, see below.

The second take away is that we can explicitly see that what underlies the binary expansion of numbers, and similarly our decimal system, is *self-similarity*. Any interval on the real line is like any other interval in the sense that one can stretch the one to fit over the other. The decimal system uses this self-similarity to break the interval $[0,1]$ into ten equal length pieces and then goes on to treat each of these ten pieces just like it treated $[0,1]$.

There is a subtlety in this, and it is one we are familiar with from learning how to use the decimal system . We recall that in the decimal system $.099999\cdots = .100000\ldots$, $.299999\cdots = .300000\ldots$, $.812699999\cdots = .812700000\ldots$ and so on. In general when a decimal expansion turns into an endless string of nines, those nines can be replaced by increasing the digit preceding them by one and then replacing them all with zeros. In the case of the binary system, it happens with strings of 1s.

Consider what happens on the FuXi keyboard for $x = 1/2$. We seem to have an option about its binary expansion. Is it $.011111\ldots$ or $.100000\ldots$? In fact they are both valid. We avoided the problem in our definition of the doubling map by partitioning $[0,1]$ so as to have $1/2$ in the second part, and not in the first. This issue only involves expansions that ultimately end in an endless string of 1s. Fortunately there are only a *countable* number of these (the endless string of 1s will occur after the first, second, third, ... place in the string), whereas there are an uncountable number of states both in $[0, 1]$ and in $\{0, 1\}^\infty$. That is what we learned from Cantor's diagonal argument.

This type of issue is common in discrete dynamics. The two systems that we have just discussed, the doubling system and the shift system on the alphabet $\{0, 1\}$, are not absolutely perfectly aligned: the mapping φ does not cover every possible infinite string and the mapping ψ is not one-to-one—it can map several strings to the same number. But apart from a countable number of states (or more importantly we shall see that apart from a set of states of measure zero) they are the same. The two systems are said to be *conjugate* when they can be dynamically identical except for such minor deviations.[8]

7.5.5 FuXi's keyboard, dictionary sets, and the transfer of structure

Now that we have made the connection between the line interval $[0, 1]$ and the shift system based on the alphabet $\{0, 1\}$, we can see another feature of FuXi's keyboard. It gives a visual interpretation of the dictionary sets. At the very bottom of the keyboard we have the simple division black-white corresponding to the two dictionary sets $[0], [1]$. Each of these bifurcates so that at the second line we have the corresponding dictionary sets $[00], [01], 10], [11]$. And so it continues. Our chosen point x is thus seen to be in the dictionary set $[010110]$. Each level up shows the interval divided into finer and finer parts.[9]

One virtue of conjugacy is that it allows us to pass features from one system to the other: what is called *transfer of structure*. For example, we know that there is an invariant probability measure $\mathbf{p}$ on the zero-one shift system $\mathcal{S}$ which gives equal probability to the occurrence of zero and one, and probability $1/2^n$ to each dictionary set $[a_1, a_2, \ldots, a_n]$. The conjugacy of $\mathcal{S}$ and $\mathcal{D}$ suggests that there must also be a corresponding invariant measure ℓ on the doubling system. There is, and it is the standard Lebesgue measure of length. Looking at the FuXi keyboard yet again, we can see it before our eyes. The width of any one of the black or white rectangles, each of which stands for a dictionary set, is identical to the probability of being in that cylinder set. For instance the bottom left black rectangle has width $1/2$ and stands for $[0]$ which has probability $1/2$ in the shift system. Here the connection between measures is visually apparent, and it is interesting to see how it matches up geometric information in the form of lengths with discrete information in the form of various finite strings of zeros and ones.

We know that the probability measure of the finite shift is invariant under the dynamics, see §6.6.5. It follows that, unlikely as it might appear, the Lebesgue measure ℓ of length ought also to be invariant under the action of the doubling map. This is true. The details are in the Endnotes.[10]

In other cases conjugacy can be used to realize the existence of parallel structure that is far less obvious. To see the power of this idea, we introduce another type of discrete pattern system.

7.6 The logistic pattern system

7.6.1 Logistic systems

This system, and variants of it, often go by the name of *discrete logistic systems*. The underlying idea, as well as the terminology, was established by Pierre François Verhulst (1804–1849) who introduced it as a way of modeling population dynamics. His model was set in the realm of continuous dynamics, but discrete versions have been extensively studied since the latter part of the twentieth century. These models are applicable to many processes in biology, chemistry, and physics, and also show a wide range of behaviors from relatively tame to unexpectedly complex.

We will concentrate our attention on the wildest one of them all, and we treat it as being a mystery that needs to be resolved. We will find that:

- invariance is sometimes hidden and can be seen only by looking at long-term behavior;
- systems can be perfectly deterministic, but at the same time chaotic;
- systems that are seemingly very different may actually be very alike;
- alikeness can lead to transferring understanding gleaned in one system to another that would otherwise remain obscure.

The Verhulst logistic system is based on a discrete-step, iterative, dynamical system that was originally inspired by a simple model of population growth. Suppose that we observe a population of some life form (bacteria, lemmings, rabbits, people). We are interested in how this population changes over equal steps in time, say hourly, monthly, annually—whatever is appropriate. The most obvious characteristic is the population *growth rate* r, so that if the population is x now, then at the next step it will be rx. We can write this as

$$x \mapsto rx\,.$$

By itself, this simply leads to exponential growth, so starting with a population x, the population goes through the successive stages

$$x \mapsto rx \mapsto r^2x \mapsto r^3x \mapsto r^4x \mapsto \cdots\,.$$

Our human population can be seen to have followed this sort of pattern. It cannot continue forever, of course. If, for instance, $r = 2$, then we double the population at each step, so even in just ten steps the population will be increased by over a thousand-fold.

A more realistic variant is to include a *carrying capacity* of the environment. As the population reaches some level, available resources and space are bound to produce a

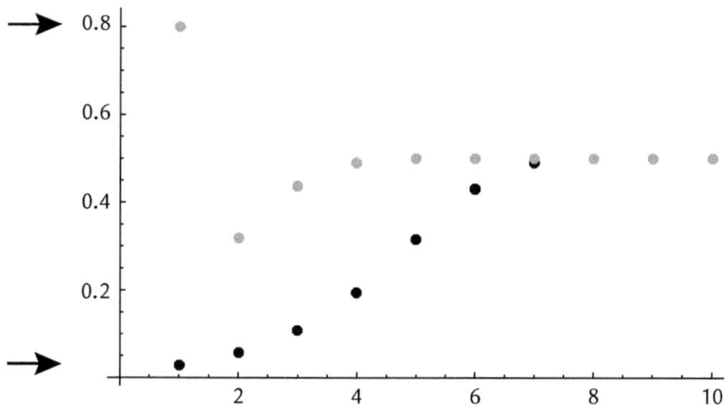

Figure 7.7: We see here plots of ten successive steps of the function $f(x) = 2x(1 - x)$ from two different starting points, 0.03 and 0.8, indicated by the arrows. Step numbers are indicated along the horizontal axis and the corresponding populations in terms of the carrying population of 1 on the vertical axis. It seems clear (and it is true) that the population tends rapidly to $1/2$ indicating that the stable population is at half the carrying capacity.

barrier beyond which the population cannot grow. The logistic system introduces such a barrier by using modified growth rule

$$x \mapsto rx(1 - x). \tag{7.4}$$

It has one parameter, namely the positive number r, which we assume is the growth rate and is assumed fixed at the outset. To keep everything as simple as possible, the measurement of population is described in units which make the carrying capacity equal to 1, so that we measure population in terms of a fraction of its carrying capacity. As x gets larger, the remaining capacity $1-x$ that can support the population gets smaller, and this cuts the population down. We can think of $1 - x$ as a survival rate, which decreases as the population grows. If x ever reaches its full carrying capacity 1, then $x = 1$ and $rx(1 - x) = 0$. This is total extinction. After that, as one can see, 0 is a stable state. Once extinct, always extinct.

In Fig.7.7 we plot what happens when the growth rate is 2, but now with the carrying capacity included in the picture. Rather than exponential growth, no matter what non-zero state from which we begin the population rapidly tends to $1/2$ (meaning one half of the carrying capacity), and this is a stable state:

$$\frac{1}{2} \longrightarrow 2(\frac{1}{2})(1 - \frac{1}{2}) = \frac{1}{2}.$$

7.6.2 A wild model

In spite of the tame behavior that we see here when $r = 2$, the way in which the population evolves under the logistic process is astoundingly dependent on the choice of r,

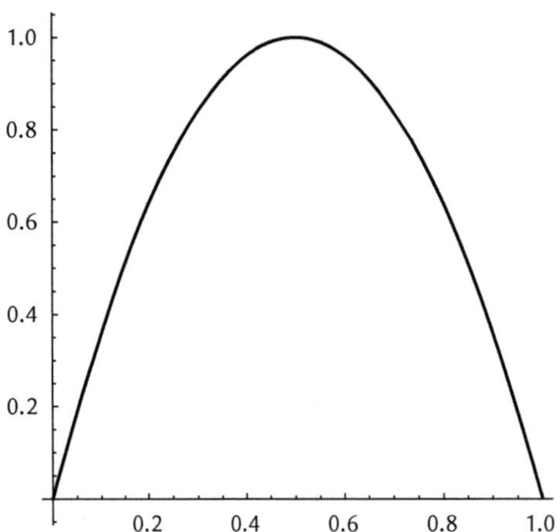

Figure 7.8: The curve is a plot of the function $f(x) = 4x(1 - x)$. The range of values in the vertical direction is between 0 and 1, so we can think of f mapping states to states. Fig.7.9 shows how this fact is used through an iterative process.

and becomes increasingly wild as r gets close to 4. There is a reason that the number 4 appears here. In any logistic system, the population $rx(1-x)$ reaches its maximum value $r/4$ at $x = 1/2$. This value cannot exceed the carrying capacity, which is 1. So the largest that r can be and still represent a population model is $r = 4$. It is this extreme example that we intend to treat in some detail.

We can put any of these logistic systems into the framework of pattern systems. All the action takes place on the set of real numbers between 0 and 1 (with 0 and 1 included), so we take $[0, 1]$ as our state space. Think of the state of the system at any moment as the population, measured in terms of the carrying capacity.

The dynamics is defined by a single operator f which, as it should, tells us how states are associated to other states thus indicating change. These step-wise transitions are given by the simple polynomial function f, which now we have decided to look at the growth rate $r = 4$, reads as

$$f(x) = 4x(1 - x).$$

So the transitions are all of the form $x \mapsto f(x) = 4x(1-x)$, or following our usual operator notation (see §6.3) we might write it as

$$x \mapsto f \triangleright x = 4x(1 - x).$$

A plot of f is shown in Fig.7.8.

We have given the state space and the dynamics. There is more to come, but as it stands now we denote this still not fully defined pattern system as

$$\mathscr{L} = ([0, 1], f).$$

Starting at any state, that is, any number in $[0, 1]$, we can follow the sequence of states through which it progresses by the repeated application of the operator f. This sequence of states,

$$x,\ f \triangleright x,\ f^2 \triangleright x,\ f^3 \triangleright x,\ \ldots$$

is, following our usual terminology, a *trajectory*. Each state x initiates its own trajectory. We can write this more suggestively as

$$x \xrightarrow{f} f \triangleright x \xrightarrow{f} f^2 \triangleright x \xrightarrow{f} f^3 \triangleright x \xrightarrow{f} \ldots .$$

It is easy to get fooled into thinking that the polynomial $4x(1-x)$ is so simple that this can hardly turn into anything complicated. But consider what happens when we apply f twice to some number x: we get

$$x \xrightarrow{f} 4x(1-x) \xrightarrow{f} 4(4x(1-x))(1-4x(1-x)) = 16x - 80x^2 + 128x^3 - 64x^4,$$

since each occurrence of x has to be replaced by $f(x)$. And this is just two steps! The point is that we keep plugging the latest value back into the formula again and again. The formula that gives what happens to x after just four steps is:

$$\begin{aligned}256x &- 21760x^2 + 731136x^3 - 12899328x^4 + 137592832x^5 - 963149824x^6 \\ &+ 4656988160x^7 - 16066609152x^8 + 40324038656x^9 - 74281123840x^{10} \\ &+ 100327751680x^{11} - 98146713600x^{12} + 67645734912x^{13} - 31138512896x^{14} \\ &+ 8589934592x^{15} - 1073741824x^{16}.\end{aligned}$$

Recursion is very powerful!

Fig.7.9 gives a graphical interpretation of what is going on as we repeatedly apply the operator f, so we can get a more geometrical idea of how the dynamics progresses. It bears saying again: at each step, the present output from f becomes the new input for f. Given the complexity of the process, it is amazing that we can say anything at all about the evolution of this system.

In Fig.7.10 we show plots of the beginnings of two trajectories arising from the logistic system with $r = 4$, one shown in grey and the other in black. These trajectories start from two different but very close points. There are two noteworthy features about these plots. In both cases the points seem to be scattered in a very disorganized way, and in addition even minute differences in the starting points seem to lead quickly to complete discrepancy between the resulting trajectories. Together they make us wonder if there is any real discernible pattern in what emerges. The rapid divergence of the trajectories starting from points that are very close is the sign that the system is chaotic. In mathematics, *chaos* does not have the same sense of complete unmanageability that we think of in common language. It is usually applied in situations like this in which there is a perfectly prescribed process going on—here the simple repetition of a very simple calculation—yet the outcome is incredibly sensitive to the minor differences. Here we

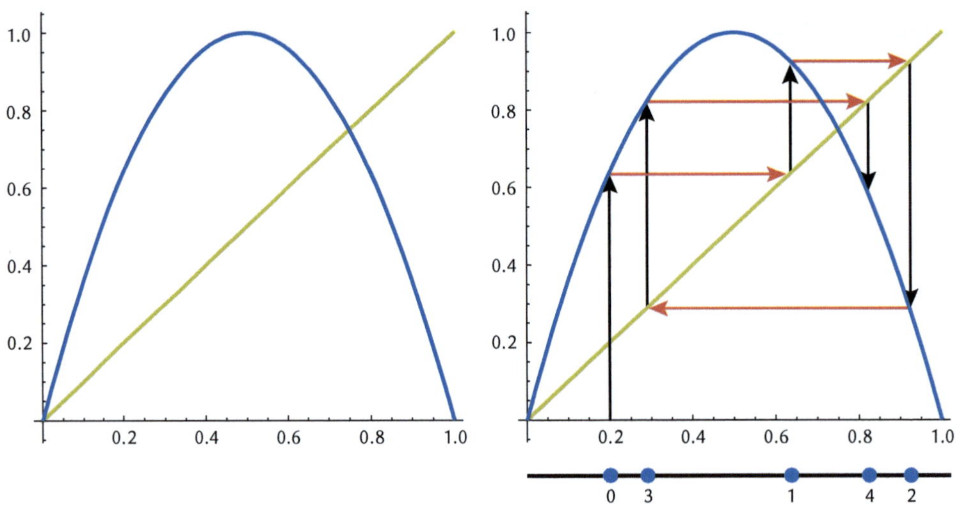

Figure 7.9: On the left we see a plot of the function f and also a plot of the straight line $y = x$. This is the line of points where the first and second coordinate are equal. On the right we see how we can visualize the making of the first part of a trajectory of f starting from some arbitrary point. We have labelled the actual sequence of points by $0, 1, 2, 3, 4$ to indicate the order in which they appear in the iteration. What the image shows is how each value $y = f(x)$ becomes the next "x" to be fed into f. A horizontal line is drawn from the latest value y of f so as to meet the line $y = x$, thus giving us the next x at which to apply f for the next step. One begins to appreciate how things get bounced around. These types of plots are often called cobweb plots (although this seems very unfair to spiders).

started with two points that differ by 0.001, yet after less than a dozen steps the trajectories are nowhere near the same and now appear unrelated. This is not because we carefully chose the points to make this happen. It happens no matter what pair of closely spaced points we choose, and no matter how close they initially are.

There are several other points to note. In spite of its apparent chaotic behavior, this (and any) logistic system is *deterministic*. The present state determines precisely all future states on its trajectory. This is not to say that it is reversible. Every state y certainly has at least one possible precursor, some state x out of which y may have come: $y = f \triangleright x$. For instance we see that $8/9$ has $1/3$ as a possible precursor, that is $f \triangleright (1/3) = 8/9$. But also $f \triangleright 2/3 = 8/9$, so $2/3$ is also a possible precursor. So if we are looking at state y just by itself, we cannot be sure what its precursor or predecessor was. All of this is visible from the symmetry of the plot of f in Fig.7.8. The one exception is $1/2$, which is uniquely its own precursor. The system is deterministic, but not reversible.

Another important point is that even if we were to follow a trajectory indefinitely, we can already see that no single trajectory will ever get to every point of the state space. Any single trajectory will have only a countable number of points on it, since it is just created by iterating f a countable number of times $(f, f^2, f^3, f^4, \ldots)$. But we already

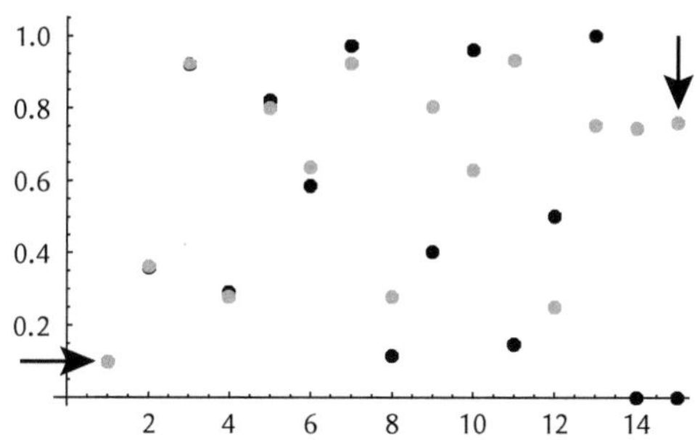

Figure 7.10: The figure shows the progression, left to right, of two trajectories, one starting at 0.1 (grey) and the other at 0.101 (black), over 14 steps of the logistic process. The positions of the points in the state space [0, 1] are shown in terms of the vertical axis. The horizontal axis represents the successive steps of repetition of the operator f. Although initially indistinguishable (shown by the horizontal arrow), by the time that we have made 14 iterations (vertical arrow) the two trajectories seem unrelated. This is the signature of chaos.

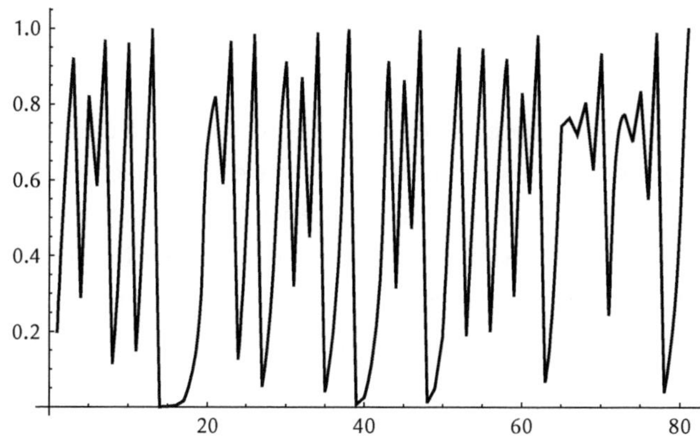

Figure 7.11: The first 80 values of the trajectory in the Verhulst logistic system starting at 0.2. It is drawn as a line plot: the values of x are indicated in the vertical direction, the iteration numbers are written horizontally. The lines connect successive values of x. It is hard to see how any order is to be found in this.

know that [0, 1] is an uncountable set. This is a typical sort of statement that we can make that reveals the usefulness of the notion of countability.

Does this logistic system display any pattern, at least in the common sense of the word? A plot of the first 80 points of the trajectory of 0.2 under iteration of the function f discussed above is given in Fig.7.11. We can stare at the output for a long time without seeing that there is a very definite pattern hidden within it.

However, suppose that we try a different tack. In Fig.7.12 we have partitioned the interval [0, 1] into 50 equal length non-overlapping intervals. Each successive state on a trajectory is some number in [0, 1] and so falls into one of these 50 intervals. Thinking of each of these little intervals as a bucket into which the values of the trajectory fall,

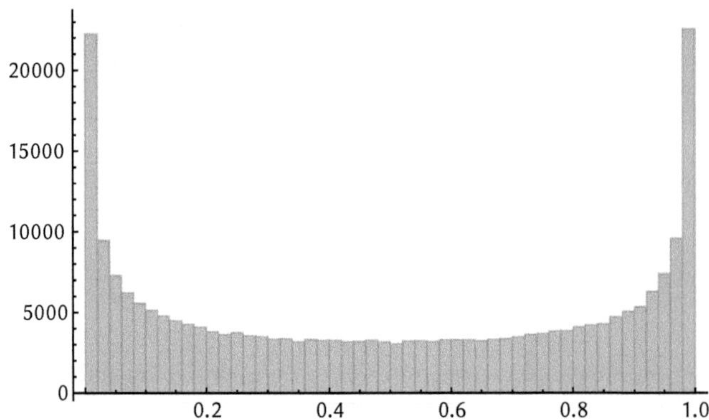

Figure 7.12: Order manifests itself in the $r = 4$ logistic system as we increase the number of buckets and the length of the trajectory: 50 buckets and 250, 000 consecutive steps along the trajectory makes for a convincing demonstration of internal patterning. The vertical axis indicates the number of hits that each bucket received. The pattern is not fully formed, but its prominent shape is clear. In fact this "ultimate pattern", which the histogram seems to suggest, really does exist and is independent of the starting point. We will be able to describe it exactly, see Fig.7.14.

we keep count of how many states along the trajectory fall into each bucket. The result is what is called a *histogram*. Instead of trying to trace the trajectory, we watch how its states are distributed numerically across the interval $[0, 1]$. Fig.7.12 shows the outcome of a run of 250, 000 consecutive states along a trajectory of our logistic system.

The result is striking. What is noteworthy is how the pattern only emerges dynamically, even though this emergence is slow. What we see here is the outcome of a quarter million steps along one trajectory. There is nothing special about this particular trajectory. Almost every trajectory produces a very similar histogram.[11] The other noteworthy feature is the way in which we have used the *frequency* of events to filter out the distribution of the states of a trajectory.

We are quite justified in thinking that the shape of this distribution has something to do with the system itself, not just the state we began with, and that if we were to make yet finer partitions and take even more iterations then similar and even smoother versions of this pattern would emerge. We can already see how we might interpret this as some sort of probability distribution. If you were to bet into which buckets a randomly chosen state out of a trajectory would most likely fall, which would you choose? Thinking about it from the perspective of some poor population of creatures evolving through this dynamics, the outlook looks pretty grim. It favors vast leaps between tiny populations to ones that almost exhaust the entire carrying capacity. Presumably the end result is extinction, since the population size is discrete and cannot be reduced in size indefinitely: zero is the only stable (everlasting!) state that the system has.

The histogram is one thing, but even if we were to draw it with a much finer partition and were to take many millions of iterations and get a much nicer picture, the question of knowing exactly what this distribution actually is would still elude us. Eventually we will see how this distribution arises from an invariant measure. But this raises an important question: just how can we find something that only emerges as a long term outcome of the dynamics? In fact there is the larger question. Our recognition of hidden order in this system is the result of changing the event space, looking at accumulation in buckets rather than looking at successive values. How do we find ways of looking at processes that reveal deeper order?

There are no genuine answers to these questions. In this particular case we will find an answer, and we will do so by using an idea that we have not yet looked at: comparing this pattern system with another one for which we know the answer. Now that we have a general definition of pattern systems, we can ask questions about how alike two different systems are. If there is an underlying similarity perhaps we can use it to infer information about one from the other.

7.6.3 An unexpected connection

The Verhulst system that we have been studying is based on the state space $[0,1]$ and the simple iterative operator

$$x \mapsto f \triangleright x = 4x(1-x)\,.$$

Surprisingly, it is a factor of the doubling system $\mathscr{D} = ([0,1], d)$ that we have discussed in §7.5.3. Both $\mathscr{L}$ and $\mathscr{D}$ are defined on the same state space $[0,1]$, but the dynamics look entirely different. Yet they are related. Explicitly there is a mapping from $\mathscr{D}$ to $\mathscr{L}$ that intertwines them:

$$\begin{aligned} [0,1] &\xrightarrow{\psi} [0,1] \\ y &\overset{\psi}{\mapsto} (\sin(2\pi y))^2 = \sin^2(2\pi y)\,. \end{aligned}$$

This looks complicated, and it is certainly surprising to see a trigonometric function appear. (Beware that the notation $\sin^2(2\pi y)$, meaning $(\sin(2\pi y))^2$, is very common in mathematics, but it is potentially confusing! It does not mean $\sin(\sin(2\pi y))$.)

But ignoring the symbols, Fig.7.13 offers a visual way to interpret this. The important thing is that this mapping respects the dynamics, or in symbols there is the commutative diagram

$$\begin{array}{ccc} x & \xrightarrow{\psi} & \sin^2(2\pi x) \\ {\scriptstyle d}\downarrow & & \downarrow{\scriptstyle f} \\ d \triangleright x & \xrightarrow{\psi} & \sin^2(2\pi 2x)\,, \end{array}$$

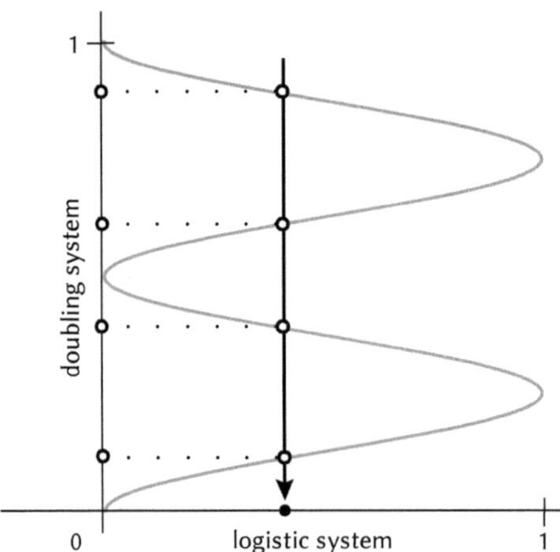

Figure 7.13: This figure shows how points are mapped from the doubling system (seen on the vertical axis) to the logistic system (seen on the horizontal axis). The plot is that of the function $x = \sin^2(2\pi y)$, though turned sideways so as to emphasize how things look from the point of view of the logistic system. Both state spaces are the interval $[0, 1]$. What is evident is how several different points of the doubling system end up at just one point of the logistic system. With just two exceptions the mapping is 4-to-1—four points map to one point in passing from $\mathscr{D}$ to $\mathscr{L}$.

which shows that the dynamics of the doubling operator d of $\mathscr{D}$ matches the dynamics of the logistic operator f of $\mathscr{L}$. Once again, "commutative" means that the same result arises no matter which of the two ways we go along the arrows from the top left to the bottom right.

This comes down to saying that $f(\sin^2(2\pi x)) = \sin^2(4\pi x)$, which is actually a fairly straightforward trigonometric identity. The details are best left to the Endnotes.[12]

The visual interpretation of the mapping ψ from $\mathscr{D}$ to $\mathscr{L}$ given in Fig.7.13 shows at once that the mapping is, with a few exceptions, a four-to-one map (four different states map to the same state).

The logistic system is some kind of shadow of the doubling system, with the dynamics of $\mathscr{L}$ following the dynamics of $\mathscr{D}$, but not in the same detail. The example clarifies the difference between *isomorphism* (equal shape) and *homomorphism* (similar shape). The doubling system is essentially the same as the shift system $\mathscr{S}$ (isomorphism, though we have called it by its usual name, *conjugacy*), whereas the logistic system appears here as only a shadow of the doubling system (homomorphism, or as we have called it a *factor* of the doubling system). Since we have

$$\mathscr{S} \xrightarrow{\varphi} \mathscr{D} \xrightarrow{\psi} \mathscr{L},$$

we can see also the logistic system as a factor of the standard shift system (again as a homomorphic image).

Just as we did before in §7.5.5, where we transferred measure from the shift system to the doubling system, we can do the same here and transfer the invariant measure of the doubling system, which is just the usual measure of length on the $[0, 1]$, over to the logistic system and create an invariant measure there. The idea behind the transfer works in just the same way: to find the measure of a subset B of $[0, 1]$ in the logistic system, we pull it back to the doubling system, getting $\psi^{-1}(B)$, and then measure this set with the Lebesgue

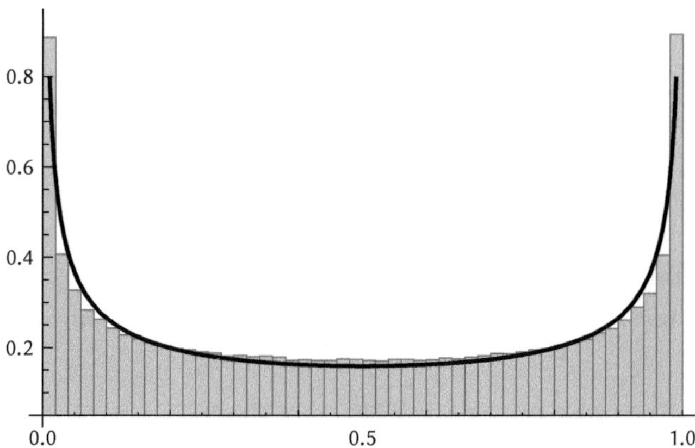

Figure 7.14: The plot shows our original histogram that we created from a long trial on the Verhulst-logistic system, and on top of it a plot of the probability distribution that arises out of transferring the length measure of the doubling pattern system to the the logistic system.

measure $\mathfrak{l}$. As a general principle this is straightforward (see the Endnote below), but in detail here it is more complicated because the mapping ψ that links the two systems is itself complicated. Still one can work through it, and the outcome is remarkable. It is a probability measure, and when it is pictured visually it reveals Fig.7.14.[13]

The probability distribution that we had foreseen from the histograms is now a reality and we have its precise form. Furthermore it is invariant under the dynamics of the logistic system.

The concepts here, which involve inverting mappings at the level of subsets, whether or not they are invertible at the level of states, are more than we need go into here, but they are general and structurally important. We have laid out the ideas behind this in the Endnotes.[14]

7.6.4 The wider view of relating pattern systems

This detailed look at the Verhulst logistic system is interesting in its own right because it is a famous example that arises in the study of chaos. But that is not our only interest in it. We see that another feature is emerging, and that is how pattern systems can relate to each other. How they can look different, but actually prescribe the same dynamics, or how one can shadow another as a factor. We have presented this in the context of mathematics, but as we have often said, pattern is not just about mathematics. It is a fundamental way in which we see and live in this world.

An essential part of pattern is that there does not have to be an exact match for us to recognize a form of patterning as something similar to what we have seen before. We are good at working with analogy and learning to pick out features that help us to function appropriately in new situations where those features appear. We are good at using the dynamics of one context to illuminate the dynamics of another and much of our language actually depends on this device. The pulling back of form in order to describe it through a more familiar or a more physical and less abstract setting is a part of how we use language. The value of the mathematics is to formalize this process and to strip it of its particular (and thus restricting) context. No matter what two pattern systems we have, if one is a factor of the other we can use the parent to articulate the dynamics of the factor, and the factor to reveal ways in which it can be seen in a wider context.

7.7 Insights from itinerary pattern systems

7.7.1 The universal nature of shift systems

What we have learned from §7.5 is that for any pattern system based on a discrete iterative operator and a partition of the state space, there is a shift system that models it, made by using its itineraries. Furthermore if the original system has an invariant measure, there is a corresponding measure on the itinerary shift system. The itinerary pattern system is a *factor* of the system that produced it. The boxed text is the formal version of this, which we call the *itinerary mapping theorem.*

Let $\mathcal{X} = (X, u, \mathcal{P}, \mathbf{p})$ be a pattern system with simple iteration u, a finite partition $\mathcal{P} = \{P_1, P_2, \ldots, P_n\}$, corresponding event space $\mathcal{E}$, and an invariant measure $\mathbf{p}$. Let $I = \{1, 2, \ldots, n\}$. Then there is an itinerary pattern system $\mathcal{I} = (I^\infty, s, \mathcal{C}, \mathbf{p}_I)$, which is a shift system on the alphabet I, and a pattern system mapping

$$\mathcal{X} \xrightarrow{\varphi} \mathcal{I} ,$$

which arises by mapping each $\mathbf{x} \in X$ to the part of $\mathcal{P}$ that it lies in. $\mathcal{I}$ is a *factor* of $\mathcal{X}$ by this mapping.

The event space of the itinerary system is the standard one defined by the set $\mathcal{C}$ of its dictionary sets. In this process an invariant measure $\mathbf{p}_I$ is created on I^∞ by the process pulling back each dictionary set C of I^∞ to a corresponding set of states in X whose itineraries fall into that set:

$$\mathbf{p}_I(C) := \mathbf{p}(\varphi^{-1}(C)) .$$

If $\mathbf{p}$ is a probability measure ($\mathbf{p}(X) = 1$) then also $\mathbf{p}_I$ is a probability measure since

$$\mathbf{p}_I(I^\infty) = \mathbf{p}(\varphi^{-1}(I^\infty)) = \mathbf{p}(X) = 1 .$$

Furthermore, the full entropy $h(\mathcal{I})$ of the itinerary pattern system equals the entropy $h(\mathcal{X})$ of the evolution of the pattern $\mathcal{P}$.

Conceptually this matches the way we most often approach pattern systems in Nature. We pick out particular features that we can measure or observe and then watch their progress in time. We may not be able to describe the full detail of the underlying pattern system, or even understand it, but we can follow the itineraries of the features that we can describe. The itinerary system is a model of the real thing.[15] This result shows the *universal nature* of shift systems in discrete patterning. Although the shift system was the first discrete pattern system we looked at, and it is one of the simplest to think about, we can see now that it lies at the heart of all discrete iterative patterning. For every discrete iterative pattern system there is an itinerary pattern system that shadows it. The important statement about equality of entropy is stated here for future reference, see §8.

We might note that although the itinerary process is

$$\text{state} \longrightarrow \text{trajectory} \longrightarrow \text{itinerary} = \text{string of symbols}\,,$$

there is no guarantee that every possible sequence of symbols in the resulting shift system will actually appear as an outcome of the trajectories of the source system. Each trajectory produces an itinerary, which is a string of symbols, but it is not necessarily true that there is a trajectory for every possible string of symbols. However, the way in which the induced measure $\mathbf{p}_I$ is defined actually takes care of this issue in its own way. If C is a dictionary set of I^∞ that does not occur as part of any trajectory, then $\varphi^{-1}(C)$ must be empty, and $\mathbf{p}_I(C) = \mathbf{p}(\varphi^{-1}(C)) = \mathbf{p}(\emptyset) = 0$. So the entire collection of empty dictionary sets gets assigned the measure 0. They are in the pattern system $\mathscr{I}$, but they amount to a set of measure zero.

7.7.2 The Markov property

The notion of an itinerary system allows us to write a succinct definition of what we mean by a Markov pattern system. The intuition behind a Markov system or process is that the future depends on the present, but not on how the present came to be. The drunken man's position at his next step is dependent only on his present position, not how he got there.

More formally the Markov condition is stated in terms of the *law* of a pattern system and its corresponding itinerary system. At the level of states, a pattern system $\mathscr{X} = (X, u, \mathscr{P}, \mathbf{p})$ is determined by the operator u, and there is no question of what happens next. However, at the level of events, passing through a sequence of particular events does not in general tell us which event will follow next. We can ask about the probabilities of future events given the present event and those that have passed. The Markov condition is that this probability only depends on the present event, and not the past events.

More precisely, if the corresponding itinerary system is $\mathscr{I} = (I^\infty, s, \mathscr{C}, \mathbf{p}_I)$ then the system $\mathscr{X}$ is said to have the *Markov property* if the probability $\mathbf{p}_I(E_1\, E_2\, \dots\, E_{n-1}\, E_n)$ of some event E_n following after the sequential occurrence of the events $E_1\, E_2\, \ldots\, E_{n-1}$ depends only on the last step, E_{n-1}. Formally this is written using the notation of *conditional probability*: the probability of a "this" happening given that a "that" has happened: $\mathbf{p}_I(\text{``}\textit{this}\text{''}|\text{``}\textit{that}\text{''})$. The Markov property then reads:

$$\mathbf{p}_I(E_n|E_1\, E_2\, \ldots\, E_{n-1}) = \mathbf{p}_I(E_n|E_{n-1}) \text{ for all } E_1, E_2, \ldots, E_{n-1}, E_n \in \mathscr{C}\,. \tag{7.5}$$

In this case we say that $\mathscr{X}$ is a *Markov pattern system*, or has the *Markov property*. There are similar versions for continuous systems.

7.7.3 Pattern mapping and pattern equivalence

The value of itinerary pattern systems is that they allow us to look at a dynamical system at two different levels. There is the level at which we have the system with all its states and the level at which we can actually make statements about the system—the level defined by its sameness/difference relation, the partition. The evolution of the partition is the event space, and this may be the limit to which we can actually distinguish things about the system and its dynamics.

It is worthwhile pointing out again that when it comes to the real world of processes and change, we will always be limited by what we can experience through our senses and our instruments of measurement. We are almost bound to be at the level of some itinerary system.

Suppose that $\mathscr{X} = (X, s, \mathscr{P}, \mathbf{p})$ and let $\overline{\mathscr{X}}$ be the corresponding itinerary system, with the overline suggesting that it is some sort of shadow of $\mathscr{X}$ itself. In keeping with this we will write $\overline{X}$ for the state space I^∞ of the itinerary system. Then the itinerary mapping theorem says that there is a mapping

$$\mathscr{X} \longrightarrow \overline{\mathscr{X}} \tag{7.6}$$

and this mapping makes $\overline{\mathscr{X}}$ a factor of $\mathscr{X}$.[16]

We can use the idea of these shadow itinerary pattern systems to extend the relationships that two pattern systems $\mathscr{X} = (X, \mathscr{P}, s, \mathbf{p})$ and $\mathscr{Y} = (Y, t, \mathscr{Q}, \mathbf{q})$ can have to each other.[17] The idea behind this is that although two pattern systems may not be conjugate to each other, they may be so at the coarser, and perhaps more accessible, level of itineraries.

In saying that $\mathscr{X}$ is a *factor* of $\mathscr{Y}$ we mean that there is a mapping of Y to X that intertwines the dynamics and for which the inverse image of each part P of $\mathscr{P}$ consists of the union of some parts of $\mathscr{Q}$.

At the coarser level, we will say that $\mathscr{X}$ is a *pattern factor* of $\mathscr{Y}$ if $\overline{\mathscr{X}}$ is a factor of $\overline{\mathscr{Y}}$. The difference is that now we are not dealing with states of X mapping to states of Y, but rather itineraries of X mapping to itineraries of Y. We will denote this by writing

$$\mathscr{Y} \rightsquigarrow \mathscr{X}\,. \tag{7.7}$$

In itself this does not require that $\mathscr{X}$ is actually a factor of $\mathscr{Y}$. That is a relation that lies at a more refined level. The shadow follows the process that casts it, but it is not its equal. If $\mathscr{X}$ is actually a factor of $\mathscr{Y}$, then it is automatic that the shadows follow along, and $\overline{\mathscr{X}}$ is a factor of $\overline{\mathscr{Y}}$. Then we can express the result as a commutative diagram

$$\begin{array}{ccc} \mathscr{Y} & \longrightarrow & \mathscr{X} \\ \downarrow & & \downarrow \\ \overline{\mathscr{Y}} & \longrightarrow & \overline{\mathscr{X}}\,, \end{array}$$

which visually suggests the shadowing of the factor mapping of $\mathscr{Y}$ to $\mathscr{X}$.

In the same way that we talked about conjugacy as a way of expressing that two pattern systems are the same, we can say that $\mathscr{X}$ is *pattern equivalent* to $\mathscr{Y}$ if $\overline{\mathscr{X}}$ is conjugate to $\overline{\mathscr{Y}}$. This time we will write this as

$$\mathscr{X} \leftrightsquigarrow \mathscr{Y}\,.$$

This is a weaker relationship than to say $\mathscr{X}$ is conjugate to $\mathscr{Y}$, though again shadows follow their masters, and the latter does imply pattern conjugacy.

> $\mathscr{X} = (X, s, \mathscr{P}, \mathbf{p})$ and $\mathscr{Y} = (Y, t, \mathscr{Q}, \mathbf{q})$ are pattern equivalent if there is a one-to-one correspondence between $\mathscr{P}$ and $\mathscr{Q}$ through which their itinerary systems are conjugate.

Pattern equivalence emphasizes the identical nature of the dynamical evolution of the two systems *as seen from the perspective of events.*

7.7.4 Itineraries in continuous pattern systems

The idea of pattern itineraries carries over easily to pattern systems based on continuous temporal dynamics. Before we look at the details we should note that what we are about to describe is one of the most common ways we have for looking at dynamical systems. The Moon, for instance, moves continuously and so the particular ways in which we see it also change continuously. However, that is not the way in which we ultimately come to speak of it, nor was it the way in which its nightly visits were first described. Far more typical is to follow its continuous waxing and waning on a daily, or perhaps we should say nightly, basis. We tend to describe the Moon in terms of its eight phases—new moon, waxing crescent, first quarter, waxing gibbous, full moon, waning gibbous, last quarter, waning crescent. Many things in Nature appear to be continuous but we make tables of measurements and use those to build our models, and in doing so model the continuous with the discrete.

Suppose then that we have a continuous pattern system $\mathscr{X} = (X, T, \mathscr{P}, \mathbf{p})$, where T represents time parameterized by the variable $t \geq 0$ and $\mathscr{P}$ is some partition that we intend to use as our source of distinguishing. There are two obvious (and useful) ways in which to discretize $\mathscr{X}$.

The first is to discretize time into intervals of fixed length, say of length 1 second. Then, for each state $\mathbf{x} \in X$ we follow its trajectory stepwise, second by second, writing a list of the corresponding parts of $\mathscr{P}$ in which the trajectory falls in. Again we are switching from trajectories to itineraries. If $\mathscr{P} = \{P_1, P_2, \dots, P_n\}$ then, just as before, we can use the alphabet $I = \{1, 2, \dots, n\}$ and obtain a mapping

$$X \xrightarrow{\varphi} I^{\infty}.$$

This process, mapping one pattern system onto another which is simpler, is what we do all the time. What is a movie but an example of this? The weakness of this form of discretization is that it can quite possibly miss what is really important in the dynamics. This version of an itinerary model can at best match the original faithfully at the level of dynamics proceeding in steps of one second (or whatever the unit of time is). But if the appropriate time scale of the original pattern system is more like milliseconds or nanoseconds, then we are missing all the detail, some of which might be important. A state may oscillate between two states a thousand times in a second, and our model will miss it all.

The second way to discretize of a continuous pattern system is more in tune with the qualitative dynamics. Again we have a mapping $X \longrightarrow I^{\infty}$, but now we follow each trajectory of each state continuously in time, listing *the successive parts* of the partition into which states flow. This is dynamics measured in terms of *events*, in the sense that change is now interpreted as being the occurrence of a new event. The relationship to time is weakened to the sequence of differences measured by changing events rather than what happens in steps of equal time.

A familiar example is the way a GPS guidance system works. When we search out directions from one location to another, the mapping system comes up with a sequence of steps which amount to a sequence of events: drive south on Highway 230 for 2 miles, turn left at the stop sign and drive 0.4 miles until you reach Cross Street. Turn left and the destination is on your right. This converts what will unfold in continuous dynamics into a sequence of steps that are marked by distinct events (arriving at Cross Street, etc.). How long it takes, whether we stop for gas or coffee along the way, all of this is irrelevant to this way of looking at the dynamics.

The two different approaches to discretizing the dynamics of a continuous system raise an interesting question. The first, based on watching the system on the basis of equal steps in time, preserves the clock-based sense of time, but may lose sight of what is really happening. The second reveals what is happening, but loses all sense of time in the sense of clocks. This raises an ancient philosophical issue: is time an attribute of reality that flows on independent of actual distinguishable events—Newton's sense of time—or is time born out of the essence of the passing of events, actual changes? Pattern systems can accommodate either view. We will come back to them in §8.4.11. Interestingly, the writing of what we refer to as the Cosmos is written in classical and modern Chinese as 宇宙 (yuzhou), which literally combines time and space, not making the clear separation that we have inherited from classical physics.

7.8 Versions and shadows

It seems clear that pattern systems can relate to one another, and that intuitively we are quite disposed to seeking out such relations. If we know one system is like another then we can make inferences, perhaps not perfect but still highly valuable, about the one from knowing the other.

The details of the explicit mappings of pattern systems into versions, factors, or shadows of them are not relevant here. We will see examples as we go along that illustrate the idea. What is important is to realize that it can be formalized and it is an important part of the patterning world. Two pattern systems can be alike to the extent that each is effectively the same as the other, but they just look different. The shift system and the doubling system are such a pair. More commonly they can be alike in the sense that one is a version, though perhaps not as refined, of another, or they both can be seen to be related to some common pattern system. That is the value of moving from a pattern system to an itinerary system, using a partition to coarsen the detail of individual states.

One of the goals of this book is to show how our concept of pattern is applicable to any situation in which there are difference and change. As we will see in succeeding chapters, there is no limit to this. There may be vast differences, but also there are similarities, some striking, some hidden, which are revealed by appearance, morphology, environment, mutual interaction, and so on. Perhaps the most common of these is cyclical repetition and its variations—seasons, lunar cycles, economic cycles, menstrual cycles, and so on. Here, as the very words suggest, there is a common feature of circularity, and they all can be related to the simple pattern of traveling around a circle.

7.8.1 Paradigm patterns

Plato is famous for his theory of ideal forms, the perfect ideal non-material forms that the material world, known to us through sensation, can only emulate but never achieve. Nowadays we have a keen respect for what we can know from our senses, and ideal

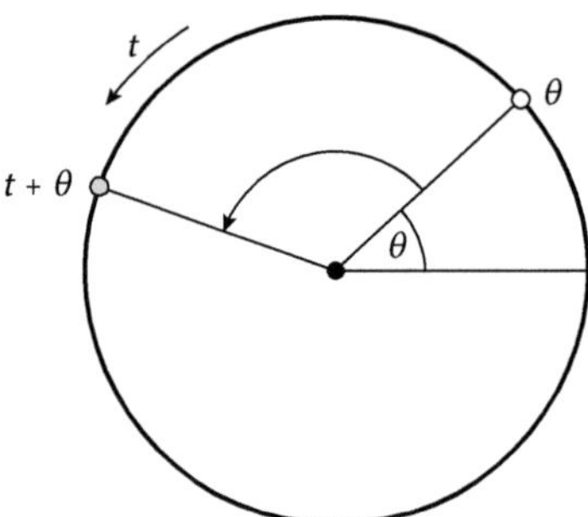

Figure 7.15: A circle in which we have labeled its points by the angle they make with respect to a horizontal axis. The action of time t is to increase the angle by the value of t.

models of the perfect heavens with their perfect epicycles are long gone. Nonetheless, ideal forms are familiar and important concepts, and we all have an intuitive appreciation for what Plato was talking about. Our shift systems with their infinitely long states are examples. Here we suggest one mathematical interpretation of this type of concept, though here we shall talk not about ideal patterns, but rather *paradigm patterns*.

A simple place to start is with the familiar idea of cyclical change. The paradigm for this is circular motion. The origin of the word is the Greek κυκλος (*kyklos*) = circle. Closely related is periodic change, again with a suggestive Greek root, namely *periodos* = *peri* + *hodos* = a going around + a journey, a way.

In its ideal mathematical form cyclical change is expressed as a point moving around a circle. We will discuss periodic motion in considerable detail in Ch.11, but here it suffices to describe the simple dynamical system consisting of a circle C, whose points we will take as comprising the state space, and the action of time t acting as rotation upon it. The position of a point on C (a state) can be described simply by using its angle relative to some pre-chosen reference point. Thus, referring to Fig.7.15, we see the angle θ identified with the point on the circle which is then also labelled θ. The action of time is simply to add to the angle, in effect producing rotation:

$$\begin{aligned} \mathbb{R}_{\geq 0} \times C &\longrightarrow C \\ (t, \theta) &\mapsto t + \theta\,. \end{aligned}$$

We have parameterized time by the non-negative real numbers, i.e. starting at time equal to 0 and going forward from there. The details are quite unimportant—they just confirm that we can state this in mathematical language. Let us refer to this system as $\mathscr{C} = (C, \mathbb{R}_{\geq 0})$.

Although no partition, event space, or invariant measure is offered here, this is a pattern system—one of the most basic ones of all. Any pattern system that involves cycling in some sense can be interpreted in terms of this simple system or, more mathematically, mapped onto this system. An example shows how.

7.8.2 The limit cycle

An important form of pattern system is what is called the *limit cycle*. It is most easily described as a dynamical system in the plane in which every trajectory (except for one) tracks towards one circle C on which we see the circular dynamics that we have just discussed. Fig.7.16 illustrates what is going on. The trajectories spiral around the origin. Outside the stable circle, trajectories shrink towards it; inside the stable circle, they expand towards it. The exception is the fixed point at the origin, an equilibrium point unaffected by time. Each point in the plane is described by two variables $r > 0$ and θ: its radius from the center and its angle with respect to some reference point.

The change of r is such that r is decreasing if r is greater than the radius of C, increasing if r is less than the radius of C, and unchanging on C. The effect of this is for the circle C to be a trajectory in its own right, and to *attract* all other trajectories towards it. The one exception, the center of the circle is an equilibrium point, but *unstable equilibrium*—the slightest deviation and the state will start tracking towards the limit cycle.

This system models an important property: a stable cycle that is able to re-establish itself after perturbations. The limit cycle is an *attractor* in the sense that trajectories head towards it. Disturb the cycle and the system will re-establish it. It is not hard to see the significance of such cycles in life forms, which depend on being able to recover from unforeseen disturbances.

An early mechanical application of this idea of the limit cycle is the *centrifugal governor*, a *feedback* mechanism invented by Christiaan Huygens and later used by James Watt to keep steam engines running at a constant speed, see Fig.8.1 in §8.2. For some mathematical details about limit cycles, see the Endnote.[18]

The limit cycle definitely embraces the idea of cyclic and periodic motion, but is considerably more sophisticated. Still we can see the simple cycle as a *factor* of the limit cycle system: if the radius of the circle is equal to 1, then

$$(r, \theta) \mapsto (1, \theta). \tag{7.8}$$

We can visualize this as a standing at the center of the circle and projecting everything we can see on a ray emanating from the origin onto the point where that ray meets the stable circle C, see Fig.7.16.

What we see here is applicable to all pattern systems that have cyclical dynamics.

> The simple circle with its rotating dynamics
> is a *paradigm* for what cyclical dynamics means.

We can formalize that to mean that every pattern system that has a cyclical aspect to it has the basic cyclical system $\mathscr{C}$ as a *factor*. Details will vary, but the underlying idea is the same.

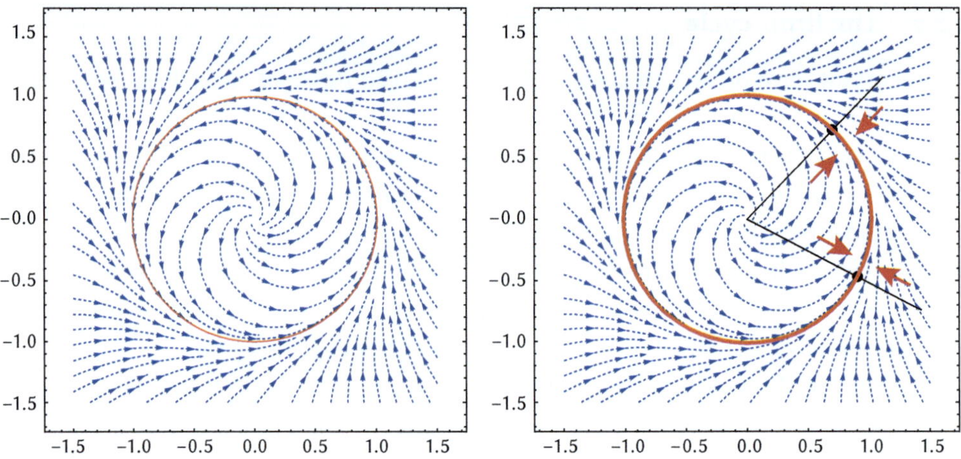

Figure 7.16: On the left we see a "vector field" picture of the dynamics of the limit cycle, see the Endnote 18 (7.7). The arrows indicate the direction of movement at enough points to make the effect of the dynamics quite visible. Pictured in red is the circle C which is the limit cycle. Trajectories starting outside and inside the limit cycle converge towards it. On the right is one version of how this system can be mapped onto the simple system of a point moving around a circle. Computed on XPP, a freely available application written by Bard Ermentrout.

Another example is what we have seen as the conclusion to § 7.7. Every dynamical system with discrete iterative dynamics and a partition with n parts maps to the simple shift system on an alphabet of n letters. If we regroup the partition into just two parts, then it maps onto the most famous of all shifts, the zero-one shift that is so common a theme in this book. It is the grand *paradigm* of discrete iterative dynamics, and as such is a constant source of information about how discrete iterative dynamics works.

7.8.3 Paradigm patterns defined

With these examples in mind, we try to formalize what we are saying when we say that one system is a paradigm for others. It is easiest to start rather abstractly. Suppose that we have a family (= set) $\mathscr{F}$ of things, each of which has some properties that come from some list of properties L. Each element $a \in \mathscr{F}$ has some set of properties from the list L, some more, some less. If $a, b \in \mathscr{F}$ then we write $a \succeq b$ if a has all the properties from L that b has. It may be that a has more, but it at least has all those of b. Let's say that two elements $a, b \in \mathscr{F}$ are *equivalent* if they have exactly the same properties: $a \succeq b$ and $b \succeq a$. For some pairs (a, b) neither $a \succeq b$ nor $b \succeq a$ may hold.

We say that an element $c \in \mathscr{F}$ is a *paradigm* of this family if $a \succeq c$ for all $a \in \mathscr{F}$. In other words, c has properties that are *common* to all the elements of the family $\mathscr{F}$. There is no guarantee that there is such an element, and if there is, whether or not it is unique. But notice that if $d \in \mathscr{F}$ is also a paradigm of $\mathscr{F}$ then c and d must be equivalent by the very definition. In this abstract setting, we can see that paradigms epitomize what is common to all the things (elements) that are in the family $\mathscr{F}$.

Now we can apply this simple idea to pattern systems. If we have a family $\mathscr{F}$ of pattern systems and $\mathscr{X}$ and $\mathscr{Y}$ are two members of this family, we will write $\mathscr{X} \succeq \mathscr{Y}$ if there is a pattern mapping from $\mathscr{X}$ to $\mathscr{Y}$, or to say the same thing, $\mathscr{Y}$ is a pattern factor of $\mathscr{X}$. Then any pattern system $\mathscr{Z} \in \mathscr{F}$ which is a factor of *every* element of $\mathscr{F}$ is a paradigm of this family. It has properties that are shared by all members of $\mathscr{F}$, so is a representative of their commonality, and hence a paradigm of what they collectively have in common.

We do not specify what this list of properties may be. In the context of pattern systems, factors are to be thought of as simplification and the dropping of details. The paradigm patterns of a family of pattern systems have shed all details except those that are common to the whole family.

This paradigm property may not be anything spectacular, or even anything that is particularly noteworthy. The definition does not promise that. However, it encompasses the idea of a shared common property of a family of pattern systems, and it is for that family a sort of paradigm for something that they share. In the case of dynamical systems arising from differential equations, often the most important features have to do with qualitative aspects of the phase space, attractors, points of equilibria, limit cycles, etc. These often depend more on some easily specified aspects of the defining differential equations than they do on their more meticulous details. Our description of the differential equations describing the limit cycle in Endnote 18 is an example which can be considered as a paradigm model for the idea of limit cycles. There are countless versions of this idea. In the classical theory of dynamical systems via differential equations such examples are called *normal forms*, see [86].

7.9 The Poincaré recursion theorem

> I do not know its name. The word is uttered: "Dao"
> If we must, call it "great." Great, say that it is passing.
> Passing, say that it goes far. Going far,
> say that it returns. [42, *Daodejing*, Ch.25]

If you ask people to say what they think pattern is about, you find that many of them will make some reference to repetition. The whole idea of pattern as something that we *recognize* already has this implication of repetition. This is a pretty good answer. As the famous statement from the *Daodejing* that is the epigraph for this section suggests, repetition stands at the basis of the flow of things. If it were not so, the consistency we so depend on in this world of constant change would not exist, and neither would we. This idea of return seems to lie deep in the workings of Nature.

One of the famous results from the theory of dynamical systems is a statement that says that this idea of return or recurrence can actually be formulated mathematically, and follows directly out of the axioms that we have sketched out in §6.1 and formalized in §6.7. It's interesting that our formalism of pattern, which does not suggest any idea of return, should match this instinctive intuition: once we see a pattern in a dynamical situation, we will almost surely see it again. This is the *Poincaré recursion theorem*.

We will state it in the context of discrete iterative dynamics, assuming that we begin with a dynamical system $\mathscr{X} = (X, s, \mathscr{E}, \mathbf{p})$, where $\mathbf{p}$ is an invariant probability measure defined on the event space $\mathscr{E}$. Our idea of pattern is based on the idea of events, the *subsets* of X about which we can make statements about sameness and difference in this dynamical context. The theorem says that if A is an event with $\mathbf{p}(A) > 0$ then the trajectory of any randomly chosen state of A will almost surely pass through A again. We can think of it as saying if an event happens then it will almost certainly happen again. Return.

There is that now familiar proviso, "almost surely". More precisely, except for a set of states of measure zero, if $\mathbf{x} \in A$, we will find that its trajectory

$$\mathbf{x},\ s \triangleright \mathbf{x},\ s^2 \triangleright \mathbf{x},\ s^3 \triangleright \mathbf{x}, \ldots$$

returns to A; there is some value $k > 0$ for which $s^k \triangleright \mathbf{x} \in A$. In fact, almost surely the trajectory will return to A infinitely many times. In probabilistic terms, the probability of not returning is zero and the return is endless.

The theorem does not say how many iterations of s are required or how long we have to wait before a trajectory of a state from A will return to A. Nor does it give us any way to distinguish which states might return and which ones will never return, though it does say that it is with probability zero that a random choice from A will be a non-returning state. As we can see, this is a *purely qualitative* result. It is these qualitative insights, which are derived as genuine mathematical statements, that allow us to connect intuition with formal mathematics, bringing meaning to both.

Theorem 7.9.1 *[The Poincaré recursion theorem] Let $\mathscr{X} = (X, s, \mathscr{E}, \mathbf{p})$ be a discrete iterative pattern system where $\mathbf{p}$ is an invariant probability measure, and let A be any event, i.e. $A \in \mathscr{E}$. Then the trajectories of almost all the states that start from A will return infinitely often to A.*

There are very few proofs of theorems in this book. Proofs don't make for easy reading and are slow to digest. However, the actual proof of this result is remarkably short and does not use anything except the basic definitions we have already introduced. Since the result is so famous, so qualitative rather than quantitative, and since it brings together all of the concepts that are relevant to our quest for understanding of pattern, we have included the standard proof of the theorem as an endnote. Of course it is okay to skip the proof. But if you do follow it, you can truly feel a wonderful sense of what you have accomplished in reading this far into the book. This is a famous result and to follow its proof is a significant achievement for anyone who has not had a lot of mathematical training. You may also find that there is something truly wondrous in this proof—the beauty of all these ideas dovetailing so exquisitely without one of them failing to be used to its full capacity.[19]

7.10 Conclusion

Everyone is already a fluent expert in the world of pattern: we all have the right intuitions built in. Formalizing pattern in terms of systems, bringing together various aspects of context, change, likeness (qualification), and measure (quantification), allows us to treat them as integrated unities. Pattern systems can interact with each other, be different yet alike, and can be models for transporting ideas from one setting to another. Since they are based on the fundamental principles of difference and change, they have a certain universal nature to them. Perhaps most importantly, once formalized they offer new insights into the world we see around us. While Ch.6 was about the formalization itself, this chapter has been about assimilating this new perspective.

Central to this has been the introduction of two new discrete iterative pattern systems, the doubling and the logistic systems, showing how they relate to each other and to the familiar zero-one shift system. The doubling system turns out to be effectively the very same pattern system as the zero-one shift system, and in understanding that we also see what underlies the remarkable way in which we connect points on a line interval to numbers in our number system (the decimal system, or here, the binary system). See also ‡Ch.16.

The logistic system, a sort of wild model of population dynamics, starts as a mystery. Just looking at trajectories of points, even though they are completely determined by iterating a very simple formula, leaves us bewildered as to what pattern can possibly lie behind such chaotic behavior. Yet there is a very definite patterning going on, one that is revealed by the evolution of the dynamics, even if it is hard to quantify. The surprising revelation is that the dynamics of this system is actually a perfect shadow of the dynamics of the doubling system. Eventually we find that the ordinary probability measure that is by now so familiar in the shift system actually transfers across to both the doubling system and to the logistic system. These are insights that would be hard, if not impossible, to infer without the benefit of the abstract formalization.

A second important concept raised in this chapter is the idea of moving from pattern systems where the dynamics rules at the level of states, to one where the dynamics is understood on the basis of changing events. This not only simplifies things, but seems to correspond with the reality of how we actually experience the world. What we witness is not the intricate dynamics, but the passage of events that we can actually physically experience. In the end we are able to see that event-based pattern systems are in reality just versions of shift systems, thus showing the universal nature of shift system in the theory of discrete pattern systems.

In the case of trying to discretize continuous pattern systems, which is important for any computer modeling, there is an interesting question of whether event-based patterning should be taken on the basis of equal steps in time or on the basis of the changing of events. It comes down to the philosophical question of whether our ideas of time are based on the changing of events or on some fundamental continuum of flow that exists outside the human mind.

The final part of the chapter is our first look at the interesting philosophical idea of return. The idea of return is an important and recurring theme in classical Chinese thought where it can be understood in the context of emergence of form and its return into emptiness (lack of form). The universality of this is something that we all come to learn, often to our deep regret. It is an overarching theme of Chinese poetry.

In our conception of pattern we have often noted the importance of recognition as a foundational idea in pattern, and recognition in effect implies return, a return to something experienced and remembered from before. Recognition/return is not explicitly built into our idea of a pattern system, but it is a mathematical consequence of those ideas. This fact is the famous Poincaré recurrence theorem, which is a basic theorem of the theory of dynamical systems. It says that if an event occurs, then with almost surety it will occur again (and extrapolating from this, again and again and again).

This chapter begins with some ideas about how the mathematical world deals with the intangible quality of the infinite. It begins with something simpler: the notion of cardinality. We are so familiar with counting that we rarely think about what lies behind our qualitative idea of numerical quantity. Once cardinality is seen for what it is—a quality based on one-to-one correspondence—then it is possible to explore cardinality at the level of infinite sets. Our main objective here is to show that between the two infinite sets with which we are very familiar, the set of positive integers and intervals of real numbers (say the interval between 0 and 1) there is an amazing difference in cardinality. The latter is somehow incomparably larger than the other. Just how this comes about is both a rather enjoyable story in itself and a source of some of the great paradoxes of the infinite.

This difference between countable sets (the ones that can be listed in the sense of being put in one-to-one correspondence with the positive integers) and uncountable sets (the ones that cannot) is one that we often have occasion to point out as we go on. A discrete iterative dynamical system based on the state space $[0, 1]$ (e.g. the doubling or the logistic system) reveals this in a very stark way. Every trajectory of states in the system must be either finite or countably infinite, but the state space itself is uncountable. This means that individual trajectories come nowhere close to passing through all the states of the system. This actually brings up a serious scientific issue. If observation of a discrete system can come nowhere near seeing all the states of the system, how can we expect to deduce genuine understanding of the system through observation? We will come back to this and give an important insight into this in Ch.8.

CHAPTER 8

Entropy and synthesis

8.1 The question of synthesis

When we (the authors) were young students the fields of geology, biology, chemistry, physics, mathematics, and neuroscience, and even their various sub-disciplines, were typically separated and jealously guarded compartments of study. Perhaps the greatest change that we have witnessed in our lifetimes has been the dissolution of these unnatural borders with the increasing appreciation of Nature as an immense and intricately interwoven unity of which we are a part and in which no part exists or functions as an isolated entity. The familiar criticisms of compartmentalism and extreme specialization still arise, though we would know little were it not for the lifetimes people have dedicated to understanding details that underlie the physical and biological world. What is new is the emergence of a holistic story of how the world we live in came to be, how and why it has changed and will continue to change, and the seriousness of the complex questions whirling around our human experience and our synthesis into the extraordinary mosaic of life on this very finite planet. Important for us is that it is a *system-based view.*

Strangely, this conceptual view—a systems view of reality, and in particular a systems view of life—was not a central concept in the western development of science or philosophy. From a creationist point of view there is no compelling reason to see the living world as self-organizing, self-creative, and self-sustaining. Although there have been constant reminders through writers like Rachel Carson, Loren Eiseley, Jane Goodall, Robinson Jeffers, or Gregory Bateson, it is only relatively recently that we have had such a deeply instructive exposition of the systems view as that presented in Capra and Luisi's *The Systems View of Life* [29]. We are slowly (very slowly) reawakening to what is in actuality an ancient truth. It was completely obvious to the indigenous people around the world that they lived in a deeply interconnected and deeply interdependent world. Before the domestication of animals and development of agriculture, our dependency as creatures of Nature was clear. Nowadays, for someone growing up in a city, Nature can be seen as something as either convenient or inconvenient according to the circumstances, but not something of which they are a part.

Although it is not expressed directly as such, the systems point of view has been a central feature of Asian thought for millennia, particularly both in Daoist and Buddhist thought. In this chapter we want to explore the systems idea in the context of pattern and patterning. Our own language here is largely scientific, but we also want to include a few passages from ancient literature as well. They may not be "scientific" in a modern sense, but there is an immense depth of thought and insight there that correlates well with the systems view and is increasingly relevant to the predicaments of the modern world.

From our perspective of patterning, the dual aspects of analysis and synthesis of pattern systems are a natural and inevitable pair. Of the two, analysis of the parts of a divided whole has always been the easier side to follow. It's easier to break things into parts than to put them together again. There is truth to the familiar statement that the whole is more than the parts. In fact the whole is often something *quite different* from the sum of its parts. But this raises questions about what we even mean by *synthesis*. In what ways might we express a concept like synthesis that would match an encompassing view of pattern and patterning that we have been developing?

The purpose of this chapter is to suggest that the synthesis (or sometimes we will say, integration) of pattern systems is connected with the concept of entropy. It is true that entropy is an elusive entity, hard to define and carrying with it connotations of disorder and decay. But as with any concept, it arises along with its opposite, and that suggests that entropy must also be about order. The idea of synthesis of pattern systems would seem to suggest that when put together their totality has a form of order that was not there before, and at their disintegration there would be a corresponding increase in disorder—hence an increase in entropy. This is the idea that we want to follow.

The importance of entropy in relating dynamical systems and in pattern recognition is nothing new to this book, see for instance [117]. However, it stands here as a central feature of our conception of pattern and patterning. A good part of this chapter is devoted to getting a better understanding of entropy and then seeing how it enters into the idea of the synthesis of pattern systems. Along the way various philosophical questions inevitably arise and, as we have said, many of these go well beyond mathematics, or even beyond the usual bounds of science itself. This is a place where we can see possibilities for exchange between mathematical/scientific ideas and more general ideas that we have about the world.

The chapter ends with a striking example of this interchange which presents two very different ways of seeing as being essentially equivalent: one based on following trajectories of a pattern system and the other based on the totality of the system itself—

the Birkhoff ergodic theorem. There is a clear metaphorical interpretation of this result that we can see as validating our deep-seated faith in the principle of induction—the inference of pattern though repeated observation.

8.2 Systems

8.2.1 Origins of systems thinking

We have adopted the name *pattern system* to describe what we think is the essence of pattern. The Merriam–Webster dictionary defines a *system* as a regularly interacting or interdependent group of items forming a unified whole. This definition brings three aspects together: entities that can be articulated, dynamical interaction of these entities, and coherence in the resulting process. It is neither purely about things nor purely about process and change. It is about these seen as a flowing stream that generates its own sense of unity. The most significant point of our approach to pattern is that it is a *systemic* idea. Pattern is a physical or conceptual unity that is system based, hence the name *pattern system.*

In their book *A systems view of life*, Capra and Luisi, [29], suggest that there is an ongoing tension in Western thought:

> From the systems point of view, the understanding of life begins with the understanding of pattern. [As we have said,] there has been a tension between two perspectives—the study of matter and the study of form—throughout the history of Western science and philosophy. The study of matter begins with the question, "What is it made of?"; the study of form asks, "What is its pattern?" Those are two very different approaches, which have been in competition with one another throughout our scientific and philosophical tradition.

To what extent this tension has been explicitly about systems, as such, is not clear, but we agree totally that the understanding of life, and Nature itself, necessarily has to be systemic in nature.

The systems idea is certainly not unique to Western thought. From classical times Chinese philosophers held a systems view of Nature, even if they did not explicitly say so. The designation for the entire world was traditionally designated as tiandi, 天地. That is commonly translated as "heaven and earth" though there is no "and" in the term. The combination itself is significant. In the *Yijing*, heaven is given the name *Qian* and earth is given the name *Kun.* Qian-Heaven is said to initiate while Kun-Earth nurtures. Neither can complete the entire world alone. Together, the two form a system.

In the commentary for Kun, the *Yijing* states:

Perfect is Kun's greatness.
It brings forth all things
And accepts the source from Qian.

Earth brings forth all things after accepting what Qian initiates. Heaven and earth are co-equal: each needs the other. The passage goes on:

Responding in its richness sustains all beings;
Its virtue is harmony without limit.
Its capacity is wide, its brightness is great.
Through it all beings attain their full development.

Qian is the source, but Kun fulfills. The passage goes on:

A mare is a creature of earthly kind.
Its moving on Earth is boundless,
Yielding and submissive, advantageous and steadfast.

The theme is one of development, movement, and harmony, and an implied sense of a *self-generating* and *self-sustaining system*. The mare represents life and its potential for freedom and expression. In the *Daodejing* this self-generating idea appears as a specific concept that relates to Nature: *ziran*, 自然. These two characters literally mean "self-so"—"occurrence by itself"—or "self-generating". Ziran appears in the famous verse of Ch. 25:

People accord with Earth
Earth accords with Heaven
Heaven accords with Dao
Dao accords with 自然.

We leave the last word untranslated. It's a good example of the difficulties of translation. It would be misleading to translate the last line as "Dao accords with Nature", for that would set Nature apart as an entity separate from Dao. Dao is understood as the dynamic of the total system of Heaven, Earth, and life. The line is usually understood as reading "Dao accords with itself", or "Dao follows itself". Another interpretation:"cycles Dao back to its processes of becoming" [135, Ch.25].

And the moving Way is following
The self-momentum of all becoming,

and thus avoids the temptation of objectivizing Dao as a separate metaphysical entity.

There is no hint here of a *Cartesian view*, where mind is totally separated from the corporal body and all animals (save for humans, of course) are mere automatons. Nor is there any sense of *vitalism*, with its beliefs in a fundamental divide between life and the material world. It is about form, and form in a very dynamical sense. Although the Chinese philosophers did not talk of systems as such, they recognized the principle. Indeed they imagined the human form as a microcosmic expression of the cosmic whole and so came to see the human body and mind also as a holistic system. This concept underlies traditional Chinese medicine and increasingly is found in integrative medicine in modern times.

8.2.2 Relevance of the Daoist viewpoint

Things That Do Not Linger for a Moment:
A boat with hoisted sails.
People's age.
The four seasons.

Shi Shōnagon

There is a significant divide in the world of physics that became particularly clear with the creation of quantum theory. Generally before that time physicists were confident that they were describing reality as it *really is*. Physics aimed at objective knowledge of absolutes. But quantum theory, especially as framed by Neils Bohr and Heisenberg, suggested that the human observer is always a part of any description of Nature that we might make and reality cannot be articulated in any absolute sense. The famous mathematical physicist Eugene Wigner said, "Physics does not describe nature. Physics describes regularities among events and *only* regularities among events"[50].

We will take this up again in Ch.13, but it is a relevant question to keep in mind. Is science the search for absolutes or is it the framing of Nature through models that are simply instrumental, in the sense that they work? Einstein wanted absolutes that were not limited by human conceptions; Bohr thought that was not possible.

It is safe to say that neither of these two can be "proven" correct or false: they represent different philosophical positions. What is clear though is that the Western tradition has greatly favored absolutes and the capacity of human beings to articulate them. This philosophical position was the great driver of the scientific revolution. What is also clear is that Eastern philosophies, at least those deriving out of Chinese thought, claim that any articulation of the nature of reality is necessarily provisional and temporary. Nature can be spoken of, but never in some absolute way. The famous opening words of the *Daodejing* read

> Dao can be spoken of,
> but not enduring Dao.
> 道可道非常道[1]

As it stands now we seem to be at a crossroads where science more and more seems to tend towards the Eastern view while the great majority of the public strongly favors the old tradition on which our Judeo-Christian-Islamic cultures are based. Not only is that tradition strongly attached to absolutes, but it is also strongly teleological —assuming that the evolution of Universe and life within it is directed towards something higher. (These days higher consciousness is popular).

This book is evidently more committed to the Eastern view, and we want to emphasize (perhaps over-emphasize) it because there is a fairly strong feeling in academic philosophy of the West that there is really nothing to learn from the philosophy of the East. But there is, and it's deep and subtle, for it is a philosophy in which there are no absolutes to fall back on, change is not teleological, and human beings co-exist with all other living forms, all of which can experience the world in their own unique ways.

In the long run these two views—a world designed for humans and human destiny versus a world in which all beings, living or otherwise, are passing manifestations of a dynamic and fecund unity—will lead to very different outcomes for our planet.

In this book we quote the *Daodejing* quite often (for a list of the chapters or verses quoted, see the Index). As a work, the so-called received version of the *Daodejing* is a compilation of an ontological position and associated wisdom that represents a school of thought dating from at least the fourth century BCE and, as we have seen, originating many centuries earlier. Though impossible to translate its title uniquely, we can think of it as *The classic of Dao and its potential for manifestation.*

The excerpt in §7.5.2 from Ames' and Hall's *A Philosophical Translation of the Daodejing* echoes the thoughts of Capra and Luisi quoted above, offering an interesting insight into the differences between the philosophical insights of the European tradition in Greek thought and the insights of the Asian tradition in Chinese thought.

The latter present the systems point of view: things are not things in a substantial sense; they are a part of the flow. They are persistencies, or we might say local invariances that survive over sequences of events, long enough to become "things" within a particular context. Ames and Hall continue:

> [A consequence of] this acknowledgement of the reality of both change and the uniqueness that follows from it is that particular "things" are in fact processual events, and are thus intrinsically related to other "things" that provide them context.

The outcome of the Chinese understanding was that change is the way of Nature and that the universe is timeless, neither externally constructed nor internally directed towards some end. What humanity, as all life, has to do is to learn from it and abide within it.[2]

Such views are not uniquely Chinese, of course. Epicurism dismisses any teleological explanation of things, see for instance Lucretius [108, pp. 4.823–4.857], Roger Bacon cautioned against it, Kant and Hume had serious doubts, Thomas Huxley and Richard Dawkins rejected it, and Steinbeck makes a long case against it [160, Ch. 14]. The whole concept of our book is that there is an over-arching unity to the Cosmos within which change is fundamental and systemic, and patterning its manifestations. The fact that we have been drawn to such an understanding and can explicate it in a formal way is an outcome of the vast enterprises of science and mathematics over the course of centuries. We have come to understand that the Cosmos is not directed towards some expressible end, that law and chance are the perpetual dance of creation and annihilation, and that in spite of the narrowness of our perceptions, this all takes place within an extraordinary unity in which things only exist in the dynamical context of everything else.

Perhaps most convincingly of all we can see that these conclusions have been impressed upon us by sheer evidence. These were not ones to which science and mathematics were particularly directed; in fact quite the opposite. Chance and law are far more subtle than we tend to think. Later in §8.5.4 and in §14.5 we will see how both mathematics and quantum theory can offer us deeper insights into what they mean.

8.2.3 Consilience and cybernetics

The Chinese did not develop the systems idea in our contemporary scientific sense. That was to emerge slowly and it is not until recent times that its importance was fully recognized. The history of this development is told more fully in Norbert Wiener's *Cybernetics*[189] and the aforementioned book of Capra and Luisi [29]. Still, we can see the stirring of this sort of thinking in the ideas of the influential and remarkable scientist William Whewell (1794–1866). Any close observation of the natural world, no matter at what scale, reveals the intricate and exquisite synthesis of many forms and processes. Often quite different in nature, nonetheless, together, they could be seen to manifest coherent wholeness. To Whewell this convergence of form, which was increasingly being revealed by science, was a source of wonder. He called it *consilience.* To Whewell, who coined this word in 1840, consilience is what gives us confidence in the truths suggested by science. Newton's physics was his exemplar of consilience, with its continuity of principle across wide domains of space and time.[3]

We pick up the story again with an explosion of thought after the second world war that came to be called "cybernetics". Norbert Wiener, who was one of the founders and actually coined the word, describes it in the following way:

> Thus, as far back as four years ago [1947], the group of scientists about Dr. Rosenblueth and myself had already become aware of the essential unity of the set of problems centering about communication, control, and statistical

Figure 8.1: The governor mechanism of a Boulton and Watt steam engine of 1788 is a classic feedback system. The engine spins the mechanism and the two balls. Increasing speed means increasing centrifugal force, and that spreads the balls apart. In turn this mechanically produces feedback to the engine slowing it down, thus reducing the centrifugal force. In effect the separation of the two balls is governed by a limit cycle, see §7.8.2. Dr. Mirko Junge, Science Museum, London; photo by CC BY 3.0 Deed.

> mechanics, whether in machine or in living tissue. On the other hand, we were seriously hampered by the lack of unity concerning these problems, and by the absence of any common terminology, or even a single name for this field. ...We have decided to call the entire field of control and communication theory, whether in the machine or in the animal, by the name *cybernetics*, which we form from the Greek kybernetes or steersman [189].

The new field gathered a number of the luminaries of the time, among them Norbert Wiener, John von Neumann, Claude Shannon, W. S. McCulloch, Walter Pitts, and later Gregory Bateson, and Margaret Mead. The underlying theme was the principle through which mechanical and living systems provided their own feedback and so their own forms of *self-governance.* Certainly, as we can read, the initial impetus was more on the mechanical side of control and self-controlling systems, with the biological side being drawn in later. The idea of self-governance was epitomized by the governor mechanism that was invented by Christiaan Huygens and later used by James Watt to keep steam engines running at a constant speed, see Fig.8.1.

The cybernetics movement was contemporary with the beginning of the new age of electronic computers. It was an insight of von Neumann that the instructions which a computer executes should co-exist within the computer's working memory, thereby allowing it the ability to alter its own code during execution. Along with this there has been increasing understanding of biological processes and the functioning of the brain, with their own abilities of self-organization. It became clear that it was not just self-governance of systems that was important, but their potential for *self-organization* and *self-emergent* pattern. This was eventually to transform cybernetics into what is now called *information theory* and *complexity theory* [29].[4]

For us in the twenty-first century, the unification of science across the domains of physics, chemistry, biology, genetics, evolution, climate, geology, ecology, neuroscience, psychology, and the social sciences is truly impressive. And if our thinking is correct, the ideas of pattern and patterning that we are discussing are a form of a consilience that can link these facets of Nature and the mind with corresponding advances in mathematics.

8.2.4 Pattern systems as systems

This raises the natural question of what is involved in trying to understand a pattern system. It would seem that very often we are quite aware that we are in the midst or presence of some system, but at the same time we know that we do not understand it. In ancient days, early humans were certainly aware of the great cycles of day and night, the phases of the Moon, and the passing of the seasons. The whole was the totality of their experience. What was necessary was to create conceptual models to explain it, to act in harmony with it, and to somehow control it. In a more modern context, the situation with COVID-19 has been a good example. The cycle of infection, sickness, recovery or death, and then re-infection, along with its potential to create variants, overwhelm hospital systems, and upset the global economy, is a very complicated and hazardous *system*. We have an instinctive understanding that there is a system, a whole, and an instinctive feeling that if we are to live and survive with it we need to reduce it into simpler components and their interactions. In the case of COVID this has included understanding the methods of infection, learning ways in which the disease can be treated, uncovering the biological structure of the virus itself and its potential for change, learning how it can be prevented from reproducing itself, devising hospital and societal protocols, and so on. The difficulties in this were all too obvious.

The natural process of conceptualizing a new pattern system is one of first *perceiving* it as a system. Jan Zwicky explains it in terms of gestalts:

> Understanding of a pattern begins with the perception of, or a heightening of attention to, what appears to be rhythm or rhyme—that is, the perception of structural echoes. After this comes the coalescence of a gestalt, the sudden arrival of an image of a whole with interacting aspects. Ultimately, we perceive pattern as the physical manifestations of a system. This is not simply dividing into simpler parts, but being compelled by the sense that there must be an integrative unity. There ought to be a gestalt [203].

It is penetration into the interior structure that can precipitate insight into a gestalt. In her book *The experience of meaning* [204] she quotes Konrad Lorenz (1903–1989):[5]

> I strongly suspect that, at the time when a set of phenomena seriously begins to fascinate me, my Gestalt perception has already achieved its crucial function and "suspected" an interesting lawfulness in that particular bunch of sensory data .... [I]ncreased observation accelerates the input of sensory data until, when a sufficient redundancy is achieved, the consciously perceived lawfulness detaches itself from the background of accidentals, an event which is accompanied by a very characteristic experience of relief ..., the sigh: 'Aha!' .... [I]f our conscious effort at cognition really had to start at the level of miscellaneous, unprocessed sensory data ...[i]nductive procedure would, I think, be impossible.

Though Lorentz speaks from a rather personal point of view, we expect that this is how it is in general. There is a *creative cognitive* process. Perceived features begin to be seen as manifestations of a previously unrecognized system. A creative act of the mind brings unity and understanding where previously there was only confusion and disorder. There is no guarantee, of course, that our conceptual creation of a model is in some way an absolute aspect of reality. In fact, it is most assuredly not. As Heisenberg cautioned, "What we observe is not nature itself, but nature exposed to our method of questioning." Our models reflect our *human* nature and concerns.

Think of the explanation of the seasons of the year. Certainly by the second century CE Claudius Ptolemaeus (Ptolemy) (c. 100–170), basing his astronomy on centuries of Babylonian observation, presented an excellent mathematical model to explain the seasons in his famous book *The Almagest*. His model was incorrect on fundamental grounds (Earth centered with everything moving in circles and epicycles), but he really did understand the seasons in terms of the tilting of the Earth's axis of rotation with respect to the plane of the ecliptic, and his understanding was enlightening—a genuine "aha" moment.[6]

Summarizing we have the important conclusion:

> Although the world impresses itself upon us in physical ways,
> its realization as an integrated unity, a system, is a feature of the mind.
> Systems are mental constructions.

This creative cognitive process is only as good as it faithfully models future continuing experiences. It can never be taken as absolute.

8.2.5 Subsystems and supersystems

Perception of some patterning as a system often reveals that the interacting parts are subsystems themselves. For instance, attempting to understand the brain leads us to perceive it at many levels. At the level of neurons, which are certainly basic neural entities, we find ourselves at the level of living cells that are entire systems in their own right. Within these cells we see other systems, like the complex sodium-potassium ion exchange that lies behind the electro-chemical process by which neurons function as neurons. There are countless subsystems in a living cell, and within them sub-subsystems. No conceptual model is likely to grasp the entirety of what makes it what it is, but neither is that what we necessarily want to know. The level of distinction that we take is relative to the gestalt that we experience and wish to elucidate.

> Great fleas have little fleas upon their backs to bite 'em,
> And little fleas have lesser fleas, and so ad infinitum.
> And the great fleas themselves in turn have greater fleas to go on;
> While these again have greater still, and greater still, and so on.
>
> Jonathan Swift and Augustus de Morgan

As this little piece of doggerel suggests, the nesting of systems within systems naturally leads to questions in the reverse direction. How do systems interact to become larger systems, *super-systems*? This brings us back to our original question: what does it mean for systems to become parts of a larger integrated whole, even a whole that is something beyond what seems to be implied by the parts? This is where entropy enters.

8.3 Information and entropy for finite partitions

8.3.1 First ideas about entropy

Entropy is a word that most of us have heard about, mostly in negative terms, but find very hard to define. It brings to mind disorder: entropy is said to be always increasing. Entropy points to the natural processes that we see around us all the time where things left to themselves seem to fall apart, wither away, and become progressively disordered. The second law of thermodynamics says that in a *closed* system entropy can only increase.

Information is a far more familiar word, mostly with positive implications, but just as hard to define. We usually use it in the sense that it brings us new knowledge about the world or circumstances in the world. But it is complex, because it is highly dependent on context. For instance "2 4 6 8 " is a piece of information if we are told that it is the address of our friend's house on Main St., a different piece of information if it is the key code for a lock, and different yet again if is telling us how the first four even numbers are written in the Hindu-Arabic system of numbering. Without a context, what can we make of it? Information cannot be detached from some context by which it becomes information. The surprising thing is that, at least in the domain of science, entropy and information are the flip sides of the same coin.

The scientific ideas of entropy date back to the late nineteenth century, mainly to the work of Rudolph Clausius (1822–1888) on the theory of heat. This was the great age of steam engines, and the physics of heat was central to understanding the way in which they worked. It was the current hypothesis that heat is a material substance (called *caloric*), and it had been argued that if the body of the working substance, such as a body of steam, is returned to its original state at the end of a complete engine cycle, no change had occurred in the condition of the working body [53]. Clausius argued otherwise and insisted that there was change, and this change was a measurable quantity for which he coined the word *entropy*. He derived this word from the Greek εν τροπια (*en trope* in roman letters) = "in transformation", or more usefully, "transformation content".

Later, once it began to be understood that heat was not a material thing but actually an indicator of change (the movement of atoms and molecules), attention moved from heat engines to understanding the emergence of the macroscopic features of physical things around us from the movements and interactions of the countless microscopic entities of which they are composed. Through the insights of such famous physicists as Ludwig Boltzmann (1844–1906, Austrian), Josiah Willard Gibbs (1839–1903, American), and James Clerk Maxwell (1831–1879, Scottish), the same idea of entropy appeared, but now as a foundational concept in statistical mechanics. It is this version, which directly links to probability and the distribution of events, that lies at the heart of entropy as we shall see it in pattern systems. For us, entropy will be used as a way of indicating the degree of order/disorder in a system, or the degree to which two or more pattern systems can be viewed as being integrated into a larger pattern system.

The modern scientific conception of information as a fully developed idea is due to Claude Shannon. The basis for this has already been laid out in Ch.4 and our discussion of shift systems as models of communication systems. Shannon was certainly aware of the entropy as established by Boltzmann, and perhaps also of the work of Leo Szilard (1898–1964, Hungarian-American) who had introduced the same unit of information in a paper of 1929 while studying paradoxes around the undoing of the doing of entropy.[7] In Shannon's paper, entropy and information are fully understood as being flip sides of the same coin, and importantly for us, he established these concepts in a way that easily extends to the theory of discrete dynamical systems, and from there to our pattern systems.

We will go into more details of this in what follows, but we can explain the general idea in a few lines. Think of a pattern system consisting of a state space, some dynamics, a partition (leading eventually to an event space), and a measure. If we think of the partition as our basis of distinguishability of states and if we think of the measure as defining the relative weights that we give to these classes of distinguishability then the entropy/information is a single number that summarizes the distribution of this distinguishability. This can be interpreted as uncertainty/certainty over the potential outcomes of the events described by the partition. More importantly here, it can be taken as a numerical indicator of the internal disorder/order of the system.

Initially we will just take the partition and ignore the dynamics. When we include the dynamics, which here we take to be simple iterative dynamics, we can take into account the increasingly complex levels of distinguishability that emerge from the evolution of the system, that is, the event space. This leads to refined versions of entropy, and finally a limiting version that we can consider as taking account of the distribution of disorder/order of the entire pattern system.

In the end, whether we look at it dynamically or not, the entropy of a pattern system is just a number, always greater than or equal to zero, that expresses the average uncertainty associated with the system. Information is its opposite—the removal of uncertainty—and is given the very same numerical value as entropy. Depending on the situation, it is sometimes more useful to take information with opposite sign to entropy, and it is spoken of as *negentropy*.

In a sense, entropy is like our concept of temperature, which produces one number to indicate the average movement of a vast collection of atomic particles. A single number does not sound like much, but entropy is important and we will use it as the basis for defining the consolidation or synthesis that occurs when two or more different pattern systems are joined together into a larger integrated system.

Two pattern systems will each have their own entropy, but if they are configured so as to interact the new combined system will have its own entropy. The maximum it can be is the sum of the entropies of the two parts, and this amounts to saying that the two systems are really behaving independently of each other. There is no genuine synthesis. Anything less than this maximum value may be considered as indicating that the two systems, now seen as a coupled pair, no longer behave as separated entities. Entropy has decreased, the amount of order has increased. This is what we will call *synthesis*, and it is the first step in understanding what it means to combine pattern systems into new pattern systems.

This is the story we want to tell in this section. There are two sides to it which we should separate, at least initially. There is a conceptual side, which is what we really care about, and a technical side, which defines what is going on. The technical side is vital but should not override the intuition. We keep the technical side rather limited in the main text, and leave details to the endnotes.

8.3.2 The basis of entropy

In understanding the way in which entropy and information are measured in pattern systems, it is conceptually easiest to start from the point of view of information. It is also easiest to start without dealing with the dynamical side of things. That will come later.

So, just as we have done with pattern, we begin in the setting of the very simplest sort of information: the distinction between two things. And just as we have done before, we take our model to be the pair $\{\mathbf{0}, \mathbf{1}\}$. (In this section we shall write the zero-one elements of the state space in boldface type $\mathbf{0}/\mathbf{1}$ in order to keep them clearly distinguishable from the numbers $0/1$ that also appear as probabilities.) We are not dealing with a shift system here—just the simplest level of difference possible: yes/no, yin-yang. The only thing we can know about a state taken from this space is whether it is $\mathbf{0}$ or $\mathbf{1}$. In the common parlance of today, we say that there is *one binary bit* of information available here. In finding out whether the state is $\mathbf{0}$ or $\mathbf{1}$, we have learned one bit of information. The binary bit is the basic unit of information, distinguishing one thing from a pair of things.

Two binary bits can distinguish two pairs of things, thus distinguishing the states of the larger state space $\{\mathbf{00}, \mathbf{01}, \mathbf{10}, \mathbf{11}\}$. So two binary bits can distinguish four things, and following on this three binary bits can distinguish eight things. Quite generally n binary bits can distinguish 2^n things. The basic measure of information is binary bits, and the basic relation is

n binary bits can articulate 2^n states.

This is the first indication that information involves logarithms. Just as

$$n \mapsto 2^n$$

is the process of exponentiation, its inverse

$$2^n \mapsto n \tag{8.1}$$

is the process of taking logarithms (base 2). The information that is available in a state space with 2^n states is $\log(2^n) = n$ bits.

People love to talk about exponential growth these days, even though all they mean is that something is growing very quickly. Literally exponential growth means that things are growing according to some *power law*: in our case, expressed in terms of the number of bits n, the growth is 2^n. The logarithm says what the power (or *exponent*) is, i.e. n. This is the underlying idea behind entropy/information—a measure of the complexity of the system in terms of the number of parts into which it is partitioned. As the number of parts grow, $2, 4, 8, 16, \ldots$, the entropy grows as their logarithms, $1, 2, 3, 4, \ldots$, which are the number of bits involved in distinguishing these parts. When we add in the dynamics, later on in §8.3.3, this will become more apparent.

This way of looking at things leads to what seems like a correct intuition about how information should work. If we have two state spaces X and Y, which have say 8 and 32 elements respectively, and put them together so that we look at all possible pairs (x, y) with $x \in X$ and $y \in Y$ (the combined state space $X \times Y$), we have altogether $8 \times 32 = 256 =$

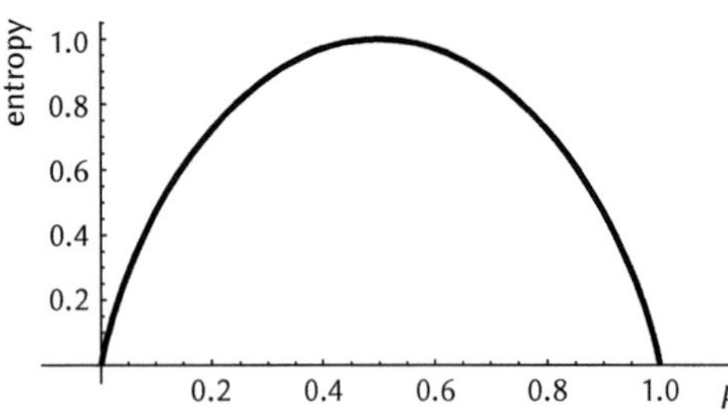

Figure 8.2: With probabilities p for **0** and $1 - p$ for **1**, the information/entropy of the state space $X = \{\mathbf{0}, \mathbf{1}\}$ varies continuously in p. It is 0 at both ends (no uncertainty) and is 1 bit at $p = 1/2$ where there is the maximum uncertainty/information. Written in the form of (8.2), this is a plot of the function $-(p\log(p) + (1-p)\log(1-p))$.

2^8 states. Now according to our discussion above, the information embedded in X is 3 bits, that in Y is 5 bits, and that in $X \times Y$ is $8 = 3 + 5$ bits. In other words, the information available in joining the two systems together is the sum of the information available from each system separately. It makes sense: we have three bits of information and five bits of information, so altogether eight bits of information. This is the characteristic feature of logarithms: they replace multiplication by addition.

Looking at this same thing from the point of view of entropy, being told that a state is coming from X and nothing else is total lack of information about what the state is (other than it comes from X) and so leaves an uncertainty of 3 bits. If lack of information is considered to be confusion or disorder, then that disorder amounts to 3 bits. Again, the disorder in $X \times Y$ is the sum of the disorder of X and the disorder of Y, which in this example is 8 bits.

However, this is not the end of the story. Let us return to our simplest state space $\{\mathbf{0}, \mathbf{1}\}$. Suppose now that there is a probability associated with these two states, say p and $1 - p$ respectively, *and we know this in advance.* This affects our knowledge of the system, and so the amount of information that we can derive from it. If **1** has very little chance of appearing, then the system is not as unknown as it might be. For instance, suppose that $p = 1$, so **1** has no probability of occurring. Then the system is not really capable of making distinctions. Given a state from it, we are certain that it will be **0** and we gain no information from it. We have no doubts about what to expect. We get the most information and the most amount of uncertainty by having the probabilities of **0** and **1** be equal. Entropy/information is defined so as to take into account the relative weightings or probabilities assigned to the various parts of the partition. In fact the entropy is completely defined by knowing the partition and the relative weights associated to its parts.

The exact description is given by the formula of (8.2), which we will come to below. Fig.8.2 shows how the entropy of the simple $\{\mathbf{0}, \mathbf{1}\}$-system varies as the probability p of **0** occurring varies from 0 to 1. Not surprisingly the entropy is symmetric (**0** and **1** play symmetrical roles), takes the value 0 at the extremes when $p = 0$ or $p = 1$, and is maximum when $p = 1/2$, where see that it is 1 (meaning 1 bit).

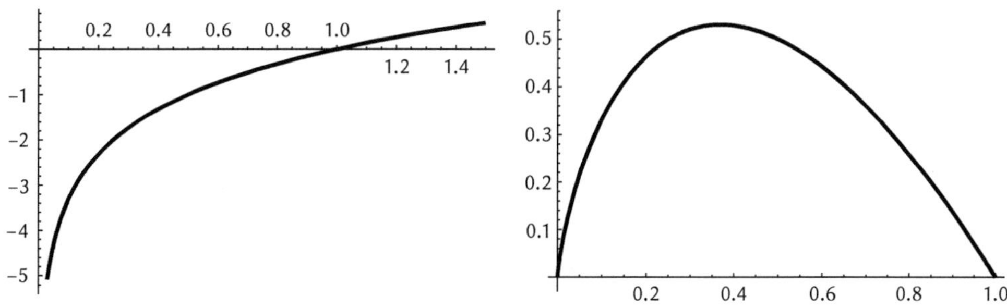

Figure 8.3: On the left is a plot $\log(x)$, showing that in the range $0 < x < 1$ it is negative. Here the logarithm is taken base 2 which is the natural one in the context of information. On the right is a plot of the function $-x\log(x)$ over the range $0 \leq x \leq 1$. This is the function which underlies the definition of information and entropy.

The formula for entropy

Although we do not need to deal with entropy in any technical detail, it is still valuable to see what the formula for entropy looks like and get a feel for how it actually produces numbers. Entropy involves the logarithm function, and its defining relation is given in the expression (8.1). There we have only used values for n that are positive integers, but this relation defines the logarithm for any real number n. Remembering that $2^{-n} = 1/(2^n)$, with $n = 0, -1, -2, -3, \ldots$ we have

$$2^n = 1,\ 1/2,\ 1/4,\ 1/8, \ldots ,$$

so the logarithms (base 2) are

$$\log(1) = 0,\ \log(1/2) = -1,\ \log(1/4) = -2,\ \log(1/8) = -3, \ldots .$$

In mathematics, log is defined for all positive numbers, see Fig.8.3 and also the Endnotes[8].

Notice that log gives *negative* values to all numbers between 0 and 1. This is important because we will be applying the logarithm function to the values of probability of the parts of a partition, and these probabilities are always in this range, i.e., the logarithms are negative or at most equal to zero (or more succinctly, non-positive). The reason for the prominent minus sign in formulas for entropy is to make the entropy come out to be positive or at least equal to zero (non-negative).

At this point we can look at pattern systems in general. Ignoring the dynamics, we have a pattern system $\mathscr{X} = (X, \mathscr{P}, \mathbf{p})$ consisting of a state space X (all the potential states that the system can be in), a partition $\mathscr{P}$ (the groupings together of various states indicating the level at which we distinguish different states), and a probability measure $\mathbf{p}$ (that gives the expected relative expectations associated with the various parts of the partition $\mathscr{P}$). There are just two things that enter into the definition of the entropy of this system: the partition and the measure. Here is the key definition: for each part P of the partition $\mathscr{P}$

the information attributable to P is defined to be $-\log(\mathbf{p}(P))$ bits.

Remember that the minus sign is there to make the value of $-\log(\mathbf{p}(P))$ be *non-negative.*

Now if $\mathscr{P} = \{P_1, P_2, \dots, P_N\}$ then each of these parts has its own value $-\log(\mathbf{p}(P_j))$ of information. The information of each part is now taken *according to the probability* $\mathbf{p}(P_j)$ that a state will actually lie in that part, so finally the *entropy*, or *information*, embodied in the system is defined as

$$H(\mathscr{P}) = -(\mathbf{p}(P_1)\log(\mathbf{p}(P_1) + \mathbf{p}(P_2)\log(\mathbf{p}(P_2) + \cdots + \mathbf{p}(P_N)\log(\mathbf{p}(P_N)) \,, \qquad (8.2)$$

or in the more compact form using summation notation

$$H(\mathscr{P}) = -\sum_{j=1}^{N} \mathbf{p}(P_j)\, \log(\mathbf{p}(P_j))\,.$$

We will use varying notation for this: $H(\mathscr{X}), H(X), H(\mathscr{P})$, according to what seems most natural, but in any case it is the partition that is important. For more on the appropriateness of this definition of entropy see [184].

The N parts of the partition set our levels of sameness and difference. What we learn from a single observation of the system is which part it is in. How much we learn from this depends on not only which part we find, but the probability of that part being the outcome anyway. On the one hand the formula for entropy tells us how much information we should expect to get if we take a state of the system at random, and on the other hand how much uncertainty we have before we take a random sample. These are the two complementary sides of knowing and not knowing.

Entropy of triples of binary bits

Let's see how it works in a simple situation. Suppose X is the system consisting of all zero-one triples :

$$\{X = \{(\mathbf{0},\mathbf{0},\mathbf{0}), (\mathbf{0},\mathbf{0},\mathbf{1}), (\mathbf{0},\mathbf{1},\mathbf{0}), (\mathbf{0},\mathbf{1},\mathbf{1}), (\mathbf{1},\mathbf{0},\mathbf{0}), (\mathbf{1},\mathbf{0},\mathbf{1}), (\mathbf{1},\mathbf{1},\mathbf{0}), (\mathbf{1},\mathbf{1},\mathbf{1})\}$$

with the partition consisting of these eight one-element sets and the probability assigning $1/8$ to each. This is the system that comes from three binary bits with equal $\mathbf{0}/\mathbf{1}$ probabilities, or equivalently the tossing of three fair coins. From what we saw above the information embedded in this system is 3 bits. We expect that the entropy will be 3.

Now let's look at our formula. Each part consists of a one-element set and its probability of occurrence is $1/8$. So each part contributes the same amount and we have the entropy

$$H = -\left(\frac{1}{8}\log(\frac{1}{8}) + \frac{1}{8}\log(\frac{1}{8}) + \cdots + \frac{1}{8}\log(\frac{1}{8})\right) ,$$

there being eight equal terms added here. We saw above that $\log(1/8) = -3$, so

$$\frac{1}{8}\log(\frac{1}{8}) = -3/8\,.$$

Adding eight of these and putting a minus sign in front gives us exactly 3. This is what we had expected, but now we see it arising in the context of parts and probabilities.

The checkerboard revisited

To get a deeper sense of the way in which entropy/information works, we consider again the 8×8 board game of §5.2 with its black and white checkers. We looked at this in two ways. Initially we just looked at the 2^{64} possible states of the board—all ways of filling the board with black and white checkers. All of these states are equally likely, so if we treat the system just as we did with the state space of $2^3 = 8$ equally probable states, we find that the entropy/information in this system is 64 bits.

However, we also examined this board simply from the point of view of the number of black or white checkers, disregarding exactly where they are. Now we have grouped the 2^{64} states into 65 parts, ranging from 0 black checkers to 64 black checkers, and the distribution of probability for these states is the binomial distribution which is highly peaked around the relatively equal numbers of black and white checkers. The entropy for this can be worked out by (8.2) and the result (to five decimal places) comes out to be 4.04707.

Thinking of entropy in terms of disorder, the original setting, where everything is equally likely, has no order at all and the entropy is as high as it can be, namely 64. In the second setting, we are viewing our system from a far more organized point of view and the entropy has dropped enormously. If all those 65 states were equally likely, the entropy would be at its maximum for 65 states, namely $\log(65) = 6.02237$. As it is the states are not equally likely, but highly favor board positions with a similar number of white and black checkers.

Entropy zero

By the way it is defined in (8.2), the entropy is always a non-negative number, i.e. it is either zero or positive. What, though, does it mean to say that the entropy given by a partition $\mathscr{P}$ is zero? Since every summand in (8.2) is of the form $-\mathbf{p}(P)\log(\mathbf{p}(P))$,which is non-negative, the only way that they can all add up to zero is for all of them be equal to zero. There are two ways for $-\mathbf{p}(P)\log(\mathbf{p}(P)) = 0$ to happen. Either $\mathbf{p}(P) = 0$ or $\log(\mathbf{p}(P)) = 0$. The former means that the part P has no probability of occurring. The latter happens only if $\mathbf{p}(P) = 1$ (see Fig.8.3). Since $\mathbf{p}$ is a probability measure, the total measure of X is one. Putting this together we see that all of the parts of $\mathscr{P}$ have measure

0 except for one, which has measure 1. So in effect the partition of X is just a partition with one part, namely all of X up to a set of states that have probability zero of occurring.

This is an important fact. If the partition effectively only has one part, then there is no distinguishing—we need at least two parts for that—and the amount of information and the amount of ambiguity about where states are is reduced to zero. If there is any real distinction then the entropy will be positive.

8.3.3 Dynamical entropy

The full concept of entropy is based not only on the way in which we assign difference and sameness on our state space, that is to say the way in which we partition the state space, but also *how that difference/sameness relation evolves through the dynamics.* If, now, we include the dynamics we are looking at a fully developed pattern system $\mathscr{X} = (X, s, \mathscr{P}, \mathscr{E}, \mathbf{p})$. The dynamical version is based on the average amount of uncertainty or information that arises *per step* of the evolution of the system. Entropy is measured at the level of partitions, and the way that the dynamics enters is at the level of the *dynamical evolution* of the initial partition $\mathscr{P}$. This brings us back to §6.5.8, where we defined the evolution of a partition. It is that evolution that we use here.

From this point on we shall assume that the dynamics of $\mathscr{X}$ is simple iteration, the measure $\mathbf{p}$ is a probability measure, and $\mathbf{p}$ is invariant under the dynamics.

Recall that the evolution of a pattern partition $\mathscr{P}$ is the sequence of refinements $\mathscr{P}^{(k)}$ of that partition through the action of the dynamics partition of X. Each part of $\mathscr{P}^{(k)}$, $k = 1, 2, 3, \ldots$, gathers together all the states of x that follow one particular itinerary over their first k steps. There is one part for each such possible itinerary. These parts, the subsets of X that arise in this way, are *events*—they are in $\mathscr{E}$, by the very definition of what it means to be an event space. The measure $\mathbf{p}$ assigns numerical values to each of these parts, which we can think of as their probability or frequency of occurrence. This allows us to define the entropy of $\mathscr{X}$ using the refined partitions $\mathscr{P}^{(k)}$ instead of the initial partition $\mathscr{P}$.

As k increases the partitions become finer and finer, each partition $\mathscr{P}^{(k)}$ being a refinement of the previous one $\mathscr{P}^{(k-1)}$. The basic formula is exactly the same as given at (8.2): it is the partitions that change, and as they do this leads to the sequence of values

$$H(\mathscr{P}^{(1)}),\ H(\mathscr{P}^{(2)}),\ H(\mathscr{P}^{(3)}),\ \ldots\ .$$

This sequence of numbers is increasing—each is at least as large as its predecessor, and in general actually larger. There is an intuitive reason for this. If the original partition has n parts, then $\mathscr{P}^{(2)}$ has $n \times n = n^2$ parts, $\mathscr{P}^{(3)}$ has n^3 parts, and so on. Some of the parts might be empty (no state makes that particular itinerary) but the message is clear: the complexity is increasing. In terms of information or entropy we would expect something like twice as much in $\mathscr{P}^{(2)}$, three times as much in $\mathscr{P}^{(3)}$ and so on. In other words, we might expect an information gain of about k bits in k steps. That doesn't necessarily happen. The system evolves, and as it does so it manifests its own internal shape. We may well anticipate some increasing internal order as the process continues. The *dynamical entropy* is defined as the limit of the average gain in entropy per step: that is,

$$\frac{1}{k}H(\mathscr{P}^{(k)}),$$

as k gets ever larger. Commonly it is denoted by $h(\mathscr{X})$, or $h(\mathscr{X}/\mathscr{P})$ if we wish to make the underlying partition $\mathscr{P}$ that defines the entropy explicit.[9]

It is perhaps useful to digest this definition by looking at the familiar zero-one shift system $\{\mathbf{0}, \mathbf{1}\}^\infty$ associated with the tossing of a fair coin. The basic partition is $\mathscr{P} = \{[\mathbf{0}], [\mathbf{1}]\}$ and the associated probability measure $\mathbf{p}$ gives equal probability to the occurrences of **0**s and **1**s in the various dictionary sets. There are four dictionary sets of length two, and they are all equally probable with probability $1/4$. So the entropy of the partition $\mathscr{P}^{(2)}$ has four equal summands and we get

$$H(\mathscr{P}^{(2)}) = -4 \times \frac{1}{4}\log(1/4) = 2\,.$$

In the same way, with the dictionary sets of length three

$$H(\mathscr{P}^{(3)}) = -8 \times \frac{1}{8}\log(1/8) = 3\,,$$

and so on. So we conclude that if we continue this then in the limit we will have

$$h(\mathscr{P}) = \lim_{n\to\infty}\frac{1}{n}H(\mathscr{P}^{(n)}) = \lim_{n\to\infty}\frac{1}{n}\,n = \lim_{n\to\infty} 1 = 1\,.$$

This says that the entropy is increasing at 1 bit per step. Each step, shift to the left, we learn exactly one new bit of information, just as we would expect.

Warning: Typically in the literature the *entropy* of a dynamical system is defined as the supremum of these limits as $\mathscr{P}$ runs over *all* possible partitions of $\mathscr{X}$. However, in our context of pattern systems, the partition is part of the definition of $\mathscr{X}$ and no other partition is involved in describing the dynamics. If the event space is actually the evolution of the initial partition, which is our usual assumption, then the dynamical entropy as we have defined it is actually the same that would be achieved by using the supremum over all possible partitions. This is a well-known theorem of Kolmogorov and Sinai [153, Lecture 14].

The dynamical entropy $h(\mathscr{X})$ of $\mathscr{X}$ is defined as the average gain in entropy per step of the process.

Two examples: Bernoulli shifts and Markov processes

Bernoulli shifts are standard zero-one shift spaces which are treated as random processes in which successive steps are considered to be probabilistically independent of one another and there are probability values p and $1-p$ set on the outcome of each successive symbol being a **0** or a **1**. For the simplest Bernoulli system, where **0** and **1** appear with equal probability, the entropy gain per step, we have seen that the entropy gain is 1 bit per shift, and the dynamical entropy is equal to 1. If the probabilities are set at p and $1-p$, the independence of each step correctly suggests the dynamical entropy is the same as that of a single step, namely,

$$h(\mathcal{X}) = -(p\log p + (1-p)\log(1-p)).$$

This is what is illustrated in Fig.8.2.

For a Markov process, the outcome of each step is influenced by the outcome of the previous step, though not on how that outcome came to be, see §5.4. The next state of the system *does* depend (probabilistically) on the present state, but not the past states. From this we can see that it is the initial partition $\mathscr{P}$, and the transition partitions that make up $\mathscr{P}^{(2)}$ and their corresponding probabilities, that are important. The probabilities associated with all subsequent transitions are already implicit in these. If $\mathscr{P}$ has n parts $P_1, \dots, P_n$ with probabilities p_i and $\mathscr{P}^{(2)}$ has n^2 parts P_{ij} with probabilities p_{ij} representing all the potential transitions, then the dynamical entropy is [184]

$$h(\mathcal{X}) = -\sum_{i,j} p_i p_{ij}\log(p_{ij}). \tag{8.3}$$

This condensed looking formula is not as hard to fathom as it seems. Each term in the sum formulates the probability of a state being in a particular part P_i (i.e. p_i) followed by the probability of it transitioning into part P_j (i.e. p_{ij}). Each of these contributes to the entropy, and their entire sum as both i and j range over their entire range 1 to n is what we are looking for.

The checkerboard game of §5.2, where we keep track of the number of white and black checkers (but not where they actually lie), is a Markov process. Applying this formula, we find that the dynamical entropy is 0.98864. This is very different from the entropy 4.04707 of this system that we derived above. But that was entropy before taking account of the dynamics. When we take account of the dynamics, internal order emerges. We know that at each step of the checkerboard game, there are only two things that can happen: either the number of white checkers increases by one or the number of black checkers increases by one. That is essentially one bit of information. Actually the information gain works out to be slightly less than 1. We know that the dynamics emphasizes the middle values where there is neither a great preponderance of white or black checkers and the outcome of the next step is roughly balanced between white and black. But the degree of uncertainty about what will happen at the next step is much lower when there is a highly *unequal* number of white and black checkers, and this lowers the entropy below 1. Not a huge amount, because highly unequal numbers of white and black checkers happen with very low probabilities.

It is interesting to compare this with the second checkerboard game that we discussed in §5.2.1 in which a checker is chosen at random, and whatever color it is, say white, another checker of the opposite color, black, is chosen somewhere at random on the board and replaced by a white checker. This is also a Markov process. If the entire board is white then it will stay white forever more. Similarly with black. These are equilibrium states at which nothing further happens. The game may go on for a long time, but the process promotes imbalance and will almost surely end up at one of the two equilibrium states, with probability $1/2$ for each. The rate of gain of information per step is ultimately *zero* since the system is almost sure to end up in one of the two equilibrium states, whereafter nothing changes. The entropy is zero.

8.3.4 Thoughts about entropy

The particular importance of the concept of entropy to this book is the way in which we can use it to understand the synthesis of pattern systems. Pattern systems do not stand alone in this world. What we experience at every moment of life can be seen as interactions of pattern systems that through these interactions become, in effect, larger pattern systems. Most often these compound pattern systems have emergent features that simply don't exist in their components. This is what we think of as *synthesis*, and we will find that entropy offers a way to express what synthesis means in some quantitative way.

Entropy has its foundational ideas in statistical mechanics, but those very same ideas can be applied here. Vast numbers of atoms are moving around in some range of possible states, each state being the collective list of positions and velocities of all of them. The usual principle is to assume that every state is equally likely. But even so, just as with the checkerboard game, not all the physical outcomes *seen at the macro-level* are equally likely. It may be that all the atoms in your cup will suddenly move in the same direction, in which case something very visible and unusual will happen, but the chances of this are impossibly small. Entropy arises by assessing the distribution of states as seen from the perspective of some explicit partition of the state space and the evolution of states through it.

Although it is transparent in the definition, it is not often emphasized that entropy is really an outcome of partitioning, and so ultimately goes back to the question of what we do and what we do not distinguish. In mathematical models we can look at the individual states of the state space and work at any level of partitioning we please, from that most detailed perspective up until we end up with the partition with just one part, the entire state space. This total freedom in choosing the level of our distinguishing is simply not the case in the physical world. In order to perfectly model a vast array of atomic particles we would need a model that could take into account the positions and velocities of every one of these particles. In reality the world disperses into a mist of events that simply cannot be brought to the level of individual states. We are bound to be dealing with some sort of partition of a proposed state space.

This partitioning may be considered as the origin of one of the most profound outcomes of entropy—that it is the source of the arrow of time. We take this up in §8.4.11, but in brief, the equations of physics that describe the process of change are all time-reversible. Change the sign of time t from $+$ to $-$ and the equations will run the universe backwards. But this is based on a concept of continuous time that can be parameterized by the real numbers which can be determined with infinite precision. It has been enormously successful, of course, but nonetheless it is pure idealism. The real numbers are infinitely remote from anything that can ever be experienced. The reality is that partitioning is inevitable, and along with it that lack of resolution that prevents the past from being recoverable from the present.

The riddles of time are far more complex than this, dominated as they are by the theories of special and general relativity. But the natural limits to precision and entropy must play a crucial role in this basic arrow of time, see §8.4.11.

8.4 Synthesis of pattern systems

After this discussion of entropy, we can return to the central question of synthesis of pattern systems. What, then, is it about a pattern system composed of smaller pattern systems as components that constitutes synthesis? If we take the view that synthesis is related to *order* (or lack thereof), and take entropy as a measure of disorder, or lack thereof, it seems natural to concentrate on entropy. In this discussion of synthesis, entropy plays the starring role, with information appearing as its negation. Key words are:

synthesis order entropy information

We begin with an example that arose out of a rather popularized idea called the Gaia hypothesis.

8.4.1 Daisy Worlds

James Lovelock is famous for his *Gaia hypothesis*, the hypothesis that the Earth itself, along with all its numerous forms of life, may be considered as a self-organizing and self-sustaining living entity.[10] One of the criticisms of Lovelock's idea was that it seemed impossible that the Earth could possibly create and regulate conditions for its own existence without some sort of teleological input. In response Lovelock, along with his collaborator, Andrew Watson, created a simple toy model, called *Daisy World*, that was able to offer a paradigm (they call it a parable) for answering one of the puzzles about the Earth's surface temperature, namely that it has remained relatively constant over a vast stretch of time during which the Sun's luminosity is believed to have increased by 30%. Using a simple model in which two species of daisies cover the world, temperature control arises as an emergent feature. Indeed, the relative constancy of the Earth's temperature is now thought to be a result of complex bio-geological feedback loops.

Here we briefly discuss Lovelock's Daisy World, because it is an example of a dynamical system whose behavior is intuitively clear without involving any detailed mathematics and it is one from which emergent pattern appears by itself out of the integrated interaction of two rather simple systems.

The basic idea behind a Daisy World is that of a planet with just two species living on it: one species of white daisies and one of black daisies. The soil of the planet is fertile and scattered all over with seeds of both species. The daisies can only survive in a small temperature range [5°–40°C]. Heat arrives from a sun very much like our own, with an expected life cycle like ours in which the amount of radiation from it will gradually increase until the point at which it self-destructs.

The main feature of the Daisy World is its ability to keep its atmosphere at an essentially constant temperature so that the daisies can continue to survive, even though the heat from the sun is slowly increasing in time. The dynamics is based on the differences between the albedo of the black daisies, the white daisies, and the bare ground. Albedo refers to the amount of reflection. The model starts off at 0°C. Black daisies absorb more heat (lower albedo) so survive better at a lower temperature than the white daisies: they absorb heat better and also tend to warm up their local environment. The first thing to appear as the temperature rises are black daisies in a ring around the equator. As the temperature rises further the black daisies start to occupy the more temperate zones above and below the equator and white daisies appear around the equator. The white daisies work in the opposite way: they reflect more heat and so tend to cool themselves and their local environment. It is easy to guess the further evolution as the black daisies move to the poles and the white daises move into the temperate zones. Eventually the black daisies die out, and beyond that it finally becomes too hot for even the white daisies to survive and they peter out starting at the equator and outwards from there.

What is relevant about this evolving dynamics is that the surface temperature of the planet is evidently affected by the total albedo, which itself is a function of the populations of the daisies and their distribution. Computing this out, Watson and Lovelock arrived at the graph of Fig.8.4, which plots the planet's surface temperature with and without daisies. The self-correcting nature of the system is rather evident, and so too is

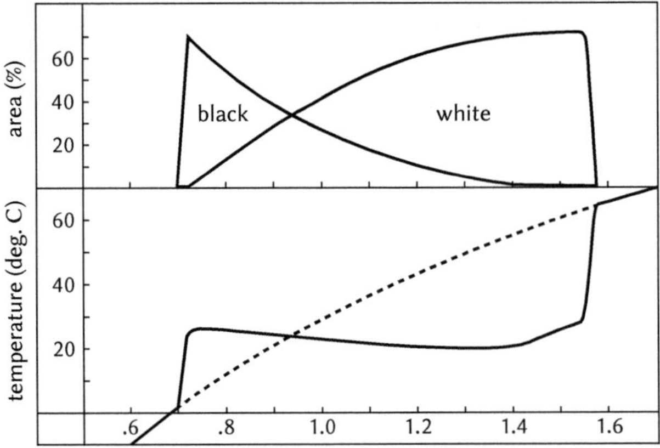

Figure 8.4: Daisy Worlds. At the top, a plot showing the area of the planet occupied by black/white daisies. Below a plot showing the average surface temperature of the planet in terms of the luminosity of the sun (dotted curve). The key point is that the temperature is held relatively constant over a long period of time during which the luminosity of the sun is always increasing. Figure from [186].

the drastic rise in temperature once the daisies become extinct.

From the perspective of pattern we can, for instance, imagine two state spaces, one for each species, each driven by the temperature of the sun. If we have only black daisies in the system, they will heat the surface up and then kill themselves through over-heating. If we have only white daisies, we have to wait much longer for suitable conditions for them to appear, after which they too eventually disappear. The self-emergent self-regulating dynamics created by integrating the two daisy systems ultimately acts as a control that stabilizes the temperature of the planet over a great stretch of time.

Although the actual mathematical analysis required to fully describe the process is not so easy, the intuition behind this model is rather clear: the synthesis of two pattern systems can create something new. It perhaps gives a renewed sense of how profoundly climate, environment, and the rich variety of life forms that accompany them are integrated in the natural world around us.

The Gaia hypothesis was not widely accepted when Lovelock first presented it. Times have changed. We have a much greater appreciation of the vast inter-connectedness of the biological and physical systems of our planet, and this includes human life. Now, standing as we are at the edge of the sixth great extinction of life, it has to be considered as a plausible, even vital, way of thinking. A very readable discussion of the Gaia hypothesis in terms of complex systems is given in the important book by Capra and Luisi on the systems way of thinking [29, *The systems view of life*]. For a passionate and informed discussion of living through this tumultuous period see [25, *Learning how to die*]. Jeffery Bennett has written a good science-based introduction to global warming [19, A global warming primer]. The original Gaia hypothesis paper itself is [186]. Since then the model has been studied extensively and in many variations, see the review article [195].

8.4.2 Combining two pattern systems

We now take a more formal approach to synthesizing or integrating pattern systems. We begin then with two pattern systems

$$\mathcal{X} = (X, s, \mathcal{P}, \mathcal{E}, \mathbf{m}), \qquad (8.4)$$
$$\mathcal{Y} = (Y, s, \mathcal{Q}, \mathcal{F}, \mathbf{n}).$$

We will assume that the dynamics of both is that of discrete iteration, in other words the dynamics is discrete and for each there is a single operator. To keep the notation simpler we are using the same symbol s for the operator for $\mathcal{X}$ and for $\mathcal{Y}$, with the context distinguishing to which it applies. We might think of the iteration as being discrete incremental steps in time.[11] Each system has its own state space, its own partition that defined sameness/difference, its own dynamics, and its own compatible event space that arises out of its dynamics. Each has its own law or measure which we will take to be a *probability measure.* The two systems need not be related in any particular way. At the outset we will consider the situation without the dynamics. It is easy to include them once we have the idea of what information synthesis means in this simpler situation.

Individually, each pattern system has its entropy/information, as we have previously described it. Thus,

$$H(\mathcal{X}) = H(\mathcal{P}) = -\sum_{P\in\mathcal{P}} \mathbf{m}(P)\log(\mathbf{m}(P)),$$

see §8.3. Remember that in spite of the minus sign, the entropy is always greater than or equal to zero. The entropy depends on the partition, but we will usually find it more convenient to write $H(\mathcal{X})$, with the partition being implicitly understood to underlie the situation. Similarly we have $H(\mathcal{Y})$.

As we have noted before, entropy is really a quantitative measure of the distribution of probability within the partition. The corresponding information, or capacity for encoding information is *negentropy*, and in this section we will indicate this by making it negative explicitly: $-H(\mathcal{X})$. This has no connotation of meaning or knowledge, but is more an indication of the capacity to encode, say in binary bits, the sorts of things that we would consider as information in the ordinary sense of the word. To the degree that one *can* know them, there is a corresponding degree of uncertainty when one does *not* know them. That is why uncertainty is also called "disorder": uncertainty amounts to disorder: disorder amounts to uncertainty.

The sum of the entropies of the two systems is $H(\mathcal{X}) + H(\mathcal{Y})$, and if the two systems have no mutual interaction with each other, this is what we would expect to be their combined entropy. However, if $\mathscr{X}$ and $\mathscr{Y}$ are actually in some relation of mutual interaction this must mean that only certain states of X exist in conjunction with certain states of Y, or the probabilities that certain states of X coexisting with certain states of Y are not what they would be otherwise.

Thus we are looking at certain pairs (x, y), which we will call *compatible pairs* of states, that can coexist in this relation and their probabilities of occurrence. In general we will find that each state of X is associated with some states of Y and reciprocally each state of Y is associated with some states of X. However, these *pairs* of states may exist with modified probabilities. It is like a horse and cart. Not connected, each has its own range of movement. Once linked together, their positions are quite literally tied to each other.

We should make it clear that we take a very wide view of what it means for two systems to be interacting. In any situation where the states of one system are affected by the states of another, that we will call interaction. In a communication system, the source and receiver are interacting. Even if the receiver is entirely passive, its states are being affected by those of the source and so in our sense there is interaction. If you are perceiving any pattern system (and that includes just about any perception of the world) then a part of your sensory system and the neural systems of your brain are in interaction with that pattern system. *There is no objective view: the observer is always in interaction with what is observed.* This all goes under the name of *interaction*, and from our perspective the two systems together form a new pattern system. That is what we call synthesis of pattern systems.

Every *interaction* is a synthesis.

At the extreme opposite to no interaction between $\mathscr{X}$ and $\mathscr{Y}$, we have the case where $\mathscr{Y}$ is a pure copy of $\mathscr{X}$ and the interaction is to pair identical states in each and no others. Then what is gained by synthesis is *redundancy*: the information of the combined system is no different from that of $\mathscr{X}$ itself. Everything is shared. These then are the two extremes: one of total lack of interaction and one of total duplication. Both are legitimate, but presumably the more important situations are those in which there is a mixture of sharing and independence, so that the combined system is indeed something new.

Tossing a fair coin twice

Thinking in these terms suggests that synthesis has to do with sharing information and decline in the total entropy. Fig.8.5 gives some idea of what this means. Here we consider a fair coin that is flipped twice, but with different assumptions made about how the two tosses are related to each other. Heads are indicated by 0 and tails by 1 and the resulting entropy by H. Synthesis corresponds to *decreasing* entropy.

H = 2	0	1
0	1/4	1/4
1	1/4	1/4

H = 1.918	0	1
0	1/6	1/3
1	1/3	1/6

H = 1	0	1
0	0	1/2
1	1/2	0

H = 1	0	1
0	1/2	0
1	0	1/2

Figure 8.5: Four variations on probabilities for tossing a fair coin twice (row: first toss; column: second toss). The probabilities for a head (0) or tail (1) remain $1/2$ for each toss (add rows and add columns). It is the probabilities for the *pair* of tosses that changes. Top left is the standard case of independence of tossing of the two coins. Top right is for people who believe that the outcome of the second toss is more likely to be the opposite to the outcome of the first toss. The bottom pair are situations where the outcome of the second toss is determined completely by the outcome of the first. The *decreasing entropies H* indicate the *increasing level of synthesis.*

8.4.3 Synthesis of pattern systems

Now let us consider this in full generality. We start off with the entire set of all possible pairs

$$Z := X \times Y := \{(x, y) \ : \ x \in X, \ y \in Y\},$$

and along with it the partition $\mathscr{R}$ whose parts are made up of all possible choices from the parts of $\mathscr{P}$ and $\mathscr{Q}$:

$$\mathscr{R} := \mathscr{P} \times \mathscr{Q} = \{P \times Q \ : \ P \in \mathscr{P}, Q \in \mathscr{Q}\}.$$

This provides us with the entire set of possible states that are available to the combined system, and the full extent of sameness/difference that is available for making distinctions.

So far we have just made the obvious steps that would lead to what is usually referred to as a *direct product.* What is missing is a probability law $\mathbf{p}$ which should give the probabilities for pairs (x, y) to appear in the various parts $P \times Q$ of the partition $\mathscr{R}$, i.e.

the values $\mathbf{p}(P \times Q)$. This is the point at which the synthesis really comes in: the law that governs which combined events are most likely, which less likely, and which never occur at all.

We should not expect that there will be some unique way of choosing or determining the law $\mathbf{p}$. Even in the simplest pattern systems like the example shown in Fig.8.5 there are infinitely many choices for $\mathbf{p}$—just replace the pair $1/6, 1/3$ by any pair of positive numbers that add to $1/2$. This is what gives synthesis so many ways to produce something new, something emergent, not implied by the two systems standing by themselves. Although the most obvious thing to do is to take $\mathbf{p}$ to be what is called the *product measure* $\mathbf{m} \otimes \mathbf{n}$ defined by

$$(\mathbf{m} \otimes \mathbf{n})(P \times Q) = \mathbf{m}(P)\,\mathbf{n}(Q)\,, \tag{8.5}$$

this is exactly the "unrelated" situation that we described above—in effect it is the statement that the two pattern systems behave independently of each other, each acting as though the other were not there. For real synthesis, $\mathbf{p}$ will have to be something more creative than this!

In integrating the two systems we expect neither new states nor finer degrees of distinction within the individual systems. However, we do expect the compatibility of pairings (x, y) to be affected, and it is the measure $\mathbf{p}$ that can control this. With this in mind we shall make only one assumption on $\mathbf{p}$, namely that it is compatible with the measures $\mathbf{m}$ and $\mathbf{n}$ in the sense that $\mathbf{m}$ and $\mathbf{n}$ are what are called the *marginal measures* of $\mathbf{p}$. Saying that the marginal measures are unchanged simply means that if we take all the pairs (x, y) but totally ignore the second components, the overall frequency of first components is just as it originally was. In words, the overall probability of x being in the part P is the same before or after pairing. Similarly for y and Q. Written mathematically the meaning is that we assume that

$$\begin{aligned}\mathbf{p}(P \times Y) &= \mathbf{m}(P)\,,\\ \mathbf{p}(X \times Q) &= \mathbf{n}(Q)\end{aligned} \tag{8.6}$$

for all $P \in \mathscr{P}$ and $Q \in \mathscr{Q}$.

The example of Fig.8.5 has this property. It is the statement that the sums of the rows are the probabilities for the first toss, and the sums of the columns are the probabilities of the second toss—both are $1/2$. The first toss still has probability $1/2$ of being a head and $1/2$ of being a tail. The same goes for the second toss. No synthesis is the case when all the pairs have probability $1/4$.

This assumption might easily be questioned. Conceivably a subsystem of a system, when studied on its own outside of the entire system will not function with identical probabilities as it does when studied on its own inside the full combined system. However, as long as the subsystem is studied in a context that includes it as being part of the larger system, then it will display the marginal probabilities of the larger system. That is what we are assuming here: the subsystem is truly a subsystem.

Note that this says nothing about what the probabilities $\mathbf{p}(P \times Q)$ of the individual parts may be. That is something for which there is usually a vast range of possibilities, and it is their values that are critical to the resultant system.

For instance, $\mathbf{p}(P \times Q) = 0$ is in effect the statement that the pair (x, y) is not a compatible pair if $x \in P$ and $y \in Q$. This is very convenient in discussing the synthesis. Instead of having to explicitly say which pairs are compatible and which not, we use the probability measure $\mathbf{p}$ to do this for us: $\mathbf{p}(P \times Q) = 0$ becomes the statement that the pair (x, y) is not a compatible pair if $x \in P$ and $y \in Q$. It does not mean that there are no pairs $(x, y) \in Z$ with $x \in P$ and $y \in Q$. They are all there, but the totality of all such pairs has $\mathbf{p}$-measure 0: "almost surely" the pair does not arise in the dynamics of the process.

Thus, in its bare essentials, a *synthesis* (or sometimes we will say an *integration*) of the pair (8.4) is a new pattern system

$$\mathcal{Z} = (Z, s, \mathcal{R}, \mathbf{p}), \tag{8.7}$$

where Z and $\mathcal{R}$ are simple product sets, as described above, and $\mathbf{p}$ is a probability measure with $\mathbf{m}$ and $\mathbf{n}$ as its marginal measures. As for the dynamics, we shall take it to be the step-wise action on both components simultaneously:

$$s \triangleright (x, y) := (s \triangleright x, s \triangleright y). \tag{8.8}$$

It should be remembered that we are using the same symbol to represent the actions of the dynamics on the two different pattern systems. Now we use the same symbol for the new operator that operates on Z.

Red-green, yes-no

Here is another example of a synthesis of two systems that is really a variation on the coin tossing example we discussed above. Consider two simple shift systems, one based on the alphabet $\{0, 1\}$, which will stand for no, yes, and one on the alphabet $\{r, g\}$, which will stand for red, green. In both cases the states are strings of these letters. Suppose at each second on a clock there is either a red or green flash on a monitor (r or g), and at the same time either a click sound or silence (yes or no). Suppose that in both cases these events occur with equal probability and that they are totally unrelated. Combining them we have the system consisting of strings of all possible pairings of these two alphabets: $\{(r, 0), (g, 0), (r, 1), (g, 1)\}$, and each pair is equally probable, namely $1/4$.

But now suppose that the setting is an ophthalmologist's office and a patient is seated in front of a monitor on which tiny red or green flashes are displayed, one each second and with equal probability, and is given a clicker and told to click it each time they see a red flash, but not if they see a green flash. Now a click or not a click depends on the patient and the probabilities are changed. The resulting combined system should put higher probability on the pairs $(r, 1)$ and $(g, 0)$ and lesser on $(g, 1)$ and $(r, 0)$. The two systems are integrated by the action of the patient and have become something different. The level of synthesis tells us something about the level of concentration of the patient and also whether there is some form of red/green color blindness. Uncorrelated the entropy is $1+1=2$. Correlated it would reduce to 1 in the case of a perfect score, and somewhere between 1 and 2 for anything else.

8.4.4 Entropy and synthesis

Now we come to the basic point: synthesis means more internal order, and more internal order means less entropy. If we go back to where we defined entropy in the first place (§8.3.2) we recall that one of the basic requirements of the definition was that the entropy $H(\mathscr{X})$ and $H(\mathscr{Y})$ of two systems is simply added when we form the direct product:

$$H(\mathscr{X} \times \mathscr{Y}) = H(\mathscr{X}) + H(\mathscr{Y}) .$$

The direct product is the least integrative way to put the two systems together. Effectively it means that there is no synthesis. Real synthesis happens only when the entropy of the new system of (8.7) is less than this simple sum:

$$H(\mathscr{Z}) < H(\mathscr{X}) + H(\mathscr{Y}) ,$$

so the entropy drops (more organization) and the two systems in some sense share information.

As we have already noted, information can be thought of as neg-entropy and written with the opposite sign. In this section entropy is always non-negative, i.e. ≥ 0, and information is always taken as non-positive, i.e. ≤ 0. We will flip back and forth between speaking of entropy and information, their intuitive meanings forming a complementary pair.

The situation of minimum synthesis occurs when $H(\mathscr{Z}) = H(\mathscr{X}) + H(\mathscr{Y})$, maximum entropy and no shared information. The *maximum* synthesis occurs when the entropy of the combined system is whichever is the larger of $H(\mathscr{X})$ and $H(\mathscr{Y})$. This is the minimum entropy (amount of disorganization) that we could possibly expect, since both $\mathscr{X}$ and $\mathscr{Y}$ are subsystems of the combined system and presumably the whole system cannot be any more organized than the least organized (largest entropy) of them. Soon we will see more precisely what this means.

The simplest measure of synthesis is the difference between $H(\mathscr{Z})$ and the sum $H(\mathscr{X}) + H(\mathscr{Y})$. We denote this difference by Ω (Omega) and call it the *consolidation*. Writing this out explicitly we have:

the *consolidation* of the synthesis $\mathscr{Z}$ of $\mathscr{X}$ and $\mathscr{Y}$ is:

$$\Omega := H(\mathscr{X}) + H(\mathscr{Y}) - H(\mathscr{Z}) . \tag{8.9}$$

Equivalently

$$H(\mathscr{X}) + H(\mathscr{Y}) = H(\mathscr{Z}) + \Omega .$$

We should note a few things here.

- The first is that $\Omega \geq 0$. $H(\mathscr{Z})$ cannot be more disordered than $\mathscr{X} \times \mathscr{Y}$, so entropy is reduced and the consolidation Ω is increased. The second form of the equation can be viewed as a sort of *conservation of entropy*. The reduced entropy of the synthesis is "stored" as order. That order, Ω, is released when the two systems are disintegrated.
- The actual numerical value of Ω depends on the base that we use for the logarithms. We assume base 2 throughout, but in any case changing the base only amounts to an overall scaling factor. We will only use the value of consolidation in the form of comparisons.
- Ω is a number that has to lie between 0 and the smaller of $H(\mathscr{X})$ and $H(\mathscr{Y})$, i.e.

  $$0 \leq \Omega \leq \min\{H(\mathscr{X}), H(\mathscr{Y})\} .$$

  This is already indicated by the fact that the entropy of the synthesis $\mathscr{Z}$ can never fall below the larger of these two numbers, and hence the consolidation can never be more than the lesser of them. We see more about this below.[12]

8.4.5 Relative entropy

So far we have viewed synthesis from the point of view of putting two systems together and looking at the potential drop in entropy as a result of this. There is an alternative way of thinking about it. In the combined system $\mathscr{Z}$, the two subsystems $\mathscr{X}$ and $\mathscr{Y}$ still exist. We are bound to have at least as much entropy as each of these systems separately (or in terms of information, the combined system must be able to carry at least as much information as each of its subsystems). Suppose that we take the view that we are combining system $\mathscr{X}$ with the already existing system $\mathscr{Y}$. Then we can ask the question how much more entropy is added by the synthesis of $\mathscr{X}$ with $\mathscr{Y}$ *given that we have already the entropy of* $\mathscr{Y}$ *within it.* We call this *relative entropy* and denote it by $H(\mathscr{X}|\mathscr{Y})$ and express its meaning formally by:

$$H(\mathscr{Z}) = H(\mathscr{Y}) + H(\mathscr{X}|\mathscr{Y}) . \tag{8.10}$$

Read it as saying that the entropy of the synthesis is made up of the entropy of $\mathscr{Y}$ plus whatever extra entropy comes from $\mathscr{X}$. The smaller $H(\mathscr{X}|\mathscr{Y})$ is, the less entropy is coming from the synthesis of $\mathscr{X}$ *given that we have taken into account the entropy of the system* $\mathscr{Y}$. The smaller $H(\mathscr{X}|\mathscr{Y})$, the more $\mathscr{X}$ is compatible with $\mathscr{Y}$.

As it stands (8.10) is nothing more than a definition. However, $H(\mathscr{X}|\mathscr{Y})$ is a well-defined and directly computable number, see then Endnote 13. It also appears as an important tool of the mathematics of pattern recognition where it is called the *Kullback–Leibler divergence* [117].

8.4.6 The meaning of zero relative entropy

The condition $H(\mathscr{X}|\mathscr{Y}) = 0$ implies maximum consolidation of the two pattern systems $\mathscr{X}$ and $\mathscr{Y}$. It says that $\mathscr{X}$ contains nothing in the way of further entropy (or further information) than what may be inferred from $\mathscr{Y}$. What that means turns out to do with a special way in which the events of X are associated with the events of Y. The relative entropy $H(\mathscr{X}|\mathscr{Y})$ is zero if, and only if, each part of the partition $\mathscr{Q}$ of $\mathscr{Y}$ is matched exclusively with just one part of the partition $\mathscr{P}$ of $\mathscr{X}$. This is something that is better understood by looking at an example, see Fig.8.6. We can think of it as saying that $\mathscr{P}$ *looks like* some sort of coarsening of the partition $\mathscr{Q}$, so that one or several parts of $\mathscr{Q}$ are matched exclusively with one part of $\mathscr{P}$; or to put it the other way around $\mathscr{Q}$ looks like a a refinement of $\mathscr{P}$. This happens in such a way that the probabilities match.

We say "looks like" because in actuality the partition $\mathscr{P}$ is a partitioning of the state space X and the partition $\mathscr{Q}$ is a partitioning of the state space Y, and these two state spaces are not in any prior form of connection. The only connection between them is through the probability measure $\mathbf{p}$. The measure tells us which part-pairs $P_i \times Q_j$ have positive probability in the integrated system. Zero relative entropy requires the positivity of the probability be assigned to pairs that match in the way we have just stated. For more details see the Endnotes.[13]

The overall conclusion of this is that $H(\mathscr{X}|\mathscr{Y}) = 0$ means that $\mathscr{X}$ adds nothing to the integrated system that is not already in it due to $\mathscr{Y}$. In some sense $\mathscr{X}$ is a *shadow* of $\mathscr{Y}$. Such things are not uncommon in Nature, notably in instances of duplication or redundancy. It is frequent in the brain for instance. Individual neurons are in themselves pattern systems, their basic feature being to fire or not to fire. This firing depends on the complex networks through which they are connected to other neurons, and often it is the averaged firing rates of a collection of neurons or neural unit that is relevant. Typically an individual neuron is rather noisy, in the sense that there is inherent degree of randomness in it firing. However, averaged over the neurons of a neural unit in which all the individual neurons do much the same thing, the redundancy greatly reduces the effects of individual randomness, see Fig.10.11 in §10.2.2.

An example of shadowing that is not duplication can be seen in the example of the way that the logistic system of §7.6.3 is related to the doubling system. The logistic

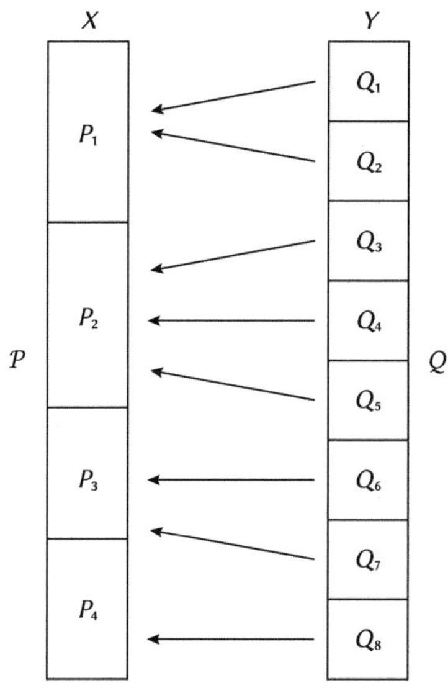

Figure 8.6: When $H(\mathscr{X}|\mathscr{Y}) = 0$ there is an association at the level of the partitions that associates to each part $Q \in \mathscr{Q}$ exclusively one part $P \in \mathscr{P}$. Several parts of $\mathscr{Q}$ may be associated with the same part in $\mathscr{P}$, but not the other way around. In effect there is a grouping of the parts of $\mathscr{Q}$ to make a coarser partition which then aligns with the partition $\mathscr{P}$. Intuitively $\mathscr{Q}$ has all the disorder of $\mathscr{P}$ and more. This correspondence dictates which pairs $\mathscr{P}_i \times \mathscr{Q}_j$, and then in turn which pairs (x, y), are compatible. Further, probabilities must match: e.g. here $\mathbf{m}(P_3) = \mathbf{n}(Q_6) + \mathbf{n}(Q_7)$.

system is a shadow, but not a duplication, and we cannot recover the doubling system from this shadow.

If both $H(\mathscr{X}|\mathscr{Y}) = 0$ and $H(\mathscr{Y}|\mathscr{X}) = 0$ then the matching of partitions $\mathscr{P}$ and $\mathscr{Q}$ becomes one-to-one and their measures also match. This is the basis of pattern equivalence, see §7.7.3.

8.4.7 Synthesis of pattern systems in a dynamic setting

So far our discussion of synthesis and consolidation has not taken into account the dynamics. We have been working only on the base level of the partitions and their probabilities, but not how states move on trajectories. Following the ideas of §8.3.3 it is actually quite easy to move into dynamical forms of entropy. We continue to consider the same two pattern systems $\mathscr{X}$ and $\mathscr{Y}$ of §8.4.3.

In addition we shall suppose that the probability measures $\mathbf{m}$ and $\mathbf{n}$ are invariant under their respective dynamics, and the synthesis measure $\mathbf{p}$ of (8.7) is also invariant.

In §6.5.8 we introduced itineraries which partition the state space into the sets of the form

$$P_{i_1} \cap s^{-1} \triangleright P_{i_2} \cap s^{-2} \triangleright P_{i_3} \cap \dots \cap s^{-(k-1)} \triangleright P_{i_k},$$

which consist of those elements of X that begin in P_{i_1} and evolve successively through the parts $P_{i_2}, P_{i_3}, \dots, P_{i_k}$ of the partition defined by $\mathscr{P}$ in the course of $k-1$ iterations by s. For each value of $k = 1, 2, 3, \dots$, we have the resulting partition $\mathscr{P}^{(k)}$ that consists of all the various ways in which states can progress from their initial part of the partition through subsequent parts as the process proceeds. Since each $\mathscr{P}^{(k)}$ is another partition of X in its own right, we can define the entropies $H(\mathscr{X}^{(k)})$ that emerge from the dynamics of $\mathscr{X}$ when the initial partition $\mathscr{P}$ is replaced by the ever finer partitions $\mathscr{P}^{(2)}$, $\mathscr{P}^{(3)}$, and so on. The dynamical entropy $h(\mathscr{X})$ is based on the average entropy gain per step of the process. Ultimately it depends only on the itinerary pattern system created from the dynamics of $\mathscr{X}$ as seen in terms of the partition $\mathscr{P}$.

We can do the very same thing with $\mathscr{Y}$ and its partition $\mathscr{Q}$ with corresponding entropies $H(\mathscr{Y}^{(k)})$, and this leads to the relative entropies $H(\mathscr{X}^{(k)}|\mathscr{Y}^{(k)})$. We still have the same combined system $\mathscr{Z}$ with the state space $X \times Y$ and the same dynamics, but now the evolved partitions $\mathscr{P}^{(k)}$ and $\mathscr{Q}^{(k)}$ are taken into consideration, and these lead to the corresponding dynamical entropies $h(\mathscr{Y})$ and $h(\mathscr{X}|\mathscr{Y})$.

With these we can define the corresponding versions of the *consolidation* based on these more refined partitions in (8.7) and on the average values of these as k gets ever larger. In this way we lift our previous discussion of consolidation to the dynamical level. The key feature is the imposed measure $\mathbf{p}$ which we need to assume is an invariant probability measure.

The situation starts to look complicated, but in practice things may be far simpler. We already have seen in § 4.4 how Shannon made his case about the power of probability to shape the written words of the English language using the refined probabilities of longer and longer letter strings, the range increasing from $k = 1$ to $k = 5$. But much of his analysis was at the level of Markov processes (the case $k = 2$). Many physical processes are modeled on Markov pattern processes where the models depend only on the part of the partition occupied by the *present state* to determine the probabilities of its being in the various parts on the next step. If we know these transition probabilities then by (8.3) we know the dynamical entropy. In this way the dynamical entropies of $\mathscr{X}, \mathscr{Y}, \mathscr{Z}$ relative to their partitions $\mathscr{P}, \mathscr{Q}, \mathscr{R}$ are known and the dynamical consolidation can be defined as

$$\Omega = h(\mathscr{X}) + h(\mathscr{Y}) - h(\mathscr{Z}).$$

8.4.8 Thoughts from Hua-yen Buddhism

The idea of a system and its subsystems, the whole and its parts, and how they seem to define each other is particularly emphasized in certain branches of Buddhist philosophical thought, notably in the tradition of Mahayana Buddhism. All Buddhist thought emphasizes change within a context of the total inter-connection and inter-dependence of all things, and from this arises its stress on the emptiness of self and a sense of compassion for all sentient beings. Within the Chinese cultural sphere, Buddhism and Daoism naturally affected each other, with Chan (Zen) Buddhism often seen as a fusion of these two philosophical systems. Overall Daoism is more Nature oriented than Buddhism, which tends to be more mind oriented.

It is this intense interest in the mind and the centuries of experience in witnessing the workings of the mind through meditation that makes Buddhist philosophical interpretations interesting in terms of modern science. Of course no one was talking in terms of systems and subsystems, but putting aside the vocabulary it is easy to see that there is a strong connection to the ideas that we have been developing here. This is particularly clear in the philosophical ideas of the Hua-yen school. Although it is nowhere near as well known in the West as forms of other Mahayana schools like Chan, Vipassana, Tibetan Buddhism, or Pure Land Buddhism, the Hua-yen school, which developed in seventh-century Tang dynasty China, is often considered to be the most profound philosophical exposition of these ideas. Hua-yen (this is Wade–Giles, Hua-yan in pinyin) means *flower garland* (also *flower adornment* or *flower ornament*) and derives from the *Avatamsaka Sutra*, or *Flower Garland Sutra*.

Foundational to Hua-yen thought is a pair of terms *li* 理 and *shì* 式 (or *shih* in Wade–Giles, see §15.3). Here *li* is the same *li* of the Frontispiece, its meaning here being *principle(s)*. The word *shì* means *event(s)*. So *li* is understood as the world of abstract principles, the immanent reality, which we can associate with our idea of laws of physics, and *shì* , or *shih*, refers to phenomena, the world of events, the world of difference and change. In Hua-yen thought these two are inseparable: the system is experienced as phenomena—the phenomena are expressions of the system.[14]

> [**Shih**] This is the realm of phenomena, in which all things are seen as distinct and different objects or events. A river flows, a tree grows, birds fly, fish swim, the fire is hot, and the ice cold—all the multitudinous phenomena which occur in the empirical world are of this realm. Things and events are looked upon here as distinct and independent.
> [**Li**] This is the realm in which only the abstract principles which underlie phenomena, and the immanent reality that upholds all dharmas, are seen. It is a realm beyond sense perceptions, a realm grasped only by intellect and intuition. All the principles and laws that dictate the events of the phenomenal world belong to this category [32, Chang, p. 142].

This is all to be interpreted at the level of totality—the full scale from the infinitesimal to the cosmic immensity, in both space and time. Within this context *li* and *shì* are understood to be an inseparable unity, much as heaven-earth are a unity in the *Yijing*: *shì* is at the empirical level, *li* at the mental level. Each implies the other, even to the extent that any event, however apparently insignificant, is considered to be equivalent to the totality. The understanding is that inter-relationship and inter-dependency make everything connected and in some way every event is dependent on every other conceivable event. No event can be that event without the full extent of totality being implicated.

This idea is often described through the metaphor of Indra's Net. The great god Indra lives in a palace, above which is magically suspended a vast net, strung as countless jewels, extending without limits in all directions. Fazang, one of the major figures of the Hua-yen school, describes it as follows: Because of their brightness and transparency, the jewels reflect one another, so that in a single jewel there appear images of all the rest .... Thus they are multiplied to infinity, and all these infinite images appear with vivid clarity in every single jewel [132].

In other words, every part reflects every other part, and in some way every part of the Cosmos is co-equal to the whole: mutual identity and mutual inter-causality. It has been suggested [132] that the non-locality, now well established in quantum theory, is representative of this idea. At a more accessible level, we are all learning that our life on planet Earth is entwined with all other life and also with the entire geo-physical processes of oceans, skies, and continents.

What is relevant to our discussion here are Fazang's thoughts on the relationships between the parts and the whole. Thus Fazang speaks of six characteristics, three pairs, that express these relationships:

> First, the names: they are the characteristics of universality, particularity, identity, difference, integration, disintegration. "Universality" means that the one includes many qualities. "Particularity" means that the many qualities are not identical, because the universal is made up of many dissimilar particulars. "Identity" means that the many elements [that make up the universal] are not different, because they are identical in forming one universal. "Difference" means that each element is different from the standpoint of any other element. "Integration" means that [the totality of] interdependent organization is formed as a result of [the collaboration of] these [elements]. "Disintegration" means that each element remains what it is [as an individual with its own characteristics and is not disturbed [in its own nature]. [35, Ch.6]

These seemingly mutually contradictory statements put beside each other are typical of Hua-yen thought. But they hold an important truth: we always see things from some point of view. Change the point of view, and the appearances of reality change. *There is no universal point of view*, and with that understanding are swept away all our hopes for absolute resting points on which to build theories of reality. There are none. All of this is thoroughly explored in the texts that came out of the Hua-yen school.

We can see ideas that parallel our systems ideas, wherein the system is seen as a universal, and events (the distinguishable manifestations, the patterns) are what makes the system what it is. The system is inferred (by the mind) as the compass of the parts, while the parts can only be described within the evolving context of the system. The part cannot be a "part" in any absolute sense all by itself, but only has the meaning of a part relative to the entire context (and a dynamical one at that) through which it becomes a part. That context is the system. Our discussion of marginal measures in §8.4.3 touches on this: the marginal measures of the system are required to be those of the subsystem, or, putting it the other way around, seeing the subsystem truly as a subsystem already includes information about its relation to the system.

Thus Hua-yen takes it as fundamental that change is the innate fabric of *being*, and being as "things" is not being at all, but rather a ceaseless flow of becoming [35, p. 40]. At the end of this chapter in §8.5 we will describe a mathematical result which shows that in any ergodic pattern system the dynamical sequence of events that arise from any single event effectively offers all that can be described about the system. This is pure mathematics deriving entirely from our ideas of patterning. Taken metaphorically, though, it gives further intuition to the idea that a single event somehow encompasses the whole.

8.4.9 Shannon's noisy channel theorem

When we discussed Shannon's theory of communication of information in §4.4.4, we left aside the important question of what happens when the communication channel is noisy and binary bits are randomly changed, replacing zeros for ones and vice versa. This surely degrades the rate of information flow. The natural question is by how much?

This is a far more important question than one might first suppose. The existence of noise down a communication channel is a plain matter of fact. How much we are willing to accept depends on the use to which the channel is put. Telephone transmission is often only mediocre when it comes to sound quality—acceptable for voice, good enough for our brains to make sense of it, but terrible for music. This is deliberate—it makes things cheaper. When it comes to storing or transmitting digital sound for music, the requirements for accuracy are far higher. When it comes to the transmission of computer programs it needs to be perfect. Perfection cannot be achieved by removing all noise—instead it is sought by adding extra bits of error correcting code to make up for the potential errors.

Suppose then we have a binary channel that sends data from a transmitter $\mathcal{X}$ to a receiver $\mathcal{Y}$. The question of transmission of information centers around our ability to reconstruct $\mathcal{X}$ given $\mathcal{Y}$. This formalizes to the now familiar equation

$$H(\mathcal{X}) = H(\mathcal{Y}) + H(\mathcal{X}|\mathcal{Y}), \tag{8.11}$$

where the relative entropy $H(\mathcal{X}|\mathcal{Y})$ is (the numerical value of) the extra information we need to add to recover the information that was transmitted from the information that was actually received.

Shannon clarifies what is involved here with a simple example of a communications channel in which we are sending zeros and ones from $\mathcal{X}$ to $\mathcal{Y}$, but in the sending of each binary digit there is a probability p that it will be sent incorrectly. He asks then what happens to the information rate. He points out that the first guess that might come to mind is that the rate of information would go down by p, so instead of transmitting at some rate of T bits per second the information rate would be reduced to $T(1-p)$ bits per second. This might sound reasonable, but it's wrong. For if $p = 1/2$, so half the symbols are sent incorrectly, then, since we don't know which are right and which are wrong, we can no longer get any information at all from the signal. We do not get half the information we would get with a perfect channel: we get none at all.

In fact the information loss is based on (8.11). From the perspective of what is sent and what is received, at the level of individual bits there are four possible outcomes $0|0, 0|1, 1|1, 1|0$, where we have used the vertical stroke to separate the sent and received bits. Suppose that we have an error rate of p per symbol, and equal frequencies of zeros and ones. Assuming that we receive a 0, the probability that a 0 was sent is $1-p$ and the probability that a 1 was sent is p. The parallel situation occurs when we receive a 1. Using the formula for relative entropy, the amount of uncertainty associated with the event turns out to be

$$H(\mathcal{X}|\mathcal{Y}) = -((1-p)\log(1-p) + p\log(p)),$$

whereas $H(\mathcal{X}) = 1$ since we are sending one bit at a time.

Fig.8.7 is a plot of this expression over the range $0 \le p \le 0.5$, along with a plot of the resulting degree of certainty. At the left end we have the value 0: there is no uncertainty. When $p = 1/2$ we have the value 1: there is total uncertainty (one bit of uncertainty per bit sent), so no certainty at all.

So $H(\mathcal{X}|\mathcal{Y})$ is a measure (in the informal sense of the word) of how much information is lost in a binary channel in the presence of noise. The solution to noisy transmission is to encode extra error-correcting bits along with the message being transmitted. The creation of such error-correcting codes is an entire field in itself. While no error-correcting code can establish perfect transmission in the presence of random noise, still the level of degradation of the message can be reduced to any desired level of acceptability by transmitting sufficiently many additional error-correcting bits to the message. The question is, how many? The result is Shannon's famous *noisy channel theorem* that says that in probabilistic terms it is the very quantity that we have been looking at: $H(\mathcal{X}|\mathcal{Y})$ bits per symbol transmitted.

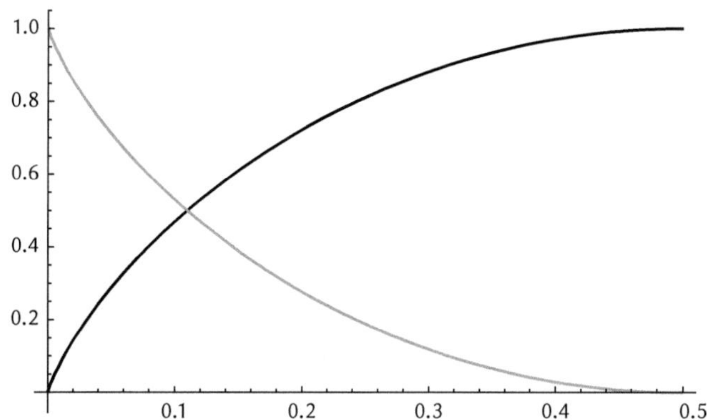

Figure 8.7: In the sending zeros and ones down a noisy channel, random errors reduce the amount of information. These plots show relative entropy $H(\mathscr{X}|\mathscr{Y})$ (increasing, the amount of information lost) and $H(\mathscr{Y})$ (decreasing, the resulting information derivable from the received data) as the error probability p per symbol (on the horizontal axis) rises from 0 to 1/2. They add to $H(\mathscr{X}) = 1$. We can see that even a modest error rate of $p = 1/8$ results in a degradation of the information about a half, and if $p = 1/2$ the degradation is complete.

Theorem 8.4.1 (Shannon) *By adding an additional capacity equal to $H(\mathscr{X}|\mathscr{Y})$ bits per symbol, it is possible to encode the correction data so as to correct all but an arbitrarily small fraction of the errors. This is not possible if the additional capacity is less than $H(\mathscr{X}|\mathscr{Y})$.*

This result has been of fundamental importance to the communications world for it sets the theoretical limit that ought to be strived for. It is important to realize that the theorem in no way presumes that the system knows in advance where the errors are, or how many there are, or that the additional correction bits themselves are free from the same chances of error in transmission. That's what makes it amazing. However, there is a strange thing about Shannon's theorem: its proof is based on probabilistic arguments that show only that such error correcting processes must exist. It doesn't say how to actually come up with a single one of them! So although the theorem tells us that it is possible to create error-correcting code to make the transmission of data in the presence of noise as accurate as we wish (though never perfection itself), it gives us no clue how to find it. There has been a great deal of mathematics created to come up with practical solutions to this question.

8.4.10 Information and pattern

"Information" is a word whose meanings and significance have changed rapidly now that we are in the "Information age". The common definition of information is understood as communication of knowledge or intelligence. This is the primary definition in the Merriam–Webster dictionary and other dictionaries that we have consulted. This idea of information is not the one we have in mind. Information in the scientific sense has nothing to do with knowledge or intelligence. DNA can be considered as information. What else could it be? But knowledge and intelligence point to mental attributes.

A definition that better fits our discussion is:

> Information is any type of pattern that influences the formation or transformation of other patterns [148, 30].

This is interesting because it says that information is pattern, so pushing off the definition to another word which is the very word that this book is trying to define! This is the game of vish again.

However, there is another concept buried in this that becomes rather obvious when we work in a mathematical setting. We know that the same information can be manifested in very different forms. The contents of this book, no matter how you are reading it, whether physically in print, or on a monitor where it is being displayed from a digital file that in turn was most likely transferred to your device from some form of electro-magnetic signal, is information. And presumably it is the same information independent of how you have come to receive it. Very different pattern systems, but the same information. Information is not something attached to a single pattern system, but something that can be "passed" from one pattern system to another.

Let's go back a bit. We can summarize our conception of patterning as:

> Pattern is that which can be articulated qualitatively and quantitatively about a system and the evolution of its states as they undergo change.

The etymology of the word "system", which stands at the root of this definition, is the Greek word *systema* (σύστημα) composed from *syn* = together and *histania* = cause to stand. Its use in English begins in the 1610s meaning "the whole creation, the universe" and soon after that as "a set of correlated principles, facts, ideas" (see the Online Etymology Dictionary www.etymonline.com).

As we have previously defined it—the standard definition in the physical sciences—information is a numerical value that relates to the probability and partitioning in a pattern system. Let's call it the *numerical information*. Now we are seeing information defined as the pattern system itself.

However, we know that the same information can be carried in very varied forms. Information is not just one pattern system, but it can be represented as a pattern system. In fact that is the *only* way in which it can be manifested. It is a similar to the question we raised about the meaning of the number "five". In the end we found that it is an abstraction for any of the numerous ways in which it can be exemplified. In the case of the number 5, the issue is resolved by defining cardinality, putting together all instances where there is a one-to-one correspondence with some standardized instance, that is $1, 2, 3, 4, \mathbf{5}$. In other words, it is an abstraction for a set of instances which are all equivalent in this special way.

We can do the same thing with information. Going back to the definition given in §7.7.3, we say that

> two pattern systems carry the same information
> if they are pattern equivalent.

Returning to information, we conclude that information is not a pattern system but a class of pattern equivalent pattern systems, any one of which is a representation of that information. We call this the *systemic information*:

> Systemic information is a class of pattern equivalent pattern systems.

Transfer of information involves two pattern systems in which the receiving pattern system is a factor of a pattern system representing the information. We should think of this in the form of two pattern systems integrating into a larger system $\mathscr{Z}$ which incorporates the transfer of information. In the present context we are assuming that we are dealing with dynamics, so we phrase things in those terms. Thus if $\mathscr{Y}$ represents the source of the information and $\mathscr{X}$ the receiver then $h(\mathscr{Z}) = h(\mathscr{Y})$ represents the situation $\mathscr{Y} \rightsquigarrow \mathscr{X}$ so $\mathscr{X}$ shadows $\mathscr{Y}$ and is in effect a receiver of information. If in addition $h(\mathscr{Z}) = h(\mathscr{Y}) = h(\mathscr{X})$ we can say that the transmission is perfect. Then we have $\mathscr{Y} \leftrightsquigarrow \mathscr{X}$, which is the meaning of equivalent pattern systems.

8.4.11 Time

Time is one of the most mysterious of concepts. We have already seen three different ways of thinking about time. The first is the standard concept of an absolute time that flows continuously and inexorably. This is the time that we have seen in our pattern systems with continuous dynamics. The other two arise in the context of pattern systems with discrete dynamics. In modeling continuous temporal systems as discrete systems we can mark off the continuous time evolution in discrete steps of equal length, say in seconds, sampling the state space once each second, resulting in a discrete iterative system. Quite different is to mark off time in terms of changing events. Instead of a steady ticking clock marking off seconds, we follow the trajectory of a state by following the successive parts of the partition that it lies in. This may seem more irregular, but perhaps it is more fundamental. This way of looking at things is based on the idea that time ultimately comes back to the sequencing of events, not to some absolute time that endlessly flows on in the background.

The idea that it is change that is fundamental and that time is a quality drawn out of our perception of change is an old one. Aristotle (384–322 BCE) in his *Physics* puzzles about time and tries to get some grasp on what it is.

> There would be no time if there were only a single now, rather than different nows, and by the same token, if the difference between nows is not noticed, the time between them does not seem to exist ...clearly time does not exist without change. [6] IV.11.

Lucretius (see §12.2.1), with his highly empirical, physical, and experiential outlook, writes

> Likewise time has no independent existence: rather from events themselves is derived a sense of what has occurred in time past, of what is happening in the present, and what is to follow in the future; and it must be admitted that no one has a sense of time as an independent entity, but only as something relative to the movement of things and their restful calm [108, p. 1.459].

Given the closeness with which Lucretius follows Epicurus (341–270 BCE), we can assume that this was also his view.

It was Newton who specifically set time as a continuous, absolute, and measurable quality of the universe, which along with continuous, absolute, and measurable space set the frame in which his physics worked. His model has been so successful that we all think of time and space in these terms, even though we know that it is not quite right. In fact, not really right at all. Einstein's special theory of relativity and his general theory of relativity destroyed any idea of absolute time and knit space and time into a malleable unity that is inseparable from gravity.

Yet even in Einstein's view time/space is a continuous in nature. Newton's theory was successful because it was so closely able to model and so able to predict observable events. Einstein's theories arose out of a deeper understanding of the physics of electromagnetism and the increasing failure of Newton's theory to explain observable events, notably subtle astronomical deviations and the entire world of physics at the atomic level. Quantum mechanics puts question to the idea that time and space are continua in the ways that we have come to believe in them. In current physical theories that seek to find all-encompassing models of physics, the continuum idea of time breaks down at the scale of Planck time (approximately 5.4×10^{-44} seconds). What that might mean remains an open question. Is time ultimately divided into minute discrete steps or is it continuous? Or, perhaps, is it neither?

> And time is a dictator, as we know it.
> Where does it go? What does it do? Most of all, is it alive?
> Is it a thing that we cannot touch and is it alive?
> And then, one day, you look in the mirror — you're old–
> and you say, 'Where does the time go?' [152, Nina Simone]

Time remains a mystery and there is little agreement amongst physicists, even those whose entire careers center around such ideas. However, the ancient idea that time is really a human concept derived from our experience of change is still quite convincing.

The story is eloquently told in Carlo Rovelli's *The order of time* [138]. After demolishing other ideas about time, Rovelli writes

> None of the pieces that time has lost (singularity, direction, independence, the present, continuity) puts into question that the world is a network of events. On the one hand, there was time, with its many determinations; on the other, the simple fact that nothing *is*: things *happen* [his italics].

These are philosophical questions. Our point is that we can think of time as driving a dynamical system with change as a consequence, or we can see the changing events as driving it, with time as a consequence. Pattern systems will work either way. These are human concepts derived ultimately from our senses and the tools we have created for explaining reality.

Entropy and the arrow of time

> The statement that entropy never decreases is the only mathematical relation of physics that knows any difference between past and future. Behind this unusual [relation], a whole world lays hidden. Revealing it will fall to an unfortunate and engaging young Austrian, the grandson of a watchmaker, a tragic and romantic figure, Ludwig Boltzmann. [138], Ch. 2.

The equations of change that form the basis of modern physics are all time-reversible. From their perspective, potentially time can run backwards just as well as forwards, the future returning to the past.[15] Of course this is completely counter-intuitive to the way in which we experience the world. The present cannot be undone to recover the past. Watching a movie played backwards rarely looks believable for even a few moments. How can it be that the laws of physics cannot manifest the direction of time, but our actual experience of time is so very much one directional?

Time remains a philosophical and physical riddle, but the discussion of entropy does shed some light on why we can only experience it as flowing in one direction. The laws of physics are deterministic and reversible, but only at the level of ultimate precision. These laws are typically framed in space and time coordinates that are exact real numbers. At the practical level there is an inevitable blur or lack of perfect resolution in our ability to separate states, and this is why the partitions of the state space (and ultimately the event space) appear. A little bit of fuzziness or blur would not be such an issue if it were not for the fact that the evolution of a pattern system typically makes it worse. This inexactitude can be measured as entropy, and the dynamical entropy measures the average rate of increase of entropy. Only if this dynamical entropy is zero do we have any hope of recovering the past. In the end, this is not just a practical issue. In what sense does infinite exactitude exist in an apparently finite universe?

This discussion on the arrow of time can be given more mathematical content by considering a simple shift system $\mathcal{X} = (X, s.\mathcal{P}, \mathcal{E}, \mathbf{m})$ relative to some sameness/difference partition $\mathcal{P} = \{P_1, P_2, \dots P_n\}$. We need not assume anything about how large the alphabet underlying the shift system is or what particular invariant probability measure is placed on it. The dynamical form of entropy that we have introduced comes from the evolution of the partition $\mathcal{P}$. In order to think of the system in temporal terms we will suppose that the shift s operates in uniform time steps, so what we witness as we watch the shifting of a state—an infinite string of symbols from the alphabet—is the time evolution of that state.

Now we ask a curious and interesting question: how much would we be able to know about the present if we knew the future? How much the future determines the present is really the question of what running time backwards entails. It turns out there is an exact statement about entropy that throws light on this. To understand what it means we have to look again at our idea of evolution of the partition $\mathcal{P}$ under the shift action. The key idea is that as the dynamics progresses we watch the various itineraries that the parts of $\mathcal{P}$ can produce, namely the succeeding refinements $\mathcal{P}^{(2)}$, $\mathcal{P}^{(3)}$, ... that we defined in §6.5.8. Recall, for instance, that $\mathcal{P}^{(3)}$ is the partition of X made from all possible sets

$$P_i \cap s^{-1} \triangleright P_j \cap s^{-2} \triangleright P_k \,,$$

that is, all the states that start in P_i and then pass to P_j and then to P_k, where i, j, k run independently over all possibilities $1, 2, \dots, n$. This goes from present to future. What we want to know is how the future affects the present when phrased in terms of entropy. For this we look at these same sets, but push them one further step into the future , so for instance $s^{-1} \triangleright \mathcal{P}^{(3)}$ is the partition of all possible sets

$$s^{-1} \triangleright P_i \cap s^{-2} \triangleright P_j \cap s^{-3} \triangleright P_k \,.$$

This is the set of all the states that will next be in P_i, then pass to P_j, and then to P_k, where i, j, k run independently over all possibilities $1, 2, \dots, n$. This is the future of the itinerary. The question becomes "what does that tell us about $\mathcal{P}$?" Here is the result [184, Thm. 4.14]:

$$\lim_{n \to \infty} H(\mathcal{P} \,|\, s^{-1} \triangleright \mathcal{P}^{(n)}) = h(\mathcal{P}) \,, \tag{8.12}$$

or read in words, the *entropy due to the partition $\mathcal{P}$ that is unaccounted for by knowing the entire future evolution of the system is exactly the same as the dynamical entropy of the system.* In the sense that entropy is a measure of uncertainty, the present cannot be recovered from the future (and here we mean the entire future!) unless the entropy of the system is zero. That amounts to saying that the past cannot be recovered from the present.

We can interpret this as supporting the idea that dynamical entropy underlies the arrow of time.

Entropy saves us from deterministic fate?

It is easy to lament the effect of entropy. We are told that entropy always increases, and we viscerally feel it when we witness everywhere the inevitable collapse of form that it entails. But again, zero dynamical entropy has an implicit meaning, this time as we pass from the present to the future. As we have seen in §8.3.3, the dynamical entropy measures the average step-wise increase in entropy as the system evolves. That is, it measures the step-wise uncertainty that accompanies each step of the system into the future. If this entropy turns out to be zero, then that means that there is ultimately no uncertainty, at least as measured by entropy. In effect, nothing new can happen. A world with no entropy is a world in which creativity (taken at the physical level) would come to an end. We might say that far from being totally negative in its effects, it is positive entropy that allows the present to be different from what is dictated by the past. Entropy is what saves us from a determinism.

But as we just said, it also allows us to reinvent history! If we look at a system with positive entropy then the future leaves us with uncertainty about the present, which amounts in particular to saying that the present does not determine the past. We can see both a tree branching into the future and a complexity of roots spreading back into the pasts (!) that can account for the present. The present is a conduit which channels the past into the future. Of course reinventing the past is a favorite pastime of politicians and nationalists of all persuasions. We all like to produce plausible stories for our present situation. Perhaps entropy can offer us some insights into just how this works!

8.4.12 Metaphor

Most of us coming out of high school classes in English literature know metaphor as a poetic device. It is used to bring deeper meaning to a concept by identifying it with another, often quite different, concept. The metaphor is usually understood as a direct assignment that "this" is a "that". In referring to a warrior as a lion in battle, we bring to him the image of strength, power, courage, and even dignity. However, contemporary study of language and cognition has come to understand metaphor in a far wider sense where it becomes an essential aspect of abstract thought and human cognitive experiences. The principle underlying this wider sense of metaphor is often referred to in the literature as one of "mapping" one contextual system onto another, where one can use previous understanding of the latter to give meaning to the former [64].

This has significant philosophical importance. As Mark Johnson puts it "Philosophy's debt to metaphor is profound and immeasurable. Without metaphor, there would be no philosophy. ...Where a philosopher stands on this key issue [philosophy's debt to metaphor] can be determined by their answer to one question: are our abstract concepts defined by metaphor, or not?" [91]

Thus metaphor is increasingly understood to be far more than simply an intentional device that we use to enhance meaning. It is understood as a basic feature of our cognitive processing. This is particularly the case with abstract ideas. Our fundamental association with the world is through the senses. When we come to abstract ideas that are not directly associated with the sensory world, we find that in one way or the other they ultimately link back to the physical world. Take for instance the simple sentence:

> It is <u>surprising</u> to <u>realize</u> the <u>extent</u> to which <u>metaphor</u>
> <u>pervades</u> <u>language</u> and <u>shapes</u> our <u>understanding</u> of the <u>world</u>.

Each of the underlined words is an abstraction whose etymology traces it back to the physical world. The word *metaphor* itself, for instance, arises as a combination of the Greek words *meta* = between and *phero* = carry, thus linking the idea to the carrying of something between two places. This is exactly what metaphor does. Metaphor is an abstract concept, but the word *metaphor* literally refers back to the physical world and thereby carries meaning.

This is not to imply that these roots convey the entire meaning that we now understand by particular words. The idea is to suggest how metaphor is bound into the cognitive processes of our brains. As concepts build on each other, we may use metaphor to link new abstractions to older ones that we already know. Thus, for instance, the word *application* in the sense of an *app* used to perform functions on a computer derives from the common sense of *application* as putting something to use. In turn this word derives from the verb *apply* that derives from the Latin *ad + plicare*, which means to fold, bend, or roll up, and ultimately has its roots in weaving or plaiting. Metaphor in this sense is the act of inferring meaning by comparison to, or even identification with, situations or patterns of process that are already understood.[16]

This wider understanding of metaphor has emerged very much in parallel with the increasing understanding of the neural structure of the brain and accompanying advances in cognitive science. In our terms we can see metaphor as the matching of pattern in one place with pattern in a second place. Often this pattern matching has both temporal and spatial components. One of the great metaphors that is universally common in human societies is that of life as a voyage. All the joys, difficulties, unexpected discoveries, terrors, perseverance, and good fortune that accompany travel can be linked to our journey through life.

> Months and days are eternal travelers, as are the years that come and go. For those who drift through their lives on a boat, or reach old age leading a horse over the earth, every day is a journey, and the journey itself is their home. Many people in the past have died on the road, but for many years, like a fragment of a cloud, I have been lured by the wind into the desire for a life of wandering [2, Basho, *Oko no hosomichi* (Narrow Road to the North)].

Here the more abstract pattern of our evolving life is placed into the more empirical pattern of travel. By making the connection between one pattern system and another we are able to explore commonality that they share, and through that connection, consciously or unconsciously, give greater depth to our understanding.

With this in mind we can see that there is an aspect to metaphor that we can understand at the level of pattern systems: the process of mapping one conceptual system (pattern system) to another. In our language this is what homomorphism does, see §7.6.3. As a reminder, the word homomorphism literally means similar shape, in this case meaning similar structural shape as seen from the point of view of pattern systems. The point here is that one system, which is essentially a system of relations, is matched to another system and its perceived relations. The metaphor assumes that at least to some degree this matching is effective enough to carry useful semantic content. It does not mean that the two systems are alike at the level of states, even remotely so, but rather the systems are alike at the level of the structural relationships—there is *structural resonance*. The metaphor is valuable to the extent that it works. Thus for simple purposes it is a satisfactory metaphor to imagine atoms as miniature planetary systems though for serious science it is impossibly flawed.

Beyond this there is another level at which metaphor is interesting in the mathematical context: mathematics itself is based on metaphorical associations. This idea, that mathematics itself might be highly dependent on metaphor for meaning, is one that seems strange at first, and even completely wrong. But very good cases have been made for the truth of this. As a simple example, consider the way we have spoken of one pattern system being a shadow of another. This makes use of a metaphor that takes into account that the shadow system is somehow a version, though perhaps with less detail, of the original, and even more, that like a shadow it follows the same dynamics. The word thus gives meaning to a relation that is properly defined only in abstract terms.

More significantly, it has been pointed out by Nuñez [122], and we can see it in our entire discussion of dynamics, that it is only through metaphorical transference that the idea of actual action and change is encompassed in the mathematics. In discussing dynamics there is nothing that literally changes; there are no evolving processes as we might imagine in a rolling wave or a moving tape. All that we have to indicate motion is an assignment of pairs (x, y) indicating an association of states, but not the actual movement. There is no suggestion of how x got to y, what changes of molecules were involved, or how the process was accomplished. We talk of the trajectory of a state, but states don't actually move. The collection of states is just the same as it was before the change. We can look at the symbols on the page, but they are not moving. Instead we have to learn how the formalism of the mathematics matches actual dynamical process by using metaphorical association based on our own experiential knowledge of process and change. Once we do that we can immediately understand what we mean by periodicity, repetition, or iteration in discrete time steps and then give the appropriate mathematical definitions that fit those words.

The history of mathematics suggests that this particular metaphor was essential, but that it is far from obvious. Mathematics in one form or another is ancient, and often very sophisticated, but it was only in the seventeenth century in Europe that effective ways of assimilating movement and change into mathematics were found. Much of the book to this point has been devoted to establishing this metaphorical connection in the context of the dynamics of pattern and change.

We do not go so far as some to say that all of mathematics is derived from metaphor. There are numerous wonders of mathematics that are the outcomes of pursuing directions that emerge from the mathematics itself, and it is hard to see their origin as being metaphorical. In ‡§16.7 there is a discussion of the prime numbers and their distribution. The set of ordinary natural numbers $1, 2, 3, 4, \ldots$ seems very worldly, deriving from the simple necessity of counting. Yet the mathematics of the set of all natural numbers is one of austere beauty derived through long chains of reasoning that depend neither on faith nor empirical observation. Towards understanding its mysteries many mathematicians, both now and for the past few thousand years, have dedicated extraordinary effort. It's safe to say that the vast majority of them have seen themselves as working in a world of ideas in which there are absolute truths. We have included ‡§16.7 partly to give some insight into this. The physical world does not come to us in the form of end results of long chains of logical deductions based on unassailable axioms.

All this being said, we might look back at the development of dynamical systems and pattern systems, both as part of the history of science and as we have explored them in this book. The outcome is an abstract system that can serve as a metaphorical device for modeling any of a vast variety of our experiences in this ever-changing world. Once set up though, it has its own theory, its own consequences, its own principles, and its own revelations. Thus it becomes a part of the mental realm, seemingly free of its origins in the physical realm. Yet its discoveries can be returned, using the metaphor in the opposite direction, to give meaning to the physical world that we did not have before.

Although we have not attempted to create any mathematical theory of metaphor, we can see that the potential for metaphor exists whenever there is an synthesis of two pattern systems $H(\mathscr{X})$ and $H(\mathscr{Y})$. To the extent that there is positive consolidation, there is commonality between the two systems and so the potential for using the features of one to stand for features observed in the other.

8.5 The Birkhoff ergodic theorem

8.5.1 The process of induction

The way in which we humans come to understand the processes behind recurring events is to observe them over time and draw conclusions on the basis of repeated observation. This is how any animal learns to live within its environment. It is called is the principle of *induction*—the inference of generalized conclusions from particular instances.

It has been argued numerous times, especially since the writings of David Hume (1711–1776), that no logical conclusions can truly be drawn about general principles from the observation of particular instances. The history of science is replete with examples of the truth of this, yet induction is the only way we have to draw conclusions about this world. If we were not able to do this, human existence would be impossible. The Birkhoff ergodic theorem, which is the subject of this section, offers insight into induction, assuring us that we can draw reliable information from our models of reality in spite of the apparent gap between observation and proof.

Suppose that we have a pattern system $(X, s, \mathscr{E}, \mathbf{m})$ with discrete iterative temporal dynamics. We will assume that the law of the process, the measure $\mathbf{m}$, is an invariant probability measure. In temporal terms its invariance says that the probabilities of occurrence of events are not affected by the particular time we begin or end our measurements.

We know that it is the measure $\mathbf{m}$ that ultimately controls the shape of the dynamics of the system, the principles on which it evolves. But here is the issue: often we don't know $\mathbf{m}$. This is a common enough situation: we can observe the dynamics and watch what happens, frame a hypothetical system, but not know an actual measure that controls it. The most natural way to observe a system is to make numerical observations about how it looks in various states. Typically this involves making measurements which collectively describe a function on the state space—recording temperature, wind velocity, brightness, frequency, and so on. These are typical sorts of observations that we can make on a system. In doing so we are, in effect, describing the system through the values of a function

$$f : X \longrightarrow \mathbb{R},$$

where for each state x, $f(x)$ is the observed value at x. In fact such functions are actually called *observables* in physics.

We can record our observations along the trajectory of a single state as it progresses in time. Presumably this is all influenced by the dynamics of the system, which in turn is shaped by the measure $\mathbf{m}$. The mathematical problem is then to uncover the law $\mathbf{m}$. But what mathematical reason might there be that tells us that long-term observation can do this?

At this point we can see how the intuition that we have for evolving systems is typically used. We want to know about the whole system, but the best we can do is to look at the evolution of an individual state of the system, or the evolutions of some finite number of individual states of the system. The trouble is that there is no way that in starting at one state (or even any finite number of states) and watching it evolve over time (even if we had an infinite amount of it) we will ever see all the states. We are only going to see states that lie on the particular trajectories we choose to follow.

The story, which we have already alluded to, arises out of statistical mechanics and the work of Ludwig Boltzmann. The dynamical systems for him were systems of atoms, say a gas in a container, free to move at random. The questions arose around the observable outcomes of such systems, which are the outcomes of astronomical numbers of moving particles. Boltzmann faced the problem of whether measurements, which are at the macroscopic level, could adequately lead to correct conclusions about the dynamics. He wanted to believe that no matter at what state in a physical system one began, that state would eventually evolve through *all* the other states of the system so that over time the physical system would, all by itself, explore all of its possibilities. But it is evident (especially since we now have much more mathematics at our disposal than Boltzmann had) that this, taken literally, cannot be true. In most pattern systems, the time trajectory of a single state does not pass through every state of the dynamical system—not anywhere close. We can see this with the simple zero-one shift. Given any state whatsoever, simply applying the shift s repeatedly to it will not eventually yield every possible state (that is every possible infinite string) of the system, far from it. Repetition of the shift certainly produces new states, but even if continued forever, it can produce only a *countable* number of new states, whereas we know that the state space of this shift system has *uncountably* many states. Even if we were to repeat the process endlessly there would still be a vast and uncountable number of states that would not been seen. Put this way it seems that the task is impossible.

This issue was used to criticize Boltzmann and it was at least partly to blame for the mental anguish that led to his suicide. There are other things, too, to worry about. For example, let's go back to shift systems and the tossing of a coin. Recall that one trial is considered to be an infinite string of throws, so each outcome is a long (infinitely long) list of zeros and ones (tails and heads). The state space consists of all possible outcomes (trials). Of course we know what the statistics are supposed to look like, assuming a fair coin. Boltzmann's problem is that of a person who gets to look at just one trial, and even at that, only a finite extent of it. Infinite though it ideally may be, it is just one of uncountably many trials, and we could get unlucky. An infinite string of heads is very unlikely, but actually no less likely than any other particular string. Even an imbalance of a few percent on average, say 1% more heads than tails, would completely ruin the statistics we expect. But there are infinitely many trials that have this property. Do we have any right to expect to get right answers in this way? Intuition says it ought to be okay, and in fact we have seen that the law of large numbers tells us that in this particular case it will "almost certainly" be okay. Still, the question remains, how much can all of this tell us about the measure $\mathbf{m}$?

What we want to introduce here is Birkhoff's beautiful and subtle theorem that actually includes the law of large numbers as a special case, but is more intuitive and more transparent about what we are actually saying.

8.5.2 Some background

We start with a *discrete* dynamical system $(X, s, \mathbf{m})$ and a function $f : X \longrightarrow \mathbb{R}$ which we think of as an observable. Pick some state x and then follow its trajectory $x,\ s \triangleright x,\ s^2 \triangleright x,\ \dots$ and the corresponding values $f(x),\ f(s \triangleright x),\ f(s^2 \triangleright x),\ \dots$. We are interested in the average value of the observable f over this trajectory, which reads

$$\lim_{N\to\infty} \frac{1}{N} \left(f(s^0 \triangleright x) + f(s^1 \triangleright x) + \cdots + f(s^{N-1} \triangleright x) \right) . \tag{8.13}$$

Ignoring the limit, we see the average of the first N values of f along the trajectory. We then look at how this average looks as we let N get larger and larger, in other words the limit.

Now the issue is this: how well does this average resemble the real average of f over the entire state space X? This average involves both f and the measure $\mathbf{m}$. This mathematical notation for this average (see §11.10, §11.10.3) is

$$\int_X f \, d\mathbf{m} . \tag{8.14}$$

The principle of induction (in the philosophical sense that we have been talking about) suggests that the two quantities of (8.13) and (8.14) should be equal. This is the question that the Birkhoff ergodic theorem addresses.

It is important to realize just how different these two quantities are. The sum represents the poor observer (that's us) who wants to understand an unknown system. We don't know what X is like and we don't know what the invariant measure $\mathbf{m}$ may be. We can make lots of experiments with different observables f. What the system is really like is expressed by the overall averages of these observables when weighted by the measure $\mathbf{m}$ and taken over the entire space X, that is, (8.14). What we get to observe is the system starting at some particular state (which is probably unknown) and watching what happens as that state evolves over a long series of iterations. We can hope that the average value of our observations is actually converging to the real overall average, but is that really likely to happen? The good news is that the answer is yes—but as usual there is some fine print.

First of all, we can't expect this to work every time because we know that there really are "bad" trials. The saving point turns out to be that although they can occur, the "bad" trials *in totality* have zero probability of happening. If f is an observable and B is the set of all "bad" states, ones that have an average different than the one given by the real average of f over X, then $\mathbf{m}(B) = 0$. This is the same as the "almost surely" proviso in that we have seen in the law of large numbers. Here it is expressed by saying that except for a set of states of $\mathbf{m}$-measure zero, we will get the desired equality of two quantities given by (8.13) and (8.14).

There is a second issue that we might not first think of. This can be exemplified by the infamous VW diesel emissions scandal. Suppose that we test a car to see if it passes the local exhaust-emission standards set by the state. We take a number of tests and get excellent results. Later we find out the manufacturer has rigged the engines to detect that a test is being made, and only then does the car turn on the full emission controls.[17] Then the reality is that unexpectedly we are looking at two systems, and depending which one we are in we get totally different answers.

Mathematically this is avoided as an explicit condition: the pattern system is called *ergodic*, or the invariant measure $\mathbf{m}$ is called ergodic, if it cannot be decomposed into two totally disjoint systems, each invariant under the dynamics, and each of which has *positive probability* of occurring.[18] For instance the standard zero-one shift system $\mathcal{S}$, viewed as a coin tossing system with a fair coin, is ergodic (as is any Bernoulli shift). It certainly contains invariant subsystems—the set of all states that ultimately just repeat with all ones is such a system—but their combined probability of occurrence is zero. Quite different is the case of the simple pendulum, where the dynamics splits into three very different regions—there are librations, rotations clockwise, and rotations counterclockwise, each of which is an invariant subsystem. It is not ergodic.

Ergodicity is ultimately a property of the measure $\mathbf{m}$, and so we can equally speak of $\mathbf{m}$ as being ergodic or not ergodic.

8.5.3 The Birkhoff ergodic theorem

With all of this we finally arrive at the Birkhoff ergodic theorem. Here we will give the version of the theorem that deals with discrete dynamics. There is also a continuous version where we begin with a pattern system in which the dynamics is the usual flow of continuous time.

Suppose that $(X, s, \mathbf{m})$ is a discrete dynamical system where $\mathbf{m}$ is an invariant ergodic probability measure. Then for any measurable function $f : X \longrightarrow \mathbb{R}$ (see (11.31) in §11.10.1) and any $x \in X$,

$$\lim_{N\to\infty} \frac{1}{N} \sum_{k=0}^{N-1} f(s^k \triangleright x) = \int_X f\, d\mathbf{m} \tag{8.15}$$

almost surely.

The "almost surely" means that the result is true for all starting states $x \in X$ except for a set of states whose totality is of $\mathbf{m}$-measure zero. The word measurable is explained in §11.10.1, but it should not be considered as a major concern since any function likely to occur is surely going to be measurable—in fact it is a real challenge to construct one that is not measurable.

Although this is stated at the level of functions, it does imply that the measure itself is revealed inductively, for any measure is completely specified by the average values it gives to functions. Two different measures are bound to give different average values on some functions. Even more, although it is not explicitly stated, under quite general conditions the same evolution of a state x can be used for any f with the desired convergence.

In the case of continuous dynamics the trajectories are continuous flows and the average value of f along a trajectory has to be written as an integral (effectively a continuous sum). The definition of ergodic remains the same. Then the theorem reads:
Suppose that $(X, \mathbb{R}_{\geq 0}, \mathbf{m})$ is a dynamical system where $\mathbf{m}$ is an invariant ergodic probability measure. Then for any measurable function $f : X \longrightarrow \mathbb{R}$ and $x \in X$,

$$\lim_{T\to\infty} \frac{1}{T} \int_0^T f(t \triangleright x)\, dx = \int_X f\, d\mathbf{m} \tag{8.16}$$

almost surely.

G. D. Birkhoff proved this theorem in 1931. The proof is not particularly easy, but the theorem is impressive. Its formulation seems deeply physical, deeply informed by metaphor, and subtly aware of the surrounding difficulties. In some way it also justifies the ways in which we seek knowledge and gives more confidence in the principle of induction as a means of obtaining knowledge. We can never do more than sample the system, but given enough time we can expect to get appropriate answers. It also gives some insight into how we can say that the parts imply the whole (see §8.4.8), for we can see that the evolution of even one state somehow implicitly implies the entirety.

This is straight pure mathematics—the actual formulation of the dynamics and all the mathematical entities within it and its various proofs are completely formalized in mathematics. But it says something that is greatly relevant to the way we learn about the world.

Boltzmann's work dates to times before measure theory or a properly developed theory of dynamical systems existed. At the time he could not have phrased the statement of Birkhoff's theorem.

In §5.1.6 we discussed the law of large numbers in the setting of an independent identically distributed set of trials of a random variable, and its almost sure outcome. It is easy to derive the law of large numbers from the discrete version of the Birkhoff theorem, see the Endnotes.[19]

8.5.4 Laws and chance

The question that naturally arises in thinking about the Birkhoff ergodic theorem is the pre-supposed existence of an invariant ergodic measure. How big an assumption is this? As it turns out the existence of at least one invariant measure is guaranteed under relatively tame assumptions, and if there are invariant measures then there are guaranteed to be *ergodic* invariant measures. This result is known as the Krylov–Bogolyubov theorem [118].[20]

There is an interesting consequence to this. We have seen that the Birkhoff theorem gives us reason to support the process of induction, which we might say is how we use observation to conclude correct information about the laws governing dynamical processes. But there is another way to think of it. Given that under mild conditions there will be some invariant measure, some law implicit in the process, we can see the Birkhoff theorem as working in the other direction and making up a law by telling us what it should look like. It even tells us how to interpret the resulting system in a probabilistic way, since having derived the measure from the averages, naturally the law "predicts" those averages.

It seems like a slight difference, but perhaps an important one. Rather than our observations leading us to "discover" a law that we now see as some pre-existent universal ingredient responsible for the unfolding of the process, instead we can say that the process, when seen from a certain perspective, can lead us to *interpret it* as unfolding with respect to some law.

Often one hears about how subtly the various variables underling our models of reality have been set, and that with even slight differences life as we know it could not have existed. Underlying this way of thinking is the idea that there is some wonderful set of equations with various variables, and that setting these just right produces the universe we see. Then it seems miraculous indeed that things have been set just right for us to be here at all [4].

But what if we look at this from the other way? We can say that it is the workings of the Cosmos that *create* the invariances we conceive, so it is quite natural that they will turn out to be aligned with what we see. This is not to say that the Cosmos is not wonderful, nor that the laws we have derived are not both beautiful and amazing. It simply takes inspiration from the Birkhoff ergodic theorem, suggesting that the right way to see things may be to see that *the laws are an interpretation of the process*, not the other way around. Naturally then these laws are seen to be in marvelous alignment with reality! The laws, marvelous as they are, are still conceptual ideas that we are led to by observations and mathematical formalism. They are our creations, not ideal entities written in some transcendent book of laws.

We will see these ideas come up again in quantum theory in §14.5.2 when we look at ideas of chance and locality in quantum theory.

8.6 Conclusion

Entropy starts off in our minds as another name for disorder. The purpose of this chapter has been to see it as something more constructive, a tool by which we can establish a relationship between pattern systems. Especially important is the relationship that we have called pattern synthesis. When pattern systems are placed in relationship, their integrated totality becomes another yet another pattern system, and this may be something quite different than its constituent components suggest. Whatever dynamics arises out of combining pattern systems must in itself be patterning. There is nothing especially unusual about this. We are quite familiar with the ideas behind analyses that seek to explain complex situations by looking at simpler component parts. Analysis into simpler components is essential, but it is only truly effective when it is combined with synthesis to reveal the workings of the whole.

Synthesis of two systems is based on the idea that what we are really talking about is coherence of states—about how what happens in one system is reflected, at least in some way, in the other. The overall effect of coherence is the lowering of the total disorder, and it is this that can be made explicit by thinking in terms of entropy. If entropy represents disorder, then coherence ought to mean decreasing entropy. What may be unexpected is the way in which synthesis leads to ideas about shadowing, pattern equivalence, redundancy, determinism, and metaphor. All of these are about relationships between pattern systems. Relationship in this context means the synthesis of pattern systems, and it is entropy that measures the degree of coherence between the two systems and hence, reciprocally, broad implications about what the relationship signifies.

The other side of entropy is the slippery concept of information. The definition of information put forward by Shannon and commonly used in information theory and physics is that information is numerically the same thing as entropy, though often taken with the opposite sign and called *negentropy*. Entropy is about uncertainty, in the sense that there is potential for many events to occur and uncertainty about which of them will occur. Uncertainty produces the potential for uncertainty to be removed and so the potential for information. Information is thus seen as the removal of uncertainty. In a situation of no entropy there is no uncertainty and so no potential information either.

However, this idea of information, which amounts to a numerical value, is not one that captures our full idea of information. Information in the usual sense is not just a number, but rather elucidation of events in a context of events. Thus we see information as needing systemic context in which it can be expressed. A more intuitive idea of information is that it is pattern or patterning, and that is the way in which it is sometimes expressed. Of course if information is pattern we are left with the question of what is pattern. That is the question we are trying to answer. However, even so, this cannot be the whole story because we all realize that the "same" information can appear in many forms. This leads to the idea that information should really be described as pattern equivalence classes.

Beyond the mathematical ideas of this chapter there is an under-current of philosophical ideas about how we understand the world. The discussion of entropy and information leads us beyond scientific or mathematical contexts. These are words that relate to the ways in which we experience, explain, and predict the events of reality. Our experiences of reality are based on events and the elaborate connective pathways of mental structure that give them coherence and meaning. The latter part of the chapter has dealt with some of these philosophical issues that naturally arise when we think beyond the pure formalism. The value of the formalism is not just itself, but the increased abilities it offers in articulating and understanding how the world works—the "what's the go o'that?", as James Clerk Maxwell would have said.

The last part of the chapter brings in a very definite mathematical idea that has deep links with another way in which we see the world. The most basic way in which we (and no doubt countless other animals) infer the regularities of the world is by repeated experience. This is the principle of induction: the inference of repetition of events based on past experience. The Birkhoff ergodic theorem sounds technically forbidding, but its basis can be viewed as deeply metaphorical and its conclusion offers remarkable insight into why induction works. We will see later in Ch.10 that the neural processes of Hebbian learning reinforce it. The Birkhoff theorem shows that within the pattern system context repeated occurrence does offer confidence of continuing recurrence. At the same time the Birkhoff theorem gives us a way to rethink our understanding of the law of a process, seeing laws more as creations arising out of experience rather than transcending entities that direct it.

Any true understanding of the world, both in its natural processes and increasingly in the additional processes that human society imposes on it, must take account of their implicit systemic nature and their inherent interconnectedness. All things fit together. Difference and sameness are woven into a seamless brocade. We finish this chapter with a few lines from the famous Chan (Zen) Buddhist text, the *Cantongqi* (*Sandokai* in its

Japanese form), which expresses these ideas in its own beautiful and subtle way.

> Grasping at things is surely delusion
> according with sameness is still not enlightenment.
> All the objects of the senses
> interact and yet do not.
> Interacting brings involvement
> Otherwise, each keeps its place.
> ...
> Fire heats, wind moves,
> water wets, earth is solid.
> Eyes and sight, ear and sound,
> nose and smell, tongue and taste.
> Thus for each and every thing,
> depending on these roots, the leaves spread forth.
> ...
> Each of the myriad of things has its merit,
> expressed according to function and place.
> Phenomena exist, like box and lid fitting together;
> principle accords, like arrow points meeting.
> Hearing the words, understand the meaning;
> don't set up standards of your own.
> If you don't understand the Way right before you,
> how will you know the path as you walk?[21]

Parts are parts, but see the whole. The whole is the whole, but don't see it as such. The experiences of our senses are themselves, yet they interact. Everything is in exquisite perfection, yet our own visions of it cannot be held as universal standards. Things are the natural outcomes of the entirety of conditions from which they emerge. This is the deeper meaning of what systems entail.

Unfortunately we do not live in a culture in which such things are self-evident.

Chapter 9

Symmetry and invariance

9.1 What is symmetry?

If anything appeals to our natural sense of pattern it is surely images such as those shown in Fig.9.1, Fig.9.2, and Fig.9.3. There is no question as to the aesthetic sense of beauty that we feel when we see these patterns, which are typical samples drawn from the great Islamic and Celtic artistic traditions. They seem to embody a local sense of artistic playfulness held within a rigid formality that speaks to a profound order that underlies the nature of things. Somehow we feel that if we could just get the pattern started then the rest would follow automatically. These patterns seem to derive from a basic motif that in itself may be as complex as we please, which is then taken over by some deep internal laws so as to fill out an entire surface. There are instances of translational repetition, rotational and reflective repetition, and even repetition at different scales. Yet the whole thing is held within some formal structure, and that structure speaks to our sense of symmetry.

This section introduces the study of symmetry and its relation to the forms of pattern we have seen so far. There are perhaps two key features in what follows that may come as revelations. The first is that it is not just in art and decoration that symmetry is important. First, symmetry is deeply embedded in the fabric of Nature, to the extent that all modern theories of physics are built around symmetry and without it cannot be properly expressed. The second is that there is a mathematical formalization of symmetry (called *group theory*) that not only rationalizes and even classifies the forms that have been discovered instinctively by artists and designers, but brings to light a vast arena of unexpected hidden symmetries that have their own ethereal beauty and even manifestations in Nature.

Let's start with the word "symmetry". Like pattern itself, symmetry is a word that is called upon to fulfill a number of similar, but actually quite different, tasks.

Figure 9.1: The ceiling of the Lotfollah mosque, Isfahan. Assuming that this is just a rectangular section of what is really a circular figure (which is actually the case) there is obvious rotational symmetry and many reflective (bilateral) symmetries. For example, the one around a vertical mirror through the center and the one around a horizontal mirror in the center are rather obvious, but there are sixty-two others. In addition there is an expansive form of scaling symmetry as we move outwards from the center. Phillip Maiwald, CC BY-SA Wikipedia, en.wikipedia.org/wiki/File:Isfahan_Lotfollah_mosque_ceiling_symmetric.jpg .

Merriam–Webster's definition of symmetry runs like this:

- balanced proportions; beauty of form arising from balanced proportions;
- the property of being symmetrical, especially correspondence in size, shape, and relative position of parts on opposite sides of a dividing line or median plane or about a center or axis;
- a rigid motion of a geometric figure that determines a one-to-one mapping onto itself;
- the property of remaining invariant under certain changes (as of orientation of space, of the sign of the electric charge, of parity, or of the direction of time flow)—used of physical phenomena and of equations describing them.

All of these are perfectly valid and extensively used meanings of the word, but gradually we will want to assume meanings closer to the last two of the definitions, which are probably less familiar.

The first definition goes back to the original Greek conception of symmetry and is

Figure 9.2: This is but one of many fabulous tilings found in the Jameh Mosque (Masjid-e Jameh) in Isfahan, Iran. In the mosque the particular tiling tiles a relatively small square area—just the square presented in the center of the figure. It is left to the viewer to realize the way in which this tiling can be extended indefinitely through simple vertical and horizontal repetitions of the square shown in the center. Only in this expanded view does the remarkable symmetry of this tiling become fully apparent. See §9.32. B.O. Kane/Alamy Stock Photo.

Figure 9.3: The triquetra, an ancient symbol found in many ancient cultures including Persian, Celtic, and Japanese. Excluding the circular ring, the over and under structure also represents the simplest knot, the trefoil knot. The 3-fold rotational symmetry is obvious, but there is no reflective symmetry at all—this is a chiral figure. AnonMoos, Public Domain, Wikimedia Commons.

still used today in this sense. The second of the definitions is the other commonly used meaning of symmetry: the bilateral symmetry so prevalent in the animal world and so familiar in architecture, formal art, and decoration. The third definition is close to the standard one of mathematics, and the fourth is its evolution into the world of fundamental physics in which symmetry comes to stand for invariance in the presence of change. It is not hard to imagine that this must be important for our ideas of patterning.

These definitions show the fluid range of meanings carried by the word "symmetry", which have evolved over the centuries, and probably will continue to evolve, along with mathematical and scientific thought.

At the start of his classic little book *Symmetry* [187], H. Weyl states: "In one sense symmetric means something like well-proportioned, well balanced, and symmetry denotes that sort of concordance of several parts by which they integrate into a whole." Similarly Robert Bringhurst [24] writes: "The general sense of συμμετρος (*symmetros*, and adjectival form of συμμετρια (*symmetria*), is well- balanced, shapely, nicely proportioned. It is very close indeed to what we typically mean by harmonious." There is also the idea, appearing at the opening of Book X of Euclid's *Elements* where he defines commensurate magnitudes. The word used here for "having a common measure" is συμμετρα (*symmetra*). The idea of symmetry being related to having a common measure is also used in a more general sense: "The term also comes up in the *Nichomachean Ethics* of Aristotle, where the theory of money is discussed. What money does, he says, is make things συμμετρα. That is, it gives them a common measure". [24]. This is compatible with our non-technical sense of the word "commensurate", meaning that things or parts, or ideas are in concordance.

Now the concordance of parts into an integrated whole is a major theme in our discussion of pattern, even to the extent that the parts have no meaning except in the context of that whole, so in this sense symmetry is something that underlies much of our thought. However, the word symmetry has acquired more specialized meanings in science and this is where the other parts of the definition come in. Trying to capture in a simple phrase what this scientific view of symmetry is all about is hard, but we like the three-word definition of R. Healy [76] and later of Frank Wilczek (Nobel Prize for Physics, 2004) [190], who describe it in a delightfully paradoxical way:[1]

Symmetry is the study of change without change.

What this means, and how it links in with our thinking about pattern is our task ahead. However, before plunging into this we might pause for a moment on the question of aesthetics that we brought up in looking at the examples of Islamic and Celtic art. The human mind has a natural affinity for symmetry (in any of the senses above), and people have always felt that this symmetry derives from something deep within the nature of things, whether that be religious, scientific, or philosophical. It is a hallmark of the thinking of scientists, and especially modern physicists, that the universe unfolds along principles that resonate with our sense of beauty, and that beauty is particularly evident in terms of symmetry. Wilczek's book [190] is dedicated precisely to this idea, and it is well known that Einstein was guided by a sense of beauty in his work. The discoverer of anti-mattter, Paul Dirac (1939), wrote:

> "the mathematician plays a game in which he himself invents the rules while the physicist plays a game in which the rules are provided by Nature, but as time goes on it becomes increasingly evident that the rules which the mathematician finds interesting are the same as those which Nature has chosen", and therefore that in the choice of new branches of mathematics, "[o]ne should be influenced very much ... by considerations of mathematical beauty."

Perhaps we should not be surprised that scientists have found Nature is beautiful. It is.

9.1.1 Reflective symmetry

> The reason why the right side of our body appears in mirrors on the left is that, when the image reaches the plane of the mirror and strikes against it, it does not turn about and so remain unaltered, but rebounds straight back. ...It is possible too for an image to be transmitted from mirror to mirror, so that as many as five or six reflections are produced ...and whenever the left side is passed on, it becomes the right and then changes back again, reverting to the same position as before.—Lucretius [108, Bk.4, 293–310]

Bilateral symmetry is the symmetry of the second of the definitions introduced in §9.1: *correspondence in size, shape, and relative position of parts on opposite sides of a dividing line or median plane or about a center or axis.* This is reflective symmetry with the dividing line or median plane being the "mirror". What is on one side is the mirror image of what is on the other.

When we look into a mirror we see another world in which each object on our side of the mirror appears at an equal distance away from the mirror but directly opposite, on the other side. The image is a virtual one. But the mathematical way of looking at reflections is a little different. We start by thinking of the mirror as taking points on "our" side of the mirror to points directly on the opposite, virtual, side of the mirror and at the same distance away from it. So for each point x on our side of the mirror we have its reflection $r(x)$ on the opposite side of the mirror. Fig.9.4 shows this. The mathematical idea of a reflection then continues by saying that the reflection also takes points on the "virtual" side of the mirror over to our side of the mirror in the same way. So the mathematics forgets about "our" and "virtual". It treats both sides of the mirror equally and simply

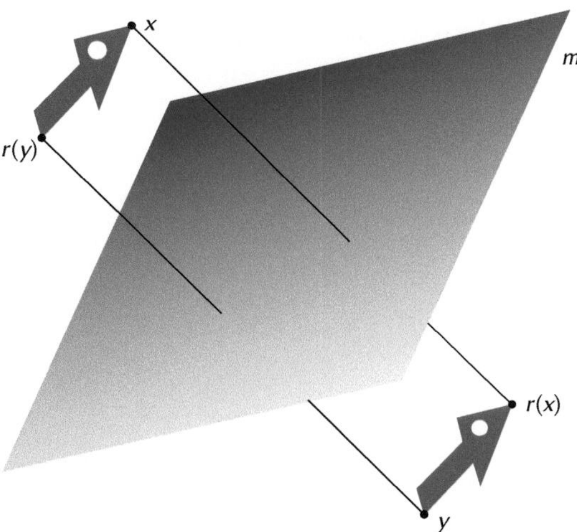

Figure 9.4: A mirror m and the reflection r in it. The mathematical reflection interchanges points on the two sides of the mirror. Just as x on one side goes to $r(x)$ on the other, so y on the other side goes to $r(y)$ on the first. Every point on m stays fixed. Evidently r^2 (i.e. r applied twice) is the identity mapping ϵ that leaves every point fixed: $r^2 = \epsilon$. It also reverses left and right. Notice the position of the white dot on the two sides of the mirror. They seem to be both on the left side of the arrow, but that is because our view point is looking down from left to right. If we look down from right to left, we see that the dot will be on the right side of the arrow.

moves points from one side to the other. As for points that are actually on the mirror, they stay fixed. So in the end we have a reflected version $r(x)$ of every point x of the plane (in 2D) or of the space (in 3D). This is shown in Fig.9.4.

Leonardo's famous *Vitruvian man*, Fig.9.5, is an example of bilateral symmetry (reflection) in art. In Fig.9.1 and Fig.9.2 there are several rather obvious symmetries by mirror reflections; but in the Celtic figure Fig.9.3 there are none.

We now move one more step into abstraction and think of r itself as an entity in its own right, an operator or mapping that moves points x into their reflected versions $r(x)$. Instead of thinking of symmetry as being a result of a passive process, like standing in front of a mirror, think of it as an active one in which points are shifted around in the plane or in space by the action of a reflection operator or mapping. This way of thinking links us to our main theme: pattern as dynamical in nature. This time the operator or mapping is acting on two- or three-dimensional space. As we have noted before, the words operator, map, mapping, transformation all signify the same thing in mathematical language, and we will use them interchangeably as suits the context. Ultimately they refer simply to the pairing of each state with some other state. In the case of a reflection the pairing is to pair each point x of the space with its reflected image $r(x)$. The usual terminology is that $r(x)$ is the *image* of x under the reflection r.

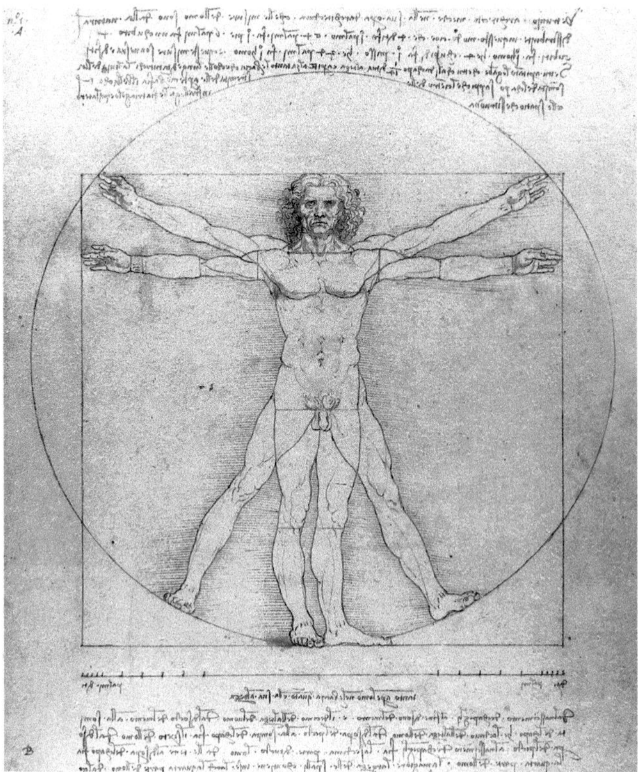

Figure 9.5: Leonardo da Vinci's famous *Vitruvian Man*, 1497. The overwhelming feature of the picture is its bilateral symmetry (reflection in the vertical line through the center of the figure). Of course, there is the artistic touch— this symmetry is not actually perfect—the feet for example are not reflected images of each other. Setting this artistic asymmetry aside, if r is the reflection and F is the figure of the man along with the accompanying circle and square, then the mathematical version of saying that the figure is symmetric is that $r(F) = F$. Applying the reflection moves points of F but in total still gives us back the same figure F. *Change without change.* Public Domain.

In keeping with our earlier notation for operators, we will sometimes write

$$r \triangleright x \quad \text{instead} \quad \text{of}\, r(x)\,.$$

This gives us a more visual separation between two different types of things: states or points, and operators that act upon them.

One cannot help noticing that if the reflection operation is applied twice then each point is returned to where it originally was: $r(r(x)) = x$ for all points x, or more simply, $r^2(x) = x$ (or $r^2 \triangleright x = x$) for all x. We can move even more firmly into thinking in terms of an "algebra" of operators, by recalling the identity operator, ϵ, namely the mapping that maps each point of space to itself. Then we have the simple, but important, fact that $r^2 = \epsilon$ for reflections.[2]

Summing up, the mathematical way to describe the mirror symmetry in a "mirror" m is by defining the operator r on the space in question that transforms each point x of the space into the point $r(x)$ which is exactly opposite to x on the other side of the m. This works for any dimension. The "mirror" is a line in 2D, a plane in 3D. In higher dimensions, though we can no longer imagine it, it is still possible to talk about reflections.

Under reflections all the metrical aspects of a figure are unchanged—the distance between two points and between their reflected images is the same—but of course the image is reflected. This even works when the two points are on opposite sides of the mirror: reflection is a true *isometry*, a preserver of distance.

But how about actually physically doing what a mathematical reflection is doing? In the 2D planar case this is, in a sense, possible: just turn the plane over by rotating it around its mirror (the median line). However, notice that the rotation takes place in 3D-space, so we have to step out of 2D to do it. Reflections in 3D are *not physically* possible to perform, as we all know well. We cannot step momentarily out of 3D into 4D to perform a rotation there. In 3D the very notions of left and right are interchanged by reflection. And this brings us to the next idea: *chirality*. Before launching into it we should set the mathematical stage more carefully.

9.1.2 Setting the stage

Our principal setting is either ordinary Euclidean two-dimensional space $\mathbb{E}(2)$, the *Euclidean plane* of Euclid's *Elements*, or ordinary *Euclidean space* $\mathbb{E}(3)$ in three dimensions, the space of our familiar world. Here we are thinking of the plane and three-dimensional space in purely geometrical terms with no imposed coordinate system, see below. However, we will assume that we have decided upon a scale of measurement for expressing distances—meters, miles, feet, nanometers, etc. Then for all points $x, y \in \mathbb{E}(3)$ we have a numerical distance $d(x, y)$ between them: the technical description is to say that d is a *Euclidean metric*. This is the usual distance we are familiar with, and it is unique up to the choice of scale we wish to use. Because of Pythagoras' theorem, the metric determines the notion of orthogonality, see Fig.9.6, and this is already enough to determine what it means to be a reflection in a mirror m. The point we are making here is that once we have space with the usual idea of distance, we already see that the space will allow reflective symmetries to be defined.

At times we shall also assume that we have a single fixed reference point in our space, whether that be $\mathbb{E}(2)$ or $\mathbb{E}(3)$. We will denote it by $\odot$. No point of Euclidean space is any more special than any other, but setting one of them to be the point of reference can be useful.

We will mostly be talking about symmetry in the context of Euclidean plane or Euclidean space, but occasionally we will look at one-dimensional space $\mathbb{E}(1)$. Although in

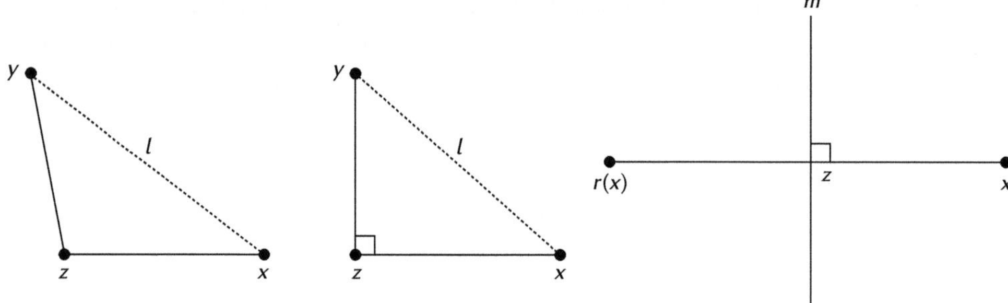

Figure 9.6: Pythagoras' theorem: the angle between two line segments zx and zy is a right angle if and only if the square of the length l of the line segment xy is the sum of the square lengths of the two line segments: $l^2 = d(x,y)^2 = d(z,x)^2 + d(z,y)^2$. Thus the distance function d (or *metric*, as mathematicians call it) already determines orthogonality. In fact, since congruent triangles have the same angles, it also determines all angles. This fact is one of the reasons why Pythagoras' theorem is so important: it is the link between two worlds, the world of geometry and the world of numbers. With the distance function, and hence the notion of orthogonality, we can see how reflection in a mirror m works: for each point x drop a perpendicular from x to the mirror and mark off the point at an equal distance on the same perpendicular but on the opposite side of the mirror.

other parts of the book we use coordinatized versions of Euclidean space, which allow for the Cartesian connection between the world of numbers and the world of geometry, they are not necessary here. A coordinate system in the plane involves setting up a coordinate frame centered on some chosen point $\odot$ and then using that to label points of $\mathbb{E}(2)$ with pairs of numbers (x, y) as coordinates. Similarly for $\mathbb{E}(3)$ with coordinate triples, leading to the familiar forms $\mathbb{R}^2$ and $\mathbb{R}^3$. However, the basic symmetries of the Euclidean plane or Euclidean space are geometrical in conception, very much in the spirit of Euclid's *Elements*, and not specifically related to coordinates.

9.1.3 Isometries

Considering three-dimensional space $\mathbb{E}(3)$, we can ask a fundamental question: exactly what are the operations that we can perform on $\mathbb{E}(3)$ that leave its intrinsic integrity unspoiled? The key feature that we have identified is distance, so the key operations are those that preserve distances. What transformations of space are possible that leave all distances unchanged? More precisely we wish to know all transformations T of the space $\mathbb{E}(3)$, i.e., all mappings $T : \mathbb{E}(3) \longrightarrow \mathbb{E}(3)$, for which the metrical distances between points remain unchanged:

$$d(T(x), T(y)) = d(x, y) \text{ for all } x, y \in \mathbb{E}(3).$$

Such mappings are called *isometries*—from the Greek iso (equal) + metron (measure). There are three fundamental types of isometries (with which, all of us, being residents of

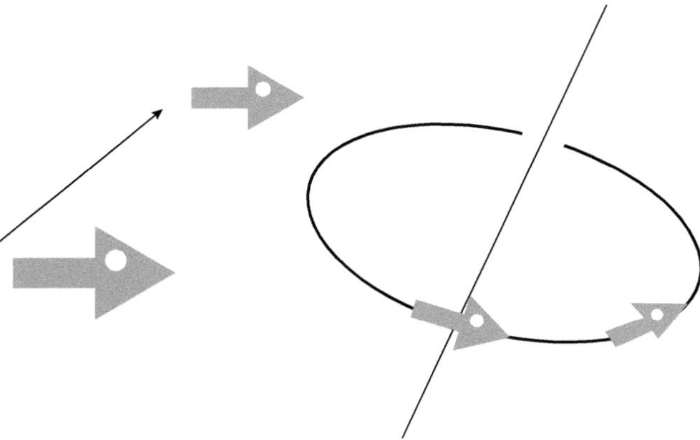

Figure 9.7: Showing a translation and a rotation about an axis in 3D. In each case we see how the mapping affects an arrow-like figure.

Euclidean space ourselves, are very familiar):

translations, rotations, reflections

see Fig.9.7 and Fig.9.4. As you can imagine from experience, any other isometry is a combination of these, i.e. the result of applying them one after another.

The modern idea of symmetry begins with the idea that these transformations lie behind our everyday idea of symmetry. The overall principle is that symmetry has to do with some aspects of a situation that do not change even though some transformational change actually does take place. The circumference of a circle does not change as we rotate it about its center. The points on the circumference move, but the circumference itself is still the same circumference as it was before. Translation moves an arrow but its length and direction do not change. Translations, rotations, and reflections in $\mathbb{E}(3)$ are symmetries of $\mathbb{E}(3)$ because they do not change the distance between points. Points move, but relative distances between them don't.

There is an important point here that needs to be explained. The usual way in mathematics of talking about the symmetry of a figure is to think of its *invariance* under some transformation of the space. We perform a transformation of the entire space and the figure "looks" just the same as before, even though its parts may have been moved around under the transformation. Saying "it looks the same" sounds vague, but mathematically it is very easily formulated. A figure F, that is, a subset of the space $\mathbb{E}(2)$) or $\mathbb{E}(3)$, is *bilaterally symmetric* if there is a reflection r of the whole space for which $r(F) = F$. The thing to be clear about is that $r(F) = F$ does *not* mean that the actual points of F stay fixed, but only that the entire *set* of points of F stay in F under the reflection:

$$r(F) := \{r(x) \ : \ x \in F\} = F\,,$$

see Fig.9.5.

This concept goes far beyond just bilateral symmetry. Subsets of the space are *symmetric* to the extent that they remain unchanged *as subsets*, under the actions of symmetries. Invariance is thus seen as co-equal to symmetry; symmetry and invariance begin to take a more formal appearance.

9.1.4 Chirality

A deep and often under-appreciated property of ordinary three-dimensional Euclidean space $\mathbb{E}(3)$ is its invariance under reflection.

Returning to the three basic types of isometries: translations, rotations, reflections in $\mathbb{E}(3)$, we can see that there are very obvious differences between them. Translations move the whole of space by some distance in some direction. No point is left unmoved by a translation. The other two types of isometry leave some points fixed: exactly one line of points (the axis of rotation) for rotations, and the whole plane of points on the mirror for reflections. These latter two types, and also products of them (one followed by the other), are often referred to as *orthogonal transformations.* This terminology usually appears in contexts in which a distinguished point has been designated as the "origin" in the space. The orthogonal transformations are the isometries that fix this distinguished point. Translations do not fix any point. Of the three types, only the translations and rotations are physically realizable. We can rotate and translate things around in space, but we cannot physically reflect them.

Thus arises a fundamental fact about three-dimensional space: from a physical perspective it has two different forms, one being the mirror image of the other. A reflection can interchange these two physically different forms, but rotations and translations cannot. This fact accounts for a pervasive aspect of our world: things come in left-handed and right-handed forms—different but related by mirror symmetry.

The difficulties around the distinction between left and right becomes rather obvious when one looks up the definition of left (as opposed to right) in a dictionary:

- Merriam–Webster: of, relating to, situated on, or being the side of the body in which the heart is mostly located;
- Oxford: on, towards, or relating to the side of a human body or of a thing which is to the west when the person or thing is facing north.

Both definitions require first setting ourselves in our physical world, and both are determined by human and culturally defined concepts (north is in effect an arbitrary choice of one of the two poles of the Earth's axis of rotation). For a mirror person living in the mirror world the heart would be on the other side and the Earth would spin on its axis the other way around; the definitions of left and right would appear reversed: if left is still defined as the side of the body on which the heart lies, left would be on the other side of our bodies.

A common way to distinguish handed-ness in three-dimensional objects is to use the left- and right-hand screws: with the fingers curled in and the thumbs pointed up, your two hands present the two forms in which a screw can appear, *chirality* being the term used to make the distinction. Rotations and translations cannot make these two types of screws come into agreement.

Life on Earth makes considerable use of the chirality of molecules, often with chiral opposites being functionally useless (even downright dangerous) and almost completely absent from the biological world. Louis Pasteur (1822–1895), who is more known for his seminal work in microbiology, vaccines, and pasteurization, was the first to explain a strange paradox. Tartaric acid taken from life-forms (notably grapes) has the property of rotating the plane of polarized light passing through it while tartaric acid derived from chemical synthesis does not. What Pasteur discovered is that the molecule for tartaric acid is chiral, its two forms being mirror images and effecting rotations of opposite chirality. In life-forms tartaric acid is biologically produced and it is of only one chirality. It is here that the polarization effect arises. Ordinary chemical synthesis produces both molecules of both chiralities and in essentially equal numbers. In this mixture the polarization effect is destroyed.

Amazingly this was all before the atomic structure of molecules was known. Pasteur actually deduced the chiral structure from a crystalline form of sodium-ammonium tartrate, where (luckily) the crystalline forms are visibly chiral. He was able to separate out the two types and show that they produced opposite rotational effects on polarized light. Serendipity—but as Pasteur said, "Maybe you will say, by chance, but remember that in the fields of observation, chance only favors the prepared minds". A nice historical overview of this is to be found in [177].

In biology shape matters! Bio-molecules of the wrong shape simply cannot match and fit together.

9.1.5 Symmetry and pattern systems

In our discussion of pattern systems, the most important idea is that they involve change. Change appears as the effect of operators that alter the state space. The most obvious agent of change is time, so much of our discussion has ranged around time, both in its continuous and discrete forms. In the discrete form we have a single operator that is used repeatedly to effect change. In the continuous version there is a continuous set of operators t, one for each moment of time, that produce change. In either case, the mathematical abstraction of this is the basic idea of a dynamical system, which we have axiomatized as the first and third axioms of a pattern system: specifically a set and a set

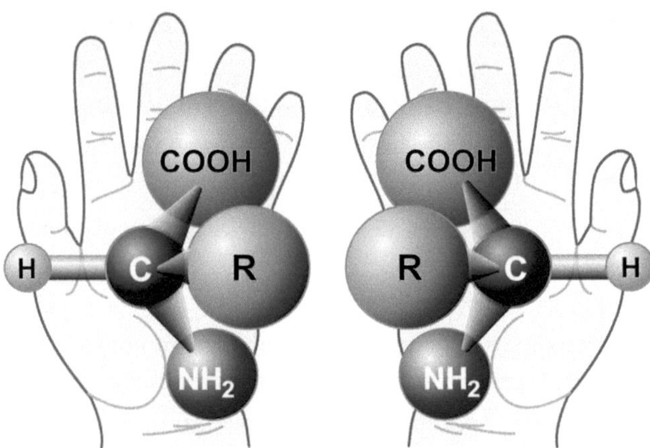

Figure 9.8: Chirality: two mirror related forms of a generic α-amino acid. They are called *enantiomorphs* of each other. In Nature only one of the two (most commonly the left-handed version) is functional [29]. Chirality matters. The infamous drug thalidomide comes in two enantiomorphic forms, one of which has tragic consequences in pregnancies. Nowadays in the drug industry enantiomers are considered as separate molecules rather than just different forms of the same drug [33, 167]. Public Domain. NASA.astrobiology.nasa.gov/news/chiral-molecules-may-have-hitched-rides-to-planets/.

of operators of change on this set, §6.1.

Looking at the ideas that we are developing about symmetry we see these same two basic features: some form of Euclidean space (one, two, or three dimensional) and a set of operators (reflections, rotations, translations) that effect change. In time we will find how the other two axioms fit in, but the principle of change is central. The interpretation may have changed from temporal to spatial, but the abstraction is just the same. This places symmetry firmly in the setting of pattern systems.

In this chapter we have been using notation like $r(x)$ for the way that the isometry r acts on the point x. This is common in mathematics. In our pattern system notation of this book we are more likely to write $r \triangleright x$. The $\triangleright$ notation is intended to be expressive of motion, and that is something we prefer to emphasize. We will use both, whichever seems clearer.

The importance of putting symmetry into a context that carries with it all these underlying currents of dynamical change is that we begin to see that symmetry too is about change. Somewhat paradoxically, though, it is really about what does *not* change, or does not *appear* to change, under change. Change without change. It is a certain type of invariance and, as we shall see, that is why it becomes so important in modern physics.

What we need to understand now is how the carriers of these changes—rotations, reflections, translations—fit and work together. The mathematics that does this is called *group theory*.

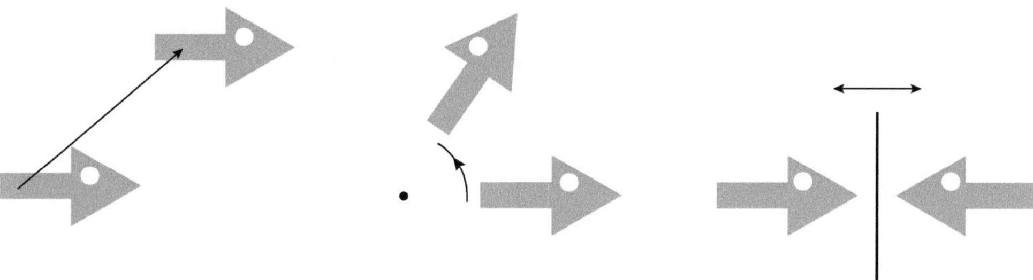

Figure 9.9: The three basic types of Euclidean motions in 2D: translations, in which the whole space is just moved a fixed distance in some fixed direction; rotations about some point; reflections through some mirror. The white dot in each arrow serves to give us left-handed or right-handed versions. Translation and rotation do not alter the handedness, but reflection does.

9.2 The group of isometries

Isometries are distance preserving transformations or mappings of space. Of course we have lots of first-hand experience with these things because when we move objects around we are witnessing such transformations. Still, in looking at them more objectively, some less familiar ideas will start to emerge.

By "space", the implication is ordinary three-dimensional space $\mathbb{E}(3)$, or 3D-space for short, but it is informative to also consider isometries in one-dimensional space $\mathbb{E}(1)$ (the line) and two-dimensional space $\mathbb{E}(2)$ (the plane) as well. Each has its own set of isometries.

9.2.1 2D-space

We begin with 2D space, the ordinary Euclidean plane, $\mathbb{E}(2)$. There we recognize the three basic forms of isometries: translations, rotations, reflections, mentioned in Fig.9.9. Although the term chirality is usually reserved for three-dimensional objects, there is a two-dimensional version of it: reflections in the plane reverse clockwise and counter-clockwise, and we will extend the use of the word chirality to refer to this too.

In §6.3.1 we already introduced the idea of following one operator by another to form their *composition*, which is a new operator. This idea is the starting point of a new form of algebraic operation, a sort of "multiplication", which ultimately becomes an entire area of algebra in its own right.

If U, V are two isometries of $\mathbb{E}(2)$ then we can apply them successively, first U and then V, to get a new isometry $V \circ U$. The process of putting U and V together is called *composition* and the little circle is used like a multiplication sign to indicate that U and V have been composed to form a new isometry $V \circ U$. In fact it is common to call $V \circ U$ the *product* of V and U. It is important to know that the convention in composing operators is to read from right to left: $V \circ U$ means first apply U and then V. As Fig.9.10 shows,

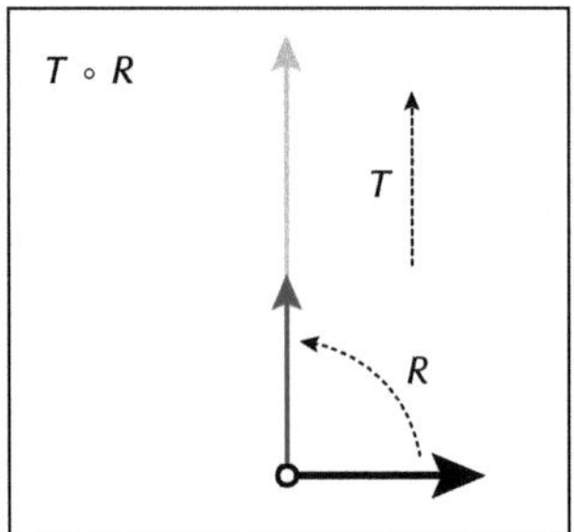

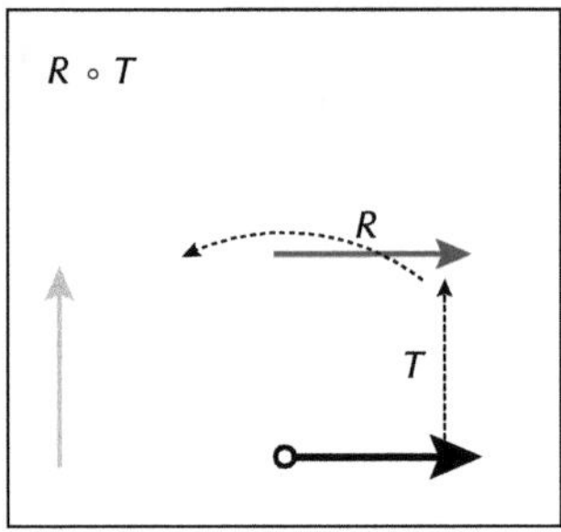

Figure 9.10: Order counts. A rotation R (centered on the small circle) and then a translation T are performed on an arrow, with the transitions shown as black -> grey ->light grey (left). If the same operations are performed with first the translation and then the rotation, again with the transitions shown in the order black ->grey -> light grey (right), the result is very different. $T \circ R \neq R \circ T$. These two operations do not commute.

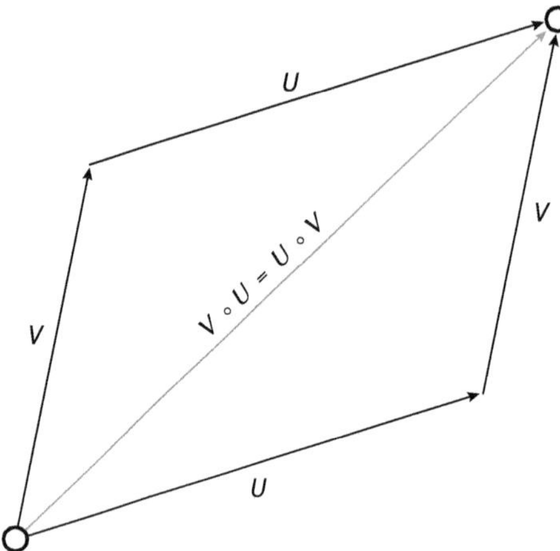

Figure 9.11: For translations, order does not matter. The composition of two translations is another one, and the result is independent of the order in which the two translations are made. They commute. The parallelogram figure shows why this is called the *parallelogram law*.

the order is crucial. We were taught to appreciate the importance of order in performing operations by thinking of first putting on socks and then shoes. Doing it the other way around produces a very different effect.

Sometimes it doesn't matter which order we do the operations: $U \circ V = V \circ U$. In this case we say that U and V *commute*. Translations are examples: one translation composed with another creates yet a third translation, and the fact that the order of composition is irrelevant is a simple geometrical observation called the *parallelogram law*, see Fig.9.11. In any case, the upshot of all of this is that we can compose isometries to get new ones.

The next important thing to note is to observe that whatever an isometry does can be undone by another isometry: just do the opposite. If R is rotation through 30° around some center p in the clockwise direction, then we have its *inverse*, denoted R^{-1}, which is rotation through 30° around the same center p in the counter-clockwise direction. The composition $R^{-1} \circ R = \epsilon$, where ϵ is the identity isometry that does nothing. It is equally true that $R \circ R^{-1} = \epsilon$, a general property of invertible transformations and their inverses. Likewise a translation through a distance in one direction can be reversed by going the same distance in the opposite direction. We already saw that reflections are their own inverses.

If R is rotation through 30° around some center p in the clockwise direction, then $R^2 := R \circ R$ is a rotation through 60° around p in the clockwise direction, and similarly $R^3 := R \circ R^2$ is a rotation through 90° around p, and so on. Thus R^{12} is a rotation through 360°, which is nothing other than the identity transformation ϵ. We say then that R has *order* 12 (not to be confused with the "order" in which we apply transformations). In the same way R^3, which is a quarter turn, has order 4. Translations cannot have finite order like this (unless the translation is through zero distance, in which case it is ϵ).

We should make note here of the fact that in our discussion of pattern system, and in our axioms for pattern systems, we do not require that the operators be *invertible*. We have stressed the fact that in the dynamics of the real world many changes cannot be undone. However, in the case of symmetry, we are not conceptually bound to the temporal world. This fact is an essential feature of the algebraic structures that we are now describing. These operators are not involved with time, which cannot be undone, but with spatial transformations, that, at least in principle, can be undone.

There is one more crucial property that composition has. It automatically satisfies the *associative law*: no matter what three operators U, V, W we choose, say acting in $\mathbb{E}(3)$, we must have $W \circ (V \circ U) = (W \circ V) \circ U$, see Fig.9.12.

This has nothing particular to do with the particulars of the space we are working in. It is just a property of composition itself: it automatically satisfies the associative law. In practice this means that just as we do with multiplication of numbers, when composing operators or transformations we can rearrange, or even ignore, the bracket arrangements. However, unlike with numbers, keep in mind that we cannot usually change the order in which the elements of a composition are written. In general $V \circ U \neq U \circ V$.

9.2.2 Defining groups

Let ISO(2) denote the set of all isometries of $\mathbb{E}(2)$. Then we have just noted four facts about ISO(2):

G(i) ISO(2) is closed under composition: if $U, V \in$ ISO(2), then also $V \circ U \in$ ISO(2);

G(ii) composition is associative;

G(iii) $\epsilon \in$ ISO(2): the identity mapping ϵ, which maps every element to itself, is an isometry and we have $\epsilon \circ U = U = U \circ \epsilon$ for every isometry U.

G(iv) ISO(2) is closed under inversion: if $U \in$ ISO(2) then there is an isometry $V \in$ ISO(2) so that $V \circ U = \epsilon$.

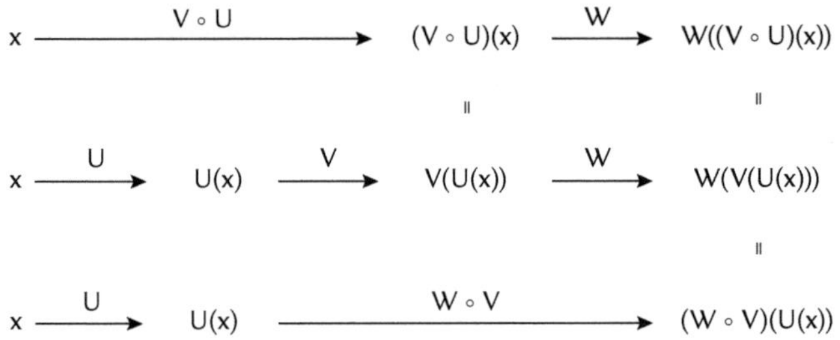

$$x \xrightarrow{V \circ U} (V \circ U)(x) \xrightarrow{W} W((V \circ U)(x))$$

$$\| \qquad\qquad \|$$

$$x \xrightarrow{U} U(x) \xrightarrow{V} V(U(x)) \xrightarrow{W} W(V(U(x)))$$

$$\|$$

$$x \xrightarrow{U} U(x) \xrightarrow{W \circ V} (W \circ V)(U(x))$$

Figure 9.12: The associative law: for all symmetries U, V, W we have $W \circ (V \circ U) = (W \circ V) \circ U$. The effect of each is simply to send each point x to $W(V(U(x)))$. It is clear that what is going on here has nothing particular to do with the types of operators involved (that they are isometries). It is just a property that all mappings have. The associative law comes for free with mappings, operators, transformations, symmetries, isometries, ... whatever we call them. The associative law means that we do not have to worry about the arrangements of parentheses.

These four properties are what make ISO(2) into a *group*. As we shall use the concept in this book, groups will almost always be sets of operators or transformations of some space or set, with composition (following one transformation by another) as the "multiplication". The four properties **G**(i)–**G**(iv) are essentially axioms that define what it means to be a group.[3] We will see many examples. Later we will look at the corresponding group ISO(3) of isometries of $\mathbb{E}(3)$.

As a matter of notation it is common to leave the composition symbol $\circ$ out and simply use adjacency of symbols (e.g. rT instead of $r \circ T$), much as we do in arithmetic when we write ab for $a \times b$.

Clearly there are infinitely many different isometries of the plane—rotations through any angle around any point, translations through any distance in any direction—so we see that ISO(2) is an *infinite group*. We will also see finite groups as we go along, §9.3.

It is unfortunate that a common word like "group" has come to have a very specialized and not very suggestive meaning in mathematics. In mathematics a group is a set with a multiplication rule on it that satisfies the four rules **G**(i)–**G**(iv).

In **G**(iv) we see that each U has a "left inverse" which we denoted by V. In fact this same V is also a right inverse and instead of V one writes U^{-1}. This and some simple consequences of definition are derived in the Endnotes.[4] Group theory is an immensely developed area in mathematics, physics, and crystallography.

Different types of isometries

Let Δ be a triangle in $\mathbb{E}(2)$, chosen with all its edges of unequal length (we don't want it to have any symmetry), and choose a point $\odot$ inside it. Then for any isometry S of $\mathbb{E}(2)$, $S(\Delta)$ will be a copy of Δ, though it may be now rotated, translated, and even reflected. If there is a single reflection involved the triangle will have reversed its chirality. If there

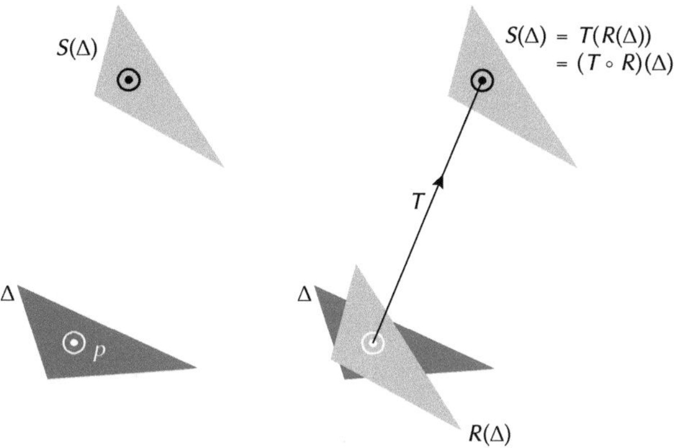

Figure 9.13: On the left we see the reference triangle with the point ⊙ (white) inside it, and its image under an isometry S that has retained the chirality. On the right we see how S can be seen as the composition of a rotation (around the white ⊙) and a translation T.

are two reflections involved, the chirality will have changed twice, so will be back to the one we started in, and so on.

Wherever $S(\Delta)$ is, S is the *only* possible isometry that could have moved Δ to exactly that position. Another way to put it is this: once you have decided where the three vertices of Δ have been moved to, you know how every point in the plane will be moved and the isometry S will already be completely known. This is the same principle as triangulation used in surveying: you know where any point in the plane is once you know its distances from three fixed points (as long as those points are not on a single line). The fact that we chose the triangle without symmetries is vital here. If the triangle were isosceles for instance, then there would be a reflective symmetry that maps it to itself, and this would mess up the uniqueness.

If S has not altered the chirality then simple experience makes it clear that whatever the isometry S is we can move Δ to its position $S(\Delta)$ by rotating Δ around the point ⊙ into the right orientation and then translating onto $S(\Delta)$. So there is a rotation R and a translation T so that $S = T \circ R$. See Fig.9.13.

If S has changed the chirality of Δ then we can rotate Δ to an orientation $R(\Delta)$ so that a reflection r in a mirror placed between $R(\Delta)$ and $S(\Delta)$ interchanges them. In this case $S = r \circ R$. See Fig.9.14.

Altogether we have established that

> every element of isometry of the plane is the composition of a rotation around ⊙ and a translation, or the composition of a rotation around ⊙ and a reflection.

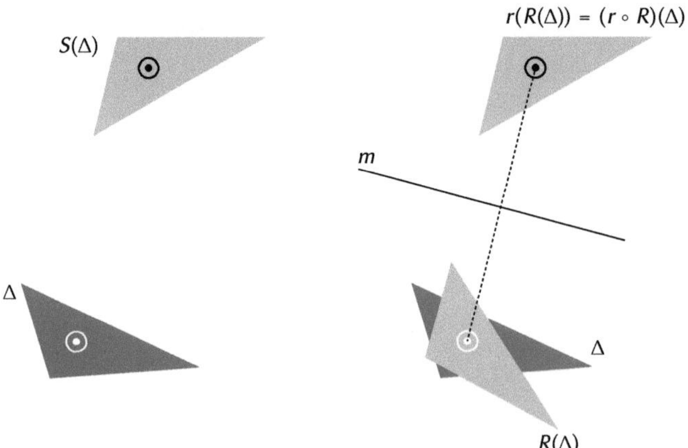

Figure 9.14: On the left, we see the reference triangle with ⊙ (white) inside it, and its image under an isometry S that places the triangle in a rotated, and reflected form. On the right, we see how S can be seen to be the composition of a rotation around ⊙ and a reflection in a mirror m. The dashed line indicates how the mirror is defined and then determines the rotation required before the reflection takes place.

Subgroups

We have seen that there are isometries that reverse chirality and ones that do not. The ones that preserve chirality form a subgroup $\mathrm{ISO}(2)^+$.

A *subgroup* of a group is a non-empty subset of that group which itself is closed under composition and inversion (and in consequence must contain the identity element ϵ).

The key point in the present instance is that the composition of chirality preserving mappings is itself chirality preserving. The chirality reversing elements of ISO(2), denote them by $\mathrm{ISO}(2)^-$, *do not* form a subgroup since the composition of two chirality reversing mappings is actually chirality preserving.

What we showed above is that every element of $\mathrm{ISO}(2)^+$ can be written as a product of a rotation around ⊙ and a translation, and every element of $\mathrm{ISO}(2)^-$ can be written as a product of a rotation about ⊙ and a reflection. In fact, if r is *any reflection* then the set of elements $\mathrm{ISO}(2)^+ \circ r = \mathrm{ISO}(2)^-$.

There are some other notable subgroups. The first is the subgroup of all translations, TN(2). The composition of two translations is another one, and the inverse of any translation is also a translation. So TN(2) is a subgroup of both ISO(2) and $\mathrm{ISO}(2)^+$.

An important form of subgroup is the set of isometries U that leave a particular point p fixed: $U(p) = p$, often called the *stabilizer* of that point. In our case we want to look at the stabilizer of ⊙, the point that we had decided to single out in the plane. The only isometries that can do this are the rotations around ⊙ and the reflections based on mirrors that pass through ⊙.

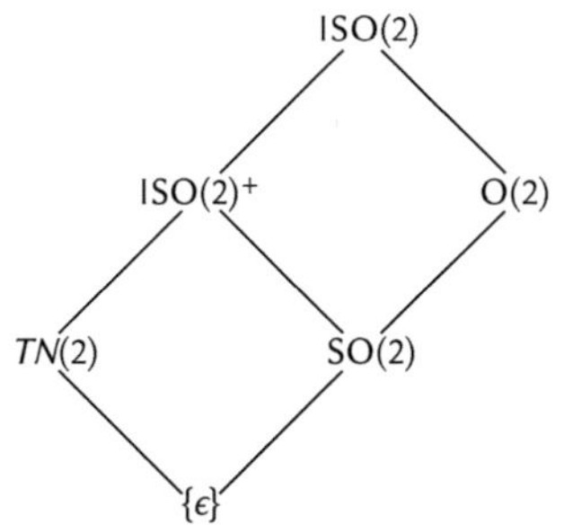

Figure 9.15: The group ISO(2) and some of its subgroups, the lines indicating subgroup relationships.

If two isometries fix ⊙ then so does their composition, and if an isometry fixes ⊙ so does its inverse. The set of all these elements is a subgroup of ISO(2). It is called the *orthogonal group*, and denoted O(2). Inside that we have yet another subgroup, namely the elements of chirality preserving elements of O(2), that is

$$\mathrm{SO}(2) = \mathrm{O}(2) \cap \mathrm{ISO}(2)^+ .$$

This group is called the *special orthogonal group*. We might also note that $\mathrm{O}(2) \cap \mathrm{TN}(2) = \{\epsilon\}$, that is to say, the only element that is simultaneously a translation and also fixes the point ⊙ is the identity transformation.

Fig.9.15 schematically shows how all these subgroups are related. In view of our discussion above, every element of $S \in \mathrm{ISO}(2)$ can be written in the form $T \circ U$, where $U \in \mathrm{O}(2)$ and $T \in \mathrm{TN}(2)$. This is often written in the self-explanatory form

$$\mathrm{ISO}(2) = \mathrm{TN}(2)\ \mathrm{O}(2) .$$

In fact this decomposition is unique.[5]

9.2.3 Euclidean space $\mathbb{E}(3)$

Remarkably, switching to three-dimensional space brings no new immediate conceptual obstacles. We can repeat everything above, and it all works the same way. Of course triangles have to be replaced by their 3D equivalents, which are tetrahedra (again we may want ones that are skewed to avoid any unwanted symmetries). The groups and sets that we have identified above are all replaced now by their three-dimensional equivalents: $\mathrm{ISO}(3), \mathrm{ISO}(3)^+, \mathrm{ISO}(3)^-, \mathrm{TN}(3), \mathrm{O}(3), \mathrm{SO}(3)$, which are inter-related by the same abstract scheme that we see in Fig.9.15. The latter two are again called the *orthogonal group* and the *special orthogonal group*. The dimension of the underlying space is implicitly understood. Moreover we have the corresponding decomposition result:

$$\mathrm{ISO}(3) = \mathrm{TN}(3)\ \mathrm{O}(3)$$

with the corresponding decomposition of every isometry into a composition of a translation and a rotation being unique.

Figure 9.16: Think of the figure as a book with a red edge, a white label, and a black edge. In the top row we see the result of two consecutive rotations, first around an axis directly out towards us and the second around a horizontal axis that sends the bottom towards us. In the lower row the same two rotations have been applied in the opposite order.

However, there are significant differences between the structures of SO(2) and SO(3). In the group of rotations about ⊙ in $\mathbb{E}(2)$ the order of composition of rotations is not important. If R_1 is rotation through angle θ_1 and R_2 is rotation through angle θ_2 then $R_1 \circ R_2 = R_2 \circ R_1$, both being rotations through an angle of $\theta_1 + \theta_2$. When the multiplication of a group is *commutative* in this sense (that the order of the elements in a product does not matter) then the group is called *abelian* (pronounced *a-bee-lian*, named after the famous Danish mathematician Niels Abel (1802–1829), which is actually pronounced *a-bell*). But when we come to $\mathbb{E}(3)$, SO(3) is nothing like abelian. Fig.9.16 gives an example which shows why. This marks a profound difference between two-dimensional and three-dimensional space.

So SO(3) is a non-abelian group. There is another significant difference between O(2) and O(3) that is best visualized by looking at Fig.9.17: the transformation that all interchanges points that lie directly opposite to each other with respect to ⊙. This unique isometry is called the *central inversion* and we denote it by I. In O(2) the central inversion is the rotation through 180 degrees around ⊙. In O(3) we still have $I \circ I = \epsilon$, but here I is chiral reversing even though it is certainly *not* a reflection: its only fixed point is ⊙.

One way to convince oneself that I is indeed chirality reversing is to imagine an inflatable ball centered on ⊙. The central inversion takes each point on the ball to the point directly opposite to it across ⊙. If we could accomplish this physically it would do something like turning the ball inside out, but in such a way that the inside still remained inside, a sort of "inside-inside". The central inversion I can be written as a product of three reflections. In a coordinatized version of $\mathbb{E}(3)$ with ⊙ taken to be the origin $(0, 0, 0)$, I would be the mapping $(x, y, z) \mapsto (-x, -y, -z)$.

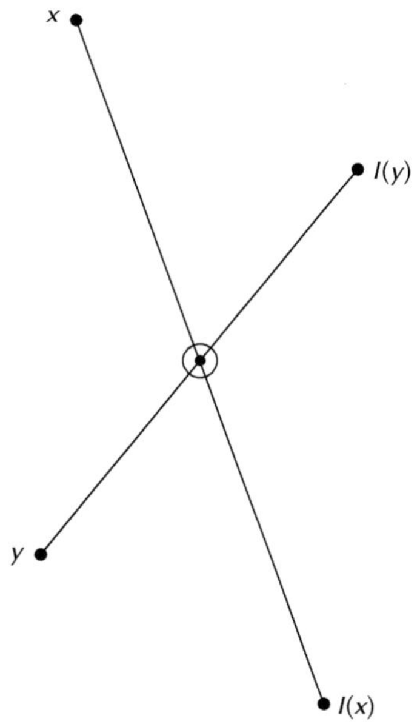

Figure 9.17: The so-called *central inversion* (relative to ⊙) works by mapping every point into the one diametrically opposite to it relative to ⊙. In two-demensional space, the central inversion is a rotation through 180 degrees. In three-dimensional space it is neither a rotation nor a reflection, though it is chirality reversing. We denote the central inversion in three-dimensional-space by the symbol I.

Although O(3) is non-abelian, nonetheless I itself does commute with every element of it: $I \circ U = U \circ I$ for all $U \in \mathrm{O}(3)$. Elements of a group with this property of commuting with everything in the group are called *central elements.* The set of all central elements is a subgroup called the *center* of the group. In fact ϵ and I are the only central elements of O(3), so the center is the two-element subgroup $\mathscr{Z} := \{\epsilon, I\}$. We can write

$$\mathrm{O}(3) = \mathrm{SO}(3) \cup (I \circ \mathrm{SO}(3)).$$

There is a more revealing way to write this, namely $\mathrm{O}(3) = \epsilon\,\mathrm{SO}(3) \cup I\,\mathrm{SO}(3)$ can be written as

$$\mathrm{O}(3) = \mathscr{Z} \times \mathrm{SO}(3), \tag{9.1}$$

suggesting how we can view O(3) being built out of the two subgroups SO(3) and $\mathscr{Z}$ in the most obvious kind of way. This is the *direct product* of these two groups. SO(3) cannot be further decomposed as a direct product of subgroups.

9.3 The finite symmetry groups in three-dimensional space

It is a natural and interesting question to ask exactly what symmetry groups are really possible in three-dimensional space. Although we are all familiar with the basic forms of symmetry, we are far less familiar with them when they are formalized in terms of group theory. If the question is not just what kinds of symmetry an object might have (rotational, reflective, translational, etc.), but how the entire collection of symmetries of an object looks as a *group*—that is, as an algebraic structure—then this becomes a genuine problem of classification. In its full generality this is a huge question, but if we stick to finite or to discrete symmetries (as opposed to continuous ones like the smooth rotations of a sphere) then there is a rather remarkable answer: there are only finitely many such "types". In the sections below we look at this question far enough to understand what it is about and to see what sort of answers we can find. It is about pattern and it is about what Nature can manifest, or alternatively depending on your philosophical position, what constraints Nature must abide by. We divide the discussion into two parts: finite symmetry groups (which do not include translations) and infinite symmetry groups (which do).

9.3.1 The symmetries of a finite object

To start, suppose that we have some material object in $\mathbb{E}(3)$, call it F. We imagine that it is some sort of object that we could pick up and turn around in our hands. A symmetry of F is an isometry t that takes F precisely back onto itself: in symbols

$$t(F) := \{t(x) \ : \ x \in F\} = F .$$

There is no need for individual points to stay fixed under t, but after applying t to all the points, F will "look" just like it did before. Imagine rotating a cube of sugar in your hand.

Let G_F denote the set of all the symmetries of F, that is all the isometries t for which $t(F) = F$. In fact, G_F is more than just a subset of ISO(3), it is a *subgroup* of ISO(3), and in particular a group in its own right. Specifically,

- for all $s, t \in G_F$, $s \circ t \in G_F$: $(s \circ t)(F) = s(t(F)) = s(F) = F$;
- the identity element $\epsilon \in G_F$: it fixes each of the individual points of F;
- the inverse of any element of G_F also takes F to itself: $t^{-1}(F) = F$, since we can always undo whatever we did. More formally, since $t(F) = F$, we get, by applying t^{-1} to each side of the equation, $F = t^{-1}(F)$.

No matter what F may be, the identity operator ϵ is always a symmetry of F and so $\epsilon \in G_F$. Sometimes it is called the *trivial* symmetry, because it fixes every point of space and tells us nothing interesting about F. It is quite possible that the trivial symmetry is the only symmetry that F has, i.e. $G_F = \{\epsilon\}$. A typical tree, for example, for all its beautifully composed structure, has no non-trivial symmetry. We then say that its symmetry group G_F is *trivial*.

No matter what F is, G_F exists and is an object with the structure of a group. *Every object has its own symmetry group.* With this we come to the heart of the mathematical idea of what it means to be symmetric or to possess symmetry.

An object F is said to have symmetry if its symmetry group is not trivial.

The most obvious example of a symmetric object is a perfectly spherical ball. A perfect ball has the full symmetry group $SO(3)$ of rotations around its center as its symmetry group, and if we are allowed reflections as well then it has all of $O(3)$ as its symmetry group. An ice-cream cone, at least an ideal one, has perfect rotational symmetry around its central axis, and even reflectional symmetry in all the planes that contain this axis.

Putting aside objects like these, which have an infinite number of symmetries, what about the case when F has only finitely many symmetries, in other words, G_F is finite? This is what we concentrate on now. We shall assume that F is some finite object, that is to say it is spatially bounded and three-dimensional, whose symmetry group G_F is finite.

A finite symmetry group always has a *center of symmetry*, a point that stays fixed under all the elements of G_F. This can be seen in the following way. Consider the smallest sphere that contains F. Since any symmetry of F that leaves F invariant as a set will also leave this sphere invariant, it must also leave the center of the sphere fixed. So the center of the sphere is a fixed point of F. There could be more than one fixed point—for instance if F simply has bilateral symmetry, like a butterfly, then there is a mirror plane of fixed points—but there is always this one which we will call the center of symmetry of F and denote by $\odot$.

Referencing the group with respect to $\odot$ means that we can view G_F as a subgroup of the orthogonal group $O(3)$, the entire group of symmetries of $\mathbb{E}(3)$ that fix $\odot$. With this we arrive at the question which we want to ask: what types of finite subgroups does $O(3)$ possess? If we limit ourselves to subgroups that are truly three-dimensional (we will elaborate on what this means later), then the answer is remarkable: there are exactly *eight* such groups. Furthermore they all can be seen to arise within the context of Euclid's *Elements*. Indeed the perception of symmetry is probably as old as humanity itself. "Groups" are a nineteenth-century idea.

9.3.2 Euclid's *Elements*

The *Elements* dates back to about 300 BCE. Little is known about Euclid. He appears to have been a Greek resident of Alexandria and his book seems to have been a compilation of the mathematical knowledge of the Greeks at the time. The book is certainly the most famous mathematical book from antiquity, and probably the most famous ever. Justly so. Mathematics is ancient, but Euclid's *Elements* is the first sustained development of mathematics and mathematical ideas *derived from a system of axioms*. Until recently it was a part of any student's mathematical education.[6] The work consists of thirteen chapters, referred to as Books. It is usual to think of the *Elements* as being about plane geometry, but in fact it also lays down the foundations of number theory (for instance the infinitude of the set of prime numbers is proved there, see §16.7.1) and tackles the difficult problems of incommensurability, or what we would now call irrational numbers.

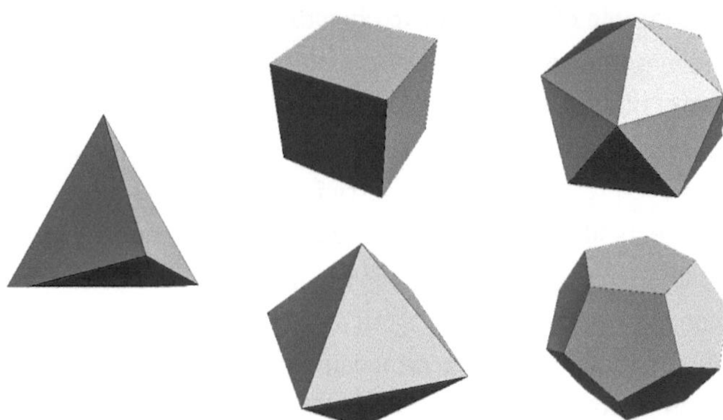

Figure 9.18: The five Platonic solids. From left to right: the tetrahedron; then the cube and beneath it its dual, the octahedron; and finally the icosahedron, and its dual, the dodecahedron. They are arranged in dual pairs (the tetrahedron is self-dual), see Table 9.1. The names derive from the number of faces: $4, 6, 8, 20, 12$ respectively.

Table 9.1: The numbers of vertices, edges, and faces of a polytope are related by the Euler formula $V - E + F = 2$. This is what the result looks like for the regular polyhedra in 3D. Notice the duality between cube and the octahedron, and again between the icosahedron and the dodecahedron, where the numbers of vertices and faces are interchanged. This meaning of this duality is made visual in Figs. 9.20, 9.21, 9.22.

polytope	V	E	F
tetrahedron	4	6	4
cube	8	12	6
octahedron	6	12	8
dodecahedron	20	32	12
icosahedron	12	32	20

The last three Books of the *Elements* deal with three-dimensional space. In particular Book 13 is devoted to the five *Platonic solids* (mathematically also called the *regular polyhedra*): the tetrahedron, the cube, the octahedron, the dodecahedron, and the icosahedron, see Fig.9.18. These are evidently highly symmetric objects, and it is the symmetry groups of these three-dimensional objects that arise in our classification of the finite subgroups of $O(3)$. The cube and the octahedron have the same symmetry groups, as also do the dodecahedron and the icosahedron (see the Figures below). So there are actually three subgroups of $O(3)$ arising in this way. Restricting to the physically realizable (i.e. chirality preserving) subgroups of these groups gives us three more symmetry groups, all of them in $SO(3)$. That makes six of the eight groups.

The Platonic solids are called such because of their appearance in *Timaeus*, a famous work of Plato which is sometimes described as his theory of everything. The actual elucidation of the five regular polyhedra and their classification as being the only regular polyhedra is believed to be the work of a contemporary, and friend, of Plato, Theaetetus of Athens (c. 417–369 BCE). He is also responsible for the theory of incommensurable quantities which is expounded in Book X of the *Elements*. Plato identified four of the Platonic solids with the basic elements of which everything was supposed to be composed (fire, earth, air, water). The fifth (the dodecahedron) he identified with a more metaphysical entity called the *quintessence* [105]. In fact our word "quintessence" derives from this, "the fifth and highest element in ancient and medieval philosophy that permeates all nature and is the substance composing the celestial bodies" [113]. Perhaps the fact that the dodecahedron most obviously involves the number five has something to do with this too.

Plato was not the first to try and frame the Cosmos in mathematical terms. Pythagoras is famous for seeking (and finding) numerical connections between the physical world and the natural numbers (positive integers). Galileo saw the world as an open book whose language was mathematics. Johannes Kepler used the Platonic solids in one of his earlier models of the solar system, in which he was still trying to fit planetary motions to spheres. Later he framed his theory in terms of elliptical orbits and his three famous laws (again mathematical in nature), which ultimately Newton was able to deduce as consequences of his theory of forces and universal gravitational attraction. It all continues today where it is an article of faith in modern physics that mathematical ideas can offer genuine insights into the nature of things.

9.3.3 The symmetries of the regular polyhedra

What is it about the regular polyhedra (Platonic solids) that makes them so special? The mathematical way to look at it is this. These are solid figures bounded by polygonal faces. The geometric entities that we see on their surfaces are vertices, edges, and faces. Each vertex is part of several edges, and each edge is part of several (actually two) faces, so what we can see is triples (v, e, f) consisting of a vertex, an edge, and a face in the relationship

$$\text{vertex} \subset \text{edge} \subset \text{face}\,.$$

Such a triple is called a *flag*, this picturesque name arising from the idea that such a combination looks like a flag on a pole, Fig.9.19.

With this definition of a flag, we have a very simple definition: a polyhedron is *regular* if for each pair of its flags there is a symmetry (actually, exactly one) of the polyhedron that takes the one to the other.

This all becomes much more evident as we look at the five regular polyhedra in turn. We start with the tetrahedron, Fig.9.20. For the time being, look only at the left-hand side of the figure. We see the faces of the tetrahedron decomposed into alternate black

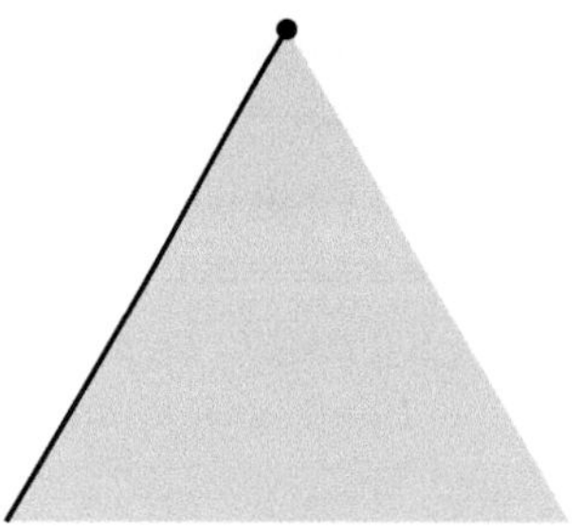

Figure 9.19: A flag of a polytope is a triple consisting of a face, an edge of that face, and a vertex of that edge. A polytope is called a *regular polytope* if its symmetry group can map any flag to any other flag. The concept applies in a similar way at all dimensions.

and white triangles. Notice that each of these black and white triangles has exactly one vertex which is a vertex of the tetrahedron, one edge which is part of an edge of the tetrahedron, and itself lies in one face of the tetrahedron. So each triangle describes one vertex-edge-face combination, i.e. one flag, and indeed the triangularization pictures the entire set of flags for us.

What makes the tetrahedron a regular polytope is the fact that its symmetry group has symmetries that can move any flag to any other flag, or to put it another way, any of the black and white triangles to any of the others. We don't have to use any mathematics to see this. Consider the most visible face in Fig.9.20. We can obviously rotate the tetrahedron so that any white (black) triangle of the face is moved onto any other of the same color on that face. We can also interchange any two adjacent white and black triangles on the face by a reflection (of the entire tetrahedron) along a mirror plane passing through their common edge and the center of the tetrahedron. Finally we can rotate any face to any other face. Combining these, we can move any triangle to any other.

There is something else we can see. Not only is there one symmetry that takes any of the black and white triangles to any other, there is *only* one. This follows because the symmetry not only involves the triangles themselves but the entire tetrahedron, and in particular its center, which is fixed by the symmetry. So each symmetry describes the movement of four non-co-planar points and is completely determined by what it does to those four points alone. Choosing one triangle, we see that there is exactly one symmetry of the group moving it to any other one. This implies that the number of triangles ($6 \times 4 = 24$) is also the total number of symmetries in the group, what we call the *order* of the group.[7] The group itself is called, not surprisingly, the *tetrahedral group*, and it has order 24. Let's denote this group by the symbol $\mathscr{T}$.

At this point we can go onto Fig.9.21 and Fig.9.22 in which we see the very same thing happening. But now something else occurs. As the figures show, the cube and the octahedron, and likewise the dodecahedron and the icosahedron, are dual to each other, with the face-centers of one matching to the vertices of the other (see the figure captions for explanations). As a consequence, they have the same symmetry groups. These groups are called the *octahedral group* $\mathscr{O}$, which has order 48 (count the triangles), and the *icosahedral group* $\mathscr{I}$ of order 120.

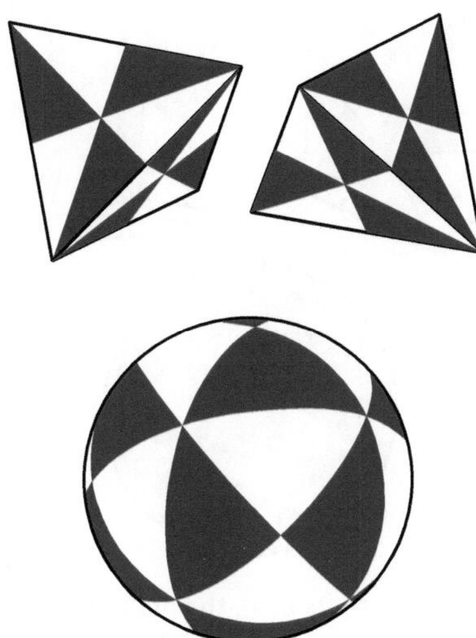

Figure 9.20: The tetrahedron, along with its dual (another tetrahedron), and a triangular decomposition of its faces. For a discussion of the duality relation between the two tetrahedra, see Fig.9.21. The tetrahedral group does not contain the central inversion (see the text).

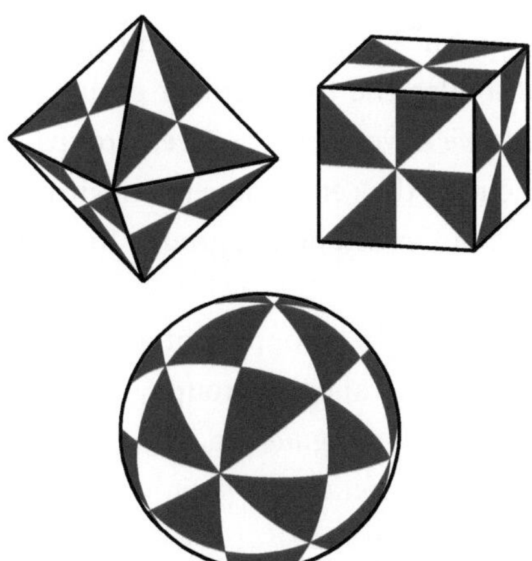

Figure 9.21: The octahedron along with its dual, the cube, and triangular decompositions of their faces. The duality between the octahedron and the cube is rather easy to see here: the centers of the faces of one become the vertices of the other, while its vertices become the face centers in the other. The figures show how the triangulations line up too. By puffing each of the polytopes into a ball, we can see the triangulation, now freed from being tied to either the octahedron or the cube. The symmetry group of all three figures is exactly the same: the same rotations and reflections work in each of the figures. This is the octahedral group $\mathcal{O}$. The symmetry of these figures under the central inversion is clear.

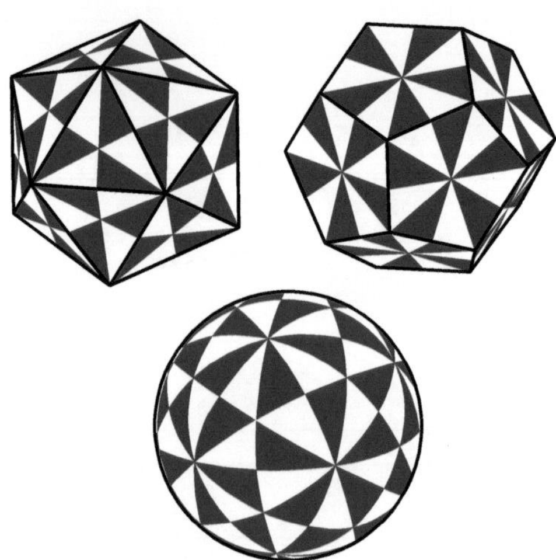

Figure 9.22: The icosahedron, along with its dual, the dodecahedron, and triangular decomposition of their faces. The duality of the two figures works the same way as in Fig.9.21.

All three of these groups involve both rotations and reflections. Half the elements are physically realizable (they are in $SO(3)$) and the other half are chirality reversing. In each case the part of the group inside $SO(3)$, the chirality preserving ones, forms a subgroup, so this gives us three more subgroups, $\mathscr{T}^+, \mathscr{O}^+, \mathscr{I}^+$ of orders 12, 24, 60 respectively. We will prefix their names with "even" (as opposed to "odd"), to make the distinction.[8]

To this list we will add in also the two-element group $\mathscr{Z} = \{\epsilon, I\}$ that arises from central inversion I. With this we have a total of seven groups. But there is still one more group, which arises by combining the central inversion with the rotation group $\mathscr{T}^+$. We find this group as a subgroup of the octahedral group. To see this we have to look at the octahedral group in more detail.

Both the octahedral and icosahedral groups contain the central inversion I but, as we can plainly see from Fig.9.20, the tetrahedral group $\mathscr{T}$ does not. Putting the two tetrahedra together produces the *stella octangula*, Fig.9.23, which now does have the central inversion as a symmetry. The stella octangula plainly has the same symmetry group as the cube, and hence the octahedral group. But if we restrict the symmetries that either do not interchange any vertices of the two tetrahedra or interchange them all at the same time, then we have the new group

$$\mathscr{T}^+ \times \mathscr{Z}, \quad \text{which stands for the set of all the elements}$$
$$R \circ J, \quad \text{where } R \in \mathscr{T}^+ \text{ and } J \in \mathscr{Z}.$$

This really is a group (it is closed under composition since both ϵ and I commute with all the elements of $\mathscr{O}$) and it is definitely chiral because it contains the central inversion. But it has no reflections in it. Instead it has what are called *rotation-inversions*, which, as

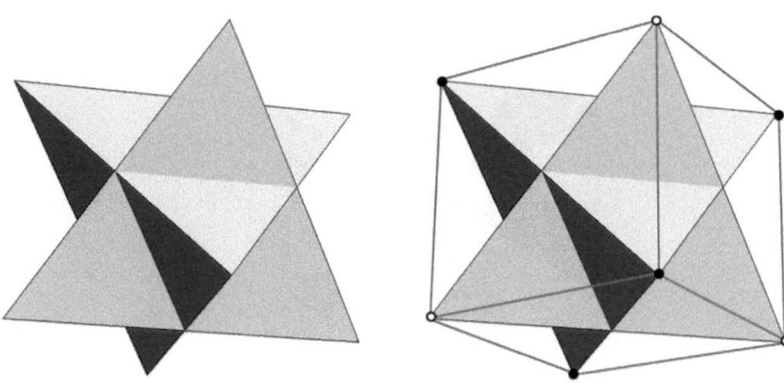

Figure 9.23: The alternate vertices of the cube (indicated with filled and unfilled dots) are the vertices of two tetrahedra. The stellated figure formed from the union of these two tetrahedra is still called by the name that Kepler gave it, the *stella octangula.*

their name suggests, are made by composing a rotation with the central inversion. The new group has 24 elements in it, but it is not the same as the tetrahedral group since that does not have any element like I that has order 2 and commutes with all the elements in the group. This group is the last of the eight groups that we were seeking. The lovely book [48] of Patrick DuVal offers more insights into these remarkable symmetries.

9.3.4 Point groups in 2D and 1D

Are these the only finite groups of symmetries in 3D space? Not quite. But all the remaining ones have the property of having an invariant plane. In effect they are 3D extensions of 2D symmetries, and so are not "three-dimensional" in the same way as the eight groups that we have just seen. As with what happens in 3D, any finite group of symmetries in 2D must have a fixed point. The obvious symmetrical figures in the plane are the regular polygons—the equilateral triangle, the square, the pentagon, the hexagon, and so on. We content ourselves with looking at one of them, the hexagon, Fig.9.24, which we show already triangulated. We can make use of the idea of flags again, but this time each flag consists of one vertex of the polygon and one of the two edges that contain it. The symmetry group can take any triangle to any other in exactly one way (black and white being interchanged by reflections). The symmetry group is called the *dihedral group*, denoted $\mathscr{D}_6$. Its has order 12 (it has 12 elements). Evidently there is one of these for each $n = 3, 4, 5, 6, \ldots$, so we have the dihedral groups $\mathscr{D}_n$ of order $2n$.

If we look at the subgroup of chirality preserving isometries here, the ones that don't interchange black and white, then we are left with just six rotations (we include the identity ϵ here). This group, of order 6, is denoted $\mathscr{C}_6$. It is an example of one of the simplest families of groups, the *cyclic groups*.

The characterizing feature of cyclic groups is that they are *generated* by just one element. In our case, the rotation R through $2\pi/6$ radians $(= 60^\circ)$ in the counterclock-

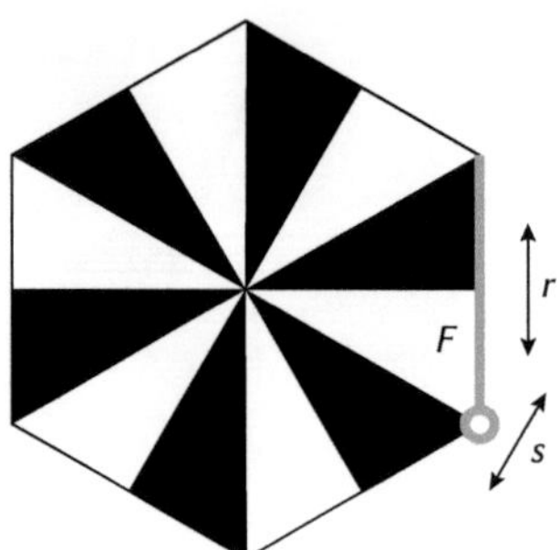

Figure 9.24: A hexagon and a triangulation of it that reveals its symmetry group $\mathscr{D}_6$ of order 12. Two reflections r and s generate the entire symmetry group. For instance $R := r \circ s$ is rotation through 60° in the counterclockwise direction. The chirality preserving part is the cyclic group $\mathscr{C}_6$ of order 6 consisting of the six possible rotations of the hexagon. We have also shown, in grey, a flag F: a vertex-edge combination. The symmetry group $\mathscr{D}_6$ can take any such flag to any other such flag in exactly one way.

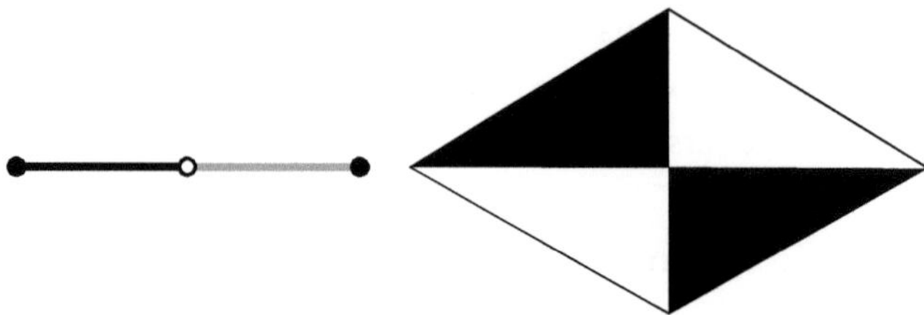

Figure 9.25: On the left, the regular polygon of 1D. Its symmetry group $\mathscr{D}_1$ has only two elements, ϵ and the reversal that flips the line over. The "mirror" for the reflection is shown as the single point in the center. On the right, a parallelogram illustrating $\mathscr{D}_2$ symmetry: two reflections and the two rotations (a half-turn and the identity mapping) making up the cyclic group $\mathscr{C}_2$.

wise direction (the usual positive sense in mathematics) generates the whole group $\mathscr{C}_6 = \{R, R^2, R^3, R^4, R^5, R^6 = \epsilon\}$. There is another generator, namely $R^{-1} = R^5$ which is rotation in the clockwise direction through $2\pi/6$ radians. The other groups $\mathscr{C}_n$ work in the same way.

Because there are no polygons with just two sides we are missing here the cyclic group $\mathscr{C}_2$, generated by a rotation through 180°, and the corresponding dihedral group $\mathscr{D}_2$ of order 4, see Fig.9.25.

In effect we have now seen the full extent of the classification of all the finite groups of symmetries in Euclidean 3D space. A complete classification can be found in Coxeter's book [38, pp. 15.4–15.5].

We have spent a lot of time on the finite groups of symmetries in 3D. Apart from the intrinsic interest of knowing more about what is actually so close to us, it actually shows us what Nature itself has to work with. Whether or not we think of these as constraints

Figure 9.26: The beautiful symmetry of a snow-flake. Although usually described as having 6-fold symmetry, we cannot but notice that there is reflectional symmetry too. The symmetry of the snowflake is D_6 of order 12. Curiously the 6-fold nature of snowflake symmetry was only noticed very late in the West. Johannes Kepler is one of the earliest to mention it (*Harmonices Mundi*, 1619). By contrast it had been noticed and documented in China over 2000 years ago. In a book from about 135 BCE, Han Ying wrote "Flowers of plants and trees are generally five-pointed, but those of snow, which are called ying, are always six-pointed", [166]. Janeklass, Creative Commons CC BY 4.0 commons.wikimedia.org/wiki/File:Snowflake_(lumehelves).jpg .

on Nature or just as the way that Nature has constrained itself, these are the only finite symmetries available and they are genuine restrictions on what sort of patterning we can expect to physically exist. Needless to say the natural world has fully explored this.

When it comes to the 3D world, we see crystalline cubes and octahedra. Many atomic structures are tetrahedral, for example methane, CH_4, which has the four hydrogen atoms at the vertices of the tetrahedron and the carbon atom at its center.

Nature often seems to favor distributions of atoms around a central atom that are as equally distributed as possible. As a result one might expect that the icosahedron, for example, would be a common feature. The most famous examples are the carbon-based buckyballs (carbon 60) and their relatives, which were predicted to exist before they were actually discovered, Fig.9.27. Indeed in the process of crystallization of liquids into solids, icosahedral clusters are common, because they are energetically favorable.

But then something very strange happens. As we shall see in §9.4.5, icosahedral symmetry is not compatible with the repetitive symmetry of lattices. Eventually in the crystallization, icosahedral symmetry succumbs to the higher priority of lowering energy to be replaced by local clusters that are fully compatible with lattice symmetry.

However, we now know that icosahedral structures are possible in recently discovered forms of long-range order called *quasicrystals*. These might be thought of as almost-

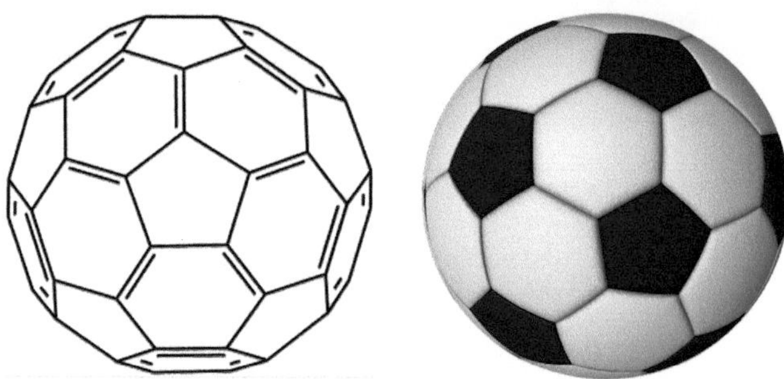

Figure 9.27: Carbon 60, also known as buckminsterfullerene, is composed of 60 carbon atoms arranged in space so as to form the vertices of the semi-regular solid shown here. It has 20 hexagonal faces and 12 pentagonal faces, the same as the traditional international soccer ball. The left-hand figure indicates the single and double bonds of carbon 60, involving the four shared electrons for each carbon nucleus. Public Domain. en.wikipedia.org/wiki/Buckminsterfullerene .

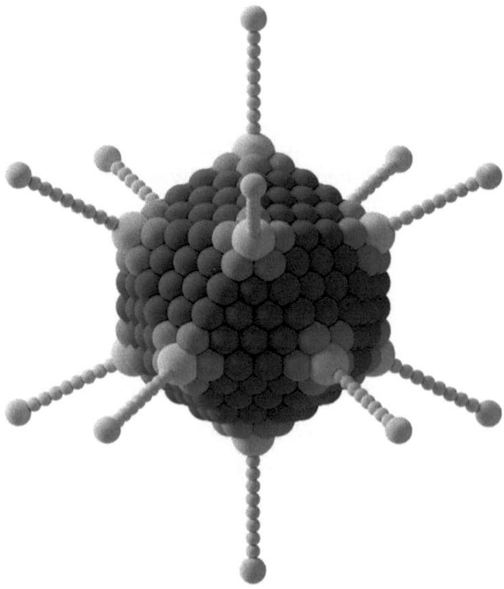

Figure 9.28: A simplified illustration of the capsid of an adenovirus, showing its icosahedral symmetry. Public Domain. Wikipedia .

lattices. We will come to them in §12.4 and ‡§18.2. As for the organic world, there are wonderful examples of icosahedral symmetries, notably numerous viruses whose resting states have icosahedrally symmetric coverings (capsids). Fig.9.28 illustrates the icosahedral capsid of the family of adenoviruses.[9]

9.4 Periodic symmetry

Everywhere you look in the man-made world you will see a recurring solution to the problem of creating physically extensive flat surfaces: brick walls, tiled floors and walls, wallpapered walls, windowed surfaces on buildings. All of these are based on simple repetition of a basic block or tile: squares, rectangles, triangles, or hexagons. In Nature too we see the same thing, notably in three dimensions, all crystals being composed of repeating copies of a basic atomic or molecular motif. In effect this is the defining property of crystals. This repeating property goes by the name of periodicity, and from the perspective of this chapter it is an outcome of translational symmetry. In this section we will place most of our attention on the 2D situation, where this sort of patterning is most often seen in the form of tilings of the plane. We call them *periodic tilings*.

In the real world tilings are necessarily of finite extent, but the underlying principle is that they can be extended indefinitely in all directions (in 2D or 3D as the case may be). Mathematically we simply take them to be infinite in the first place. Not to do so means to have to worry about the complications around boundaries—where they are, their shape, and so on—all of which may have significance, but which are irrelevant to the basic discussion here, namely the repetition that underlies them.

As soon as we do this the repetitive nature of the periodic tilings appears as *translational symmetry*. We can translate the entire tiling so as to make it coincide exactly with itself. These special translations are the *periods* of the periodic tiling. In 2D there are two independent directions of translational symmetry; in 3D there are three. In either case, the entire set of periods (including the null "translation" that does nothing at all) is a group since the composition of one period with a second one is still a period, and the inverse of a period (undoing it) is also a period. This group of translations is called the *period group*. We will use the symbol $\mathscr{P}$ for the period group, being aware that we use the same symbol for partitions!

9.4.1 City blocks

The basic intuition of periodic tilings is very familiar. Imagine the typical planned city (of which there are excellent examples from both modern and ancient times, see Fig.9.29) laid out in a regular rectangular grid of blocks, say in the east-west and north-south directions, Fig 9.30.

The block layout is, in essence, a periodic tiling. To get from any block to any other one walks so many blocks east or west and so many blocks north or south. To be economical in thinking we can restrict to just east and north and allow walking a negative number of blocks to deal with walking in the other two directions. In this language, with E for walking one block east and N for walking one block north, walking from a point at the corner on one block to the corresponding corner on any other block amounts to the walk

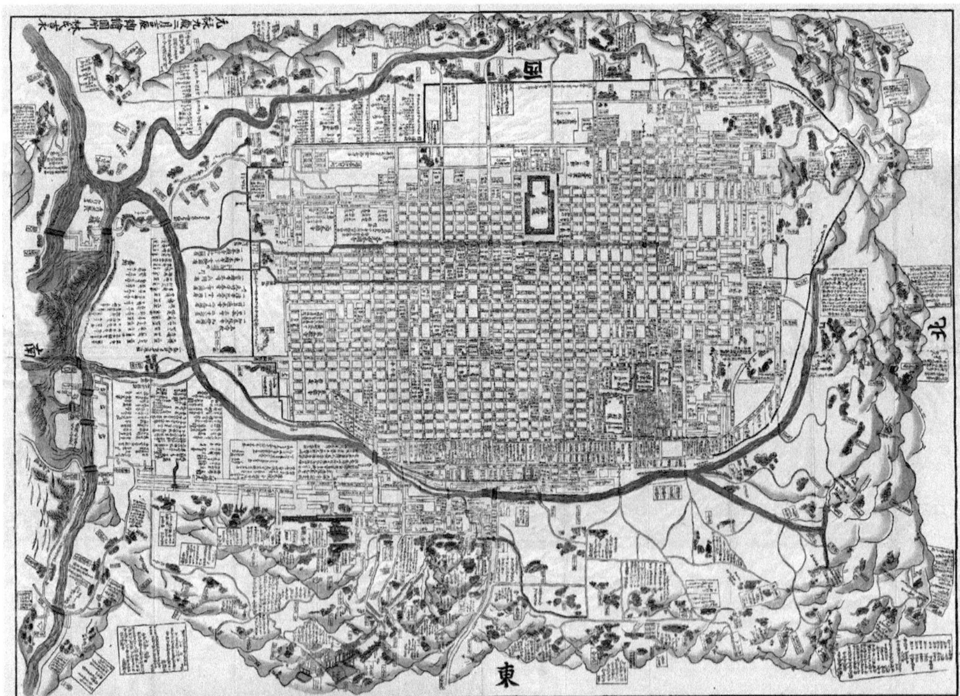

Figure 9.29: A seventeenth-century map of the city of Kyoto. The basic city plan of Kyoto was laid out in the eighth century as a scaled-down replica of the Tang dynasty city of Chang'an. Geographicus Rare Antique Maps, Wikimedia Commons, geographicus.com/mm5/cartographers/japanese.txt .

(or translation) $E^k \circ N^l$ for some integers k, l. There is something we know well about this: it doesn't matter in which order we do the east and north parts, and we can even mix them, as we often do in actually walking such paths. This is a statement that the two translations commute: $E \circ N = N \circ E$, so the Es and Ns can be performed in any order, and in particular $E^k \circ N^l = N^l \circ E^k$.

Although we are framing this in terms of walking, the larger picture is that we can shift the entire tiling one block east or one block north, or invert these and shift one block west or one block south, and these are translational symmetries of the entire block layout. Think of E and N as periods that translate the entire layout. The entire translational symmetry group, the period group, that they generate is

$$\mathscr{P} = \{E^k \circ N^l \ : \ k, l \in \mathbb{Z}\}.$$

No matter what two-dimensional periodic tiling we have, there is an associated period group. E and N are called, naturally enough, *generators* of $\mathscr{P}$, since every other period can be "generated" out of them by using the group structure of composition and inversion. The fact that we chose E and N as the generators does not mean that they are the only

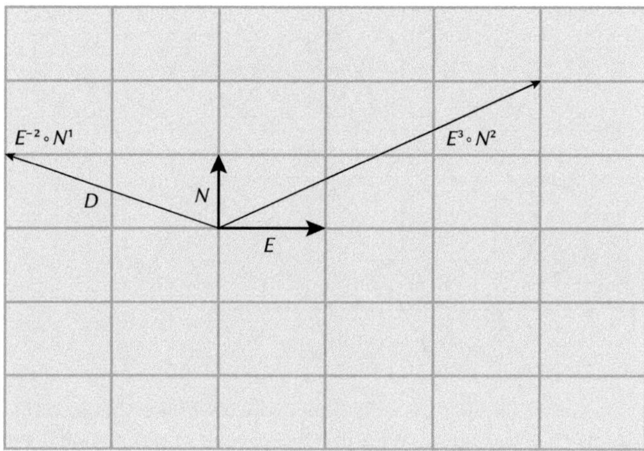

Figure 9.30: A typical city block layout, mathematically idealized to go on forever in all directions. The translational symmetries are generated by the two translations E (east) and N (north). We see the particular translations $E^3 \circ N^2$ and $D := E^{-2} \circ N^1$, where we use the convention of negative indices for the indicating translation in the opposite direction, e.g. E^{-1} is translation west. The period group $\mathscr{P}$ is the entire set of translational symmetries. It is abelian (it doesn't matter in which order the translations are performed. In walking, you can walk east/north directions in any order, but if you want the outcome to be $E^{-2} \circ N$ then the total number of eastward translations should add up to -2 and the total number of northward translations to 1.

ones. In the case of real physical city blocks, where we cannot walk "as the crow flies", we do not have the option of anything other than east-west or north-south walking, but if we are thinking in terms of translating the city landscape without physical constraints, then we could, for example, make the "diagonal" translation $D := E^{-2} \circ N$ one of the generators, see Fig.9.30, since along with E it also produces an alternate pair of generators $\{D, E\}$. Note that $N = E^2 \circ D$, so the new pair can generate the original pair $\{E, N\}$ and vice-versa. There are endless pairs of generators (an infinite number), though that may not seem at first very obvious.[10]

The fact that a group can be "generated" by a subset of its elements is important. By capitalizing on the algebraic group structure (that is, the multiplication rule and the existence of inverses) it is possible to express every element of a group using only a few elements of the group. In this case of two-dimensional periodic tilings the period group can be generated by just two elements. For three-dimensional tilings and crystals, we need three generators.

The city block picture is one of repetition of a single geometric feature, one city block. In the language of crystallography the city block itself is an example of what is called a *fundamental region* for the translation group: the action of the period group is to fill out the entire plane with copies of it with no overlap except at the boundaries where the city blocks join each other. Typically when there is a periodic tiling it is useful to find a

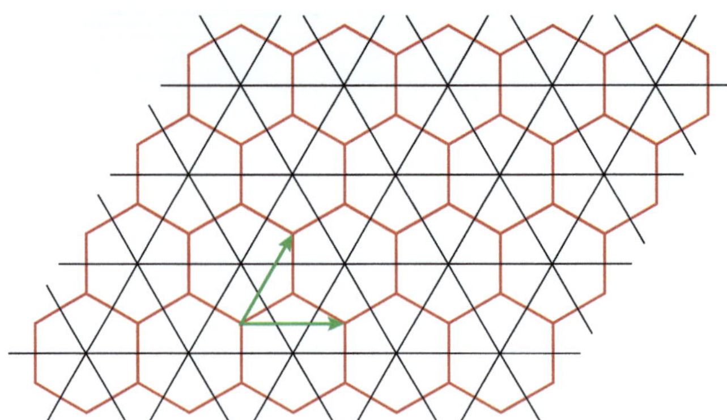

Figure 9.31: Here is another familiar periodic tiling, which works either as a tiling of equilateral triangles or as a lattice of hexagons. We still have a group of translational symmetries, and we can still pick (in numerous ways) two of these that generate them all. We have chosen two, shown by the green arrows, and these two translations can be used to generate all the other translational symmetries of the tiling.

fundamental region. They are never unique, but usually one will come rather easily to mind, or seems more natural, once one starts looking. See, for example, the discussion of the Isfahan tiling in §9.4.2.

In Fig.9.31 we see an example of the ubiquitous hexagonal tiling, which is another variation on the way simple translational repetition can be used to fill out a plane. Again we can find two translations that will generate the entire group of translational symmetries (periods) of the tiling, say two at a 60° angle to each other. Evidently the full symmetry group of the hexagonal tiling is far larger than this, including both rotations and reflections. We will look more closely at the nature of this symmetry in §9.5.

9.4.2 The symmetry underlying the Isfahan tiling

With this notion of the period group in mind, it is instructional to compare the city blocks picture of Fig.9.30 with the magnificent tiling of the Jameh Mosque seen in Fig.9.2. There we showed just a single square panel from the tiling as it appears in the mosque. Using vertical and horizontal translations we can open it up into a square tiling of the plane. Evidently the entire tiling is periodic with vertical and horizontal translations as periods. However, closer inspection shows that the translational symmetry is finer than this. In Fig.9.32 we can see that the smaller rotated square, bounded by the white edges in the figure, can be translated by diagonal translations to fill the entire plane. These two diagonal translations generate the true period group of the tiling, and the smaller square is a fundamental region for the period group—that is, the translations of the group perfectly tile the plane.

The symmetries hidden in this tiling do not stop there, see the caption to Fig.9.32.

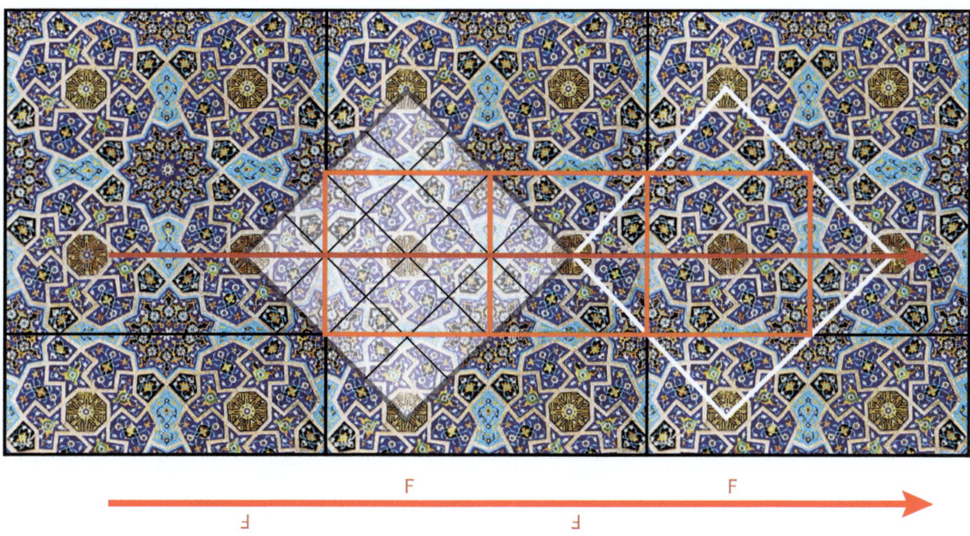

Figure 9.32: Here we go back to beautiful tiling from the Jameh Mosque, Fig.9.2. Here, the upper part of the figure consists of three copies of the original square (black boundaries). Below there are three partial copies. Looking at this in a new way in terms of the square shown which has been whitened, we see that it is a smaller region that covers the tiling by using two diagonal translations. It is a fundamental region for the period group and these two translations generate the period group (the group of translational symmetries) of the tiling. However, there are many other symmetries too. For instance the center of the square with a white border is a center of 4-fold rotational symmetry around its central octahedral figure. Looking at adjacent orange squares, we see that the chirality of their rotational symmetries are opposite (clockwise and counter-clockwise). The adjacent orange squares are not translationally equivalent: instead they are related by what are called *glide reflections*—translate and reflect along the red line—shown schematically at the bottom of the image. The entire symmetry group can be generated out of repeated applications of one of these 4-fold rotations and a glide reflection.

The Isfahan tiling is a wonder. Having seen it, one may ask what other wonders may be out there. That begins by looking at the variations in the geometry of the lattice that underlies periodicity.

9.4.3 Bravais lattices

Group theory is considered as a part of algebra. It has to do with the structures that can arise out of the algebraic operation of combining symbols subject to certain rules, namely the axioms **G**(i)–**G**(iv), see Endnote 3. As a part of what is normally called *modern algebra*, group theory is studied as a subject for its own sake. However, the way we are looking at groups here is *geometric*, and there is a difference. One such difference arises when we look at the period group of tiling. In two dimensions a periodic tiling has translational symmetry that is generated by two translations A and B. The entire group of translational symmetries then consists of the translations $A^k \circ B^l$, where k, l vary over all possible

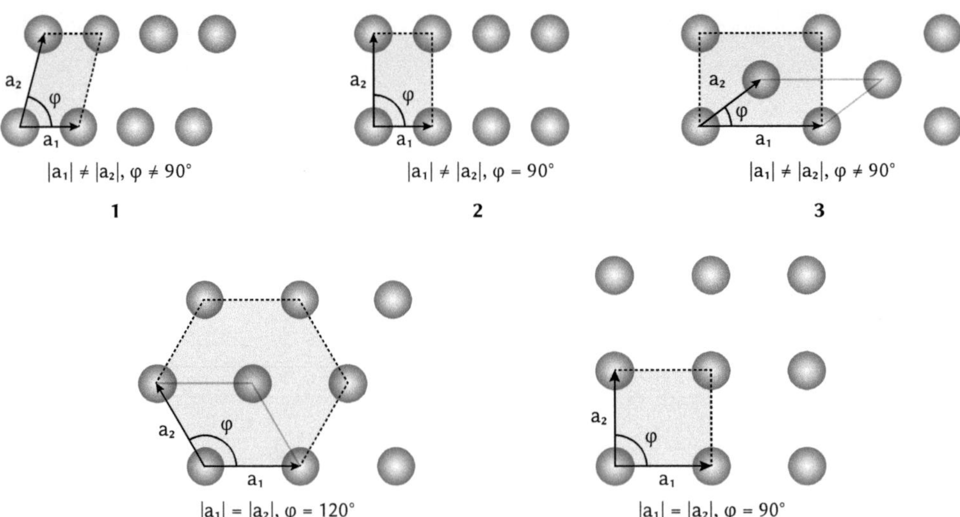

Figure 9.33: Shown here are the five basic forms that a 2D lattice can take once it is treated geometrically as well as algebraically. These are the so-called 2D *Bravais lattices.* They all have the same algebraic structure, but they are distinguished by their additional rotational symmetries. For instance the square lattice has centers of 4-fold rotational symmetry, while the hexagonal lattice has both 3-fold and 6-fold centers of symmetry. To add to the complexity, in actual tilings based on these lattices the available symmetries may or may not be present. That is a question of whether or not these symmetries survive after the repeating cells are decorated with designs. All told, there are 17 potential outcomes for the symmetry group of a periodic tiling in 2D.

integers k, l. The algebraic structure is then the simple rule

$$(A^k \circ B^l) \circ (A^m \circ B^n) = (A^{k+m} \circ B^{l+n}).$$

This is universal—it is the same for all of these period groups. That is the purely algebraic point of view. If we look at the geometry then we see differences. The generating translations may perform equal or unequal translational distances, and they may be at right angles, or sixty degrees, or some other angle.

The *lattice* of a periodic tiling is its group of translational symmetries along with the basic geometric information of the relative lengths and angles involved in the structure of its translational symmetries. There are five different scenarios, the five 2D-Bravais lattices, and they are explained in Fig.9.33.

Beyond these, taking into consideration the symmetries that may be *destroyed* due to the actual designs on the tiling, there are 17 different symmetry arrangements. Consider, for instance, the Isfahan tiling. Its Bravais lattice is the square lattice: two translational generators of equal length at right angles to each other. In principle this tiling can have both rotational and reflectional symmetry at its vertices, the mid-points of its edges, and the mid-points of its square tiles, that is D_4 point symmetries. The Isfahan tiling has none of these three reflective symmetries, but it does have rotational symmetries of order 4, 2, 4 around the vertices, mid-edge points, and center points of each of its squares.

Needless to say, this missing symmetry is not due to any deficiency of the Islamic artists. They so perfected the art of 2D periodic tilings that even without these modern perspectives they found, and used, all 17 different symmetry types. For instance, they all appear in the *Alhambra*, the famous Islamic palace in Granada. For a more extensive discussion on these symmetry groups and on symmetry in general, see Coxeter's book *Introduction to Geometry* [38].

9.4.4 Periodicity in 3D

All of this works in three-dimensional space too, and apart from the increasing number of possible symmetry types and the considerable difficulty of visualizing all the possibilities, the story is very much the same. Typically in the case of crystals, there is more than one type of atom involved, so the fundamental cell may be "decorated" with other atoms, often of a different type, and the symmetries applicable to the cell itself need to be taken into account. When all is said and done, there are 14 Bravais lattices (Bravais, 1811–1853) in the 3D situation, which result in a total of 230 space groups belonging to them. The determination of the 230 space groups—quite an amazing feat—was accomplished by Evgraf Stepanovich Federov (1853–1919) and Arthur Moritz Schönflies (1853– 1928). Arriving initially, and independently, at slightly different answers, they combined forces and finally arrived at the now accepted answer of 230. Independently of Federov and Schönflies, William Barlow (1845–1934) enumerated the 230 space groups, publishing his result in 1894. Amazingly all this mathematics was done well before the atomic theory was accepted by physicists!

In spite of the apparent complexity that arises here, the overall message is something different: there are only a very limited number of forms of symmetry that are compatible with periodic repetition. The list may seem fairly long, but this is all that can happen. Every crystalline structure in Nature must be one of the types on this list, obviously a point of tremendous importance to crystallographers. What underlies this is the combination of discreteness and repetition. In both the 2D and 3D cases there is repetition of a block of fixed size, and that means that there is an overall discreteness, that is separation, between like points. Physically we can think of every atom needing its own space and separation from its neighbors, even though they all mutually interact. To this we include all possible rotational and reflectional symmetries. The possibilities turn out to be very limited.

What is also amazing is that all 230 of these groups have been found to be manifested in crystal structures. To put it in somewhat anthropological terms, Nature has fully explored all the possible options available to it. It is this sort of fact that brings up the old question of whether or not mathematics has a Platonic ideal nature to it. These groups were discovered before atoms were accepted as physical realities. The mathematics is the mathematics of Euclidean three-dimensional space. It seems that this classification is a purely mathematical construct. Yet it is something that manifests completely in the natural world. Are these principles something that belongs to some ideal world beyond the physical domain, or are they part of the inherent principles of Nature, deeply woven into its weave?

Whatever the answers to these questions may be, there is one especially notable powerful consequence of this study of symmetry: the crystallographic restriction.

9.4.5 The crystallographic restriction

The famous crystallographic restriction says that the only kinds of rotational symmetries that are compatible with lattice symmetries in 2D or 3D are 2-fold, 3-fold, 4-fold, and 6-fold. No 7-fold, 8-fold, ..., and most notably, no 5-fold rotational symmetry is possible. So, for instance, there is no icosahedral symmetry possible in a crystal. There are some very nice ways to show this statement rigorously, but actually the intuition behind it is not hard to see by looking at the 2D situation.

Suppose that we had a mathematical crystal (a periodic tiling) $\mathscr{T}$ in 2D with period group $\mathscr{P}$, and suppose that $\mathscr{T}$ were also symmetric by a 5-fold rotation R with center p. Since $\mathscr{T}$ is symmetric with respect to all the translations of $\mathscr{P}$, not only p but all its translates by $\mathscr{P}$ would also be centers of 5-fold symmetry (remember that $\mathscr{T}$ looks identical after any of its translations is applied to it). So we have lots (in fact an infinite number) of centers of 5-fold symmetry. Since $\mathscr{T}$ is discrete, there must be some minimal separation between any two such centers, say d. So any two centers of 5-fold symmetry are at least d apart. So we take two such centers at distance d apart. The rest is explained by Fig.9.34 and its caption.

5-fold symmetry is possible in lattices in 4D, and curiously enough it can also creep into physically existing materials that seem to be crystals. Of course they cannot have lattice symmetry, but they have other properties that are so crystalline in nature that they were enough to bring the entire crystallographic community into a tizzy. These are the quasicrystals, discovered in 1982 by Dan Shechtman. We discuss them in §18.2.

9.5 The pattern idea and orbits

9.5.1 Symmetry is a systemic concept

Although symmetry is evidently concerned with pattern, we should ask how it actually fits into the conceptual idea of pattern systems. In the previous sections we have seen many examples of groups acting on sets. If E is a set and $\mathscr{G}$ is a group of transformations

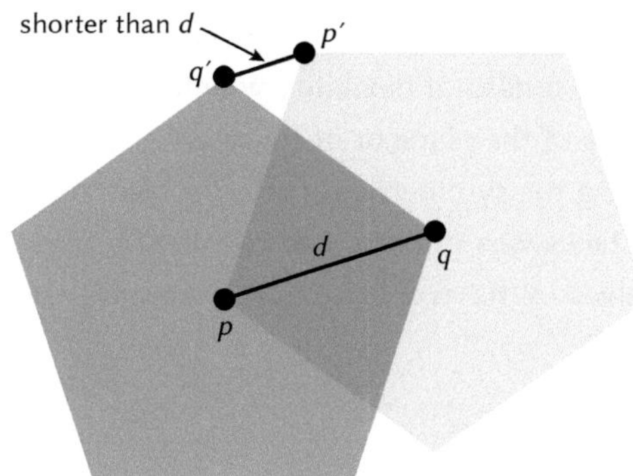

Figure 9.34: This is a proof, by way of contradiction, that there can be no 2D lattice-like periodic tiling based on centers of 5-fold symmetry. The argument is that there would be no minimal distance between such centers of 5-fold symmetry, thus violating the periodicity of such a tiling. For suppose that p and q were two centers of 5-fold rotational symmetry at a supposedly minimum distance d apart. Since the periodic tiling is completely 5-fold symmetric around p, so are all the 5-fold rotational centers themselves. So the rotated (counterclockwise) image of q around p, shown here as q', has to be another center of 5-fold symmetry. We picture this turn through a fifth of a circle in terms of a pentagon. Now doing the same with q, the rotated (clockwise) image of p around q, shown here as p', is another 5-fold center. These two new 5-fold rotational centers p', q', are closer together than d (in fact less than $d/3$ apart), contradicting our choice of p and q in the first place. Once there is a center of 5-fold symmetry there are infinitely many and repeating we see that there is no limit to how close they may be to each other: in totality they are not discretely separated. This simply cannot happen in a (periodic) crystal.

on E, then we naturally have the first two ingredients of a pattern system: a *state space* and a *set of operators* acting upon it.

In earlier sections, the nature of the operations could usually be imagined in temporal terms—the continuous flow of time or the repetition operation of time. The situation in the study of symmetry is generally different. Each symmetry is an operator and, in our present setting, it acts as a discrete operator in the sense that it affords a discrete invertible change in the plane or in space. Further, there are many of these operators and we need to compose them and their inverses, respecting the order in which they are applied, to reveal the symmetry that they hold in their entirety. Each system now has a whole group—literally a *group* in the strict mathematical sense—of operators, and it is not just the operators themselves, but also the mutual algebraic structure that they entail that is important. Even so, the abstract idea is the same: a space of states and a set of operators that produce change upon those states.

The remaining two features of a pattern system refer to the event space, that is the level to which we express differences within state space, and some sense of invariance in the form of a measure. The latter has been fairly simple so far since all the symmetries that we have discussed are area or volume preserving. In fact we have only discussed distance preserving operators (isometries), which automatically preserve area and volume, so there is a stronger invariance that is hidden in the nature of the operators themselves.

As for the events there is a natural flexibility which we have taken advantage of as we moved along. The points of the plane or of space are the most obvious choice of the basic events. But in studying the symmetries of the Platonic solids we saw that working with vertices, edges, and faces was the key, and in periodic tilings we have thought in terms of the tiles themselves. Whatever choices are relevant, we can see symmetry as systemic in nature.

9.5.2 Trajectories and orbits

In the language of pattern systems, if $x \in E$ then the *trajectory* of x under the set of transformations that make up the set of operators $\mathscr{G}$ is the set

$$\mathscr{G} \triangleright x := \{R \triangleright x \ : \ R \in \mathscr{G}\} = \{R(x) \ : \ R \in \mathscr{G}\}.$$

In words, a trajectory is all the points that one can get by applying the transformations of $\mathscr{G}$ to some point of E. In the language of group theory, trajectories are called *orbits*, and $\mathscr{G} \triangleright x$ is referred to as *the orbit of* x. We may say that orbits are what develop out of the action of the group on individual points.

A familiar example in the continuous setting is lines of latitude drawn on globes of the Earth. Under the group of rotations around the north-south polar axis, the orbit of any point on the Earth's surface is the corresponding line of latitude. The reason that latitude is relatively easy to determine whereas longitude is not, is that latitude is a direct consequence of a very physical fact: the motion of the Earth around its axis. With the equator being taken as 0 latitude and the north/south poles as $\pm 90°$, the latitude anywhere else can be measured by sighting the star Polaris or by observing the altitude of the Sun at high noon. By comparison, longitude is determined by angle of rotation away from the prime meridian, which is a completely arbitrary creation of our culture, chosen at the height of British power to be through Greenwich.[11] The solution to the navigational problems of longitude were to vex mariners for centuries until the creation of clocks that could retain their accuracy under the physical stresses of long ocean voyages.

Group orbits have nice properties, chief of which is that any two orbits of $\mathscr{G}$ in E are either equal or disjoint. In symbols this is equivalent to saying that if $y \in \mathscr{G} \triangleright x$ then $x \in \mathscr{G} \triangleright y$. This is a simple consequence that group actions are invertible: if $y = R \triangleright x$ then $R^{-1} \triangleright y = R^{-1} \triangleright R \triangleright x = \epsilon \triangleright x = x$. Since every element is in its own orbit, we see that the state space E is *partitioned* by its orbits, each orbit being one part of the partition. Put in our language of equivalence relations, the orbits are equivalence classes. All of this is visible in the lines of latitude.

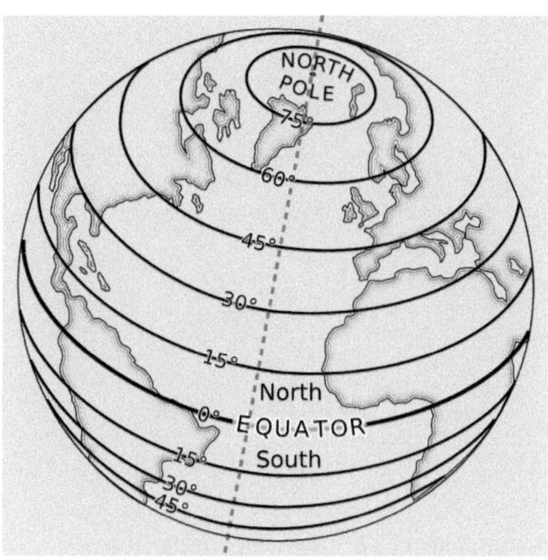

Figure 9.35: Lines of latitude are orbits of the group of rotations of the Earth's surface around its north-south polar axis. Pearson Scott Foresman Archives, Public Domain. picryl.com/media/latitude-psf-a326a2.

9.5.3 The symmetry group of the hexagonal lattice

As an example of what orbits are like in a discrete setting we look at the triangular-hexagonal lattice of Fig.9.31. Decomposing the hexagons into triangles, much as we did in looking at the symmetry group of the regular hexagon in Fig.9.24, we arrive at the tiling (or *tessellation*) of the plane shown in Fig.9.36. The tiling clearly has lots of symmetries—an infinite number, in fact, including reflective, rotational, and translational symmetries. Although it is huge, the entire group of symmetries, let's call it $\mathscr{H}$, is generated by the three reflections coming from the sides of any one of the triangles. We have selected the one marked off by dark lines in the right-hand figure. What this means is that by repeated use of just the three reflections in the sides of this one triangle it can be transformed into any other of the triangles. The triangle then serves as a fundamental region for the symmetry group $\mathscr{H}$.

In particular every point in the plane, wherever it is, can be transformed by $\mathscr{H}$ into a *unique point* of this fundamental region. Conversely every point of this fundamental region gives rise to exactly one orbit under the action of $\mathscr{H}$. These orbits partition the entire plane into disjoint subsets so that there is one orbit for each point of the triangle (including those of its interior).

Partitioning at this level is too fine to be particularly relevant in terms of our study of pattern. But it is useful to see that there are really only seven distinct types of orbit: orbits arising from each of the three vertices, orbits arising from each of the three edges, and orbits arising from the interior of the triangle. Patterns emerging from each of these are illustrated in Fig.9.36.

Orbits are certainly what we would commonly call patterns. Here we see how that sense of pattern can be interpreted in terms of change and invariance: orbits are the result of repeated transformations on a point (change), but an orbit itself is something

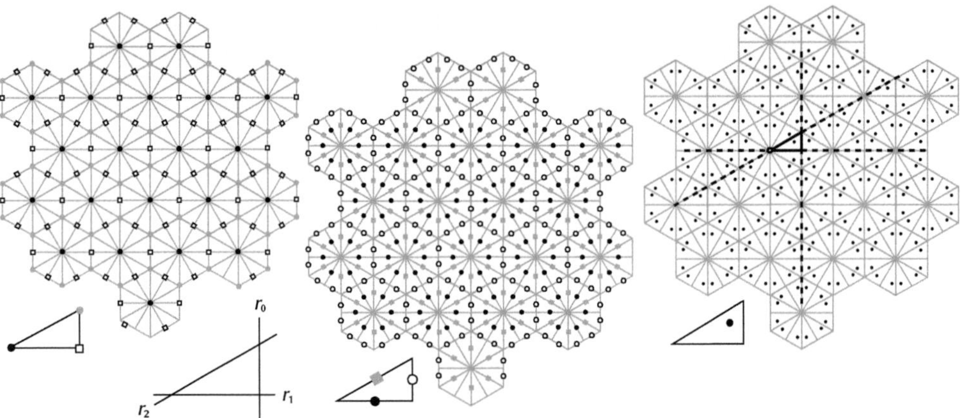

Figure 9.36: The three figures, left to right, show the appearance of the orbits under the action of the group $\mathscr{H}$ generated by successive applications of the three reflections r_0, r_1, r_2 in the three sides of one triangle. For definiteness we use the triangle and three reflections indicated by the dark lines in the right-hand figure. The three figures show patterns emerging from vertices, edges, and the interior respectively.

that does not change (it is invariant as an orbit) under these changes.

9.5.4 Invariance in physics

Physics takes this to an extreme. The function of physics is to find, to the greatest extent possible, those great features of invariance that underlie, or at least apparently underlie, the workings of Nature. In this it has been fantastically successful. Beneath the surface flux of things there are principles or laws that seem to be invariant in time or place, untouched and unmoved by petty human affairs, that allow us, at times, to accomplish awesome predictability. The physicists' approach to this, and it is something that has only come to the forefront over the past century, is that this invariance is expressible as symmetry.

The idea is actually rather easy to appreciate. The basic belief that underlies physics is that our Earth is not a privileged place. Although what happens here on Earth is evidently greatly influenced by local conditions (the local complexity), nonetheless there are basic principles that are independent of us and where we choose to look in space or time. For instance we take it to be true that gravity is a universal force that acts on other bodies in the universe just as it does here, that the speed of light is the same everywhere (in a vacuum) and is unchanging in time, that atoms are totally classified by the periodic table, and their excitations with their characteristic frequencies of their emitted photons are the same now as they were billions of years ago in parts of space billions of light years away. No doubt these are assumptions, but they are perhaps among the most reasonable that we can make—we look at the heavens and the great cycles of the planets and stars, at the procession of the seasons, and we see that there is order. We no longer assume that just because humanity is here on the Earth at this particular moment of time, that the Earth is necessarily some place especially preferred by or especially exempt from the forces of Nature that act elsewhere.

What is relevant here is that these assumptions imply fundamental symmetries. The laws of physics will not change if we simply take a different position in space or a different orientation. Thus they should be invariant under the symmetry group ISO^+. Adding in invariance under uniform motion in time, we arrive at Galilean physics, the classical physics that began with Galileo, that served as the basis Newton's development of physics, and that was later interpreted in terms of invariance under this symmetry group.

This was all based on the conception of an absolute sense of space in which velocity had an absolute meaning. When experiment and theory led to the conclusion that the speed of light was a constant of Nature independent of the translational motion of the observer, and it was understood that there is no absolute meaning to motion, but only to relative motion of independent observers, the understanding of invariance changed. This called for a more refined appreciation of the notion of invariance in which it was the time between events that remained invariant. The corresponding symmetry group is called the Poincaré group. Put in physical terms, these symmetries turn into explicit laws, for instance translational symmetry appears as the conservation of momentum and rotational symmetry as the conservation of angular momentum, time invariance as conservation of energy.

This relationship between symmetries and laws of invariance was formalized by one of the great mathematicians of the twentieth century, Emmy Noether. Noether's theorem is of great generality and assumes even more significance in quantum mechanics. In broad terms her theorem states that if a system has some form of continuous symmetry, then there is a corresponding quantity whose value is conserved [169, p. 5].

This brings in another aspect of patterning, namely *symmetry breaking*. The study of crystal symmetry is a case in point. Crystals form by loss of energy (cooling) from liquid or gaseous states. It seems that this symmetry is being formed out of nothing in front of our eyes. But the physicists suggest that what we see is better thought of as symmetry being destroyed (or broken, as they say it).

How can that be? From a physical point of view a gas is more symmetrical than a crystal, for in a gas all parts of it are essentially of the same nature. It favors no one orientation over any other, it has full rotational symmetry, and it is translationally symmetric by all translations, not just by the periods of a period lattice. During crystallization we move from continuous symmetry to discrete symmetry. As energy is taken away from a gas (it is cooled, i.e. the average speed of motion of its atoms is reduced) the atoms assume positions that are compatible with that loss of freedom to move. The optimal solutions to this problem (which depend on temperature and pressure) seem to be the

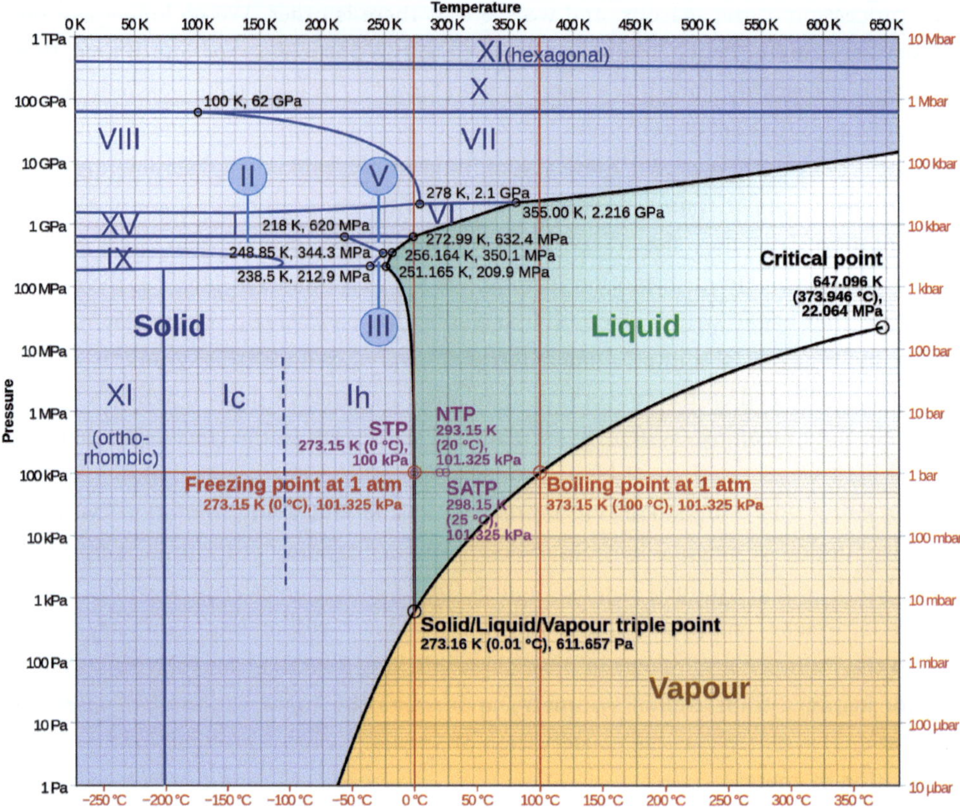

Figure 9.37: A phase diagram for ice. The horizontal direction indicates temperature, the vertical direction, pressure. The roman numerals describe different forms of ice, with crystal structures ranging over six of the Bravais classes: rhombohedral (IV), tetragonal (III, VI, VIII, IX), monoclinic (V, XIII), cubic (VII, X, XVI), hexagonal (Ih), and orthorhombic (XI, XIV), as well as some non-crystalline metastable states. London South Bank University, Creative Commons, lsbu.ac.uk/water/ice_phases.html.

periodic frameworks of crystals. There are experiments where the initial arrangements of the atoms during a cooling process leads to icosahedral clusters (being as symmetrical as possible given separation requirements), only to suddenly switch to a lower energy state with lower local symmetry but long-range periodic symmetry as the temperature drops further [96]. A fascinating scenario emerges here in which, according to its state, the system changes its symmetry. The startling complexity of crystallization is apparent from Fig.9.37, which shows the varieties of crystal formations of water according to temperature and pressure. It is safe to say that there is a great deal that is unknown about crystallization and why minimalization of energy manifests itself in this particular type of structure.

Contemporary theories of cosmology posit an early universe of much higher energy, and accompanying higher forms of underlying symmetries that have been successively broken as it cools. Just like ancient stories and histories of former more perfect times, our universe was somehow more "perfect" than it is now. Whether or not these views will prevail over time is not for us to say. Today they are important guides for cosmologists and for experimentalists in high energy physics.[12]

9.5.5 Invariance in mathematics

The notions of invariance and invariants are deeply woven into the mathematical study of symmetry. The degree to which invariants appear out of symmetry is not, at least at first, all that obvious. It is obvious that a cube is invariant under its rotational and reflective symmetries. But there is more to it than that. For example we have already talked about invariance in the context of measures. Recall that in the case of a dynamical system with temporal change, either discrete step-wise iteration or the continuous flow of time, we paid particular attention to finding stationary or invariant measures. These are measures that remain invariant, or stationary, during the process of change that the system is undergoing in time. To recall its shape, the defining equation is

$$\mathbf{m}(t^{-1} \triangleright A) = \mathbf{m}(A)$$

for all subsets A of the state space and for all times t. Since $t^{-1} \triangleright A$ is the set of states that *will* become the set A after time t, it is a statement that as the future unfolds, the measure corresponding to $\mathbf{m}$ at that future time is the same as the measure that we have now—invariance.

Similarly we have seen that the Lebesgue measures, the standard measures of length (1D), area (2D), and volume (3D), are translation-invariant, in fact ISO invariant. As we have noted before, it is precisely this invariance that underlies our whole interest in these quantities, and, up to a choice of a scale (think of acres versus hectares) these measures are entirely determined by the fact that they are invariant measures.

Observables and invariants

Our scientific and objective approach to understanding physical things is to observe them by collecting information about them, notably numerical information. Think of temperature, pressure, magnetic field strength, wind-speeds, and so on. Physically these are aspects of reality that we can measure and quantify using appropriate instruments. These measurements typically depend on location in some spatial framework. Mathematically they are functions that give numerical values to points in this space. This means that we are looking at functions defined on $\mathbb{E}(3)$, or some region of $\mathbb{E}(3)$. In the language of physics, these functions are called *observables*. They are the quantitative side of observation.

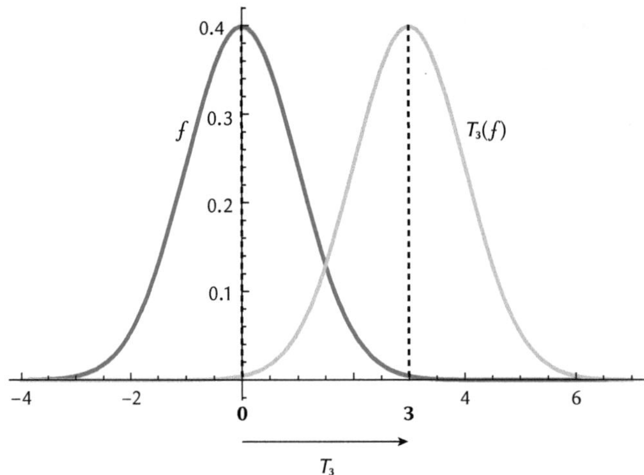

Figure 9.38: The figure shows the plot (black) of a normal distribution f centered at 0. It is a function on the real line $\mathbb{R}$ (shown as the horizontal axis) whose values are shown vertically. If we combine f with the translation T_3 that shifts the horizontal axis to the *left* by a distance of 3 then we have a new function, let's call it $T_3 f$, whose plot (grey) is shifted to the *right* by a distance of 3. Why the shift in the opposite direction? Shifting the axis to the left amounts to making f look at higher values of x, where before it was looking a lower ones. For example, since 3 is moved to 0, we have $(T_3)f(3) = f(0)$. In general $(T_3 f)(x) = f(x - 3)$. So T_3 shifts the plot of f to the right by three units.

As we go on in our study of pattern, the study of functions defined on the state space of a pattern system assumes considerable importance. We will discuss this in detail in §11.10.2, but here we want to note that the dynamics of a pattern system—its aspects of change—extend quite naturally to the observables on its state space. The idea is straightforward. An observable on a pattern system with state space X is a function

$$f : X \longrightarrow \mathbb{R}$$

that assigns numbers to points of X.[13] Its values depend on the specific points of X at which it is evaluated. If s is an operator of change on X we can ask what f looks like if we first apply the operator to states and then look at the corresponding values of f. In a way it is the same function, but the ground on which it is evaluated has been shifted, and the result is a new function. In this way the dynamics appears not only on X, but also on the functions defined on X. The dynamics extends to the observables. This is easier to see visually than when explained in words, see Fig.11.24. There is a more complete discussion in §11.10.2.

Once we have an action of change on the set of observables of a pattern system (usually referred to as the *space* of observables) we can start to ask questions about invariance again. This turns out to be of central importance, and it will occupy a good deal of our attention in the later chapters of the book.

9.6 Conclusion

Symmetry has been described as change without change [190]. At the beginning of the chapter we saw how the original ideas of symmetry related to ideas of being harmonious, well-proportioned, or bilaterally symmetric. It was this last meaning that grew into our modern conception of symmetry. Bilateral symmetry, say of the human figure, is the left-right matching of our bodies, at least at the exterior level. If we were to interchange the two sides, make a mirror reflection, then the figure would look the same. Something changes—the two sides are completely interchanged—but something doesn't change—the overall figure. The idea of symmetry has been extended over the centuries to fit more and more with the idea that the conceptual essence of symmetry is indeed change without change.

Circles and cylinders are rotationally symmetric. Crystals and wallpaper are translationally symmetric. Many flowers, like daisies, have both rotational and reflective symmetry. So too do squares and cubes. In each case there are agents of change (rotation, translation, reflection—symmetries) under which the overall visual or physical form does not change. This is invariance.

As denizens of 3D space, we are familiar with all of these forms of change, as well as combinations of them. We have no problem with rotations, reflections, and translations. What is common to all of them is that they preserve distances, they are *isometries*—things don't change their sizes or distort their shapes when they are moved around in the world. More simply, the distance between two points is the same before and after they have been moved. They are all transformations, but they all have a common invariant, namely distance. What makes an object *symmetric* is that some of these changes do not actually produce any change on the object *as seen as an object.*

In order to understand symmetry in a context beyond just individual symmetries it is necessary to consider how all the symmetries of a figure fit together. A cube has rotational and reflectional symmetry, but understood in more depth it has exactly 48 symmetries. The processes of composition (following one symmetry by another) and inversion (undoing a symmetry by reversing it) give the set of all these symmetries an algebraic structure: its *symmetry group.* The study of symmetry then changes, becoming less about individual symmetries of some object or pattern and more about the algebraic structure of its group of symmetries.

Their classification falls into two broad parts: finite groups and infinite groups, and the latter divides again into discrete groups and continuous groups. The central interest here is in the group ISO(3), which is the group of all isometries (length-preserving transformations) in 3D space. This consists of two types of elements, those that retain the chirality (left- and right-handedness) and those that reverse it. The former has all the rotations and translations, the latter all the reflections and all the compositions of a reflection and a rotation/translation. All the other groups that we look at here are *subgroups* of these. The surprising fact is how few possibilities there are for *finite* groups of isometries in three-dimensional space. We have found all of these, relating them to the very evident symmetry of five regular polyhedra of three-dimensional space. These five figures essentially capture the extent of finite symmetry in 3D, just as the regular polygons do the same for 2D.

The other form of symmetry that we have looked at is what is generally called *crystallographic symmetry*. This is the symmetry that arises from repetition of a basic motif, either in 2D or 3D. The symmetry is infinite, with the overarching feature being translational symmetry. This type of symmetry is discrete, in the sense that it is not continuous. We can rotate a circle smoothly and slide along a line smoothly without any change of overall form, so there is a sense of continuous or flowing symmetry. But when it comes to polygons, or regular polytopes, or crystals, the symmetries, the changes that result in "no change", are in the form of discrete movements, not continuous ones.

Although there are limitless ways to *decorate* translationally symmetric figures, and we have seen a bit of that with the great Islamic designs, there is actually only a finite number of different symmetry groups involved. This classification is something that we have just touched upon, but again there is an important underlying lesson to be learned: the nature of 2D and 3D space (the shape of space) puts strict limits on the possible discrete symmetry groups that are possible. There are exactly 230 forms of crystallographic symmetry in 3D. The stunning fact is that this was known even before the atomic nature of the world had been established, and that afterwards Nature was found to have manifested all of them.

A curious and important feature of translational symmetry in 2D and 3D is that it precludes any overall 5-fold symmetry, see §9.4.5. In fact the only rotational symmetries compatible with translational repetition are those of orders $2, 3, 4, 6$. That puzzle is expanded upon in Ch.12.

The intimate relationship between symmetry and Nature is manifested in the form of *invariance*. If what underlies symmetry is the concept of change without change, then in symmetry there is change but along with it something, some property or quality, that does not change. That property or quality is an *invariant* of the symmetry. Along with a group that is acting as transformations there follows the question of invariance—what is it that is not changing under the action of those transformations?

Now, this can be turned around, and in the presence of invariance we can ask if there is a group of symmetries that lies behind it. This fits into our conception of patterning, which is really about form that appears within a context of change. Nature is all about change, but within that change there seem to be unchanging principles or laws, which modern science believes are the same throughout the knowable Universe. Those principles can be seen from the perspective of invariants that arise from inherent symmetries.

If the laws of physics do not depend on our position or orientation in the Universe then they are symmetric under the group of all isometries, ISO(3). This is the realm of classical physics, before relativity and quantum mechanics. The more refined physics of today is expressed in terms of more refined symmetries (for instance space and time cannot be considered as independent of one another), but the idea remains the same.

From a strictly mathematical point of view, the appearance of invariants in the context of symmetry arises most clearly in the study of *observables.* If G is a group of transformations acting on a state space X (whatever that set may be) then the group also acts on the set (also referred to as *the space*) of all functions on that set. These functions are often called *observables* because the nature of observation of a system is most often to quantify it, in effect, creating a function on X. Now G acts on functions on X simply because it acts on X itself (§9.5.5), and the interest turns to searching for invariants among the observables. Momentum and energy are such invariants. The formalization of this search will play a considerable role in what follows in Ch.11, 12, and 13.

In this chapter the principal mathematical invariant has been that of distance, and it is this that lies behind the groups ISO[2] and ISO[3]. Later we will see other bases for invariance such as conservation of area or volume, or invariance under stretching (see the Cantor set). Details differ but the underlying principles are the same. More generally, since any pattern system involves change we can always expect to find accompanying invariances. If the changes are of the invertible kind then there will be a group involved, and hence symmetry as we have defined it here.

Chapter 10

Pattern systems and the brain

10.1 Cognition and pattern

Sensation, perception, categorization, association, attention, knowledge, action, memory. In these, and far more if we were to properly include emotional states, we recognize features of life associated with the brain. This is not just about human brains, but also those of all living beings that make up our world. Of course brains are massively complex and we are still far away from any deep understanding, especially when it comes to primates and our own human species. However the gains of neuroscience have been truly impressive and the principal features of the neural architecture are becoming clearer. Most of all from our perspective, in spite of the seeming concreteness and physical "thingness" of some of the words that we have used to characterize mental activity, all of them have temporal aspects, all are dynamical, and all can be seen in the light of pattern systems and patterning. The purpose of this chapter is to take a brief look at what some of this patterning looks like, how these basic mental processes link to patterning, and what some of the modeling of these systems looks like.

> Everything that is seen as form is the seeing of the mind.
> The mind is not mind by itself, but exists through forms [109, The recorded sayings of Ma-Tsu].

There is also a strange twist to all of this. The ultimate resting place of everything we have to say about pattern is the brain, our brains. It is true that we have come to realize the pattern of change as a *conceptual* framework that is universal in its scope, covering the entire spectrum of the physical and biological forms, spatial and temporal, and at all scales—seemingly everything we can know about the universe. However, in the end, whatever it is that we know about the world, our actual knowledge of the world, is a construct of the mind. Patterning as we have been developing it is a tightly knit collection

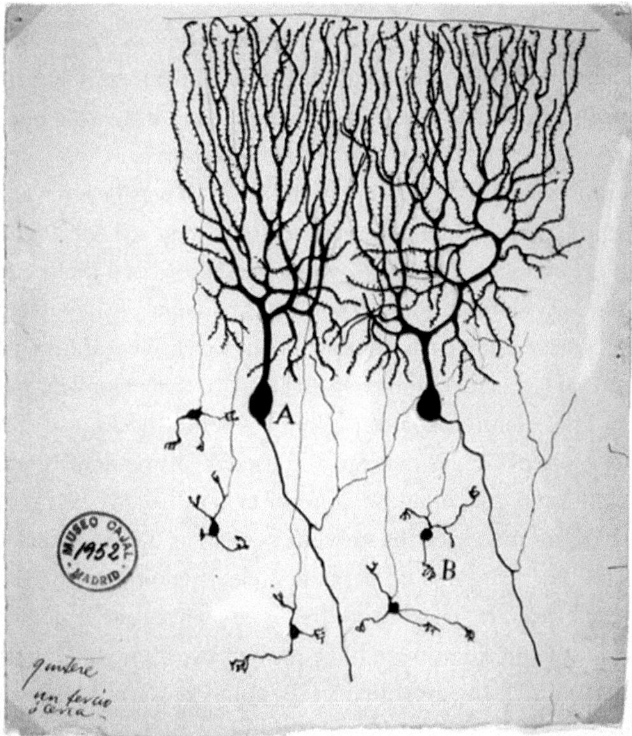

Figure 10.1: Santiago Ramón y Cajal is considered as the father of neuroscience. His lifetime study of the brain and his extraordinary powers of observation led him to formulate three powerful concepts: the individuality of neural cells, the synaptic connections between neurons, and the path of information flow from dendrites through to the axon. His beautiful drawings of neural anatomy are unsurpassed, still used and admired today. The image here shows two Purkinje cells from the brain of a pigeon. Public Domain, Wikipedia.

of concepts, and as such it is a collection of mental constructs and a property of the mind. We are not saying that the world is mind only, something we do not agree with, but that the mind is responsible for our *conceptual representation(s) of the world.* If we presume, and all evidence suggests it is true, that the mind emerges out of both structural pattern flows within the brain and its larger engagement with the body through its biological processes and sensory experiences with the world, then we come to the conclusion that our concept of pattern is itself rooted in the physical world.

There is broad agreement in the neuroscience community that neural networking is fundamental to the ability of brains to analyze incoming data, to correlate that data with past experience, and to categorize it on the basis of recognized features. Our purpose here is to look at a few examples taken from our present understanding of biological neural networks—how they distinguish different states and so encode pattern, particularly as this applies to the principal ideas of this book.

10.1.1 A new subject

We need to keep in mind that neuroscience, as it is now understood, is a relatively young subject. Ramón y Cajal, some of whose exquisite drawings we have included (see [164] for a wonderful collection of these), was awarded the Nobel Prize in 1906 for his pioneering work in understanding the neural structure of the brain. It was he who, for the first time, distinguished neurons as individual cells, realized that they are not directly attached to other neurons, but are connected through synapses, and realized that neurons are unidirectional transmitters of information from dendrites to axons. He shared the Nobel Prize with Camillo Golgi, whose principal accomplishment was a silver staining technique that he created in 1873. Using this technique nerve cells with their highly branched dendrites and single axon could be clearly visualized against a yellow background. It is ironic that Cajal made extensive use of Golgi's method and used it to advance his individual-nerve-cell understanding of the brain, while Golgi never believed Cajal's great discoveries and instead subscribed to the theory of the nervous system as a continuous single network (the *reticular theory*). We bring this up because it clearly points to the time when modern neuroscience can be said to have begun. Its history since then is full of advances that may look small at first glance, but were huge in their own time, and without which we would know very little about the most important organ in our bodies. Even so, there are still centuries of work ahead.

The following pages outline some key features of the cerebral cortex, mostly concentrated around perception, and show how they fit into our pattern system theme. We begin with a brief survey of some current conceptual ideas about the workings of the cortex. This establishes the framework in which we want to think about pattern systems. Turning to models, we look at the simplest of models around a single neuron. This leads to the emergence of the commonly observed feature of neural oscillation, modeled using just two neurons.

The later parts of the chapter broach the difficult concepts of memory, cognition, and consciousness. Science is far from any deep understanding of these concepts through mathematical modeling. Nonetheless we will find that there is no shortage of evidence of their inherent system-like natures and the physical bases that underlie them. This includes the recent ideas about system consolidation of memory, the Maturana–Varela theory of cognition, and new approaches to recognizing the markers of consciousness across the larger animal kingdom. We need to keep in mind that all animals engage in cognitive and neurological processes with the world, and use them to wonderful effect. When we speak of neuroscience here we are not just thinking about human brains or higher functions of the human mind.

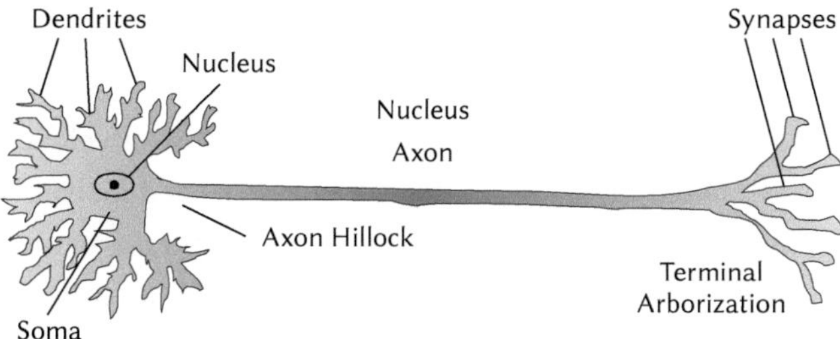

Figure 10.2: A schematic illustration of a neuron showing its principal parts.

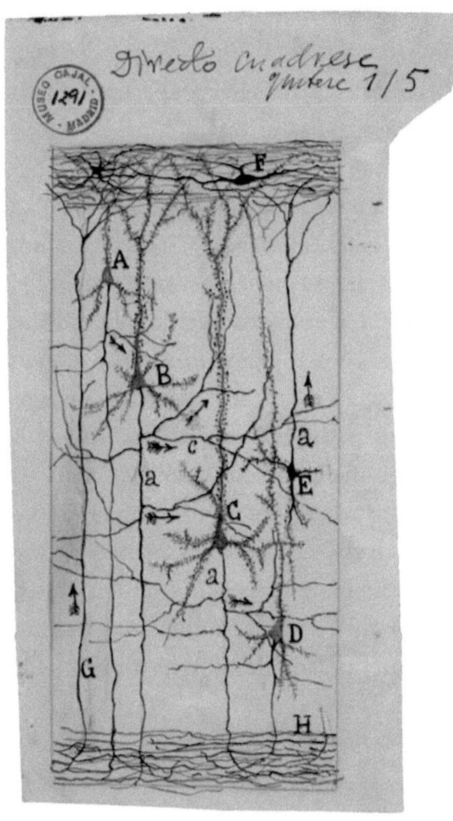

Figure 10.3: A figure drawn by Ramón y Cajal showing his realization of flow of information in neurons. He writes (1891), "The transmission of the nerve impulse is always produced from the dendrites and the cell body to the axon. Every neuron thus has a receiving apparatus, the soma, and a distributing apparatus, the nerve terminal arborization." Public Domain.

10.1.2 The neuron

The starting point for neuroscience, the basic unit or "neural atom" if you will, is the neuron. A neuron is basically a cell, but, as with many cells in the body, it is a highly specialized one. To grasp just how far up the scale of life we already are when we come to neurons, consider the words of Paul Davies [39], "The living cell is the most complex system of its size known to mankind. Its host of specialized molecules, many found nowhere else but within living material, are themselves enormously complex. They execute a dance of exquisite fidelity, orchestrated with breathtaking precision." Our starting point is already a pattern system of incredible complexity.

Neurons come in a wide variety of distinct types, but in essence each consists of a tree-like assemblage of *dendrites*, which receive input from other neurons, and a long single structure called an *axon* that extends from its main body (the *soma*) and onto which the dendrites of other neurons connect, thus providing input to other neurons in a network. The dendrites, of which there can be thousands for some types of neurons, are connected to axons of other neurons by means of structures called *synapses*.

In these few words we already see how the conception of a neural network arises. We will look more closely at the dendritic structure in §10.3.2.

The most distinctive feature of neurons is that they are electro-chemically active, with the axis membrane carrying a negative potential relative to its surrounding medium (approximately −75 mV). Neurons communicate with other neurons by means of *action potentials* during which an electrical impulse travels down the axon, away from the cell's soma, by a complicated process involving the voltage-gated exchange of ions of sodium and potassium through the axon's membrane. This transmission of electrons in neurons is chemical, so differs enormously from electrical current running through a copper wire. The generation of this electrical impulse (firing) occurs when the potential of the neuron reaches a certain threshold of depolarization (typically around 25 mV above resting potential) as a result of input received from other neurons at the cell's dendrites. The input from other neurons received at the dendrites can be either *excitatory* (which decreases the polarization, makes the potential less negative) or *inhibitory* (which increases the polarization, makes the potential more negative). Both these types are essential for the stability of the neural system. These polarizing and depolarizing signals are integrated at the *axon hillock* at the base of the neuron's axon, and if the resulting sum reaches the threshold level of depolarization an action potential is sent along the neuron's axon (more colloquially, the neuron "fires").

The experimental science together with the mathematical modeling of the electro-chemical behavior of neurons, initially carried out by Hodgkin and Huxley in the early 1950s, is fascinating and complex, and is itself a remarkable example of dynamical patterning [8]. However, we will pass this by and go on to how neurons behave and interact with other neurons.[1]

At each synapse, the places where a dendrite from another neuron (*post-synaptic* neuron) attaches to the (*pre-synaptic*) axon , the electrical impulse results in a discharge of chemicals (neurotransmitters) that affect the overall potential of the post-synaptic neuron, and this effect can be to increase the potential (i.e. move it less negative), which is *depolarization*, or decrease the potential (move it more negative) which is *hyperpolarization*. Some of the synaptic connections to the dendrites of a neuron act to excite it while others will inhibit it, but no action in the neuron occurs until the threshold is reached, whereupon a full action potential spike is generated. The neuron then quickly returns to its resting state. The change in potential due to the response of a single synapse being activated is something of the order of $1\,mV$, so the firing of a neuron is the result of numerous stimulations from other neurons. In this sense the networks are *weakly connected*.

There is an important hypothesis, called *Dale's principle*, that a given neuron is either excitatory and all of its synapses generate excitatory post-synaptic responses, or inhibitory and all of its synapses generate inhibitory post-synaptic responses. Mathematical models are usually based on this hypothesis, which is generally accepted as a rule of thumb.

This scenario (mostly derived from [86]) is only a very rough outline of neural structure. What is particularly absent is any mention of the extraordinary plasticity of neural architecture, especially in the ceaseless synaptic activities of formation, retention, and removal of dendritic spines. We will come to this in §10.3.2.

Brains typically contain huge numbers of neurons (e.g. $\sim 10^{11}$ for humans), each with thousands of dendritic connections; beyond that, there is a staggering complexity to the resulting neural structures. However, it has been noted that there is a great deal of redundancy in this structure, and often whole groups of neurons are strongly interconnected and have connections of the same types to other groups of neurons that are similarly interconnected. Individual neurons are typically noisy, in the sense that they tend to generate random firings. Redundancy is one of the simplest forms of integration of pattern systems. These larger units, each comprising many neurons, form structure on a larger scale which is less sensitive to the noise and spurious idiosyncrasies of particular individual neurons. We will call these *neural units*. With the introduction of these units as the fundamental objects, the activity of the unit becomes the averaged activity of the individual neurons that compose it. Significantly, the simple idea of the $0/1$, on/off, fire/not-fire neuron is replaced by a variable that can take a seemingly continuous range of values. These values can be used to represent such things as the averaged number of action potentials per unit of time, the frequency of oscillations, the probability of the generation of an action potential, the membrane potential, the concentration of neurotransmitters at the synapses, and so on (see [86]). Thus we shall see in our modeling that for each neural unit there is a time-dependent real variable which carries with it a numerical value corresponding to one of these measurable qualities of the unit's activity.

10.1.3 The brain and the cerebral cortex

The largest region of the mammalian brain is the cerebral cortex, and it plays a key role in memory, attention, perception, cognition, awareness, thought, language, and consciousness. This is the so-called grey matter of the brain as opposed to the white matter over which it forms a thin deeply creviced sheet. This is familiar as the highly convoluted surface that we see in standard images of the brain. It is composed of six layers of densely interconnected neurons, each layer with its own typical architecture. Vernon Mountcastle [116] proposed that the cortical layers are grouped vertically into mini-columns, six layers deep, and that these mini-columns are the basic functional units of the cortex. Apart from internal networking within each mini-column there is lateral networking within each layer which can connect local areas and also distant cortical areas. These layers themselves appear to reflect the evolutionary history of the mammalian cerebral cortex, with the lowest layers (most inward from the upper surface of the cortex) being the oldest.

Modeling neural networking of the cortex involves dealing with a very wide range of resolutions both temporally and spatially. These range from very local models, which seek to model the activity of the network at the level of ion channel activity of single neurons, to population models which look at the mean field activity or firing rate activity of large numbers of neurons or entire cortical areas. As we have already suggested, there is strong evidence that in cortical structures local neural networks are themselves connected or networked to larger ones (e.g. between cortical areas) so that there are networks of networks. The individual constituents of the networks, the nodes that form its lowest levels, as well as the networks that are made of them, and then onto the networks of networks that form the larger constituents of the brain, are all dynamical pattern systems. Thus there are natural hierarchies of dynamical systems – neurons, clusters of neurons making up neural units, and beyond them neural networks of ascending levels of complexity. One of the outcomes of this discussion is the conceptualization of these networks of pattern dynamical systems as pattern dynamical systems in themselves, as indeed they should be. This brings with it further insights into the way in which abstraction and robustness are enhanced through networking. In §8.4.2 we will take a closer look at this.

10.1.4 Cognits

In his book *Cortex and Mind* [59], the eminent neuroscientist Joaquín Fuster introduces a concept he calls a *cognit*. In his words "a cognit is a generic term for any representation

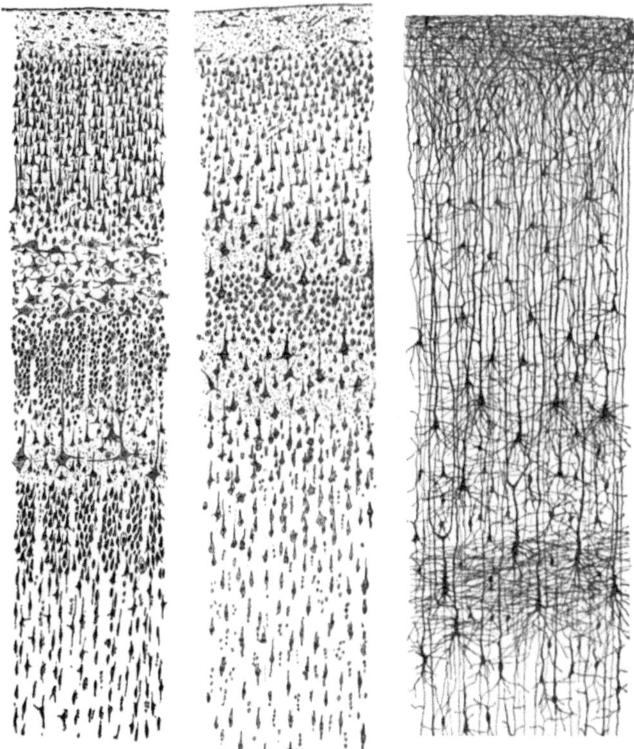

Figure 10.4: Drawings of cross-sections of the brain showing the cerebral cortex as the surface layer at the top: left to right, the visual cortex of a human adult, the motor cortex of a human adult, and the cortex of an infant. Cajal, Public Domain.

of knowledge in the cerebral cortex. A cognit is an item of knowledge about the world, the self, or the relations between them." A cognit is not so much an abstract entity: it is a physical thing, a network structure in the brain along with its intrinsic dynamical nature. In our terms a cognit is a neural structure whose internal networking and dynamics constitute a pattern system which may be construed as an item of knowledge. As with pattern systems, cognits can vary vastly in scale and complexity, and cognits can be local or global in nature (in the sense of either being highly localized or widely distributed in the brain). Cognits can be nested within each other and so form hierarchical structures, just as pattern systems can.

There is an important aspect of neural patterning which bears repeating. We may certainly think of neural networks as pattern recognizers, for that is what they do. However, given the finiteness of these networks, there must be an ultimate step in the pattern recognition process beyond which there are no further pattern recognizers to call upon. Pattern systems may be transformed, transmitted, or accumulated into other pattern systems, often with the idea that the latter ones serve as pattern recognizers of the earlier ones, but ultimately, as far as pattern recognition in the brain goes, this must come to an end where in some way pattern is actually *experienced*: "the buck stops here." One important idea from neuroscience is that ultimately this *experiencing* of pattern is nothing other than activation of a particular pre-existing neural network.[2] One particular and important understanding of neuroscience is that memory rests on re-activation of neural networks that have been created from prior experiences.

10.2 The simplest Wilson–Cowan model and equilibrium points

After this rather sketchy overview of some of the important structural and dynamical features of the brain and their relationships to pattern, we should give some indication that our ideas about pattern systems do actually lend themselves to a more detailed analysis of mental processes. Here we give an introduction to what this modeling looks like.

Within the vast system of neural units, we seek to model small subnetworks of them that are supposed to simulate some cortical function. Thus mathematically we begin with a collection of n neural units $x_1, \dots, x_n$, where each x_i is a time-dependent numerical quantity that describes the physiological state of the unit. As we have pointed out, there is a range of qualities of neural activity that these variables can quantify, but for our purposes each x_i will be a single real variable and we will refer to it as being the ith neural unit. These neural units feed each other in the form of excitatory or inhibitory inputs, the total of which determine their subsequent responses.

A neuron only partakes in the action of a network when it is capable of firing. It has to be capable of being brought into the threshold range of its action potential. The inputs into a neuron are made up of those arising from other neurons feeding into its dendritic structure, its own particular state, and finally control parameters that arise either from other parts of the neural system or from sensory inputs. These determine whether or not it will reach its threshold state. It is when the overall state of the system reaches a *bifurcation point*—a point at which the dynamics can go in two or more directions—that it is of greatest interest, for we can take this as being the onset of pattern recognition.

At its most basic mathematical level a neural network is a weighted directed graph, see for instance the right-hand side of Fig. 10.5. The nodes of the graph represent neural units and the directed edges connecting the nodes indicate the connectivity of the neural network. The direction of the edges indicates the flow of activity between the units and their weights indicate the relative strengths of these flows. Neural networks can be discrete, in which case the dynamics of the network progresses in discrete time steps and is guided by some set of time-dependent difference equations, or continuous, in which case the network is described by some system of time-dependent differential equations.

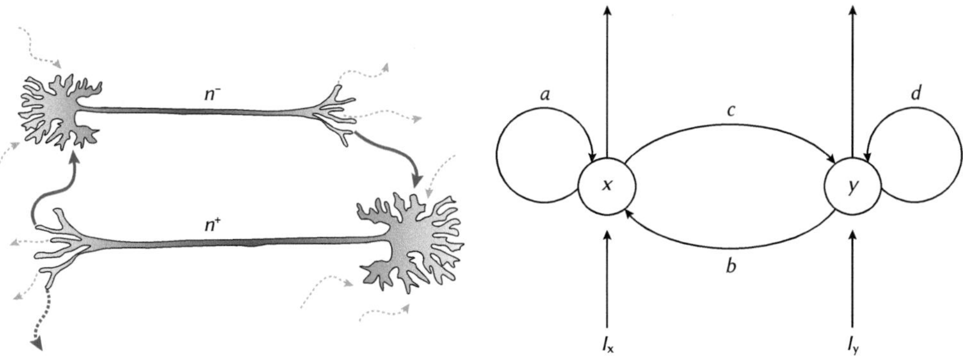

Figure 10.5: On the left, two neurons, one excitatory and one inhibitory, are mutually synapsed to each other. The excitatory neuron n^+ promotes the firing of the inhibitory neuron n^- , which in turn promotes the inhibition of the firing of the excitatory neuron. Since n^+ is inhibited it no longer stimulates the neuron n^- that inhibits it, so the neural potential of n^+ rises while that of n^- gets no further stimulation from n^+. When n^+ reaches a firing potential it repeats the cycle. This is a feed-back loop, an example of cybernetics, where the system exhibits features of self-control, see [15]. Even this simple formulation can lead to a wide variety of outcomes ranging from stability to chaos. Since each of the two neurons can be synapsed to (many) other neurons, the effect of pairing of neurons n^+ and n^- may be highly complex. Image adapted from [86]. The right-hand side gives a visual interpretation of the equations (10.2) in the form of a graph.

At this level, both the neural networks of artificial intelligence and the models of biological neural networks have the same basic structure. However, serious representations of the neural structure of the brain have to take into account the complex physiology of living neurons, the physical effects of synaptic coupling, the multi-scale hierarchical structure of the neural networks, and most importantly the actual experiential evidence that researchers have collected. Neurons, or more realistically neural units, are not simple on/off switches. Far more characteristic are spiking, oscillating, and rhythmic features that change in time and represent different patterns of behavior. The structure of a brain, even of a simple animal, is incredibly complex. Although no mathematical models can come close to describing them completely, still there are mathematical models that can simulate genuine modalities of parts of the brain and these do shed considerable light on how these modalities arise in a dynamical setting. They also point clearly to the way in which pattern can be represented and distinguished by neural networks.

Here we will take just one particular family of models that is general enough to apply to a great number of situations, transparent enough to allow us to understand the basic underlying geometry, and specific enough to allow computer analysis of their workings. These are the Wilson–Cowan models. They are based on continuous dynamics arising from non-linear differential equations. This sounds difficult, and of course, at the level of research it is. Fortunately we only need to get the feel of what these equations signify in terms of the phenomena they model, how neuroscientists think about neural networks, and how they connect into our general ideas of pattern systems. Their value to our project is to illustrate present ideas about how patterning is brought about in the brain. In some sense it is really necessary that we make this connection. Whatever pattern is, as far as we humans are concerned, it is a mental phenomenon. How the brain is able to evoke pattern-based responses is obviously something we ought to care about. The Wilson–Cowan systems of differential equations are often used to model cortical networks of neural units. They describe the dynamics of interactions between populations of very simple excitatory and inhibitory model neural units. Our discussion follows the book of Hoppensteadt and Izhikevich [86].

10.2.1 The neural oscillator

Our discussion begins with the study of a single neural unit. It is represented by a continuously varying number x, which represents the firing rate or the relative probability rate of the neuron. The Wilson–Cowan model is described by the differential equation

$$\dot{x} = -x + S(I + cx)\,. \tag{10.1}$$

If the S-term is removed then we see the simple equation $\dot{x} = -x$, which says that the rate of change of x (that is, $\dot{x}$) is in the opposite direction to the positivity and negativity of x itself. In short x decreases when x is positive and increases when it is negative. The result is convergence towards 0, which is a fixed point, unchanging in time. Fig. 10.6 shows what this means.

The Wilson–Cowan equation adds in a perturbation to this, namely $S(I+cx)$. This is based on the function S, whose plot is the sigmoid shape of Fig. 10.7. Note that $S(I+cx)$ does not mean $S \times (I+cx)$: it means that the function S is evaluated at $I+cx$.

The way the sigmoid part behaves, and thus its contribution to the dynamics, depends on two (real-valued) parameters. I is a parameter that represents some input into the system and controls how *excitatory or inhibitory* the neural system will be (the more positive it is, the more excitatory it will be and the more negative, the more inhibitory). As for c, it quantifies the amount of *feed-back* of x into the system. Neither I nor c profoundly affects the shape of the sigmoid function: the parameter I shifts the plot of S horizontally by $-I$ and c changes the rate at which the the graph of S is traced out. We still have a sigmoid function. However these two parameters do affect the dynamics. In Fig. 10.8 we look at what happens when c is held constant, but the excitatory/inhibitory parameter I is varied. What we are looking for are the points of *stability*, or *equilibrium*, where x is not changing, or in mathematical language, when $\dot{x} = 0$. According to our defining equation (10.1), we should try to see where x is equal to $S(I+cx)$.

Our first inclination might be that the stable equilibria are more desirable, but it is the unstable equilibria that are crucial. As we have pointed out, for neurons to play any

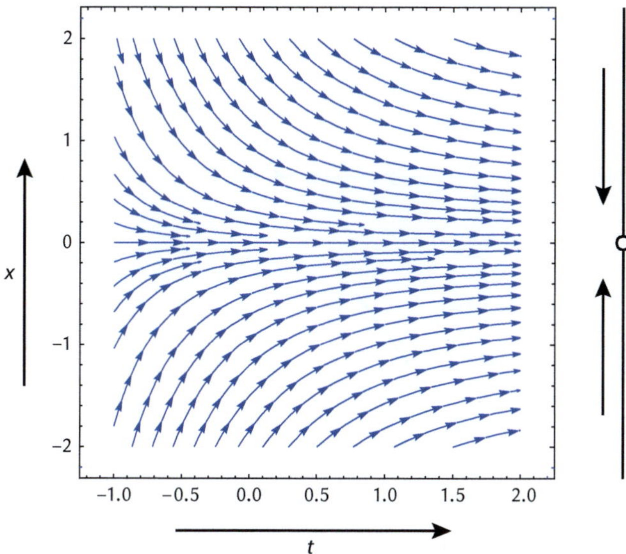

Figure 10.6: The trajectories of the numerical value of x subject to the differential equation $\dot{x} = -x$ all decrease towards 0 in time. Time is shown on the horizontal axis and the position of x on the real line is shown on the vertical axis. This scenario represents total inactivity of the neuron, or transition to its resting state.

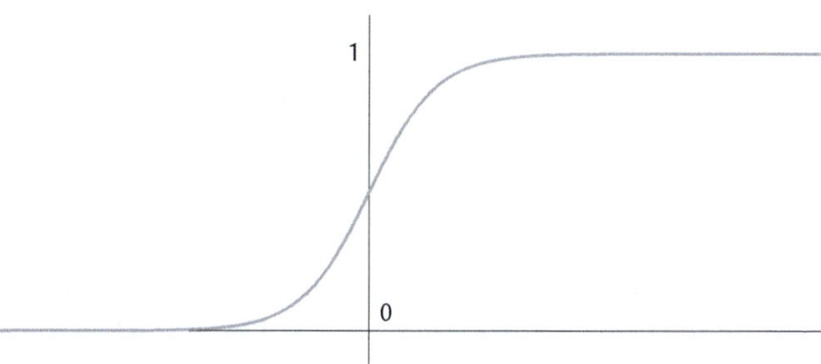

Figure 10.7: The typical sigmoid graph shape that represents the firing rate or firing probability x of a neuron or, more appropriately, a neural unit. The key features are that it is continuously increasing (left to right) and with limiting values 0 to the left and 1 to the right.

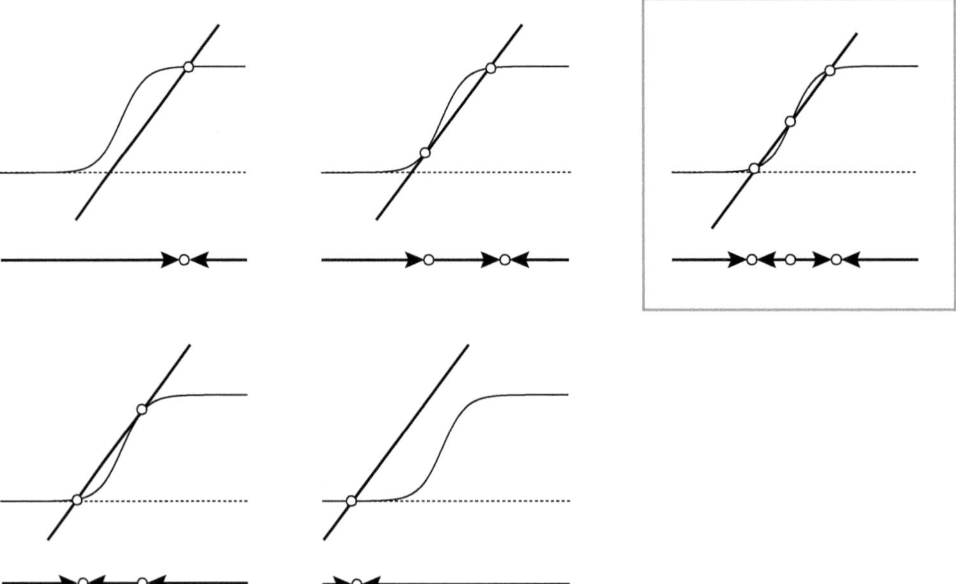

Figure 10.8: The graphs are of the line $y = x$ and $y = S(I + cx)$ for a fixed value of c and decreasing values of I. As I decreases the sigmoid curve appears more to the right. The equilibria occur at values of x at which the two graphs intersect. The arrowed line drawings below show how the values of x trend towards the points of equilibria (shown as small circles). The stable equilibria are those which are attracting (all flow is towards them) and the unstable equilibria are those which are repelling (all flow is away from them). Initially the input I is large and there is a single attractive state that is close to 1 and represents neural activity. At the end, when I is small, there is a single attractive equilibrium that is close to 0. In between we see the really relevant situation (shown boxed) where there are two attractive equilibria and an unstable one between them.

useful role in a dynamical system they must be held close to thresholds at which they fire. They should be held in states close to unstable equilibrium points, or in fancier language, *bifurcation points*. The boxed situation in Fig. 10.8 is the interesting one, where the neural unit is poised to go to either of two stable equilibria according to which side of the unstable equilibrium it finds itself in.[3]

Perception of pattern ultimately comes down to neural behavior. In this primitive example we see illustrated one of the hypothetical mechanisms of excitatory and inhibitory input by which pattern is recognized by and responded to in the brain.

10.2.2 Neural oscillators

For all this to be useful, we need a systems version of this in which other neurons that can mutually input their excitatory or inhibitory contributions to each other. The Wilson–Cowan models elaborate the single neuron model we have just seen to complex multi-

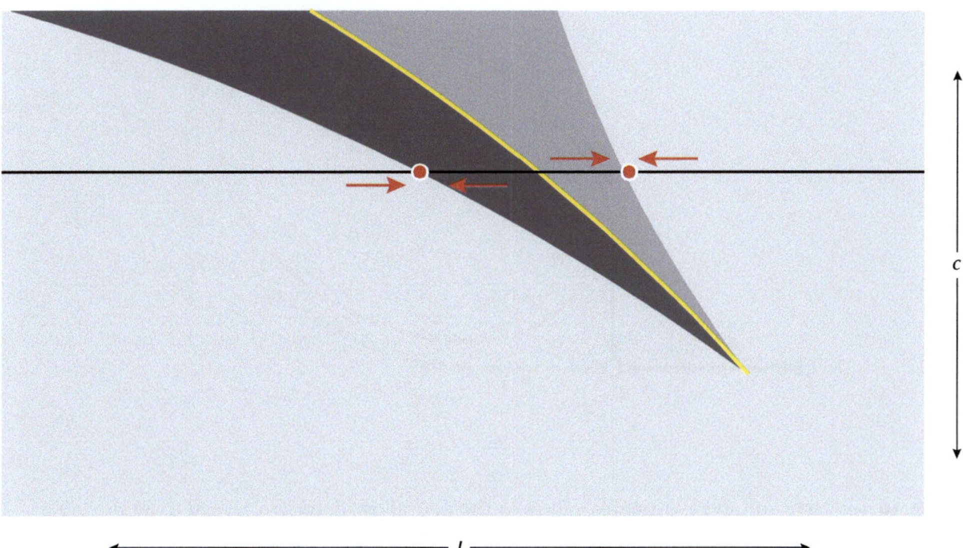

Figure 10.9: This plot over (I, c) indicates the region in which the differential equation (10.1) produces an unstable equilibrium (and two stable equilibria). The horizontal black line shows the path of the dynamical variable x as I varies, for a particular fixed value of c. Where the line enters and exits the darker region are stable equilibria. The darker region is itself divided into basins of attraction: one for each attractor, and they are separated by the line in yellow which indicates the position of the unstable equilibrium. Thus, with fixed c, the variable I can control the neuron and push it into a basin of attraction of either of the two stable equilibria. In other words there is a distinguishing mechanism here—compare with Fig. 10.8. Outside the bifurcation region there is only one (stable) equilibrium point. [86, §2.2].

neuron models, separating a homogeneous population of interconnected neural units into those of excitatory and inhibitory subtypes, according to Dale's principle (§10.1.2). The fundamental variables describe activity of excitatory or inhibitory types within the population, and are thus divided into excitatory units x_i and inhibitory units y_i.

The simplest example of this consists of a pair of neurons or neural units, one of which is excitatory and one inhibitory, which interact with each other. In spite of its simplicity this model can exhibit a great range of activities, the most important being neural oscillation. Although the basic mechanism of a single neuron is fire or not fire, normal brain activity is oscillatory, as is familiar from electroencephalography (EEG). In practice neural units, small groups of interconnected neurons, act as oscillators with specific frequencies, and it is these oscillations that are relevant to neural activity.

Fig. 10.5 illustrates a basic model of a *neural oscillator*. Mathematically it is described as a pair of equations

$$\begin{aligned} \dot{x} &= -x + S(I_x + ax - by)\,, \\ \dot{y} &= -y + S(I_y + cx - dy)\,. \end{aligned} \tag{10.2}$$

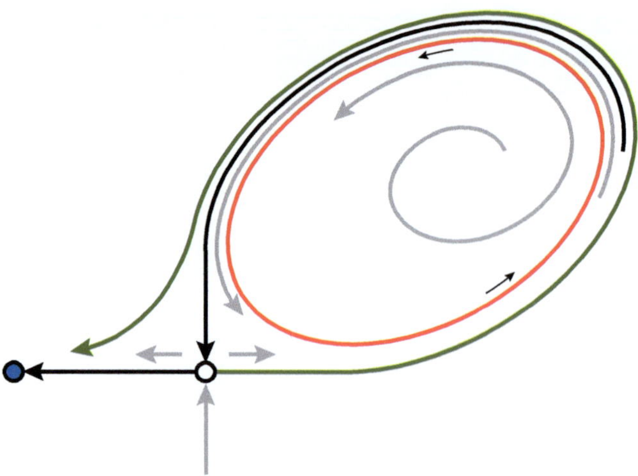

Figure 10.10: Here suitable parameter values in the neural oscillator (10.2) lead to an unstable equilibrium state (white circular dot) that can bifurcate to a stable equilibrium state of no activity (blue circular dot) or to a stable limit cycle (shown in red). This is taken to represent the transition to and out of oscillatory states of the dynamics.

Here we have two real variables x, y that are varying in time, and four non-negative constants a, b, c, d. Each equation is basically of the form of the model that we used for a single neural unit, but now the two equations are coupled, each inputing its values into the other. Notice that x gets input positively (excitatory) into the sigmoid function in the terms ax and cx while y gets input as an inhibitor through $-by$ and $-dy$. We have an excitatory neuron and an inhibitory neuron, and they feed into each other. Apart from these four controlling parameters there are the two input variables I_x and I_y that feed into their respective neural units.

If $b = 0 = c$ then the two systems are uncoupled and there is no synthesis. Each one will move towards one of its attractors depending on the values of its parameters (I_x, c) and (I_y, d). Logically it is like a pair of on/off switches. However, if there is coupling then there is a whole gamut of possible behaviors of the combined system. Of these one of the most important is the existence of a bifurcation point that stands at the threshold to two very different attractors. One is an attractor to a resting state of no or very little neural activity; the other is to a stable limit cycle where the states flow around in a periodic cyclical way, see Fig. 10.10. The importance of this is that it models a very widely observed phenomenon—neural units can turn on and off oscillatory states, see Fig. 10.11.

In general Wilson–Cowan models take more elaborate forms that we can visualize as a network of interconnected excitatory and inhibitory neurons, each synapsing to the others through weighting constants. Such models can incorporate *Hebbian learning* by the varying strengths of the interactions between units over time.[4]

One natural extension of the single oscillator model, which we have just noted can

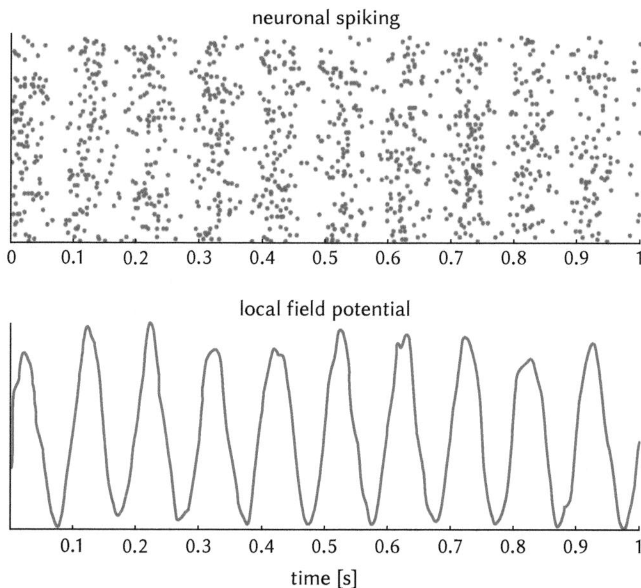

Figure 10.11: The figure shows individual spikings of neurons in a neural unit, and the corresponding averaged output which is a cyclical oscillation. The frequency shown here is about 10 Hz (cycles/second). Neural oscillations are ubiquitous and vary in a wide range from a few cycles per second to a hundred or more cycles per second. Neuroscientists divide these into broad bands: alpha, beta, gamma, delta. Higher frequencies are associated with cognitive activity. TjeerdB Public Domain en.wikipedia.org/wiki/Neural_oscillation .

lead to a stable limit cycle, is a scenario with multiple oscillators with multiple bifurcations into different limit cycles. This has been associated with the storage of multiple memories that can be triggered by variation in the input variables, which in turn might vary on association with certain neural stimuli.

10.3 Mind

Mind is a product of the brain. The latter is something material; our natural instincts are that the former is something else, something beyond the physical. But if we want to avoid stepping into the metaphysical, we should examine how the mind might be seen in systemic terms, as something arising as process out of the material. When we say "physical" we need to go beyond the mere material and include process as well. No doubt this brings us to depths that are difficult to articulate, not to mention fully understand. Yet neuroscience, psychology, high technology, and philosophical insights continue to shine new light on the remarkable physical properties and dynamical processes underlying brain function. All of them point to perceiving mind at the level of systems, and particularly seeing it as inherently dynamical. In this section we want to examine some of these ideas. It is too early for mathematics to be a serious tool in theories of the mind, but not too early to be convinced that mental processes are deeply system-based.

10.3.1 Association and gestalt

The German word "gestalt" has the meaning of shape or form, but out of the work, or under influence, of several German philosophers and psychologists during the last part of the nineteenth century and the first half of the twentieth century it came to take on a more dynamical meaning in the context of perception. The gestalt theorists were interested in how the shape or form, say of a piece of music or a visual image, emerges whole in the mind even when only a fragment of it is seen or heard. We all know the truth of this, and use it numerous times every day.

Gestalt theory came to refer to the way we perceive things, how sensations integrate into organized wholes, and how we can instantly perceive the whole from just parts—how the entirety of a perception is larger and different from any of its constituent parts. It is a sort of instantaneous completion of fragments into a meaningful whole. Much importance has been given to the way that gestalts are a way in which we give meaning to things. As poet and philosopher Jan Zwicky says, "to grasp a gestalt ... is to apprehend meaning" [204]. It must also be said that gestalten don't just happen. They are the outcome of learning processes. Some are almost universal in human experience, but others are the results of a lifetime in a field of expertise. Perhaps this is why they are often seen to be imbued with meaning.

Joaquín Fuster [59, Ch. 4] sees a parallel, or even an isomorphism, between gestalt and cognit, an equivalence between perceptual structure and neural structure. In these terms what is at work in the perception of a gestalt is the activation of an already existing cognit that then proceeds by association to bring together an entirety.

What underlies gestalt is the association of cognitive structures in such ways that activation of one part activates the others. One general principle that is thought to underlie association is the *neural plasticity*, which strengthens or weakens synaptic connections between neurons. This was first suggested by Donald Hebb (1904–1985) in 1949. Hebbian learning is commonly summarized in the catchy phrase "neurons that fire together wire together". Actually this is not quite what happens, nor what Hebb said. More accurately he suggested two principles. The first says that the synapse of neuron A to neuron B will be strengthened if the action potential of A (pre-synaptic) partakes in the action of potential of B (post-synaptic) and for that activation of B to follow directly *after* that of A. This marks a potential causality relation, and in this way it is learned that A in some way implies B. So B becomes associated with A and invocation of A can bring up the action of B, and with it an associated memory.

This has been extensively tested and it has also been seen to work in the other direction. If neural activity of B occurs immediately *before* the action of A then the synaptic connection in the direction A to B is weakened thus weakening the chance that activity in B will be invoked by the activity in A.

Thus along with associative memory, this process (spike-timing-dependent plasticity) apparently makes physical pattern out of a cause and effect relationship. This is surely not the only mechanism by which neurons become engaged with one another through learning, but still the simplicity and beauty of this idea is striking. We note though that we should really say that this learning process produces a physical pattern that we *take to be* a cause and effect relationship. This common form of physical embodiment of causal connections may be at least partially responsible for our great susceptibility to look for simple causal explanations, even when they are inappropriate or wrong.

The second Hebbian principle is that if two neurons tend to fire at the same time, and they are both presynaptic to a third neuron, then these two will tend to become associated, so that the activity of one will facilitate the activity of the other. This goes by the name of (presynaptic simultaneous) *convergence* . Both Hebbian principles may be involved at higher levels of neural units rather than just single neurons. Both are considered as forms of learning.

10.3.2 Memory

Memory connects the past to the future. Memory is such a multi-facetted part of the human mind that an assessment of memory as a forward bridge across time may seem limited at best. However, taken in the larger context of living beings in general, and not restricting memory to mental states but thinking of it more in physical terms, we may see it as ubiquitous and even essential to life.

All living forms have to face the environments in which they live, and that includes making decisions about what to eat and where to find it, what's dangerous and what's safe, how to navigate territory, knowing when to mate, and how to find and attract a mate. These and a host of others are based on memory and in this way, moment by moment, memory serves as a guide to proceeding into the future. Of course as human beings we reflect on the past, feel nostalgia, shame, or happiness from past events, and have vastly extended our memories through creative processes and inventions like painting, writing, video and audio recording, and so on. We may use memory in more complicated ways than other animals, but even so, many of our memories serve as guides to future behavior. We can assume that other animals are not preoccupied in mental musing, but they do depend crucially on memories, consciously or not, to direct their lives. Much of this memory must be phyletic, that is, passed along through the genome, effectively memory of the entire species going back in time, some of it even to the origins of life. For many animals memory acquired out of the pure experience of living is also absolutely vital. Herons have to learn how to fish. But however they are represented, memories must be physical traces of the past.

It has been suggested to us that pattern is really all about memory.[5] In this vein Fuster writes that "all perception involves remembering, in that it is an interpretation of the world according to prior knowledge" and "...every percept has two components intertwined, the sensory-induced re-cognition of a category of cognitive information in memory and the categorization of new sensory impressions in the light of that retrieved memory" [59]. In short, there is a suggestion that cognition is based on memory. If so, we would expect that memory is physically represented as neural architectural presence—a product of experience—be that at the phyletic or at the individual level.

We will elaborate on this in §10.3.3, but first we might take a look at memory at the strictly formal level of pattern systems. If pattern is about recognition, and recognition is about memory, then since our ideas of pattern are framed in terms of pattern systems, we ought to be able to see some aspects of memory within that formalism.

Let's consider again the simple shift system on the two letter alphabet $\{0, 1\}$ with its shift map s and its event space based on dictionary sets. Along with this, assume an invariant probability measure $\mathbf{m}$ which gives numerical values to events. We will see that it is this measure that encodes "memory", so it is good to remember that there is any number of such measures (uncountably many) so we are looking at something with a vast range of possibilities.

We made the point in §6.3.1 that the shift system has *no memory*, at least in the way that we normally think of it. The symbol strings shift to the left, and at each shift one letter is lost and unrecoverable. In that sense the system forgets. But if we consider a dictionary set, $[01]$ for instance, that stands for those states that begin with a 0 and at the next step become a 1, or to put it in a more suggestive form

$$[01] = [0] \cap (s^{-1} \triangleright [1]),$$

then we see that it is a statement about the future. If we look at $\mathbf{m}([01])$ we see that it represents the frequency with which the event 0 is followed by 1. And so it is in general, with $\mathbf{m}([x_1x_2 \dots x_n])$ measuring the frequency of the sequential events x_1 followed by x_2, followed by x_3, and so on up to x_n. These are statements about the future, but the frequencies themselves are, assuming that we are talking about a natural process, derived as memories of the *past*. The knowledge of $\mathbf{m}$ is in effect memory, and as such it allows predictions about the future. Knowledge of the measure is something derived out of experience and remembered from the past.

In fact it is rather easy to build explicit memories of the past into our shift systems: simply change the point of view, and instead of looking at the left-most position as the point of reference use some other position. For instance, shifting attention to the second position, we can interpret the system as having memory one step into the past. Shifting to the third position, we have memory of two steps into the past. In these terms $\mathbf{m}([011])$ is a quantitative prediction that the present 1 derives from the past steps 0 and 1, and hence offers some measure of how important previous events are to the present. Markov processes are of this type, with one step of memory.

Although in the formal terms expressed here we can speak of numerical values of a measure, its physical presence in the brain must be stored in terms of neural architecture that responds to events dynamically as they unfold. It may be that after enough study we can formulate $\mathbf{m}$ in the form of a physical law, in which case we might even think of memory as an inherent feature arising from the processes of Nature itself: sequential events to be played out again, when similar conditions arise.

Our study of entropy, which is entirely dependent on our notions of difference (the event space) and the frequencies of those events as prescribed by $\mathbf{m}$, can also give us insight into memory. In our discussion on time in §8.4.11, we saw that positive entropy has the curious implication that we cannot predict the present from the future, (8.12). Simply shifting along in time (taking the reference point further to the right in the shift system) we interpret this as saying that we cannot recover the past from the present. The probabilistic nature of a shift system that arises in the context of the invariant measure $\mathbf{m}$ serves to show that the future cannot be determined from the past. At the same time, it seems that on the basis of memory we can only make probabilistic excursions into the past. We might lament our inability to go back in time, but the mathematical formalism suggests that that would be as full of uncertainties as going forward.

Put in these terms, we can see how the pattern system formalism itself may be seen to incorporate aspects of memory, notably through its associated measures.

10.3.3 Memory and the brain

When it comes to the brain and its neural system, the processes underlying memory are vastly more complex than this, and remain poorly understood. Memory overall includes short-term memory, the sort of memory involved in keeping a telephone number in mind long enough to dial it, but then quickly forgotten. Mathematical models of neural networking that can simulate short-term memory have been created, but here we want to think about memory that is somehow laid down and retrieved much later. There is strong evidence that this type of long-term memory, at least in mammals and likely in many other life forms too, is physical, structural, dynamic, and extensively distributed in the brain—in fact, system based.

In 1904, Richard Semon (1859–1918) introduced a specific name for the memory trace, or more specifically the lasting physical changes in brain states and structure that occur in response to an event or experience: he called it an *engram*. He evidently thought of engrams as being dynamical network-based entities, capable of being strengthened or altered, and partaking in activity with other engrams as part of cognitive processes.

> After learning, Semon proposed, cuing (i.e., awakening) of the original engram would also lead to the generation of a new engram for this event. The old (retrieved) and new engrams became associated by contiguity, thereby strengthening the original memory (with the interplay between engrams described as homophony). The simultaneous retrieval of multiple engrams with similar content and their subsequent association (the "resonance between engrams") could provide a basis for complex cognitive processes, such as generalization, abstraction, and knowledge formation .... Therefore, Semon was one of the first to emphasize that the representation of an event (an engram) is not static but changes with use [150, 106, 145].

We can see here various ideas that parallel ideas of pattern systems and their potential for integration and mutual interaction. Engrams would seem to be patterns precisely in the sense of physical manifestations of a dynamical process. Of course neither the neuroscience nor the technologies of the time were in any position to substantiate Semon's ideas, and they were largely ignored and forgotten until they resurfaced again in the late 1970s. Now modern neuroscience and amazing new technologies are demonstrating the prescience of Semon's thinking, and engrams are assumed to be a primary feature of what is called *systems consolidation of memory* [150].

During an experience or episodic event a large amount of sensory information associated with that event is combined in the brain in forms that can later be recalled. This much is pretty obvious, but what actually is going on? It has been known for a long time that in the mammalian brain it is the hippocampus that is primarily involved with the initial phases of acquisition of memory, the laying down of what will become a memory engram. This laying down of memory is assumed to be based on neural synaptic plasticity and Hebbian processes that we saw in §10.3.1. But various studies of *retrieval* of memory have shown that role of the hippocampus in reviving an episodic memory decreases in time. Furthermore, damage or inhibition of the hippocampus has shown that recent memories are more susceptible to loss than older ones. This observation is known as Ribot's *law of retrograde amnesia*: memories become more resistant to decay as time goes on. It suggests that there are on-going processes of memory consolidation that lay down memory elsewhere than the hippocampus, so the hippocampus no longer needs to be involved (or at least as involved) in a memory recall. This is the origin of the idea of systems consolidation of memory.

Although the concept of systems consolidation of memory is widely held, there is considerable controversy about what other parts of the brain are involved with the acquisition and retention of memory, and how it is that so-called "silent" parts of an engram, parts that seem not involved with the recall of recent events, become involved with the recall of older ones.

Experimental studies on people cannot be invasive, so non-invasive studies on people with damaged brains have been a primary source of information on the physical aspects of memory. Thus the effect on the acquisition of memory by a damaged hippocampus has been extensively reported. But if indeed memory is a system-wide process in which the hippocampus plays a vital but nonetheless incomplete role, this offers rather limited insight. Deeper studies involve turning to the brains of other mammals, in particular mice, where invasive processes are deemed to be morally acceptable.

Fig. 10.3.3 gives some immediate insight into the complexity of the system-wide response of the mouse brain. In this experiment some mice were given what is called *contextual fear conditioning* by removing them from their home cage to another cage where they received an unpleasant electrical shock, and then returning them to the home cage. The result of the shock is seen to be a brain-wide stimulation of neural networks. Simply returning the mice some time later (24 hours in this experiment) to the cage in which they had received the shock (but with no further shock) manifested an experience of fear that was, as the figure shows, widely extended across the brain and in the same regions as the original stimulus. It seems that it is as Semon had suggested—the engram is widely distributed. In fact, in the experiments reported here at least 88 brain regions were seen to be activated in the recall. In this and other experiments the frontal cortex, the temporal cortex, and the anterior cingulate cortex appear to be especially involved in memory retrieval—and even more so in later retrieval. Although the hippocampus is no longer seen as the prime location of memory, it seems crucial for initial acquisition of memory and there is evidence that it assists in strengthening the more distributed engram.

All of this fits into basic ideas about experience and memory:

- an event is experienced through activity along neural pathways;
- memory is the strengthening synaptic connections involved in these pathways;
- recall of memory is the experience generated by invoking these neural pathways again.

Memory is the storage of information; information is pattern based; pattern is the physical manifestation of system dynamics. Since memory in the brain is the recall of experience, and experience is activation of neural pathways, we should expect that the laying down of memory of an event is the strengthening of the pathways that were originally involved with the event. Here we are using the word "event" in its common sense, but we might expect it to be appropriate in the sense of dynamical systems too.

If we follow this argument, then the next question would be, how does the neural system go about strengthening certain neural pathways and maintaining them over extended periods of time?

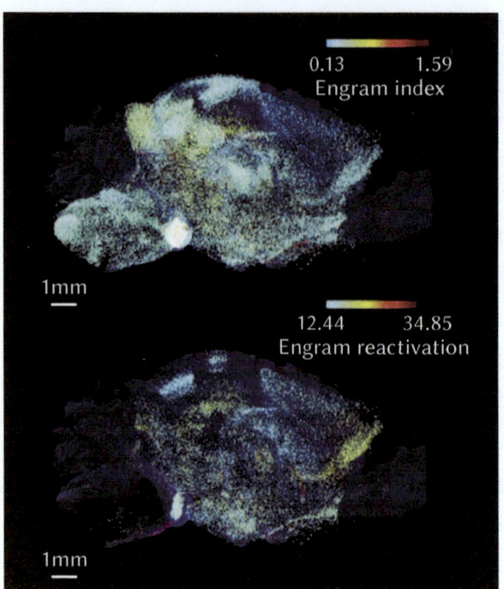

Figure 10.12: Two images showing activity in a mouse brain. The top image shows activation of the brain under an unpleasant stimulus and the lower one the activation under recall of that experience. What is clear is the system-wide involvement of the brain both in the experience itself and the subsequent response to its recall. These images come from [139] where there is a detailed assessment of the extensive list of parts of the brain involved in the engram. Image courtesy of the Tonegawa Lab/Picower Institute [123] CC BY 4.0.

There are strong indications that at the neural level the *dendritic spines* play a significant part in this. The dendrites of a neuron serve to receive information by synaptic connection to the axons of other neurons. The dendrites themselves have small spine-like extensions that make the actual synaptic connections at their heads. Ramón Y Cajal had described the spines on the dendrites and had also hypothesized that memory did not involve the growth of new neurons but rather was based on changes in synaptic connectivity. Now it is thought that these dendritic spines serve as storage sites for synaptic strength and play an active role in transmission of electro-chemical signals to the neuron's cell body. Neurons may receive tens of thousands of inputs from other neurons onto their equally numerous spines. Dendritic spines can increase and decrease in number and in density in response to various stimuli. In particular increased spine density has been connected with memory, and in the reverse direction, decreased spine density has been associated with Alzheimer's disease.[57].

The suggestion, then, is that

> long-term memory is mediated in part by the growth of new dendritic spines (or the enlargement of pre-existing spines) to reinforce a particular neural pathway. Because dendritic spines are plastic structures whose lifespan is influenced by input activity, spine dynamics may play an important role in the maintenance of memory over a lifetime [41].

There are mathematical models of the role of the dendritic spines in the synaptic processes involved with passing electro-chemical information from one neuron to another. Interestingly these are based on ideas that originated in the mathematics of telegraphic transmission by under-sea cables in the mid-nineteenth century. The extension of these

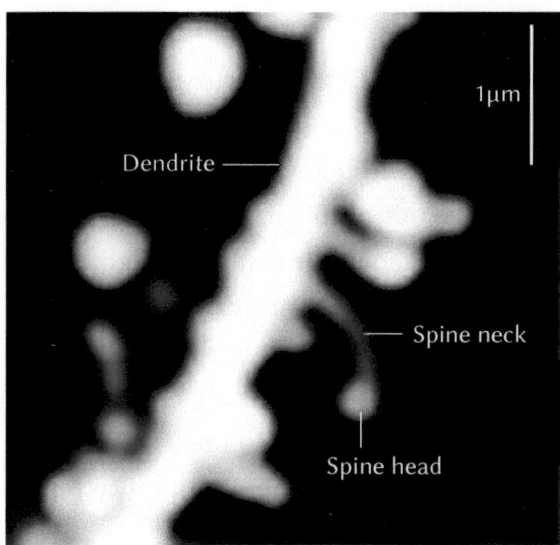

Figure 10.13: An image of a dendrite and dendritic spines branching from it. Evidence points to the dynamic nature of appearance, growth, and disappearance of dendritic spines, and their active role in strengthening neural pathways involved in memory. Timhoogland, Public Domain, en.wikipedia.org/wiki/Dendritic_spine .

cable-modeling ideas to the electrical activity in neurons began in the 1930s. Hodgkin, who we saw before in his work with Huxley, was one of the important contributors. More recent modeling of dendritic spike dynamics has largely been based on a cable theory developed by Baer and Rinzel [13].

Along with all of this there have been advances in technology. Questions about the possible electro-chemical function of dendritic spines have been greatly advanced by new imaging techniques involving a quantum mechanical process called two-photon glutamate uncaging.[6] Observations using these techniques suggest that there are separate voltage-dependent channels for sodium, potassium, and calcium. This then points to further complex dynamics at the level of dendritic spike heads.

Although we are still far from understanding memory and memory recall, there is strong evidence that memory is system-based involving structural physical alterations to the brain and the strengthening and weakening of neural pathways. As the old adage says, a path is made by the walking of it [185, Zuangzi, Ch. 2].

10.3.4 Cognition

A major shift in our understanding is the realization that mind is not a physical "thing" so much as it is outcome of physical processes. For a long time, and still for many people, mind is some sort of metaphysical entity. Instead, it is now thought that mind is physical, but in the same sort of way that weather is physical—its essence is dynamical and system-based. The brain functions on the basis of networked electro-chemical flow. The resultant dynamics, which is in effect the mind, is dependent on this flow. In turn it is a product of the physical architectural structure of the neural network and the manifold ways in which it is integrated into the physical structure of the entire body. We know that this structure is built and changes over time. This is the present understanding of mind, and the one we adopt.

Cognition, in the normal sense that we think of it, is construed as an aspect of the mind. In this sense it implies a knower and something that is known, and so implicitly, a mind. However there is a different view of cognition that is again system based but takes a wider view of cognition and is not based on the mind. In this view, a living cell, for instance, manifests cognition in its interaction with its environment. Which material entities it will accept within its boundary and how it reacts to various mechanical or chemical stresses to which it is subjected are based on some sort of recognition. There is some form of knowing, in essence, memory, but it doesn't have to carry any mental implications. On this view, cognition is seen as a necessary aspect of all life. No life form can exist without ability to interact with its environment in a discriminating way, and this discrimination is cognition.

There is a school of thought called the *Santiago theory of cognition*, developed by the two Chilean biologists and philosophers Humberto Maturana (1928–2021) and Francisco Varela (1946–2001), that has developed this idea of cognition in considerable depth. Since it ties rather naturally into our theory of pattern and offers insights into our own interactions with the world, we pursue it in some detail here. In their book *Autopoiesis and Cognition*[7] [111, p. 13], they write:

> A cognitive system is a system whose organization defines a domain of interactions in which it can act with relevance to the maintenance of itself, and the process of *cognition* is the actual (inductive) acting or behaving in this domain. *Living systems are cognitive systems, and living as a process is a process of cognition* [authors' italics]. This statement is valid for all organisms, with and without a nervous system.

Later on they called the dynamical process of such a system *autopoiesis*, and the system itself an *autopoietic system*. The word itself derives from the Latin *auto*, self or by oneself, and *poiesis*, the act or process of creation: in short self-creation. This is similar to the concept of *ziran*, see §8.1. There is an implication in their thinking that life is equivalent to the existence of this cognitive dynamics, and it is one that they develop explicitly. Whether or not there is truly an equivalence is not our point here. What is relevant is the assertion that cognition (in their sense) is fundamental to life and that autopoiesis is a systems-based concept.

The basic tenet of the Santiago theory is that autopoiesis is structural in form. There is a system, there are components, and there are *relations* between these components. The important thing is the organizational and relational structure of these components, and the dynamics that underlies their function. The autopoietic system is understood to be an autonomous identity, yet it is situated within a larger context with which it interacts—its external environment. This interaction, that is *the larger system* that includes both the autopoietic identity and its external environment together, results in continual structural changes within the system, while preserving the overall relationships of its components. These interactional responses within the autonomous identity constitute cognition.

Thinking of cognition in the context of structural systems suggests how cognition can be expressed as an emergent form of a strictly physical process. The autopoietic representation of knowledge is not one of creating an internal homologue of some external pattern, but rather *structural* outcome of interaction with the external system. In effect these structural effects and their dynamics constitute the internal representation of the external environment within which the autopoietic system exists. We can see links back to our discussion of memory in §10.3.2. As Maturana and Varela put it, "Cognition is not a representation of an independently existing world but rather a continual bringing forth of a world through the process of living" [29, §12.2.2]. The autopoietic system does not attempt to represent the outside world as such, but responds to whatever "deformations" or effects that the outside has on it. The resulting structural changes of the system due to these interactions become its knowledge of the outside world. It is in this sense that the autopoietic system brings forth the outside world. As Maturana and Varela have it, "to live is to know".

Although they tend to speak of an autopoietic system as a physical "machine", it is clear that what they are talking about is abstract and conceptual. For them it is the logical structure, not the physical structure, that is relevant. Although the brain is based on the working of a very physical neuronal network, the concept of cognition has now been brought into an abstract setting. It is at this point we might start to think again in terms of pattern systems. The idea of cognition has been reduced to the level at which distinction and classification within a context of change are essential ingredients. Maturana writes:

> Everything said is said by an observer. ...The fundamental cognitive operation that an observer performs is the operation of distinction. By means of this operation the observer specifies a unity as an entity distinct from a background and a background as the domain in which an entity is distinguished. An operation of distinction, however, is also a prescription of a procedure which, if carried out, severs a unity from a background, regardless of the procedure of distinction and regardless of whether the procedure is carried out by an observer or another entity [111, p. xxii].

What are involved here are systems of pattern and change, and it is possible to look at the Santiago system within the framework of pattern systems. Initially we can think of this in terms of two pattern systems $\mathscr{A} = (A, t, \mathscr{E}, \mathbf{m})$ and $\mathscr{X} = (X, t, \mathscr{F}, \mathbf{n})$ representing the autopoietic system $\mathscr{A}$ with its external environment $\mathscr{X}$. We take the dynamics to be in the form of discrete time steps which are iterated by the operator t. The state spaces are A, X, the event spaces are $\mathscr{E}, \mathscr{F}$, and the corresponding measures of probability of events are $\mathbf{m}, \mathbf{n}$. Interaction means synthesis or integration of the two systems into a new pattern system $\mathscr{Z}$. Its state space Z is constituted in the first place by all the pairs (a, x) where $a \in A$, $x \in X$. It has its own measure $\mathbf{p}$ that determines the probabilities $\mathbf{p}(E, F)$ of event-pairs (E, F) occurring in the synthesis, that is to say states of F that are associated through the interaction with states of E.

The *cognitive domain* of $\mathscr{A}$ is determined by the set of all event-pairs (E, F) for which $\mathbf{p}(E, F) > 0$. We then imagine the dynamics of this synthesis as representing the cognitive dynamics of the autopoietic system and interpret the cognitive content of an event E in $\mathscr{A}$ to be the union of all those events F for which $\mathbf{p}(E, F) > 0$, which is itself an event. E is an internal representation of an event from the external world. The dynamics of the entire system contains the itineraries of events in time, and these are implicitly included in this description. Thus an event E can constitute a finite sequence of events over some finite time interval and so accommodate a processual environment rather than one composed of things. There is no sense that events in some way replicate an objective external world. Its representations of the external world are the events that its embodiment in its environment bring into being.

What does it mean to say that the system $\mathscr{A}$ is autonomous? Presumably that its functioning, though influenced by the external environment via the interaction, is not simply a product of it. We could put this into terms of entropy: the relative entropy $H(\mathscr{A}|\mathscr{X})$ is a measure (in the common sense of the word) of the amount of internal ordering of $\mathscr{A}$ that is *not* accounted for by $\mathscr{X}$ in their interaction. Autonomy then means $H(\mathscr{A}|\mathscr{X}) > 0$. In (8.9) we defined the consolidation Ω by

$$\Omega = H(\mathscr{A}) - H(\mathscr{A}|\mathscr{X}). \tag{10.3}$$

It is lack of consolidation that refers to autonomy here. The maximum autonomy is when $H(\mathscr{A}|\mathscr{X}) = H(\mathscr{A})$ and the consolidation is 0. This is the situation of no integration. The other extreme, minimal autonomy, is when the consolidation is $H(\mathscr{A})$ in which case $H(\mathscr{A}|\mathscr{X}) = 0$. These seem to represent the two extremes: self and selflessness. The functioning of an autopoietic system $\mathscr{A}$ within an environmental system $\mathscr{E}$ makes sense only when there is a dynamic balance between these two extremes.

The way we have presented this here, autopoiesis is represented in the form of a prior duality between the autopoietic system and its external environment. More properly the autopoietic system represents a division of a prior unity. A more accurate scenario would be to assume that $\mathscr{A}$ is a subsystem of a pattern system $\mathscr{X}$, so that now the state space A is subset of X and its event space is a sub-Boolean algebra of the events $F \cap A$ as F runs over $\mathscr{F}$.

Since all living creatures on Earth have evolved within the environment offered by the planet, we would expect their structure to be highly aligned with the natural laws of physics and the overall nature of the physical regime around them. That being the case we might presume that some internal representations of an organism may truly be seen as mappings of external features. The perception of geometrical objects is an example.

But as we can see from this discussion, Maturana questioned the traditional epistemological idea that the internal representations of the world were necessarily mappings of the external features of the world. It was his work [with Samy Frenk and Gabriela Uribe] in the 1960s on vision in pigeons, especially in the domain of color, that brought him to this. After finding the geometry of the outer world is mapped into structural components within the brain, he goes on to say:

> ...we could not account for the manifold chromatic experiences of the observer by mapping the visible colorful world upon the activity of the nervous system, because the nervous system seemed to use geometric relations to specify color distinctions. ...After we realized that the mapping of the external world was an inadequate approach, we found that the very formulation of the question gave us a clue. What if, instead of attempting to correlate the activity in the retina with physical stimuli external to the organism, we did otherwise, and tried to correlate the activity in the retina with the color representation of the subject? ...We did this rigorously, and showed that such an approach did indeed permit us to generate the whole color space of the observer [111, pp. xiv–x].

One implication of the Santiago theory of cognition is that there is no such thing as the "objective observer". The autopoietic system simply responds to physical impressions made upon it by the external environment. To repeat, the autopoietic system brings forth the outside world: "to live is to know." The autopoietic system observes but is not objective. Certainly an outside observer may observe both the autopoietic system and the environmental system in which it functions. But that observer is then a third part of a larger system, and all forms of the observer's cognition will pertain structural aspects of his/her brain. Only to the extent that these can be aligned with other observers will there be uniformity.

The concept of an autopoietic system that interacts with its external environment emphasizes a binary conception of living beings and their environments. As Maturana later pointed out, what is missing from this is the deeper conception of autopoietic entities that interact in the form of social structures. This entails collective behavior and interaction of autopoietic entities functioning as larger systemic units. Human societies are obvious examples, and the variety of their emergent outcomes and the ways in which the individual is understood within those societies is enormous. Language in particular leads us inadvertently to a collective trend of objectifying reality in terms of the words of a common language and a cultural uniformity in interpreting the reports of our senses. There are also social insects, wolf packs, elephant families and tribes, and importantly symbiotic groupings of species. The study of social dynamics is complex and fascinating, far beyond the scope of this book. But, however it develops we can be sure that it too is systemic in nature.

10.3.5 Consciousness

> Man is but the place I stand, and the prospect hence is infinite. It is not a chamber of mirrors that reflect me. When I reflect I find that there is other than me. The universe is larger than enough for man's abode.
>
> [170, Thoreau, April 2, 1852].

The Cambridge Declaration on Consciousness [107] says "Convergent evidence indicates that non-human animals have the neuroanatomical, neurochemical, and neurophysiological substrates of conscious states along with the capacity to exhibit intentional behaviors."

Consciousness is the most unfathomable of our human experiences—unfathomable because we can only speak authoritatively of it in experiential terms, our own personal experiences. We simply cannot know how others really experience the world. How does consciousness study consciousness?

Francisco Varela, whose work with Maturana on autopoiesis we have discussed above, was deeply interested in consciousness. He came to the conclusion that the only way to make scientific progress on understanding consciousness was to study one's own self. In particular he stressed the stilling process of Buddhist meditation where conscious awareness may find the capacity to witness its own processes [178]. It is probably true that we can gain insights through meditation, but even so, if consciousness, like mind, is an outcome of physical principles, we should be able to find physical structural signs of its presence.

Unfortunately, in Western thought it has often been assumed that consciousness is restricted to mankind: animals (meaning non-human animals) are not conscious in our normal sense of the word. This is of course convenient, elevating us to a unique status in the realm of living beings and relieving us of obligations to treat other animals as thinking-feeling beings. But the evolution of life, and in particular of primates like ourselves, suggests that consciousness did not just spring up, fully developed, in *homo sapiens.* It must occur, in varying degrees and forms, in other life forms, and if this is the case we would do well to consider consciousness across the entire domain of animal life. Perhaps related, there is considerable evidence that many animals have dreams in their sleep in which patterns of behavior are rehearsed (like memorizing songs in the case of zebra finches [74]).

In this context, the Cambridge Declaration is important. It says unequivocally that the physical evidence points to forms of consciousness in non-human animals. For instance it says "Birds appear to offer, in their behavior, neurophysiology, and neuroanatomy a striking case of parallel evolution of consciousness." It not only presents this as an overall view, but also suggests ways in which we might assess degrees of consciousness in objective ways. One thing that is clear, though: we are still at a formative and controversial stage of any scientific approach to consciousness. Evidently any serious mathematical modeling of consciousness is beyond our understanding at the moment. We don't know enough. However, the "pattern of change" is not just a mathematical concept. Even without mathematical modeling it is possible to foresee that the now familiar concepts of pattern and pattern systems that we have been discussing are going to be relevant.

A very clearly laid out paper [22] gives some insight into how the larger and more inclusive path might go. The authors state at the outset that the question of consciousness needs to be seen as multi-dimensional. They then go on to ask how we may categorize different varieties of consciousness and whether it makes sense to talk of degrees of consciousness. In response to these questions they propose studying five different aspects of consciousness. These are:

(i) perceptual richness: sensual perceptions through sight, sound, taste, smell, touch, etc.;

(ii) evaluative richness: pain, fear, grief, anxiety, pleasure, comfort, love, etc.;

(iii) integration at a moment of time: the ability to integrate total experience of any moment into a unified perspective;

(iv) integration across time: "mental time travel", the ability to integrate experiences over extended periods of time, past, present, and future:

(v) self-consciousness: selfhood, the conscious awareness of oneself as distinct from the world outside.

This list suggests that consciousness is indeed "multi-dimensional", and further that it may be viewed as something that admits objective study. It also describes features that we ascribe to pattern systems, notably ideas of integration and important integrative features of gestalt. In subsequent chapters we will look at sight, sound, and touch from the perspective of the physical systems. Integration across different perceptual experiences, both in the moment and across time, suggests integration and synthesis of corresponding pattern systems. In discussing perceptual richness the authors say that "the richness of visual experience depends on bandwidth (the visual content experienced in time), acuity (the number of just-noticeable differences to which the animal is sensitive), and categorization power (the animal's capacity to sort perceptual properties into

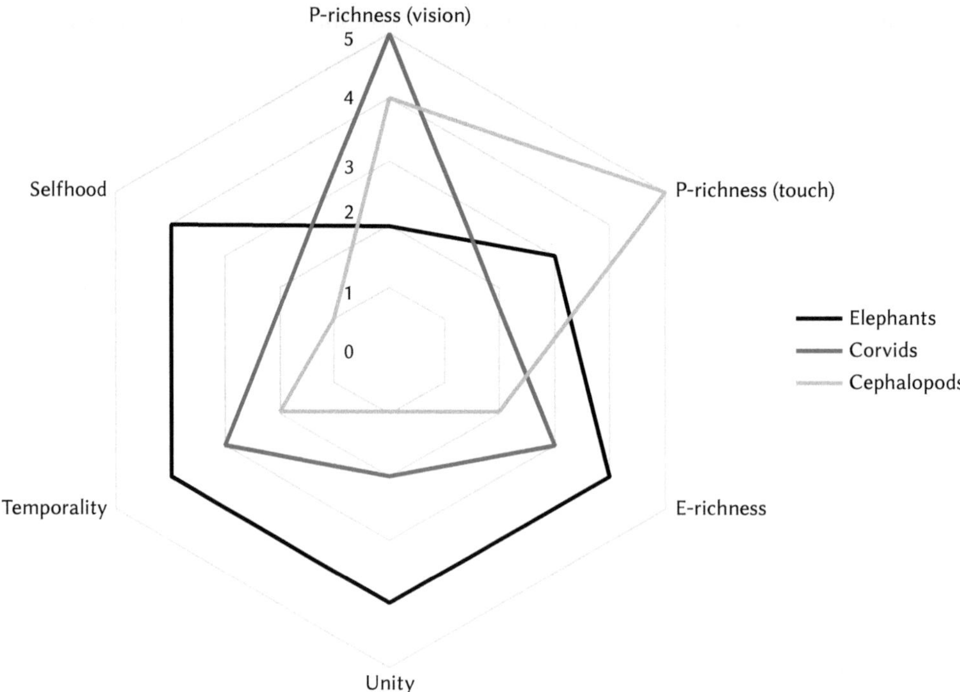

Figure 10.14: The figure shows a hypothetical profile for the degrees to which elephants, corvids, and cephalopods excel in various categories of conscious awareness. The perception category has been subdivided into vision and touch. These profiles are based on current (2020) research [22]. CC BY 4.0 Deed, pubmed.ncbi.nlm.nih.gov/32830051.

high-level categories, events). All of these are aspects of pattern systems (bandwidth to state space, acuity to states, categorization into events)".

Elephants, corvids (the crow family), and cephalopods (squids, octopuses, cuttlefishes) offer examples of animals capable of consciousness in these terms. Fig. 10.14 offers some profiles, based on current research in these three animal families.

Consciousness may be dynamic and illusive in the extreme, but its origin presumably lies in the physical world. With a deeper awareness that consciousness may exist in numerous ways throughout the animal realm, current science offers ways of looking for physical structures that are known to relate to consciousness in humans.

10.3.6 Mind in Chinese/Japanese thought

Association, gestalt, memory, cognition, consciousness—all of these are complex systemic processes related to the physical structure of the brain. They are also features that we can associate with mind, and they all tend towards the cognitive functions which most clearly distinguish us as a species. There is no question, especially in Western thought, that mind is distinguished from the body, and there is a strong tendency to treat the body as inferior to mind. Thus the famous mind/body duality or mind/body problem as it is often called, and ideas of punishing the body in order to free the mind.

Nonetheless, human behavior is greatly influenced by sensations, feelings, emotions, fear, anger, danger, doubt, shame, trauma, sexual drive, compassion, concern, love, joy, and so on. Gut feelings and body wisdom are powerful forces that can easily override rational thought. Propagators of misleading information depend on it. Heart rate, breathing, and feelings of fear, well-being, or compassion are autonomous functions of the brain and brain stem. Recent understanding of the human (and mammalian) vagal nerve system, with its multitudinous afferent and efferent connections of the brain and spinal cord with the physical organs, and its importance in emotional responses to events both inner and outer show just how much mind is embodied in the physical. It has been suggested that Descartes might better have said "I feel (taken reflexively) therefore I am ." [129]

Deeply entrenched ideas of mind as primarily metaphysical in essence and mind/body as a basic dualism are not supported by modern neuroscience. Nor were such ideas ever a part of the Chinese philosophical understanding of reality. The character 心 (xin) is ancient and remains basic in the Chinese language. It's a pictographic abstraction of the heart, but its meaning is far wider than the heart as a physical organ, though it includes it. It is usually rendered in English as *heart/mind*, and is understood to include feeling and intention. Xin is also a very rich source of additional meaning, encompassing a full range of mental states and attributes. For instance there is 心理, which already one can guess means something like mind patterns, and indeed does mean mentality and psychology. On the physical side we find 心臟, which is literally the heart, the organ of the body. Xin embraces emotional states too, like ai, 愛, affection or love, which we can see contains within it the heart radical. There are hundreds of others. Chinese thought did not put mind into some category outside of the physical being as some transcendent or metaphysical entity. Mind arises directly out of life itself. In this philosophy the famous (infamous) mind/body problem is no problem at all! The two are understood from the very start as being mutually inclusive. It is interesting that there is an extraordinary quote from Lucretius [108, p. 3.325] that fully agrees on this question.[8] But of course Lucretius' *On the nature of things* was purged, almost to extinction, for just such views.

The relationships between mind and body, between self and other, and between distinction and non-distinction don't have the sharp boundaries of opposing opposites that are so familiar to us. Superficially the fantastic success of science can blind us to the fact that it does not, and cannot, articulate some absolute or objective picture of reality. Quantum theory and the phenomena of uncertainty and non-locality, which we come to later, force us to accept paradoxical and unsuspected subtleties and to realize that our naive understandings of reality are at some level delusional. Our models of pattern systems are just that, models. To believe that they are some sort of absolutes is delusion.

The recognition of the inevitability of such "delusions" underlies Chinese and Buddhist thought.

This is rather clearly stated in the writings of the famous Japanese Zen master and philosopher, Eihei Dogen (1200–1253). Dogen himself, dissatisfied with the contemporary Buddhist teachings he experienced in Japan, sought teaching in China. He returned both to write extensively (in Japanese) and to establish one of the main Zen schools of Japan.[9] Over the course of some twenty-two years he composed his masterpiece the *Shobogenzo* (*Treasury of the true dharma eye*). Here we quote from one of its better known fascicles, *Genjo Koan* (Actualizing the Fundamental Point) [47, p. 69]:

> As all things are buddha-dharma [that is, inseparable from the Buddha's teaching] there is delusion and realization, practice, and birth and death, and there are buddhas and sentient beings.
> As the myriad things are without abiding self, there is no delusion, no realization, no buddha, no sentient being, no birth and death.
> The buddha way is, basically, leaping clear of the many and the one; thus there are birth and death, delusion and realization, sentient beings and buddhas. Yet in attachment blossoms fall, and in aversion weeds spread.
> To carry yourself forward and experience the myriad things is delusion. That the myriad things come forth and experience themselves is awakening.

The dynamics is not directed outwards from the self but is experienced inwards from the other. The writing seems paradoxical, flip-flopping between opposites, much as we find in the so-called wave-particle "duality" underlying quantum theory. It seems to leave us nowhere to stand, and in a sense that is exactly its intention. There is no absolute to stand upon. Of course to survive in this world we have to be able to distinguish dirt from rice, but we also need to be aware that our distinctions are relatives, mental fabrications, not absolutes. The history of science can be seen as a long road of discovery to exactly these ideas. Whether it is quantum theory or modern neuroscience, the path of science has undermined the very ideas which it set out to pursue—the reading of the great book of the mind of (a very anthropomorphic) God.

Continuing, Dogen writes of the dynamics about experience itself:

> To study the buddha way is to study the self. To study the self is to forget the self. To forget the self is to be actualized by the myriad things. When actualized by myriad things, your body and mind as well as the bodies and minds of others drop away.

These ideas are not specifically Buddhist or Daoist. The point is that there was a philosophical environment fostered in ancient China that assumed a universe of inherent unity and profound mutual interconnection, and this was the basis of a rich cultural, medical, and spiritual practice. Further, though these, and indeed any philosophical positions, are very human constructions, the mind/body connection itself must be a part of the experiential processes of other animals, and surely a part of the evolutionary path of our own species. As such it is ancient and deeply embodied.

10.4 Conclusion

The brain presents the greatest of all enigmas. We assume that mind and consciousness are products of the brain's functioning and everything we can know, perceive, conceive, or express is based on the workings of our brains. Even knowing what knowing means.

At the beginning of the book we presented the view that it is by pattern that we know the world, and only by pattern that we know the world. Here we have tried to assess the workings of the brain in terms of pattern systems. The point is not that reality consists of, or is even based on pattern systems, but that our *human understanding and expression of it* seems inevitably to reduce to them. In spite of all the mathematics and formalism that we have developed, it still comes down to difference and change.

In this chapter we have seen that within the brain pattern systems are recognizable at many scales, from the electro-chemical action of neurons, through simple neural circuits and neural units involving dozens to hundreds of neurons, to the vast neural structures like the hippocampus or the cerebral cortex. This networked architecture is largely built through the process of living, and ultimately grows to intimately link body and mind.

At least for some the most difficult part of the chapter may be its persistent theme that mind, cognition, and consciousness must be processes that emerge from physical entities. There is an overwhelming feeling, built through centuries in many cultures throughout the world, that the mental domain is ultimately metaphysical in nature. Often this has been accompanied by a disdain for the body and its supposed gross and unholy nature. Almost always, particularly in Western culture, access to this metaphysical world has been presumed as something unique to us and unavailable to other animal species.

Here we might use the words of Daniel Dennett, which he applied to Darwin's theory of evolution but which we can equally apply to science itself: it is kind of *universal acid* [44]. It cuts through all our cherished beliefs about ourselves, our central importance to things, our right to dominate the world, and our metaphysical hopes for the eternal preservation of self. It seems to undermine the meaning of life.

However, thinking this way we might forget that this acid comes with its own salve, removing the artificial barriers that conceal the immense unity that underlies the unfolding Cosmos. As Ames and Hall say, in this processual worldview, each phenomenon is some unique current or impulse with the temporal flow, and integrity does not mean being or staying whole but *becoming whole* in all our co-creative relationships with others. The world is ever new [3, p. 15-16].

CHAPTER 11

Waves

11.1 Introduction

> Only the strongest conviction that Light, Magnetism, and Electricity must be connected ... led me to resume the subject and persevere .... There was an effect produced [by an electromagnetic field] on a polarized ray, and thus magnetic force and light were proved to have relation to each other. This fact will most likely prove exceedingly fertile [23, Michael Faraday].

This chapter is about periodic repetition. The great cycles of days, months, and seasons are so indelibly stamped into the fabric of our beings and our existence so dependent upon them that there is no society that has not been profoundly shaped by them. Inevitably we have metaphorically extended this fundamental patterning to include the numerous cycles of birth and death, generation and decay, increase and decrease, war and peace that surround us.

It has not been hard to appreciate the power of this type of periodicity though the true depth of periodic repetition in Nature has turned out to be more profound and more prevalent than anyone could possibly have imagined. It is, quite literally, right before our eyes, in every material object that we touch and in every thought that we think. Whether it is the wave-like nature of light, or more generally electro-magnetic radiation, the frequencies associated with electron orbitals, the amazing precision and exactness of periodic structures of crystals, or the fundamental resonances that underlie our mental processes, all of them are forms of repetitive patterning. Since all of these are foundational to the way we can experience the world, and offer our deepest physical explanations of those experiences, we have to assume that periodicity is an inherent *principle* of Nature.[1]

That is our assumption here, and we can get some idea of its universality by realizing that in spite of its spectacular variety, both temporal and spatial periodic repetition are

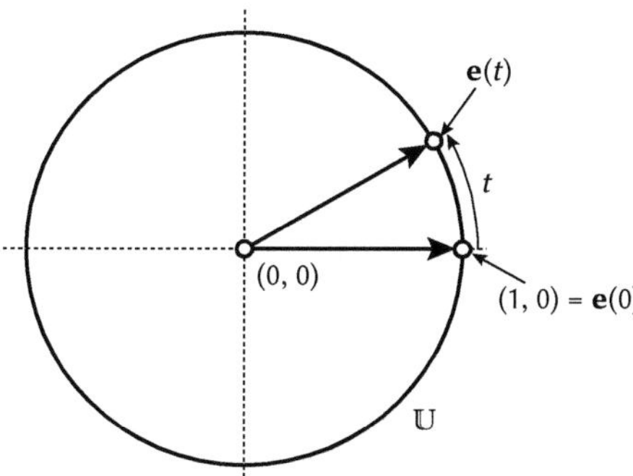

Figure 11.1: The unit circle $\mathbb{U}$ is the circle of radius 1 centered on the origin in the coordinate plane. The symbol $\mathbf{e}(t)$ denotes the point reached by going along the arc of the circle in the *counterclockwise* direction a distance of t, starting at $(1, 0)$ at $t = 0$. Negative values of t correspond to rotation in the *clockwise* direction.

based on a single common concept—waves. This is what this chapter is about.

Technically the process can look a bit daunting, mostly because of all the mathematical symbols. The notation and symbolism are important, for they allow for succinct and accurate expression. Nonetheless the intuitive understanding that we all have is that periodicity also means cycles, that is, something related to moving around a circle, and it is correct. The basic entity of the theory of waves is precisely that: a point moving around a circle. To make it completely explicit we take as our circle the circle of radius 1 centered on the origin $(0, 0)$ in the plane, often referred to as the *unit circle* and denoted $\mathbb{U}$ in this book.

The convention is that the point $(1, 0)$ is taken as the basic reference point on the circle and the symbol $\mathbf{e}(t)$ denotes the point on the circle that is at distance t along the perimeter of the circle, starting at $(1, 0)$ and going in the counter-clockwise direction. Then $\mathbf{e}(0)$ is $(1, 0)$ itself. If t is negative the direction is taken clockwise rather than counter-clockwise. This chapter is really about what we can derive from this, and from its generalization into higher dimensions when we come to repetition in things like tilings and crystals. The story falls into four parts:

(i) waves in 1D;

(ii) the complex numbers;

(iii) waves in higher dimensions;

(iv) the ideas behind linearization.

We begin with an exploration in some detail of what we call one-dimensional waves, like the familiar sine-like waves of Fig.11.3. We tend to think of these as both temporal and continuous—waves that appear in the context of a point moving around the circle at a uniform speed. But the section culminates with the realization of a profound duality that matches continuity with discreteness in the context of periodicity. For while a periodic wave is a very continuous-looking thing, it can also be described by the various frequencies that underlie it and their intensities. In this way the very continuous-looking wave also has a very discrete-looking description. This duality of description manifests as an explicit duality in the mathematics. The passing between these two descriptions is called *Fourier analysis.*

In addition, periodicity is not restricted to simply temporal repetition. In the third part of this chapter we will go on to look at higher dimensional periodicity, like wallpaper designs, tiling patterns, and most importantly crystallographic structures, where the underlying aspects of change are not temporal but spatial. Surprisingly this move upwards in dimension is quite simple. Once the forms of higher-dimensional waves are properly represented, the notation extends so beautifully that everything has just the same appearance as before. Even the Fourier analysis and the continuous/discrete duality extend without difficulty.

A significant feature that arises in the study of waves is the use of the complex numbers. We are familiar with the real numbers and their interpretation as a coordinatized line, the so-called *real line.* The complex numbers may not be so familiar, but they are incredibly relevant to the situation. Introducing these ideas makes up the second part of the chapter.

The complex numbers can be interpreted as being the coordinate plane, but endowed with an algebraic structure in which the horizontal axis is the usual real line and the vertical axis is the line of pure "imaginary" numbers that account for all the square roots of negative real numbers. This is referred to as the *complex plane.* Each point (x, y) of the plane is now interpreted as being the number $x + \sqrt{-1}\,y$, or as we usually write it $x + \mathrm{i}\,y$. The crucial connection to periodicity is that the unit circle $\mathbb{U}$, which serves as the basis of our discussion of periodicity, is now seen as a group under the multiplication of complex numbers. The group structure is simplicity itself: $\mathbf{e}(s)\mathbf{e}(t) = \mathbf{e}(s + t)$. This needs to be explained in more detail, but its essence is straightforward: if we traverse the circumference of the circle through distance t and follow that by traversing it further by s then in total we have traversed it through the distance $s + t$. This simple idea turns out to be foundational in Fourier analysis and also in the shift of focus from state spaces to linear spaces of observables.

This brings us to the fourth and final part of the chapter. Until now, our development of pattern systems has been based on the ideas of a state space, dynamics on that state space, the level of sameness/difference attached to some partition in the form of an event space, and finally some form of measure that gives numerical values to events. All of this remains, but now our attention switches to *functions* on the state space. A function defined on the state space simply gives numerical values to the states. This is exactly the type of thing scientists do in order to understand complicated systems. For this reason physicists often call a function on the state space an *observable*, since from their point of view that is what observations are largely about. We will also use this terminology.

Due to the fact that functions can be added and scaled by multiplying their values by a constant they actually take on an algebraic structure of their own. Instead of just *sets* of functions or observables we speak of *spaces of functions*. These function spaces, along with the dynamics that they inherit from the dynamics of the state space itself, greatly advance our thinking about patterning. Because these new spaces fall into a larger category of mathematical objects called *linear spaces* , we like to refer to this step forward as the *linearization of dynamics*. Although we come at it from the perspective of waves and can learn a lot from that perspective, this linearization—the concept of moving from state spaces to spaces of functions on them—is universally applicable.

11.2 Light

It is not hard to appreciate that the act of seeing is a dynamical process, involving as it does the physical structure of the eye with its rods and cones, the transmission of information through the retinal neurons of the optic nerve, and the enormous capabilities of the visual cortex of the brain that can extract pattern and color from all of this. The complex process of seeing in this sense is obviously deeply involved with pattern systems at many levels. But there is another aspect of seeing that we rarely think about that stands as the basis of our ability to see at all. Light!

The fact that we see things as things, as distinct objects distinguishable from each other, is due to the different ways in which light interacts with them. These processes involve the wavelength, transmission, refraction, reflection, scattering, and absorption of light as it interacts with physical substances, and their explanation involves an extraordinary amount of knowledge about the nature of light and the atomic structure of the material world. This is deeply related to our whole idea of pattern. We are used to the idea of visual pattern in the form of shape and contrasting luminescence and differences in color, but we rarely think of simple redness as being a form of pattern, or even patterning, just as it stands. Yet when we ask ourselves why something is seen as red, we are drawn into the complex and fascinating world of the physics of the atom and the quantum nature of light.

We might write a whole book on light, for its story is one that is both profoundly physical in its connection with the bio-chemical and bio-physical formation of life and also profoundly important in many of our metaphysical and spiritual beliefs. Even in raising the issue of "seeing" we place ourselves very much in the middle of our very physical world. In fact we confront the seemingly bizarre but nonetheless very real question of how it is that we can even see at all. The question of how bio-molecules react to certain frequencies of light, and the relationships of those frequencies to the emission of light from the Sun, is a story in itself. We have related some of this story in ‡Ch.18.

As for light itself, people have speculated since early times as to whether light consists of particles or is more akin to waves as we might see in the ocean. The Greek atomists, notably Epicurus, proposed a particle-like theory, and this is given extensive description in Lucretius' *The Nature of Things* (see §12.2.1). By the seventeenth century there started to be convincing arguments for both sides, with Isaac Newton and Christiaan Huygens representing the corpuscular and wave sides respectively. By the end of the nineteenth century, in spite of enormous advances including the spectral theories of light, the question remained wide open. Since then we have learned that light can display behaviors that seem best described as particle-like or wave-like, yet it is actually neither. Nonetheless, in the quantum theoretical model photons are represented by waves (though not in such an obvious way), and there the wave-like and particle-like aspects of light live together hand-in-hand.

The mathematics of waves itself is certainly not restricted to the electro-magnetic spectrum. Still, the electro-magnetic spectrum is by far the most obvious way in which we are affected by wave-like phenomena—think of radios, WiFi, microwave ovens, infrared night vision, heat lamps, tanning spas, X-rays, spectral analysis, nuclear magnetic resonance, and of course the vast gamut of visual experiences—so this remains the easiest way to think of waves in the one-dimensional setting with which we begin this chapter.

When we think of a *pure wave*, we are thinking of the perfect undulating wave that we see in Fig.11.3. It is a simple continuous repetitive phenomenon, the sort that we would quite naturally call *cyclical*, just as we speak of the great cycles of the seasons. In speaking of something that is cyclical we are literally speaking about something that can be thought of as being defined on a circle. And this is actually the case: the direct connection is made in Fig.11.2. So although this chapter is called "Waves", it is at its heart a study of the dynamics of circular motion.

Thinking in terms of time measured by going around a circle, the wave traces out its entire form, which then repeats itself as it goes around the circle again. There are only three things we need to know about a pure wave seen as something being traced out in time (see Fig.11.9):

- its *amplitude*—how high or low the wave gets, sometimes referred to as *intensity*;
- its *period*—how long it takes to go through one cycle);
- its *phase*—where its form is located around the circle.

Two pure waves represented on the same circle can together form a new cycle only if their combination also has a period. Such a period would have to be an *integral* (whole number) multiple of each of the component periods, allowing their forms to realign so that their combined form can begin a new cycle. This is the phenomenon of *commensurability* that appeared in Ch.2: for two periods to combine into a new period, they have

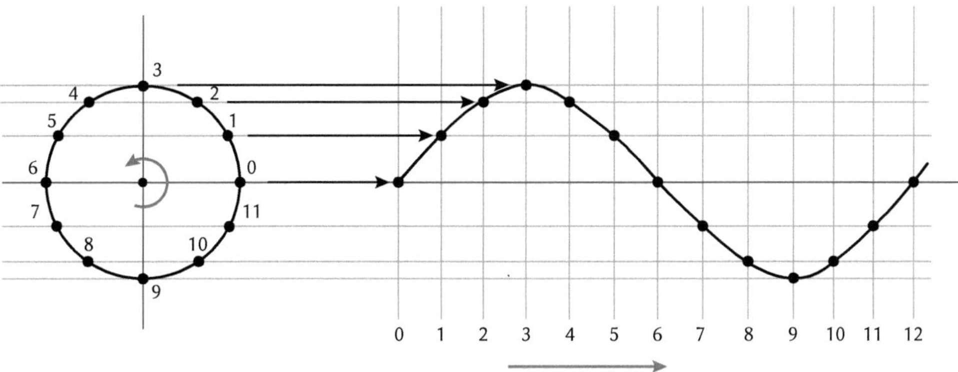

Figure 11.2: This figure illustrates how the changing values of the y-coordinate of a point moving around the unit circle trace out a wave. On the left we see twelve samples of the point as it moves along the circle. On the right-hand side we see a plot, in corresponding steps, of the heights of the point above the x-axis (i.e. the y-coordinates). The grey arrows indicate the flow of time. The wave-like plot that arises from this, is that of a sine function. This illustrates the direct connection between waves and circular motion.

to be commensurate in the sense that there is a common unit of measurement of which they are both integral multiples.[2]

The central aspects of this chapter are mathematical. If all of this is unfamiliar there may be a lot to unpack. Still, step-by-step the ideas are all here: the closer look at waves, the introduction to the complex numbers which are natural adjuncts to cyclical phenomena, and the marvelous idea of Fourier duality, which we are familiar with in the form of holograms. Waves are a basic feature of Nature. This is what comes from looking at them with a more careful eye.

Superficially light, electric fields, and magnetic fields do not seem especially tightly related. The potential connection between them was made by the discovery in 1845 by the remarkable scientist Michael Faraday (1791–1867) that the polarization of light could be affected by a magnetic field. This led him to suggest that light was some sort of high frequency electro-magnetic vibration. The quotation of Faraday that heads this chapter must be one of the greatest understatements of all time.

As an aside, Michael Faraday ought to be an inspiration for anyone who has no mathematical education. With very little formal education at all, he went on to make brilliant experimental discoveries and conjectural insights that inspired Maxwell to his mathematical formalization of electro-magnetism in the now famous Maxwell equations. In turn these were the great impetus for Einstein's theory of relativity. Those of us who are beyond the usual years of stellar achievement can take encouragement from the realization that Faraday's important work came only after he was forty. More on his story from a poor blacksmith's son to one of greatest scientists of all time can be found in [70, 23]. For more on this discovery and the difficulties Faraday went through to make it, see [56].

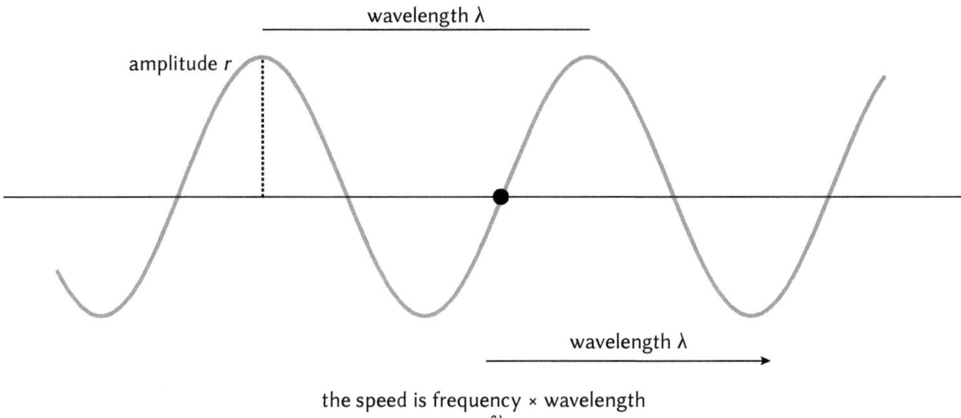

Figure 11.3: The figure shows a wave, which we can imagine moving from left to right, with the dot standing as a fixed reference point. A pure wave, as shown here, is a cyclic phenomenon, repeating with a fixed *frequency* f, the number f of complete cycles per unit of time, say per second (which need not be a whole number). The *wavelength* λ is the distance from crest to crest, so the speed of the wave is $f\lambda$. If the wave represents an electro-magnetic wave traveling in a vacuum then the speed, no matter what the wave-length, is exactly $¢ = 299,792,458$ meters per second. As a consequence wavelength and frequency are locked in a fixed reciprocal arrangement $f\lambda = ¢$: the longer the wavelength the lower the frequency.

Taking up Faraday's suggestion, James Clerk Maxwell (1831–1879) began a long theoretical study of electric and magnetic fields, leading eventually to the famous Maxwell equations. According to this theory, the electric and magnetic fields are two coupled wave-like components of a single electro-magnetic field and this field propagates through a vacuum at a constant speed, no matter what its wave-length. The fact that this speed, resulting from theory, matched the empirically determined speed of light was strong evidence that light was indeed a form of electro-magnetic radiation, just as Faraday had predicted.

Maxwell's equations do not put any limit on the wavelengths (or equivalently the frequencies) of these electro-magnetic waves, and so came the idea of the electro-magnetic spectrum which covers the full range of possibilities. It quickly emerged that previously observed infra-red radiation and ultra-violet radiation were also forms of electro-magnetic radiation that lay directly below and above the frequency range of visible light. Then Heinrich Hertz, inspired by Maxwell's theory, generated radio waves in 1886 and showed that they too travel at the speed of light and could be reflected and refracted just as light. With the arrival of the atomic age, first X-rays, and later what came to be called gamma-rays, were discovered, and these also were shown to be electro-magnetic radiation, see §12.2. In this way an array of phenomena are seen to lie in a single continuum, the electro-magnetic spectrum, of which "light" as we normally think of it covered just a small band.

As far as we know, the electro-magnetic spectrum extends, at least in principle, all the way from wavelengths on the scale of the vastness of the entire Universe to the minuteness of the Planck scale, which is many orders of magnitude below the atomic scale and lies at the very basis of quantum mechanics. "Seeing" in the strict sense means the interaction of our visual system with electro-magnetic waves in the visible part of the spectrum, a very narrow band in this apparently boundless spectrum. It is revealing to ask what is so special about this range that it should have evolved in animal life into the extraordinary sense of vision, and we explore this question in more detail in ‡Ch.18. Still, having gleaned this insight, it is only a step away to think of "seeing" using other parts of the spectrum. Actually we can feel infra-red radiation and we are fully aware of the UV side after an unprotected day in the sun. Certainly the effects of exposure to high energy radiation are all too visible—the sad lesson of the atomic bombs dropped on Hiroshima and Nagasaki. Through techniques and tools such as X-ray photography and radio-telescopes we have vastly extended our ability to see the world around us. As we continue we want to consider these notions of seeing in this larger sense.

All of it is deeply related to pattern. The pure waves that we have been describing are periodic phenomena, and in fact it is this simple cyclic repetition that is the basis of their physical nature. Cyclic repetition is one of the most basic and universal forms of pattern that we experience and recognize, and it lies at the very basis of the making of our world—photons, the carriers of the electro-magnetic force.

All this is based on what people sometime speak of disparagingly as the reductionism of science. It is true that the science that went behind all this understanding of electro-magnetic phenomena represents the long struggles of a great number of scientists who were very much reductionist. But the outcome is a magnificent unity that could not have been imagined before it was slowly revealed. As we go on, we will see how this amazing understanding leads us to the wonders of seeing. It is a false dichotomy to think of detail and wholeness as opposed antagonists. This is the point of the *Cantongqi* (Sandokai) of §8.6. The nameless and the named always arise together:

> Nameless: the origin of heaven and earth,
> Naming: the mother of myriad things.
> Empty of desire, perceive mystery, Filled with desire, perceive manifestations. The same source, but referred to differently [42, 1, Ch.1].

If you don't look, you won't know anything. Wilson's rather free translation of the same chapter of the *Daodejing* [193, Ch. 1] puts it in a more colorful way:

> With your mouth unopened, and things left undefined,
> you stand at the beginning of the universe.
> Make definitions, and you are the measure of all creation.

It is not that man is the measure of all creation in any absolute sense, but that it is in naming we bring forth the creation in forms that we can articulate. Different naming, different sense of reality. See also the passages of Dogen in §10.3.6.

11.3 Waves

Waves are by nature continuous and extended objects, the extension being both spatial and temporal. As we have seen, they are an abundant source of patterns, and indeed they produce the wonderful patterns that vision and hearing bring to us every day. Now we need to see how they fit into the dynamical context that we have been building. Doing so brings in new concepts, new tools, and new insights.

11.3.1 The space of waves

Fig.11.3 is a picture representing a simple pure wave, illustrating its wavelength and its relationship to frequency and speed. Shortly we will spell out the mathematical details of what it means to talk about a "pure" wave, but even without this we might note some of the most distinguishing features of waves:

- they are extensive in space (in this case, the 1D space of the real lines) and can move in time;
- they can be scaled up or down (amplitude);
- they can be added or subtracted;
- they can be shifted in phase;
- physically they can interfere with each other to form spatial-temporal patterns of increasing and decreasing amplitudes;
- they can diffract.

The geometric content of these six statements is suggested in Fig.11.4. We still have to give some sort of definition to what we call waves, but it is useful to draw attention to these distinguishing features because they imply that collectively a set of waves, unlike a set of points, has an implicit algebraic structure. That structure, which centers around the possibilities of scaling waves, adding them, phase shifting them, and making any combination of these, and still getting waves is called a *linear space*. Such spaces are also called *vector spaces*. The definitions of addition and scaling of functions or waves is as follows: let f, g be two functions defined on a set X (in this chapter X is the real line, the coordinate plane, or coordinate 3D space) and let c be any number. Then $f + g$ is the function defined by

$$(f + g)(x) = f(x) + g(x) \text{ for all } x \in X,$$

and cf is the function defined by

$$(cf)(x) = c\,f(x) \text{ for all } x \in X\,.$$

Notice that the addition and scaling (or *scalar multiplication* as it is usually called) of functions or waves are based on addition and multiplication of numbers. That is, $f + g$ is the function whose value $(f + g)(x)$ at x is defined as the sum of the two numbers $f(x)$ and $g(x)$ which are two real numbers. Similarly cf is defined as the function for which $(cf)(x)$ is simply c times $f(x)$.

Phase shifting is about shifting the point of reference, see Fig.11.4. Here the implicit assumption is that in talking about numbers we are talking about real numbers, and hence real-valued functions. What we have written here will apply also to *complex-valued functions*: the multiplication and addition are those of complex numbers. Eventually we will see that a context founded on complex-valued functions is the most natural since it is only by using complex numbers that the full dynamics of the wave theory is made visible, and it is in this larger context that phase shifting is most easily expressed.

As an aside, in some contexts it is also useful to multiply functions, in which case we define fg by

$$(fg)(x) = f(x)\,g(x)\,.$$

The resulting mathematical structure is called an *algebra of functions*. However, in this book it is basically the adding and scaling of functions that are important, and the objects that we deal with are linear spaces.

So at the very outset, we find ourselves dealing with new sets—sets whose elements are *functions* based on the original states that make up the state space—that carry significant structure. In this new picture waves themselves become the states, the new state space becomes the set of all waves, and this new state space forms a linear space. Although these structures are called "spaces", there is little point in trying to imagine them in terms of our familiar ideas about space. They are called spaces because their elements, which are functions, can be added and scaled, just as we have seen. In this sense they are like vectors in two- or three-dimensional space. It is another step in abstraction and best simply to accept as standard terminology. Basically we are talking about a set of functions defined on some set (like the real numbers, and later the plane, or 3D space).

The shift in emphasis from a state space to a linear space of functions (or what the physics community would call observables) defined on that state space is what we call the *linearization* of dynamics or of patterning. It plays a significant role in what follows and at the end of it we will have a whole new and more powerful way of looking at the mathematics of pattern.

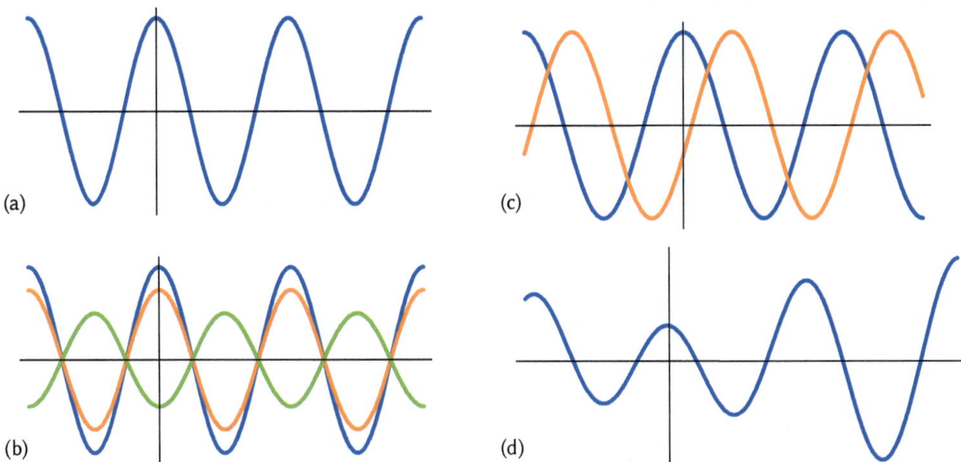

Figure 11.4: A simple wave (a) is shown. In (b) it is shown along with versions of it scaled by 0.75 (orange) and $-1/2$ (*green*). Note that the minus sign turns the wave upside down and the factor of $1/2$ scales its vertical size (amplitude) down. In (c) we see the wave shifted along (orange) – what is called a *phase shift* (which we will eventually see as a form of scalar multiplication). Combining some scaling, phase shifting, and addition, we obtain the wave shown in (d), which goes to show the variety of waves we can create by these simple operations.

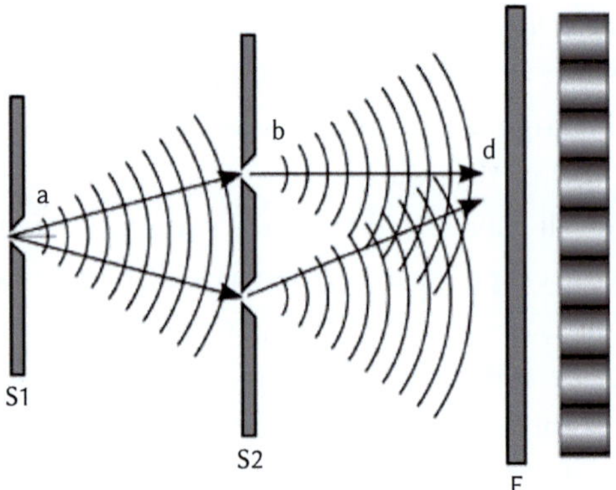

Figure 11.5: The figure shows the diffraction that occurs when a traveling wave meets a barrier with a small opening. The opening then serves as the source for a circular diffracted wave. When this diffracted wave meets another barrier with two openings, there are two new sources of circular waves and they interact (the sum of the two functions defining the waves) to produce the sort of pattern (called a diffraction pattern) of peaks and troughs with which we are familiar from water movements in ponds and quiet lakes. On the right-hand side we see an interpretation of the effect of peaks and troughs of light intensity, as might be seen by a photographic plate in the case that these were light-waves (photons).

11.4 The unit circle $\mathbb{U}$

11.4.1 Circles and waves

Periodic waves are in effect cyclical and it is in terms of a circle that they are most easily understood. So we begin with the basic concept underlying all cyclical phenomena, the circle, and look at the dynamical process of simply going around the circle at a uniform rate. In fact we refer everything back to the standard reference circle of Fig.11.1, which we have called the *unit circle* $\mathbb{U}$, in the plane.

The circumference of the unit circle is 2π since it is a circle of radius 1. In mathematics 2π shows up all over the place and we shall see a lot of it in what follows. The fact that it is usually 2π and not π is due to the definition of the Greeks that made the circumference of a circle $\pi \times$ diameter. In the development of mathematics it has transpired that it is the radius, not the diameter, which is more natural, hence $2\pi \times$ radius. But whether it is π or 2π we cannot escape their appearance in dealing with cyclical phenomena. The reason is quite simple. In temporal terms the length of time taken for one complete revolution around the circle is naturally based on its circumference. However, in terms of the outcome it is often the number of complete revolutions, the number of cycles, that seems most natural, and the relationship between the two is a factor of 2π. For instance in tuning an orchestra the pitch of the standard A is used and that is 440 cycles per second, often referred to as A440.

This relationship between continuous movement around a circle and the actual number of revolutions (a whole number) is surprisingly deep. One of the main features of this chapter is to reveal this as a deep structural connection between the unit circle $\mathbb{U}$ and the integers $\mathbb{Z}$. It is this that is the basis of Fourier analysis of periodic phenomena.

The connection between movement around a circle and the production of waves is illustrated in Fig.11.2. If we simply graph the values of the two coordinates of a point moving at a uniform rate around a circle we get the familiar wave forms shown in Fig.11.3 and Fig.11.6. These are the *cosine* and *sine* functions respectively. The unit circle combines these two numerical functions into a unity, $(\cos(t), \sin(t))$, where t denotes the angle (we use angles in radians since that equates angles to arc length, see §11.4.2). Another of the major aims of this chapter is to introduce the complex numbers, which unite these two numerical functions into a single complex-valued function.

11.4.2 The unit circle in detail

In going around the circle counter-clockwise starting at $(1, 0)$ one returns to this starting point after traversing a distance of 2π. We now introduce a variable point $\mathbf{e}(t)$ on the circle that starts at $t = 0$ at the point $(1, 0)$ and moves uniformly counterclockwise as t increases. We define $\mathbf{e}(t)$ to be the point that is distance t along the circumference of

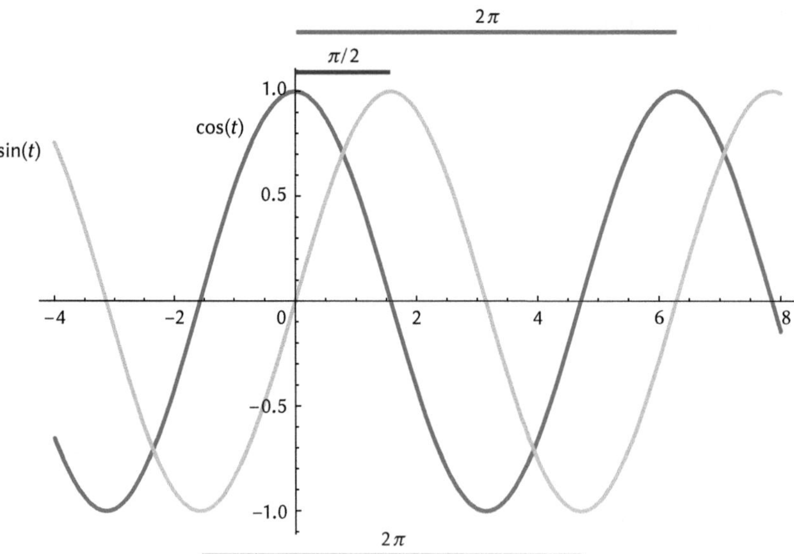

Figure 11.6: The graphs of the coordinates $\cos(t)$ (dark) and $\sin(t)$ (light) of $\mathbf{e}(t)$ in the range $-4 \le t \le 8$. They are both periodic of period 2π, and in fact identical save for the phase difference of $\pi/2$.

the circle, starting from $(1,0)$. In particular, going around the entire circumference we see that $\mathbf{e}(2\pi) = (1,0) = \mathbf{e}(0)$. If we continue around a second time then we come to $\mathbf{e}(4\pi) = (1,0)$, and in fact $\mathbf{e}(2k\pi) = (1,0)$ for all integers $k = 0,1,2,3,\ldots$. In short, it is periodic. Fractional values of t work in the natural way: $\mathbf{e}(\pi/2) = (0,1)$, $\mathbf{e}(\pi) = (-1,0)$. Some of these are illustrated in Fig.11.7.[3]

We can extend the range of t to negative values by going around the circle in the opposite clockwise direction, so for instance $\mathbf{e}(-\pi/2) = (0,-1)$, $\mathbf{e}(-2\pi) = (1,0)$, $\mathbf{e}(-4\pi) = (1,0)$, and so on. From this we can make a key observation about full cycles around the circle:

$$\mathbf{e}(2\pi k) = (1,0) \text{ if and only if } k \text{ is an integer (positive, negative, or zero).} \tag{11.1}$$

Think of it as saying that whole numbers of complete cycles around the circle return you to the starting point. This is the natural link between the integers and the unit circle.

If we equate arc length t to the angle that the arc makes (subtends) at the center, then we have the natural unit of angle, the *radian*. The relation of radians to degrees is simply that 2π radians is 360 degrees. This makes a right angle 90° equal to $\pi/2$ radians. If we use radians as the measurement of angles then we have[4]

$$\mathbf{e}(t) = (\cos(t), \sin(t)). \tag{11.2}$$

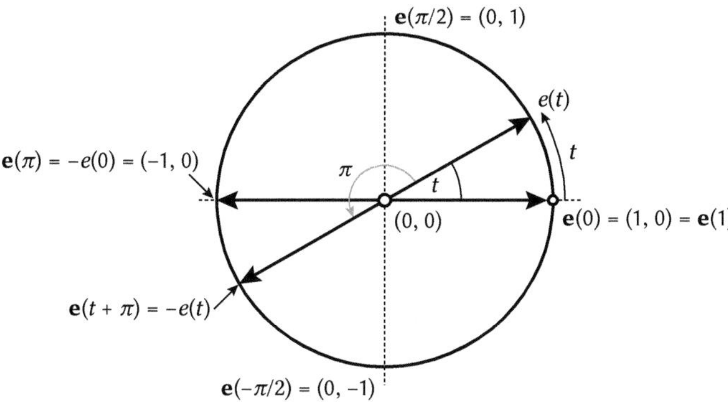

Figure 11.7: The unit circle $\mathbb{U}$ is the circle of radius 1 centered on the origin in the coordinate plane. The symbol $\mathbf{e}(t)$ denotes the point reached by going along the arc of the circle in the counterclockwise direction a distance of t, starting at $(1, 0)$ at $t = 0$. We return to this same point when $t = 2\pi$. Points $\mathbf{e}(t)$ are shown for various values of t. In the figure the point $\mathbf{e}(t)$ is also indicated as the end of an arrow that runs from the origin $(0, 0)$, which is an alternative way of interpreting it. The angle subtended by an arc of length t is t radians. In particular the angle subtended by the arc from $\mathbf{e}(0)$ to $\mathbf{e}(t)$ is t *radians.* The figure shows how adding π to the parameter t amounts to rotating through 180°, or equivalently π radians. Negative values of t correspond to rotation in the clockwise direction.

11.4.3 The dynamics of the unit circle

The idea of cyclical motion is that of going around a circle. With that in mind we should be able to think of $\mathbb{U}$ as the state space of a dynamical system in which a point $\mathbf{e}(t)$ goes round and round as t, which we can take as time, flows on. The problem with this is that in our conception of a dynamical system the dynamics does not work by one point moving around in an otherwise fixed state space. The dynamics is such that it acts simultaneously on every state of the state space. The points of $\mathbb{U}$ have to rotate as a unity. Given this, we have to observe this dynamics relative to something that is not changing (or is changing in a different way). So although it is an obvious fact, it is worth stating explicitly:

> In order to distinguish change there has to be something relative to which that change takes place.

In this case that "something" is the frame of reference of the coordinate system into which we have set the circle. The way to think about a point moving around a circle is to think of the circle moving past a fixed point. Think of the fixed point as representing "now" and rotating the entire circle around so that the future rolls past it becoming the present "now" and receding into the past. We fix the point $(1, 0) = \mathbf{e}(0)$ as representing "now", with the future in the counterclockwise orientation around the circle. Imagine that there is an observer who keeps track of which point on the circle is at the now-point as the circle rotates. Then rotating the entire circle *clockwise*, the observer sees that the future states become the present state while the present states move clockwise into the past, see Fig.11.8. The circle rotates so that the point that we have labelled $\mathbf{e}(t)$ (with positive t, representing future time) passes through the positions formerly occupied by $\mathbf{e}(s)$, $s < t$, as s decreases to zero.

If the circle rotates at a rate of 1 radian per unit of time, the action of time t will be to rotate the point $\mathbf{e}(t)$ to $\mathbf{e}(0)$ in t units of time, and quite generally the action of time s will be to bring any point $\mathbf{e}(t)$ of the state space to its former position of $\mathbf{e}(-s+t)$. This then brings us to the way in which we think of a time interval of length s acting on the points of $\mathbb{U}$—it is rotation through angle s in the *clockwise* direction, i.e. rotation through $-s$ radians. Of course we can go backwards in time by rotating counter-clockwise. In the end we see that we have an action of the real numbers $\mathbb{R}$ on the unit circle which models circular motion.

> The essence of circular motion is the dynamical system $\mathscr{U} = (\mathbb{U}, \mathbb{R})$ is rotation, with the action of $\mathbb{R}$ on $\mathbb{U}$ defined by
>
> $$s \triangleright \mathbf{e}(t) = \mathbf{e}(-s+t) \quad \text{for all } s, t \in \mathbb{R}. \tag{11.3}$$

This backwards looking effect, which indicates process towards the future, is similar to the effect one can get sitting in a stationary train beside another stationary train. Looking out the window when that train moves backwards from our perspective it often feels as though our train is moving forwards. Once one has settled that it is the other train that is moving, one can see its passing as the "future is coming to the present", or to put it the other way around, the present is evolving in time. We will see this same idea appearing when we introduce dynamics into observables. Fig.11.24 may also be useful here.

What we have just described is a way of thinking of $\mathbb{U}$ as the state space with the dynamics of arising from $\mathbb{R}$ acting on it. If we think of $\mathbb{R}$ as time, then this is time stretching endlessly from past to future. But there is another way to think of this dynamics on $\mathbb{U}$ which is perhaps more familiar, namely cyclical time. We can think of the great annual cycle of the seasons in linear time, so we might say that this is such and such a day in year 2023, but more commonly we omit the 2023. This is how we mark holidays, birthdays, and so on. It is the same with cyclical clock time, where the natural way is often to think in terms of the time of the day rather than in linear times stretching centuries into the past. In cyclical time, time s and time $s+2\pi$ may be viewed as equivalent, and it is certainly true that

$$(s+2\pi) \triangleright \mathbf{e}(t) = \mathbf{e}(-s-2\pi+t) = \mathbf{e}(-s+t) = s \triangleright \mathbf{e}(t).$$

If we *identify* s and $s+2\pi$, that is treat them as being the same, we get a new picture in

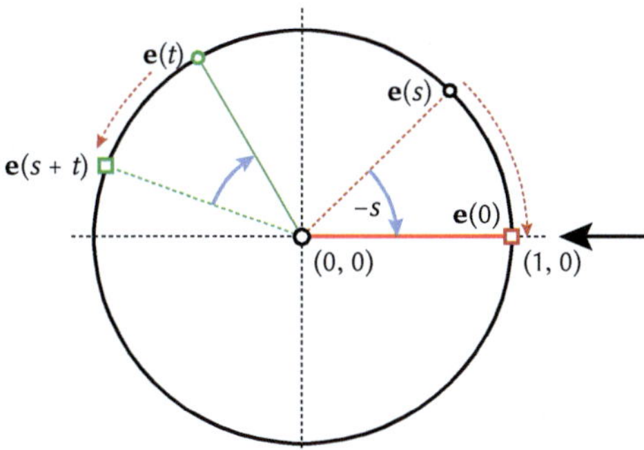

Figure 11.8: Motion, to be seen, has to be seen from a perspective of non-motion. A mathematical model of cyclical motion consists of the unit circle $\mathbb{U}$ as the state space, and $\mathbb{R}$ as the set of operators with dynamical action defined by $s \triangleright \mathbf{e}(t) = \mathbf{e}(-s+t)$ for all $s, t \in \mathbb{R}$. The effect of s is then seen to be *clockwise* rotation (for positive s) of the circle $\mathbb{U}$ through angle s (so it is shown as rotation through $-s$), which seen from the *fixed* coordinate point $(1, 0)$ makes it appear that this point has rotated $+s$ in the counter-clockwise direction since now $\mathbf{e}(s)$ appears there. In this way we can see that the dynamical action can be interpreted as progression in time.

which $\mathbb{U}$ is seen to be acting on itself and the dynamics becomes

$$\begin{aligned} \mathbb{U} \times \mathbb{U} &\longrightarrow \mathbb{U} \\ (\mathbf{e}(s), \mathbf{e}(t)) &\mapsto \mathbf{e}(-s+t)\,. \end{aligned}$$

As a more familiar example we might consider the way the set of integers acts on itself by addition

$$\begin{aligned} \mathbb{Z} \times \mathbb{Z} &\longrightarrow \mathbb{Z} \\ (x, y) &\mapsto -x + y\,. \end{aligned} \tag{11.4}$$

This example is particularly relevant because it lies at the heart of what we shall see as the Fourier duality between $\mathbb{U}$ and $\mathbb{Z}$. In both cases we see the sets serving in two roles: as a set of *operators* and as a *state space* on which they operate. In both cases what underlies this is the structure of a *group*, see §9.2.2. This will come out more clearly as we go on.

11.4.4 Pure waves

At this point we can return to the cosine and sine waves that arise out of the coordinates of the point $\mathbf{e}(t)$ as it traverses the circle $\mathbb{U}$. These coordinates define the two trigonometric functions cos (cosine) and sin (sine) respectively, Fig.11.6. The situation is familiar from the trigonometry of the right-angled triangle which has hypotenuse of length 1. Alternatively, if this is not so familiar, Fig.11.6 serves as a *definition* of sin and cos. In fact it defines sin and cos for *all* real values of t, see (11.2).

We may think of sine and cosine as producing the simplest and most basic waves of all. Most importantly, even if we form combinations of these waves (linear combinations) by adding them, scaling them, and phase shifting them, the resulting wave forms will still be periodic with period 2π. This is the basis of wave forms created by musical instruments, which we will come to in §11.4.5.

The way it is set up, the dynamics is periodic with period 2π because whatever the value of t, $\mathbf{e}(t) = \mathbf{e}(t + 2k\pi)$ for all integers k. Most of the time we will find it preferable to make the periods whole numbers, rather than $2\pi \times$ integers. To do this we scale the argument t by 2π and use the functions $\mathbf{e}(2\pi t)$. For these,

$$\mathbf{e}(2\pi(t \pm 1)) = \mathbf{e}(2\pi t \pm 2\pi) = \mathbf{e}(2\pi t)$$

so in terms of t they have period 1, and hence also periods of $2, 3, 4, \ldots$ and $0, -1, -2, -3, \ldots$: in short all integers.

From this point on, rather than dealing with $\sin(t)$ and $\cos(t)$, most often we prefer to deal with $\sin(2\pi t)$ and $\cos(2\pi t)$ which have period 1.

If k is any integer then $\sin(2\pi kt)$ and $\cos(2\pi kt)$ also have period 1, that is, they repeat every time t is increased or decreased by 1. It is true that they have even shorter periods, e.g. $\cos(2\pi 3t) = \cos(6\pi t)$ has period $1/3$, but they still have period 1. We call these higher frequency waves *harmonics* of the basic waves $\sin(2\pi t)$ and $\cos(2\pi t)$. It is more efficient and more revealing to deal with the basic waves and their harmonics in their combined form as points $\mathbf{e}(2\pi kt)$, $k \in \mathbb{Z}$, running around the unit circle:

$$\mathbf{e}(2\pi kt) = (\cos(2\pi kt), \sin(2\pi kt)).$$

It is these functions that are the focus of our attention in what follows. We call them *pure waves.* Out of them we will build *all* the periodic functions with period 1.

Fig.11.9 summarizes the key words that we will associate with waves.

11.4.5 Waves from musical instruments

Among commonly recognized waveforms are timbres that characterize different musical instruments. The sound emitted from a musical instrument does not consist of a pure frequency (a pure wave) but rather a combination of frequencies, most often arising from *fundamental frequency* along with a series of harmonics. This is the source of its individual appeal, with its own distinctive and easily recognizable voice. In producing the sound of a single note, there is a fundamental frequency or tone, and that is likely the one that we are most drawn; above it is a range of harmonics that give it its distinctive character.

Fig.11.10 shows sample waveforms and corresponding frequency charts from two familiar musical sources. The harmonics are the tones arising from frequencies that are

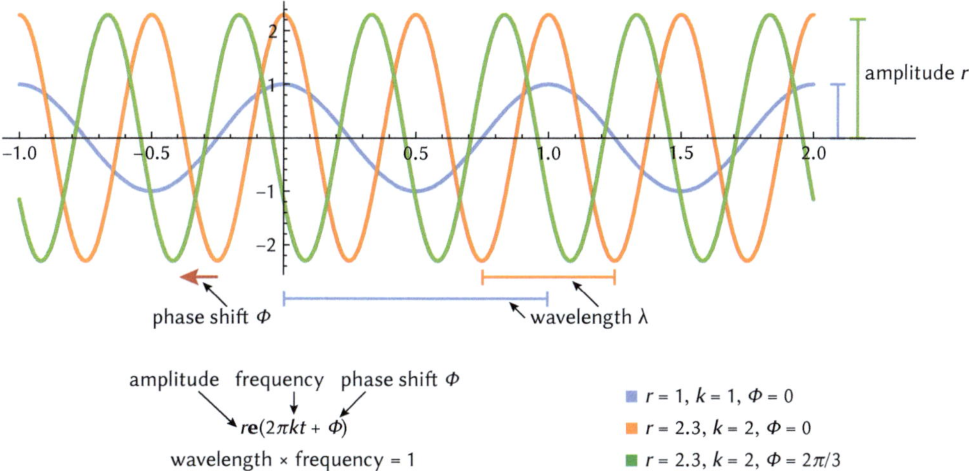

Figure 11.9: The figure shows the anatomy of a wave seen from the perspective of the circular function $\mathbf{e}(\cdot)$. There are five parts: amplitude, frequency, wavelength, time, and phase. The formal presentation of the wave is $r\mathbf{e}(2\pi kt + \phi)$, though what is illustrated here is just the first coordinate $r\cos(2\pi kt + \phi)$ (later on we will call it the *real part* of the wave.) The fundamental pure wave $\cos(2\pi t)$ (blue) has frequency 1, so we can imagine the basic wave moving to the right at one full wave per unit of time. For the harmonics, k determines the frequency, that is the number of cycles per unit of time. Here $2.3\cos(2\pi 2t)$ with $k = 2$ (orange) is of amplitude 2.3 and frequency 2: it completes two cycles per unit of time and its wavelength is $1/2$: frequency and wavelength are inversely proportional. The phase shift ϕ shifts the wave (orange plot shifts to green plot). Phase shift is often thought of in terms of angle, in this case it is $2\pi/3$ radians or 120° degrees. If ϕ is negative, the shift is in the opposite direction. The *exact period* of the wave $r\cos(2\pi kt + \phi)$ is $1/k$.

twice, three times, four times ... the frequency of the fundamental. The most notable feature here is how continuous the waves look and how discrete the frequency charts look. This points to the source of the continuous-discrete type of duality that lies at the basis of periodic phenomena.

Our immediate goal is to extrapolate this notion of frequency analysis into a more mathematical setting. At its basis there is a fundamental frequency, which, by choosing the unit of time appropriately, we can simply take to be one cycle per unit of time. Along with it are its harmonics with higher frequencies which are $2, 3, 4, \ldots$ cycles per unit of time. Each harmonic will appear with its own intensity amplitude and also its own phase shift with respect to the fundamental. The resulting wave can obviously be very complicated, but mathematically the result will have a quite straightforward appearance:

$$f(2\pi t) = \cdots + c_{-2}\,\mathbf{e}(-4\pi t) + c_{-1}\,\mathbf{e}(-2\pi t) + c_0 + c_1\,\mathbf{e}(2\pi t) + c_2\,\mathbf{e}(4\pi t) + c_3\,\mathbf{e}(6\pi t) + \cdots\,, \quad (11.5)$$

a sum of pure waves with coefficients corresponding to their intensities, and, as we will

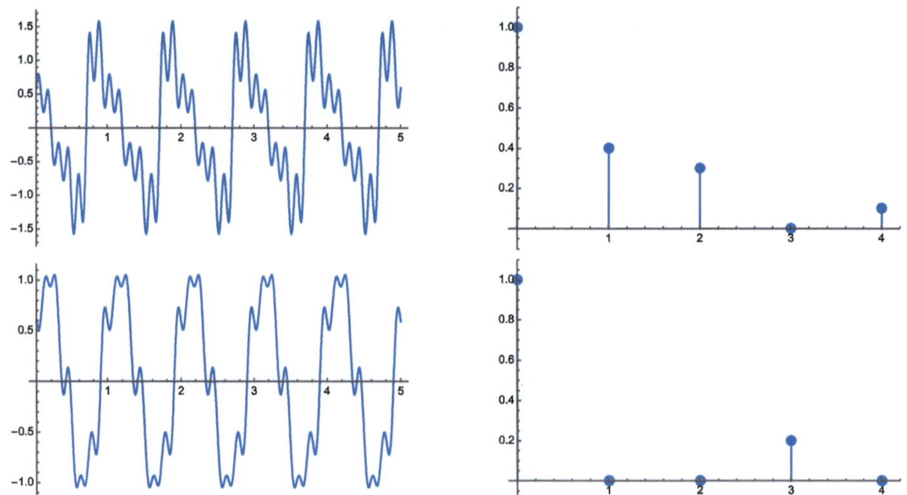

Figure 11.10: Wave form and frequency charts approximating the distinct timbres of the pipe organ and the violin. On the left the axes are time/amplitude and on the right frequency/amplitude.

see, also their phase shifts. We have included the constant term $c_0 = c_0\,\mathbf{e}(0\pi t) = c_0\,\mathbf{e}(0)$ which may not seem relevant but will be important later.

Now, it is one thing putting waves together, but another to be given the resulting outcome then trying to untangle it and determine which frequencies are present and with what intensities. This is the goal of Fourier analysis and the values of the coefficients that appear are called the *Fourier coefficients*.

There are a number of obstacles in the way of this lofty goal. The first is that the way that it is expressed $f(2\pi t)$ is actually a sum of points in the *plane* that have been scaled by numbers c_k. So $f(2\pi t)$ itself is not actually a number, but rather a point in the plane ($\mathbf{e}((2\pi t)$ etc. are points on a circle). The second, which is more serious, is that the form of the equation (11.5) does not seem to include any provision for possible phase shifting. Altering the phase of a periodic wave does not change its period, but does change the way in which it relates to other periodic waves. We could change the phase of each and every one of the pure wave forms $\mathbf{e}(2\pi kt)$ without destroying the periodicity, so we should rightly expect phase shifts to be a part of the picture.

Both of these problems are solved by the reimagining of the coordinate plane, where all these pure waves are defined, as being a single system of numbers. This is the *complex plane*. Once this is done we will be able to come back to the equation (11.5) and see how it can be untangled.

The dots that appear in the sum in equation (11.5) are there to suggest that the sum might be finite or it might go on forever, with higher and higher frequencies. We allow either because we will see that in general infinite sums are unavoidable (even though they do bring with them questions of convergence).[5] In practice the frequencies arising in a physical system cannot be indefinitely high, so it might seem that infinite sums are unnecessary. However, in the mathematical framework allowing infinite sums is crucial since only then can we express results with exactness.

With all this in mind we can pose two questions:

(1) Is every periodic function with period 1 expressible in the form (11.5), and if so,

(2) given the function f how can we find the so-called *Fourier coefficients* c_k?

In §11.6 we will answer them.

11.5 The complex numbers

Although we have pictured waves of the electro-magnetic spectrum in the form of wavy graphs spread out along the real line, it is far more relevant, conceptually, physically, and mathematically, to think of periodic phenomena directly in terms of a circle. They are, after all, called cyclical.

Now there is a wonderful way in which to treat the entire coordinate plane as a new set of numbers, the *complex numbers*. This is done by giving the plane an algebraic structure by defining addition and multiplication for ordered pairs of numbers. With this structure the plane is called the *complex plane* and is denoted by the symbol $\mathbb{C}$. In this new system the horizontal axis, the *real line*, retains the familiar properties of addition and multiplication of the real numbers. The vertical axis is called the pure *imaginary* part of the complex plane and numbers on it are called *imaginary numbers*. Since it is just a vertical line, there is nothing especially imaginary about it, but it is here that we find those seemingly outlandish "imaginary" square roots of negative numbers, hence its name; and at the same time we see the reason why the real numbers came to be called "real". Thought about without any further context, having numbers which are the square root of -1 would seem impossible, but there is nothing peculiar about the 2D plane, and we shall see that it is quite straightforward to give it the algebraic structure in which such square roots exist. In addition, in this new system of numbers the unit circle $\mathbb{U}$ becomes an algebraic object in its own right, and it is this feature that underlies its importance in studying cyclical phenomena.

We begin by recalling that there are two common ways in which to coordinatize the plane. The first is the familiar (x, y) form, which describes points in terms of their projections onto the horizontal and vertical axes. These are the *Cartesian coordinates*. The other form, the *polar coordinates*, works by writing points in the form $r\,\mathbf{e}(t)$. The points $\mathbf{e}(t)$ themselves are points on the unit circle, which we can think of as indicators of direction, namely the direction from the origin to $\mathbf{e}(t)$. We can then think of $r\,\mathbf{e}(t)$ as the

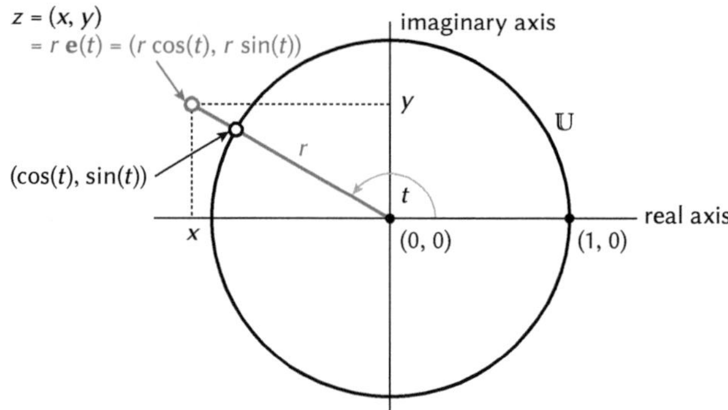

Figure 11.11: The points of the plane are interpreted as *complex numbers.* Each point z of the plane can be described both in its Cartesian coordinate form $z = (x, y)$ and in its polar form $r\,\mathbf{e}(t)$. Since $\mathbf{e}(t) = (\cos(t), \sin(t))$, we have $z = r\mathbf{e}(t) = (r\cos(t), r\sin(t))$. This point becomes $x + \mathrm{i}y$, or equivalently, $r\cos(t) + \mathrm{i}\,r\sin(t)$ when treated as a complex number.

point reached by going distance r in the direction indicated by $\mathbf{e}(t)$. This is a natural way in which we give walking directions to someone: point them in the right direction and tell them how far to walk.

With this in mind, each point z of the plane can be described in these two forms:

$$z = r\,\mathbf{e}(t) = r(\cos(t), \sin(t)) = (r\cos(t), r\sin(t)),$$

for some real number $r > 0$ and some point $\mathbf{e}(t)$ on the circle, Fig.11.11. With the exception of the origin, both r and the point $\mathbf{e}(t)$ on the circle are uniquely determined, though t itself is only determined up to differences of 2π since $\mathbf{e}(t) = \mathbf{e}(t + 2\pi)$. It too becomes unique if we constrain the angle so that $0 \le t < 2\pi$ or $-\pi < t \le \pi$, though usually we do not wish to do this. As for the origin $(0, 0)$, it can be written as $0\,\mathbf{e}(t)$ and does not depend on the value of t.

Usually r is described as the *modulus* or *absolute value* of z (and we will see this more formally below) and t is described as its *argument*. In this picture the entire plane is called the *complex plane*, $\mathbb{C}$, and the points of the plane are called *complex numbers*. The complex plane is simply the ordinary coordinate plane seen from a new point of view. Historically the idea of viewing the complex numbers as a plane was introduced by Jean-Robert Argand (1768–1822) in 1806 and one sometimes sees it referred to as the *Argand plane.*

So, each point z in the plane has its usual coordinate form: $z = (x, y)$, but also its polar form:

$$z = (x, y) = r\mathbf{e}(t).$$

In the context of the complex plane the usual x -axis is called the *real axis* and the y-axis is called the *imaginary axis.* Accordingly, x is called the *real part* of z and y is called its *imaginary part* . But both parts are just real numbers.

11.5.1 Algebraic operations

We seem to have gained little by this. However, there are two algebraic operations that make the complex plane into the algebraic object $\mathbb{C}$, and it is from these that it becomes something really new. These two operations are extensions of our familiar notions of addition and multiplication.

addition: this is very straightforward. If $z_1 = (x_1, y_1)$ and $z_2 = (x_2, y_2)$ are two points of the plane $\mathbb{C}$ then their sum is defined to be

$$z_1 + z_2 = (x_1 + x_2, y_1 + y_2).$$

Just add the real components and the imaginary components. Note that $(0, 0)$ works like a zero element for this addition, and for this reason, in the context of complex numbers it is simply denoted by 0. This never causes any problems of confusion. So $0+z = z+0 = z$ for all $z \in \mathbb{C}$.

multiplication: this is based on a new definition. The product of any two points $\mathbf{e}(s)$ and $\mathbf{e}(t)$ on the circle $\mathbb{U}$ is defined to be

$$\mathbf{e}(s)\,\mathbf{e}(t) = \mathbf{e}(s + t).$$

This multiplication mimics addition of real numbers.

With this we define the product of $z_1 = r_1\,\mathbf{e}(s_1)$ and $z_2 = r_2\,\mathbf{e}(s_2)$ by

$$z_1 z_2 = r_1 r_2\,\mathbf{e}(s_1 + s_2) = z_2 z_1 \ : \tag{11.6}$$

add the arguments and multiply the absolute values. The product is represented just as we do with symbols representing ordinary numbers simply by placing the symbols one after another. Notice that the order does not matter: $z_1 z_2 = z_2 z_1$. Notice also that the absolute value of the product $z_1 z_2$ is $r_1 r_2$, which is the product of the absolute values of z_1 and z_2.

Notice also that multiplication by $\mathbf{e}(s)$ performs rotation of the whole plane since $\mathbf{e}(s)\,r\mathbf{e}(t) = r\mathbf{e}(s + t)$. Multiplication by $\mathbf{e}(s)$ is the same as rotation through the angle s.

In particular multiplication by $\mathbf{e}(0)$ acts like 1:

$$\mathbf{e}(0)\,\mathbf{e}(t) = \mathbf{e}(0 + t) = \mathbf{e}(t).$$

In fact we simply write 1 for $\mathbf{e}(0) = (1, 0)$. We know that $\mathbf{e}(2\pi) = \mathbf{e}(0) = 1$. Indeed we get things like

$$1\,\mathbf{e}(t) = \mathbf{e}(2\pi)\,\mathbf{e}(t) = \mathbf{e}(2\pi + t) = \mathbf{e}(t)$$

so the multiplication works just fine. As for 0, $0z = z0 = 0$ for all z.

This suggests that through multiplication we are seeing the group of rotations of the circle appearing. Below we will see that $\mathbb{U}$ is not only the set of points of the unit circle, but it can also be interpreted as the group of rotations of the circle.

At this point we simply identify the entire real axis (the x-axis) of our plane as just being the ordinary real line that we are so familiar with. Thus $(r, 0)$ is identified with r itself. This fits perfectly with our definitions. Our new multiplication and addition then match the usual definitions for addition and multiplication in $\mathbb{R}$: $r + s = (r, 0) + (s, 0) = (r + s, 0)$ and

$$rs = r\,\mathbf{e}(0)\,s\,\mathbf{e}(0) = r\,s\,\mathbf{e}(0 + 0) = rs\,\mathbf{e}(0) = rs(1, 0) = (rs, 0)\,.$$

Here r and s are assumed non-negative, but we get the negative real axis by using the points $r\mathbf{e}(\pi) = r(-1, 0) = (-r, 0) = -r$, and this fits perfectly with our usual use of minus signs.

Importantly, every non-zero complex number has a multiplicative inverse: the inverse of any non-zero complex number $r\mathbf{e}(t)$ is $r^{-1}\mathbf{e}(-t)$, as we see immediately by multiplying.

To recapitulate, we have made the entire coordinate plane into a new set of numbers, the *complex numbers*, denoted by $\mathbb{C}$, and given it algebraic structure by defining addition and multiplication throughout. In defining addition we have exploited the coordinate form of points in the plane, while for multiplication we have exploited the polar form. The complex numbers contain the real numbers as the x-axis.

The complex plane has extraordinary features that the real numbers do not. Of these the most famous is the complex number i, which is *defined* to be $(0, 1)$. Since $(0, 1)$ is $\mathbf{e}(\pi/2)$, in polar form, we have $\mathrm{i} = 1\,\mathbf{e}(\pi/2)$. From this follows

$$\mathrm{i}\,\mathrm{i} = \mathbf{e}(\pi/2)\,\mathbf{e}(\pi/2) = \mathbf{e}(\pi) = (-1, 0) = -1\,,$$

or to put it in words: two quarter-turns make up a half-turn. This is the "mysterious" *square root of* -1. Using i we can write

$$z = (x, y) = x(1, 0) + y(0, 1) = x + \mathrm{i}y\,,$$

and this is the standard way in which one will see complex numbers written down. Whether we prefer to write it as $x + \mathrm{i}y$ or $x + y\mathrm{i}$ is irrelevant. In this form x is the real part, y is the imaginary part, and $\mathrm{i}y$ is the product of i and y. Note the mathematical terminology: the "imaginary part" is actually real! It is y, not $\mathrm{i}y$.

Multiplication works just as we would expect: for $z_1 = x_1 + \mathrm{i}y_1$ and $z_2 = x_2 + \mathrm{i}y_2$

$$z_1\,z_2 = (x_1 + \mathrm{i}y_1)(x_2 + \mathrm{i}y_2) = (x_1x_2 - y_1y_2) + \mathrm{i}(x_1y_2 + x_2y_1)\,, \tag{11.7}$$

where we did nothing special except use the fact that $\mathrm{i}\,\mathrm{i} = -1$.[6]

Once we think of the plane as the complex plane, we can rewrite (11.2) to get

$$\begin{aligned} \mathbf{e}(t) &= \cos(t) + \mathrm{i}\sin(t)\,, \\ r\mathbf{e}(t) &= r\cos(t) + \mathrm{i}\,r\sin(t)\,. \end{aligned} \tag{11.8}$$

These equations are basic to what follows.[7]

It is well worth getting introduced to the complex numbers because they are such a remarkable algebraic structure and they play such a huge role in mathematics. In fact, as we see here, they are rather easily defined in terms of the unit circle. Their construction may seem like a trick that we might carry out just as well in 3D or any dimension, but for all the apparent arbitrariness here, the complex numbers are special, and are a natural outcome of the real numbers themselves. They arise all over mathematics. Here we will stress their importance in the theory of waves. Later we will see how essential they are for the formalism of quantum mechanics. We have put some more details about the intrinsic properties of the complex numbers in §16.6.[8]

We make one little, but important, observation here. Of the operations on waves, adding waves and scaling them by factors are straightforward. But what about *phase shifting*? Shifting a pure wave $\mathbf{e}(t)$ by a phase ϕ produces $\mathbf{e}(t + \phi)$. Now, we see that this is just the product $\mathbf{e}(\phi)\mathbf{e}(t)$, so phase shifting by ϕ just "scales" the wave by a factor $\mathbf{e}(\phi)$ when viewed in terms of complex multiplication. To simultaneously scale a wave by factor of r *and* shift its phase by ϕ amounts to multiplying by the complex number $r\mathbf{e}(\phi)$. In this way, expanding our idea of scaling to allow scaling by complex numbers, the processes of scaling and phase shifting are made part of a common process. This is a crucial step in simplifying the formalism of periodic functions.

One final thing to keep in mind is that although we began with waves that look like we expect waves to look, the quantities $\mathbf{e}(t)$, which truly encapsulate circular motion, are complex numbers and have two components, each of which traces out wave-like paths in time, Fig.11.14.

11.5.2 Conjugation

Obviously one of the striking features of $\mathbb{C}$ is the appearance of the famous $\sqrt{-1}$. Of course there is no real number that satisfies $x^2 = -1$, but the complex plane is a much larger object than the real numbers and it does have a square root of -1. In fact it has two. The first is $\mathrm{i} = \mathbf{e}(\pi/2)$ since $\mathbf{e}(\pi/2)\,\mathbf{e}(\pi/2) = \mathbf{e}(\pi) = (-1, 0) = -1$. However, it is also true that $\mathbf{e}(-\pi/2)\,\mathbf{e}(-\pi/2) = \mathbf{e}(-\pi) = \mathbf{e}(-\pi + 2\pi) = \mathbf{e}(\pi) = (-1, 0) = -1$. This is not surprising since $\mathbf{e}(-\pi/2) = -\mathbf{e}(\pi/2) = -\mathrm{i}$, and we expect square roots to appear as $\pm$ pairs.

But this does raise an interesting point. We chose to define $\mathrm{i} = (0, 1)$, one unit up on the y-axis. We could just as easily have chosen to make $\mathrm{i} = (0, -1)$, one unit down on the y-axis. Of course we could say it is more natural to choose $(0, 1)$ than $(0, -1)$ because the latter involves a minus sign. But this ignores the fact that it is just pure custom that we choose to have the positive y axis in the upward rather than the downward direction. This choice is what makes $\mathbf{e}(\pi/2)$ into a rotation in the counterclockwise direction. If

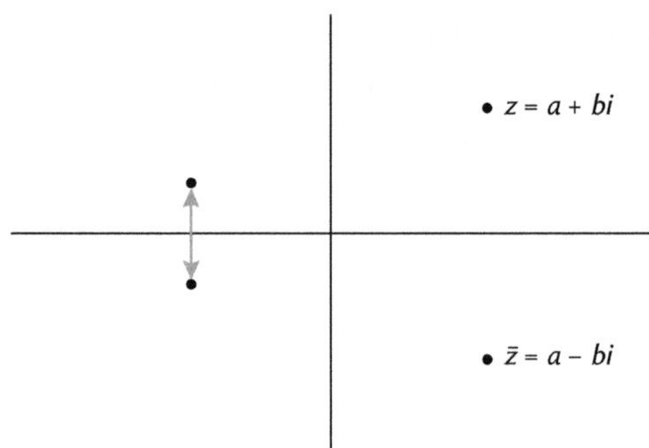

Figure 11.12: Complex conjugation is reflection in the real axis.

we had chosen the y-axis downwards, then the roles of i and $-\mathrm{i}$ would be reversed and the clockwise direction would be the positive one. This is not really an ambivalence, but rather points to a special internal feature of $\mathbb{C}$ called *complex conjugation.*

Let $z = a + \mathrm{i}b$ be a complex number. Its *complex conjugate* (or just *conjugate*) is $\overline{z} := a - \mathrm{i}b$. Conjugation replaces i by $-\mathrm{i}$. The overline here is universally used notation. The conjugation mapping

$$\overline{} : \mathbb{C} \longrightarrow \mathbb{C}, \quad z \mapsto \overline{z}$$

is visually simply reflection in the x-axis, or better stated, reflection in the real axis, Fig.11.12. As such it reverses orientation.

The conjugation mapping has a pair of important properties: for all $w, z \in \mathbb{C}$

$$\overline{w + z} = \overline{w} + \overline{z}\,,$$
$$\overline{wz} = \overline{w}\,\overline{z}\,.$$

It is what is called an *automorphism*—it is a mapping of $\mathbb{C}$ to itself that perfectly preserves the algebraic structures of addition and multiplication (in the sense that conjugation can be done before or after the operations of addition and multiplication) and its existence reflects the fact that there is no intrinsic way to choose which of the two roots of $x^2{+}1 = 0$ we should choose as i. However, once it is done, and the way we have presented it is the standard way in mathematics, we have already made that decision, the thing is settled. The conjugation automorphism itself, though, is vital.

The real numbers have nothing resembling this. The closest might be $x \mapsto -x$, but this does not match the nice multiplication property of conjugation: $-(xy)$ is not the same as $(-x)(-y)$. In fact, conjugation itself is invisible from the point of view of the real numbers, because the real line (the real axis) is the mirror of the reflection : $\overline{x} = x$ if x is a real number (and only if it is a real number).

Notice that

(i) $\overline{\overline{z}} = z$ for all $z \in \mathbb{C}$;

(ii) $z \in \mathbb{C}$ is real if and only if $\overline{z} = z$;

(iii) $z \in \mathbb{C}$ is *pure imaginary* (that is to say it has the form bi for some real number b) if and only if $\overline{z} = -z$;

(iv) $\overline{r\mathbf{e}(t)} = r\mathbf{e}(-t)$ for all real numbers $r > 0$ and t.

Now we come to the notion of absolute value in $\mathbb{C}$. Notice that if $z = a + bi$ then

$$z\,\overline{z} = (a + ib)(a - ib) = a^2 - aib + iba - ibib = \;a^2 + b^2 \geq 0\,.$$

It's always a real number, and since it is the sum of two squares of real numbers, it can't be negative. In fact unless $a = b = 0$, i.e. $z = 0$, we have $z\overline{z} > 0$. Written in polar form $z = r\mathbf{e}(t)$ we have

$$\overline{z}\,z = z\,\overline{z} = r\mathbf{e}(t)r\mathbf{e}(-t) = r^2\mathbf{e}(t - t) = r^2\mathbf{e}(0) = r^2\,. \tag{11.9}$$

The positive square root of $z\,\overline{z}$ is r, and hence (see (11.6)) is equal to the *absolute value* of z:

$$|z| := (z\,\overline{z})^{1/2} = (r^2)^{1/2} = r\,.$$

The new absolute value is defined for all complex numbers and hence also for all real numbers since they are a part of $\mathbb{C}$. Not surprisingly, given the terminology, the new absolute value is the same as the old one when applied to real numbers. For if $z \in \mathbb{C}$ is *real*, then $\overline{z} = z$ and the new definition gives us

$$|z| = (z\,\overline{z})^{1/2} = (z^2)^{1/2} = \sqrt{z^2} = |z|\,,$$

which is the precisely the definition of the absolute value of a real number.[9] Since the absolute value of a product is equal to the product of the absolute values (see (11.6)), we have

$$|zw| = |z|\,|w| \quad \text{for all } z, w \in \mathbb{C}\,.$$

The new absolute value also measures distance. The reason is clear from Fig.11.13: $|z|$ is the distance from the origin $0 \in \mathbb{C}$ to the point z. This is Pythagoras' theorem, and shows again why this ancient theorem is still so important in mathematics. Using the absolute value we can see that the distance between two complex numbers w, z when viewed in the complex plane is $|z - w| = |w - z|$.

Unlike the real numbers, there is no natural sense or ordering in the complex numbers, no natural concept of what $z \leq w$ might mean. We can certainly have $|z| \leq |w|$, but this is only about absolute values, not about z and w themselves. In particular there is no concept of $z \geq 0$ or $z \leq 0$. Having said this, even so, in mathematics one sometimes does see $z \geq 0$ where z is nominally a complex number. But this always has the implication that in fact z is actually a real number (its imaginary part is 0) and as a *real* number it is greater than or equal to 0.

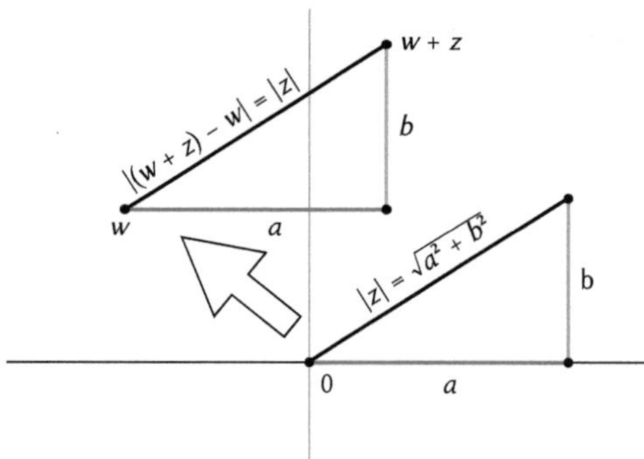

Figure 11.13: The distance between 0 and $z = a + ib$ is seen to be $(a^2 + b^2)^{1/2} = |z|$ by Pythagoras' theorem. Simply translating the two points to w and $w+z$ shows how the distance formula works in general: any two points of $\mathbb{C}$ can be written in the form $w, w+z$ for some w, z, and the distance between them is *the absolute value of their difference*, namely $|z|$.

The notation $z \geq 0$, where $z \in \mathbb{C}$, means $z \in \mathbb{R}$ and z is non-negative.

11.5.3 The circle group

All the points z in the complex plane with $|z| = 1$ are of the form $1\,\mathbf{e}(t) = \mathbf{e}(t)$ and so are precisely the points of $\mathbb{U}$. That's another reason why it's called the *unit circle*. Now $\mathbb{U}$ has one very lovely property that we alluded to before: $\mathbb{U} \subset \mathbb{C}$ is a *group* under complex multiplication. We simply need to see that $\mathbb{U}$ is closed under multiplication, it has an identity element, and each element of $\mathbb{U}$ has an inverse in $\mathbb{U}$, see §9.2.2. If $z, w \in \mathbb{U}$ then $|z| = 1 = |w|$, so $|zw| = |z|\,|w| = 1$ and $zw \in \mathbb{U}$. The inverse of z is $\overline{z}$ (since $z\overline{z} = |z|^2 = 1$) and $|\overline{z}| = 1$. So $\mathbb{U}$ now appears as the *group of rotations of the plane around the origin*: each point $z = \mathbf{e}(s)$ of $\mathbb{U}$ can be thought of as representing the rotation through the angle s, since this is precisely the effect of multiplication by $\mathbf{e}(s)$.

$\mathbb{U}$ is the carrier of the symmetry that we call cyclical repetition.

11.5.4 Recap

Although this section may seem to have accumulated quite an array of concepts, the basis of it all is simply the idea of a point moving around a circle. The ever-present symbol $\mathbf{e}(t)$ is simply the point on the unit circle, the circle $\mathbb{U}$ of radius 1 centered on the origin $(0, 0)$ in the coordinate plane, which is t units along the circumference starting from $(1, 0)$ in the counter-clockwise direction. If t increases in a continuous way then the point moves in a continuous way around the circle, repeating every time it increases by 2π (the length of the circumference).

Movement of a point around a circle certainly points to dynamics, and indeed $\mathbb{U}$ can be taken as the state space for the dynamics of rotation, see (11.3). The idea is that $\mathbb{R}$ acts on the circle so that $t \in \mathbb{R}$ acts as a rotation of the entire circle through $-t$ radians, or to put it another way, moves all points on the circle clockwise through a distance of t along the circle. Thinking in terms of time, a fixed observer at the point $(1, 0)$ sees that the point that was initially $\mathbf{e}(t)$ is now at $(1, 0)$, thus witnessing the effect of following the path of $\mathbf{e}(t)$ in time. One of the virtues of introducing the complex numbers is that rotation through $-t$ is the same as multiplication by $\mathbf{e}(-t)$, so the effect of the dynamics is

$$t \triangleright \mathbf{e}(s) = \mathbf{e}(-t)\mathbf{e}(s) = \mathbf{e}(-t + s)\,. \tag{11.10}$$

Each point in the plane has two representations: as a coordinate pair (x, y) in the usual way, or as a polar pair $r\mathbf{e}(t)$ consisting of a direction, given by a point on the circle, and a magnitude expressing how far to go in that direction, given by r. These two representations are used to define the two algebraic operations of addition and multiplication that make the plane into the complex numbers. Addition is just adding coordinates $(x, y) + (x', y') = (x + x', y + y')$ and multiplication is based on the simple rule $\mathbf{e}(t)\,\mathbf{e}(s) = \mathbf{e}(t + s)$. This is what leads to the interpretation that $\mathbf{e}(t)$ acts as rotation through an angle of t radians (Fig.11.8), or equivalently as a phase shift through t radians.

The realization that $\pm\mathrm{i}$ are both square roots of -1 and are in some sense interchangeable leads to the idea of complex conjugation, and that in turn leads to the notions of absolute value, length, and distance in the complex plane. These are no different than what we already had at the very start with the coordinate plane, but now they fit into a larger scheme of algebraic structure.

Finally $\mathbb{U}$ can now be seen as the set of complex numbers of absolute value equal to 1, and as a group under multiplication. Multiplication by $\mathbf{e}(s)$ just rotates the circle counter-clockwise, so that $\mathbf{e}(s)\,\mathbf{e}(t) = \mathbf{e}(s + t)$, and the inverse of $\mathbf{e}(s)$ is $\mathbf{e}(-s)$, which is rotation in clockwise. Using this $\mathbb{U}$ can actually be seen as the rotation group of the circle, i.e. on itself! We can think of $\mathbb{U}$ as the circle that it is, or as the group of rotations of the circle.

Oftentimes more than one interpretation can be associated to the very same mathematical formalism and this is an example. Rather than a confusion, this can be a great benefit. Being able to see things from several points of view is not a source of ambiguity but a source of enlightenment, and this applies equally well in mathematics.[10]

Still, all these features are somewhat conflated in the 1D picture. The interpretations that the circle is a 1D geometrical object, that it serves as the state space of a form of cyclical dynamics under the action of $\mathbb{R}$, that it produces wave forms, that it is also the rotation group of a circle, and that it can serve as a group of operators on itself to make another dynamical system lead to a rather bewildering picture. When we come to 2D and

3D waves, we will find that all these features are still there, but they are more separated and the situation actually becomes clearer.

> A consistently arising question in dealing with circular motion is whether to think in terms of angles (in radians) or in terms of complete cycles—that is, whether to use the mathematical form $\mathbf{e}(t)$ or the form $\mathbf{e}(2\pi t)$. No matter which approach one takes, the symbol *pi* will inevitably appear.
> We will use both without further mention, but increasingly we will prefer the second form, which emphasizes periods (complete cycles).

11.5.5 Why use the complex numbers?

We began our discussion with the mathematical objects that fit naturally with our conception of 1D waves—the graphs of the basic trigonometric functions sine and cosine, Fig.11.4. However, the idea of thinking in terms of cyclic motion around a circle and using complex waves is ultimately more revealing and more fundamental. This is not entirely obvious. If our mathematical waves are expressed as complex numbers then the more physically related waves have to be expressed in terms of their real (cosine) and imaginary (sine) parts. If we are only interested in real waves why drag around a second component? There are three good reasons for this.

(i) The first reason is simple mathematical convenience. In dealing with a pure wave we have three ingredients: amplitude, frequency, and phase shift:

$$r\cos(2\pi t + \phi) \quad \text{and similarly} \quad r\sin(2\pi t + \phi)\,,$$

with amplitude r, wavelength 1, and phase shift ϕ. The corresponding complex expression is

$$r\cos(2\pi t + \phi) + \mathrm{i}\, r\sin(2\pi t + \phi) = r\mathbf{e}(2\pi t + \phi) = r\,\mathbf{e}(\phi)\mathbf{e}(2\pi t) = a\,\mathbf{e}(2\pi t)\,,$$

where $a = r\mathbf{e}(\phi) \in \mathbb{C}$. We see that the amplitude and phase have been absorbed into one complex number $a = r\mathbf{e}(\phi)$ and its absolute value $|a| = r$ is the amplitutde.

(ii) The second reason is more subtle, but ultimately more compelling: conceptually periodicity is circular in nature, and this is made totally explicit in the complex version. The fact that $\mathbb{U}$ is actually a *group* and algebraically encodes the concept of circular symmetry is a key to wave analysis.

(iii) The third reason is necessity: the complex numbers are needed to describe the natural invariants (eigenfunctions and eigenvalues) that appear in linear analysis the arise in §11.8. They are foundational in quantum mechanics.

11.6 Functions and observables

11.6.1 Observables

Although the underlying foundation of cyclic phenomena is motion around a circle our experience with periodic systems is primarily the repetition of some type of observation. Think of the annual seasons of the year that are founded on the rotation of the Earth around the Sun and the inclination of the Earth's axis with respect to the plane of its orbit. The dynamical system here is a vast play of forces at the level of the solar system, but our experiences of the seasons are in terms of temperature, the heights of tides, the amount of precipitation, sun dials, the times of sunrise and sunset, and so on. There are numerous ways in which we measure these things but in the mathematical and scientific world its most common expression is through numerical values; in effect the study of functions

$$f : \text{state space} \longrightarrow \text{numbers},$$

which associate numbers to states. Physicists call them *observables*. By "numbers" we might immediately think of real numbers, but there is good reason to extend the idea of observables to include complex-valued functions (in effect, functions whose values are pairs of real numbers), and that will be our basic assumption. In fact, we have already been using this functional notation: when we name the points on the unit circle $\mathbb{U}$ by the complex numbers $\mathbf{e}(t)$, we are using a function, namely $\mathbf{e}$, to give numerical values to the points of a purely geometric form, a circle. This is the famous way, originating with René Descartes, in which geometry and algebra are paired. For formal definitions around functions see the Endnotes.[11]

Observables themselves lead to their own dynamical systems, and to a great extent it is these derivative systems that we directly experience, while the working of the systems that underlie them may remain obscure. We need only think of the intricate dynamics that must control the heart in contrast with the typical measurements of heart-rate or blood pressure.

So our interest switches to functions. This is the point at which the concept of *linearization* appears. The rest of this chapter explains what this is, how the ideas around pattern system extend to include it, and why it is so useful.

We begin with periodicity, which we have seen is so important to all life on Earth.

> A real or complex-valued function $t \mapsto f(t)$ defined on $\mathbb{R}$ is called *periodic* if it repeats with some period P, that is, there is a non-zero real number P for which
> $f(t+P) = f(t)$ for all t.

Since this is supposed to be valid for all t, it is also valid when t is replaced by $t+P$, so $f(t+2P) = f((t+P)+P) = f(t+P) = f(t)$, or t is replaced by $t-P$ in which case

$f(t) = f((t-P)+P) = f(t-P)$. Following this idea we can see quite generally that $f(t) = f(t+kP)$ for any integer k, including negative integers. Thus if f is periodic with period P it is also periodic with period kP for every integer k. Fig.3.5 is an example.

Saying that P is a period of f, i.e. that $f(t+P) = f(t)$ for all t, does not mean necessarily that there is no shorter period. The simple pure wave $t \mapsto \mathbf{e}(2\pi t)$ has period 1, but so does the simple pure wave $t \mapsto \mathbf{e}(6\pi t)$ which goes around the circle three times as fast. It is true that its shortest period is 1/3—we call that the *exact period*—but it also has period 1.

With this we re-introduce what we call the *pure waves*, which we also designate by w_k, one for each $k \in \mathbb{Z}$ defined by

$$w_k : w_k(t) = \mathbf{e}(2\pi kt), \quad t \in \mathbb{R}. \tag{11.11}$$

All of these are periodic of period 1 for the simple reason that if k is an integer then

$$k(t+1) - kt = k \in \mathbb{Z} \quad \text{and so} \quad \mathbf{e}(2\pi k(t+1)) = \mathbf{e}(2\pi kt).$$

This family includes the "constant" wave w_0, for which $w_0(t) = \mathbf{e}(2\pi 0t) = \mathbf{e}(0) = 1$ which is unchanging with t. It is periodic with period 1 (in fact any period you choose) by default. It may seem rather pointless function to call it a wave, but we will see that it plays an important role in what follows.

The question we wish to raise is what can we say about the entire collection of functions, or observables, that have period 1. The prime candidates are the pure waves w_k, $k \in \mathbb{Z}$. More complicated waves or observables with period 1 can be formed by taking *linear combinations* of these pure waves, that is to say finite scaled sums of pure waves. Fig.11.14 offers an example. In this case we are looking at the wave given by

$$\begin{aligned}(1/2)\mathbf{e}(2\pi t) &- (1/3)\mathbf{e}(4\pi t + \pi/4) + (1/6)\mathbf{e}(-6\pi t) \\ &= (1/2)\mathbf{e}(2\pi t) - ((1/3)\mathbf{e}(\pi/4))\mathbf{e}(4\pi t) + (1/6)\mathbf{e}(-6\pi t),\end{aligned} \tag{11.12}$$

which combines three pure waves. In this example, the second term has a phase shift, which we know can be brought out into the coefficient. In the second line we have done this, so the coefficient becomes the complex number $-(1/3)\mathbf{e}(\pi/4) = -(1+\mathrm{i})/(3\sqrt{2})$. The point is that we can normalize things so that there are no phase shifts in the arguments of the pure waves by absorbing them into the coefficients.

Quite generally we can take any *finite* linear combination of pure waves.

$$\begin{aligned}f(2\pi t) &= \cdots + c_{-2}\,\mathbf{e}(2\pi(-2)t) + c_{-1}\,\mathbf{e}(2\pi(-1)t) + c_0\,\mathbf{e}(2\pi(0)t) \\ &\qquad + c_1\,\mathbf{e}(2\pi(1)t) + c_2\,\mathbf{e}(2\pi(2)t) + c_3\,\mathbf{e}(2\pi(3)t) + \cdots \\ &= \sum_{k\in\mathbb{Z}} c_k\,\mathbf{e}(2\pi kt).\end{aligned} \tag{11.13}$$

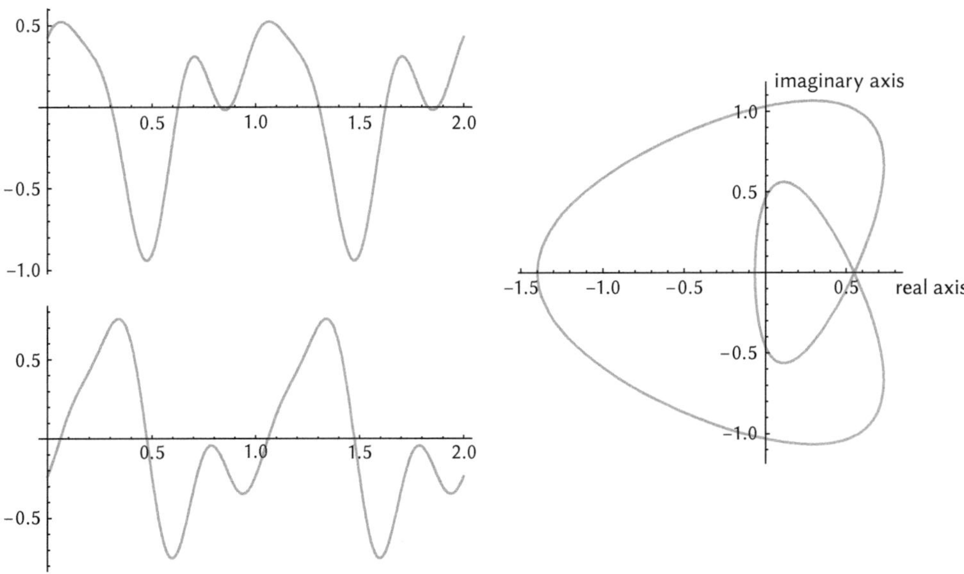

Figure 11.14: The graphs on the left are the real (top) and the imaginary (bottom) parts of the complex periodic wave form of equation (11.12), shown over two periods. To the right is a plot of the path that this wave form plots out in the complex plane as t runs from 0 to 1—one period that closes on itself. The problem of spectral analysis is to start from the latter and determine the three component pure waveforms along with their coefficients.

For each value of the real-valued variable t (say time), $\mathbf{e}(2\pi kt)$ is a complex number, c_k is a complex number, their product is a complex number, and so the entire sum, as long as we only add a finite number of these, is a complex number. So indeed this general linear combination of pure waves is a complex-valued function. Furthermore it is still periodic with period 1.

The last line of (11.13) is a more elegant way of writing the first, using the standard mathematical summation notation. Since in mathematics it is important to be careful with definitions, we should note that to be a "linear combination" the sum has to be a *finite* sum, so in the second equation we should say that all but a finite number of the c_k are zero. If we allow infinite sums we need to worry about whether the sums *converge*: do the sums tend to some finite limit as we extend them indefinitely? Actually, to get to the bottom of the question of finding all functions of period 1 then we do need to allow convergent infinite sums because they do produce functions of period 1. For the purposes of this book, and certainly for all practical purposes, we can ignore the technicalities of convergence though, to be precise, the condition for convergence is that $\sum_{k\in\mathbb{Z}} |c_k|^2 < \infty$, see (11.37) and §19.4.1.

Functions defined by (11.13) are complex-valued. The condition for a function defined by (11.13) to be real-valued it is simply that

$$c_{-k} = \overline{c_k} \quad \text{for all } k\,.$$

We will return to this later.

11.6.2 Fourier's wonderful idea

At this point we can return to questions raised at the end of §11.4.5.

(i) Is every periodic function with period 1 expressible in the form (11.5), and if so,

(ii) given the function f how can we find the so-called *Fourier coefficients* c_k?

When sunlight, which is spread over the entire visible spectrum, interacts with a physical object, the light that is scattered from it is a complicated mixture of pure waves. Whether or not we are dealing with a periodic wave, we face the same task: how can we determine which simple pure waves are involved, and with what intensities and phase shifts they appear? This process is appropriately called *spectral analysis*, and the mathematics involved is called *Fourier analysis*. It dates back to the seminal ideas of Jean-Baptiste Joseph Fourier (1768-1830), who was the originator of this field of mathematics.

This is what is going to occupy us over the rest of this chapter.

> Normally when we talk about functions we just write the function name down, e.g. f, whereas $f(t)$ literally means the value of f at t. But in what follows we need to be aware of how the argument of the function is being used so we will slightly misuse the mathematical language and talk of the function $f(t)$. Much of this has to do with the pure waves w_k which we most often write in the form $\mathbf{e}(2\pi kt)$. In reading these, read them as functions which are effectively being described in terms of what they do to the variable t.

We start with a complex-valued function $f(2\pi t)$ of period 1. We assume that we know the values that takes, what it looks like, but know nothing about its spectral components. Since $f(2\pi t)$ is of period 1, we will start with the assumption that it actually does have an expansion of the form (11.13), but we don't know what any of the coefficients c_k are. The question is how we can go about finding them.

It begins with the idea of *averages*. Given a periodic function g defined on $\mathbb{R}$ we can ask about its average value over one period. We denote this average by $\mathrm{av}(g)$. It really doesn't matter where we start on the circle since in averaging over one period then we have taken into account all the variation that the function has. We will see it appear in forms like

$$\mathrm{av}(\mathbf{e}(2\pi kt))\,, \tag{11.14}$$

where we should understand that the average is being taken over the values of this pure wave as the variable t runs from 0 to 1.

There are two straightforward things to know about averages.

- If g is a periodic function on $\mathbb{R}$, c is a number, and we scale g by c then the result is still periodic and its average over one period is also scaled by c. If we double all the values of g, then the average will double. This works no matter what number c is, including complex numbers. It reads $\text{av}(c\,g) = c\,\text{av}(g)$.
- If g and h are two periodic functions on $\mathbb{R}$ then the average of their sum is the sum of their averages. If the average number of males born each year in America is added to the average number of females born in America each year, then their sum is the average number of humans born each year in America. This reads as $\text{av}(g+h) = \text{av}(g) + \text{av}(h)$.

The technical term for the combination of these two properties of scaling and adding is *linearity*. Averaging is a *linear operation*. There is more about averaging in §11.10.3.

Now, although we do not know whether it is possible, let us *assume* that f can be put into the form that we see in equation (11.13). We start to glean information about it by asking about the average value of f. According to what we have just said about averages, averages are linear and so averaging comes down to averaging the pure waves and then putting all those averages together: the average of the sum is the sum of the averages of the components. Thus

$$\text{av}\left(f(2\pi t)\right) = \sum_{k\in\mathbb{Z}} c_k\,\text{av}\left(\mathbf{e}(2\pi k t)\right). \tag{11.15}$$

In each case we are thinking of the average as t runs from 0 to 1.

What first comes as a surprise, but which is rather obvious when we think about it, is that

$$\text{av}(\mathbf{e}(2\pi t)) = 0\,.$$

There are several ways to see this. The first is to look at Fig.11.6 which shows the two components, the real and imaginary components, of pure waves. It is pretty clear that they are negative just as much as they are positive. As t goes through one period, i.e. increases by 1, $\mathbf{e}(2\pi t)$ goes around the circle exactly once. Fig.11.15 shows quite specifically how opposite values always cancel each other out.

Alternatively we can appeal to symmetry. The simple pure wave $\mathbf{e}(2\pi t)$ moves exactly once around the circle $\mathbb{U}$ as t runs from 0 to 1. This is in all respects completely symmetrical and so the average, which is the average value of the complex numbers $\mathbf{e}(2\pi t)$ that appear as we go around the circle, must also be symmetrical. There is only one complex number that is symmetrical with respect to the center of the circle, and this is 0, the complex number corresponding to the center itself.

This also suggests that for each of the pure waves $\mathbf{e}(2\pi k t)$, no matter what integer k we take, the average will be zero. Going around the circle k times is still going to lead to a symmetrical outcome. This is true, but there is one exception. The exception is when

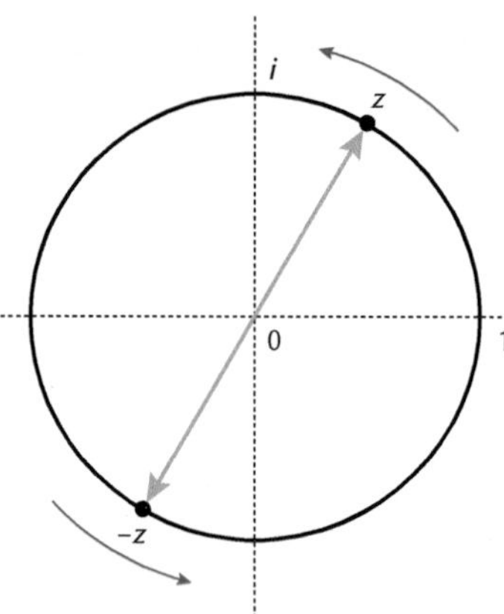

Figure 11.15: The figure shows why the average value of $\mathbf{e}(2\pi t) \in \mathbb{U}$ is zero. As we trace around the circle, $\mathbf{e}(2\pi t)$ and $\mathbf{e}(2\pi t + \pi) = -\mathbf{e}(2\pi t)$ are always diametrically opposite each other. As t goes from 0 to $1/2$, $\mathbf{e}(2\pi t)$ goes around the circle halfway and at the same time so does $\mathbf{e}(2\pi t + \pi)$: and they always add to zero. This same argument applies when the tracing is done twice as fast, three times as fast, etc., and whether or not it is clockwise or counter-clockwise.

$k = 0$, for then the pure wave is $\mathbf{e}(0)$, which does not change with t and has the constant value 1. The average value of the constant value 1 over one period is clearly equal to 1. It gets its symmetry in a different way, by simply always being 1 as we go around the circle.

So all the averages in equation (11.15), except for the one at $k = 0$, disappear and we find that the average value of our function is c_0:

$$\operatorname{av}(f(2\pi t)) = c_0 \,.$$

This might look like a humble achievement. The constant term looks like the least interesting term of all—what about all the other coefficients? But there is another way to look at it: the averaging process extracts exactly one coefficient from amongst all of them—the one in the zero position. If we could somehow slide other coefficients to the zero position, then we could extract their values too. In fact there is a very simple way to do this. Look at the case of c_1. If we multiply the function $f(2\pi t)$ of equation (11.13) by $\mathbf{e}(-2\pi t)$, the effect is to shift everything to the left:

$$\mathbf{e}(2\pi kt) \quad \text{changes to} \quad \mathbf{e}(-2\pi t)\,\mathbf{e}(2\pi kt) = \mathbf{e}(2\pi(k-1)t)$$

for every k. This does not change the coefficients, so what we see is that

$$\begin{aligned}\mathbf{e}(-2\pi t)\, f(2\pi t) = {}& \cdots + c_{-2}\,\mathbf{e}(2\pi(-3)t) + c_{-1}\,\mathbf{e}(2\pi(-2)t) + c_0\,\mathbf{e}(2\pi(-1)t)\\ &+ c_1\,\mathbf{e}(0) + c_2\,\mathbf{e}(2\pi(1)t) + c_3\,\mathbf{e}(2\pi(2)t) + \cdots \,.\end{aligned}$$

This is another periodic function and since its constant term is $c_1\mathbf{e}(0)$ we must have

$$c_1 = \mathrm{av}(\mathbf{e}(-2\pi t)\, f(2\pi t))\,.$$

Thus the method of averages allows us to compute c_1 and indeed all the c_k:

$$c_k = \mathrm{av}(\mathbf{e}(-2\pi kt)\, f(2\pi t)) = \mathrm{av}\left(\overline{\mathbf{e}(2\pi kt)}\, f(2\pi t)\right)\,.$$

All the coefficients can be computed by the use of averages. In the last part of this equation we have written $\mathbf{e}(-2\pi kt)$ in its totally equivalent form $\overline{\mathbf{e}(2\pi kt)}$ (see point (iv) of §11.5.2), which is seen more frequently in this context.

There are five points to make here.

(i) Unless one has some techniques to compute averages in general, all this is not going to help find the values of the c_k. Such techniques exist. The averaging process, for which we have relied on our intuition, is formally based on integration, see §11.10 later in this chapter, and this does lead to computable techniques. But this is not what is really important here.

(ii) Rather, what is revealed is conceptual: the Fourier coefficients form an alternative view of f in terms of a *discrete* set of (complex) numbers, indexed by integers—something very different in character from the continuous nature of f.

(iii) There is a *local-global* phenomenon here. The Fourier coefficients of f are properties that involve *all* of f—they arise from the average values of f and $\overline{\mathbf{e}(2\pi kt)}f$, over the entire circle, not just from specific values of f. This becomes the next important motif in this section: how the *global* (that which depends on the entirety of the domain of f) manifests itself as something *local*, a single complex number assigned to the integer k.

(iv) All of what we have said here is based on the assumption that f really can be expressed in the form (11.13). The argument completes by showing that effectively this is the case: functions with period 1 can be expressed in the form (11.13) using the values of c_k that come out of the averaging process. The fine print and the subtleties of this are discussed in §19.4.1.

(v) Though from practical considerations, only a finite number of Fourier coefficients are relevant, the theory requires all frequencies to be represented. That is why in the symbolism of the summation notation we see $k \in \mathbb{Z}$, which signifies that the values of k will run over all integers in general, and why, like it or not, questions of convergence arise.

11.6.3 Fourier duality

The main theorem of Fourier analysis says that functions on $\mathbb{U}$ can be written in the form (11.13), and this expression is unique. (As with most results in mathematics, there is some "fine print". The class of all functions for which Fourier analysis applies is huge, but not universal. The exact conditions are found in §11.10.4, after we have developed the concepts required to express them properly.) This leaves us with the interesting conclusion that the function f is equally described by its set of coefficients c_k. So whereas f is initially described by its point-by-point values, it can also be described by quantities that only depend on averages. Just as sound can be described by what we hear, it can also be described by the frequencies that are embedded in it, which are what our ears actually sense.[12] The first is of the instant, so to speak, whereas frequencies can only be inferred through time. That is why high-frequency production of music in digital form requires very rapid sampling of the digital signal.

So along with f there is a second function $\widehat{f}$ (it is called the *Fourier transform* of f) that is defined on the integers and gives the *Fourier coefficients*:

$$\begin{aligned} \widehat{f} &: \mathbb{Z} \longrightarrow \mathbb{C} \\ \widehat{f}(k) &= c_k = \operatorname{av}\left(\overline{\mathbf{e}(2\pi kt)}\, f(2\pi t)\right). \end{aligned} \tag{11.16}$$

What is important to notice is how the roles of t and k have been reversed. In equation (11.13) we are dealing with a function of the variable t which is expressed as a sum over simple wave functions expressed in terms of integers k. Now, in equation (11.16), we are dealing with a function of the variable k which is expressed as an average over simple wave functions in terms of the variable t. At the heart of it are the pure waves $\mathbf{e}(2\pi kt)$ which can be treated either as functions of t or functions of k.

This leads to an entirely new way of looking at the function f. It started as a function on $\mathbb{U}$ but now it has led to a function on $\mathbb{Z}$ that looks totally different but somehow encodes the same information. While f is described in terms of what it does to points on the circle $\mathbb{U}$, $\widehat{f}$ is described in terms of what it does to points in $\mathbb{Z}$. What's more, the values of this new function are the coefficients c_k and they can be used to recover f in its form given by (11.13):

$$f(t) = \sum_{k\in\mathbb{Z}} \mathbf{e}(2\pi kt)\widehat{f}(k) = \sum_{k\in\mathbb{Z}} \mathbf{e}(2\pi kt)c_k\,. \tag{11.17}$$

This is referred to as the *Fourier series expansion* of f.[13] Beneath this Fourier duality there lie the two *groups*, $\mathbb{U}$ and $\mathbb{Z}$. The first is the circle group in $\mathbb{C}$, where the group structure comes from complex multiplication. The second is the integers, where the group structure is from addition. The relation of these two groups is an example of a general concept called *Pontryagin duality*, that can be summed up in the simple equation that links them:

$$\mathbf{e}(2\pi kt)\,\mathbf{e}(2\pi lt) = \mathbf{e}(2\pi(k+l)t)\,.$$

On the left is group multiplication of pure waves using the *multiplicative* structure of $\mathbb{U}$. On the right we see addition of frequencies using the *additive structure* of $\mathbb{Z}$.

Figure 11.16: A simple tiling which can be interpreted as the doubly-periodic repetition of a square in horizontal and vertical directions. Anselm Levskaya, Computed on Eschesketch https://eschersket.ch/\protect\protect\leavevmode@ifvmode\kern+.1667em\relax.

What we have just seen is one of the great insights of Fourier analysis. We have a periodic phenomenon, but we can see it in two very different ways. In the one it appears as a periodic function on the real numbers (or more relevantly, as a function on the unit circle $\mathbb{U}$), which is a very continuous looking picture, and in the other as a function on the integers, which looks very discrete. The two are linked by the Fourier transform of (11.16) and the inverse Fourier transform (11.17).[14] The difference is that in the first the frequency k is held fixed while the time varies and in the second it is the other way around.

In looking at this formal statement of Fourier duality it is good to recall Fig.11.10 that visually illustrates what the duality is about.

11.7 Tiles and crystals

Our discussion so far has been framed in the context of temporal repetition, effectively one-dimensional periodicity in time. Time doesn't have a natural interpretation beyond one dimension but there are plenty of examples of multiple periodicity if we look instead at spatial repetition. Wall tilings and crystals abound and they all have spatial repetition. For instance, returning to Fig.9.2 we have seen that the entire tiling, in all its amazing complexity, is composed through repetition of a single square section of the tiling— endlessly left and right, up and down, in equal sized steps. It is *doubly-periodic*. Likewise in 3D space periodicity in three distinct directions is possible, and in fact crystals are ubiquitous in the material world.

This suggests that our theory of waves extends to two- and three-dimensional theories and should give us new ways to think about the patterns of periodicity. Instead of a one-dimensional wave, we can think of planar waves, like the surface of the ocean, or

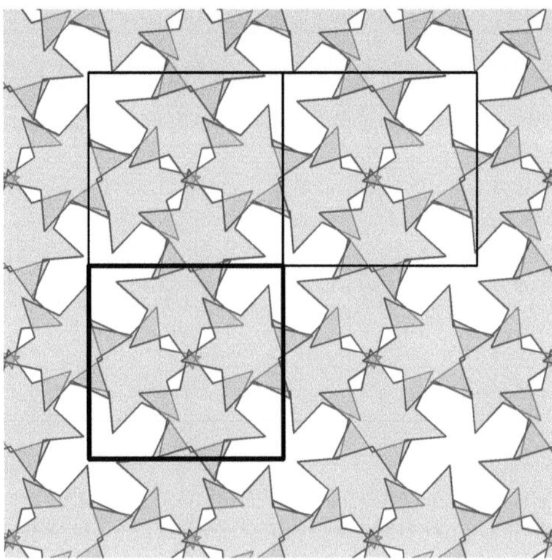

Figure 11.17: The pattern is simply repetition of the pattern in a square. Here we show three copies of this square. Looking at the bottom left square, notice that the patterning along its opposite edges is identical. We can imagine bending this square into a cylinder, matching the top and bottom edges. Continuing we can imagine bending the cylinder, matching the left and right edges of the cylinder. The resulting donut shape is a *torus*.

3D-waves like sound waves or light waves emanating in all directions from some source. Temporal dynamics is replaced by the "dynamics" of spatial translation. This extension to higher dimensions works (even beyond three dimensions). It's important. What's more, it's surprisingly easy to derive from what we have already seen in 1D. At the same time it clarifies what have so far been the rather confusing overlapping roles of the unit circle.

To make the connection we can revisit the 1D temporal situation. There a periodic temporal pattern is reimagined in terms of motion around a circle. But any 1D periodic pattern, temporal or spatial, can be thought of as a pattern laid on a circle. If instead we have a 2D pattern that repeats in one direction then we can express that repetition by wrapping the pattern around a cylinder. We saw this in §3.3.1 where the spatial repetition seen in Fig.3.7 is re-expressed on a cylinder in Fig.3.8. We can think of the cylinder as being like a paint roller with the pattern embossed on it so that as it rolls up and down on a wall it produces the full 2D repetitive pattern.

This gives a geometrical representation of 2D periodicity in one direction, where the repetition in that one direction has been represented as circular repetition. The *doubly-periodic* nature of a two-dimensional tiling can be associated with two circles. This can be readily be appreciated by looking at Fig.11.17, which is based on repetition of a single square of pattern (a fundamental region), say the lower left square in the figure. Looking at a pair of *opposite* edges of this square we see that their patterning is actually identical. This applies to both the left-right pair and the top-bottom pair.

We could bend the square into a cylinder by joining together the top and bottom sides, as we have just seen. However, the left and right ends of the roller are identical, so we can imagine bending the cylinder so as to join the left and right. And voilà—a donut! See Fig.11.18. We may not be able to use it as some sort of roller, but it has the important property that all of the periodicity has been wrapped up to form the surface of a donut. In this way all of the essential patterning has been divided into two parts: the donut encapsulates the pattern of double periodicity while the pattern that appears on its surface is the pattern that actually repeats under this double periodicity. If we were to first identify opposite edges and then opposite ends of the resulting cylinder we would get a different torus, but from our point of view it is immaterial which way it is done.

The fancier mathematical name for the donut is a *torus*, though unlike a donut, which ought to have some solidity to be enjoyable, the torus is just the surface. It does not include the interior of the donut. It is more like an inner tube or an inflatable swimming tube, on which the pattern has been painted. The torus is the two-dimensional analogue of the circle, and similar to the circle, which is one dimensional but needs two dimensions to be drawn without overlaps, the torus is two-dimensional but needs three dimensions to be drawn without overlaps. We will denote the torus by $\mathbb{T}$.[15] Fortunately the mathematics that we developed around the circle extends remarkably easily to the torus.

11.7.1 Doubly-periodic functions

Just as we can describe points on the circle $\mathbb{U}$ with a real parameter t and the notation like $\mathbf{e}(t)$ of Fig.11.6, we can parameterize the two circles involved in double-periodicity with two copies of $\mathbb{U}$, see Fig.11.18. The torus $\mathbb{T}$ is the product space $\mathbb{U} \times \mathbb{U}$, that is it is the set of ordered pairs $(\mathbf{e}(x_1), \mathbf{e}(x_2))$, with the numbers x_1, x_2 indicating how far along each of the two circles we have gone (see §8.4.3 for the notation). We have switched letter names here from the familiar t, suggesting time, to x, since we want to suggest the spatial positions of the points on the two circles.

Following what we did in the one-dimensional periodicity, we prefer to include factors of 2π in order to make the two periods both equal to 1, see §11.4. Both $\mathbf{e}(2\pi x_1)$ and $\mathbf{e}(2\pi x_2)$ are periodic in their arguments x_1 and x_2, each with period equal to 1. This means that via

$$(x_1, x_2) \mapsto (\mathbf{e}(2\pi x_1), \mathbf{e}(2\pi x_2))$$

we can treat functions that are doubly-periodic with period 1 in the two different directions as functions *defined on a torus.* Just as we can see the circle $\mathbb{U}$ as being the natural domain for studying periodic functions, so we can see $\mathbb{T} = \mathbb{U} \times \mathbb{U}$ as the natural domain for studying doubly-periodic functions—functions that are periodic in two separate variables.

Also just as we saw $\mathbb{U}$ as the state space for a dynamical action of $\mathbb{R}$ in (11.3), we can see $\mathbb{T}$ as the state space for the dynamical action of $\mathbb{R}^2$ defined by

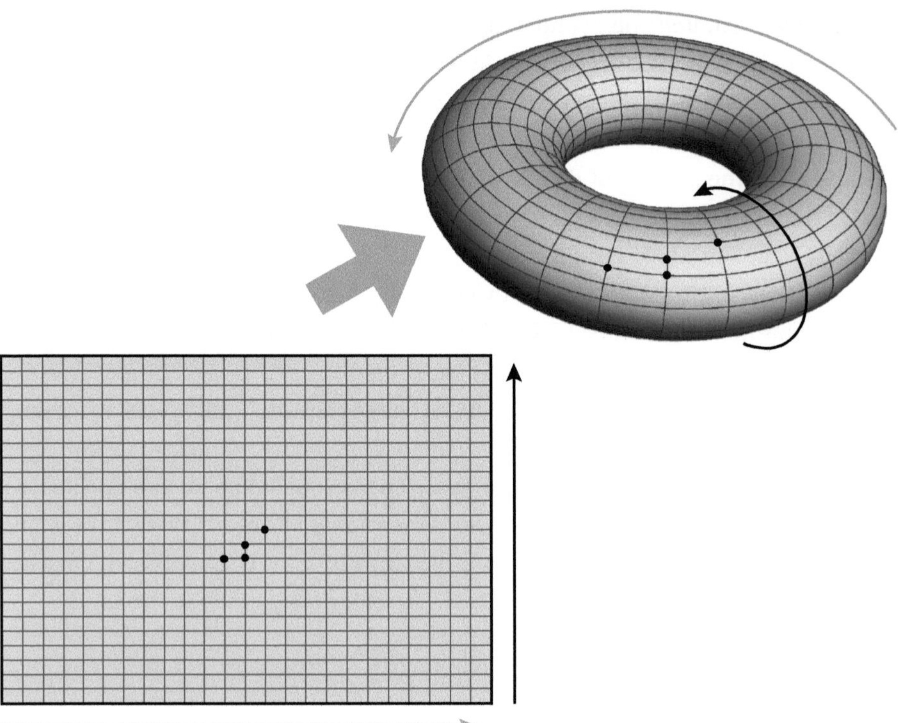

Figure 11.18: Zipping together the top-bottom and left-right edges of the rectangle produces a torus, $\mathbb{T}$. The torus may be thought of as the "product" of two circles, indicated by the black and grey arrow-headed arcs. The figure shows how the grid and four marked points appear on the torus.

$$\begin{aligned} &\mathbb{R}^2 \triangleright \mathbb{T} \longrightarrow \mathbb{T} \\ &(u_1, u_2) \triangleright (\mathbf{e}(2\pi x_1), \mathbf{e}(2\pi x_2)) = (\mathbf{e}(2\pi(-u_1 + x_1)), \mathbf{e}(2\pi(-u_2 + x_2))) . \end{aligned} \tag{11.18}$$

Everything we did in § 11.6.3 works in higher dimensions and in the end looks remarkably similar to what we have already seen, though we really need to think in spatial rather than temporal terms. Our new interest is in (complex-valued) functions of two real variables, that is, functions defined on the *coordinate plane*,

$$f : \mathbb{R} \times \mathbb{R} \longrightarrow \mathbb{C},$$

that are periodic in two different directions. We will refer to them as *doubly-periodic* functions. Here we keep things simple by supposing that these two directions are at right-angles, in the directions of the x and y axes. In other words we will assume that f is periodic in the first coordinate and also in the second coordinate. We will also assume that these two periods are equal to each other, and in fact both equal to 1. These assumptions make things easier, but do not distort the ideas that are applicable to doubly-periodic functions.

As we have noted, doubly-periodic functions can be reimagined as functions defined on the torus, see Fig.11.19. A doubly-periodic function has two aspects: its global part due to the overall symmetry of its periodicities, and its local part which is the design—its detailed appearance on one (or any) fundamental region. Seeing it as a function on a torus absorbs the repetitive part by redisplaying the function on a new space (the torus) that has a structure that implicitly enfolds that repetitiveness. We will continually shift between these two equivalent ways of looking at doubly-periodic functions.

11.7.2 Pure waves in higher dimensions

We can now turn our attention to the simplest functions we can think of in terms of the torus. For simple periodicity on the circle these were the pure wave functions $t \mapsto \mathbf{e}(2\pi kt)$, for integers k. The simplest doubly-periodic functions we can define are the new "pure waves"

$$(x_1, x_2) \mapsto \mathbf{e}(2\pi k_1 x_1)\,\mathbf{e}(2\pi k_2 x_2) = \mathbf{e}(2\pi k_1 x_1 + 2\pi k_2 x_2)\,, \tag{11.19}$$

where k_1, k_2 are *integers*, which arise simply as the product of the two component 1D pure waves. This new "pure wave" is still a complex-valued function: its values are the product of two complex numbers on the unit circle $\mathbb{U}$. Further, the multiplication rule for elements of $\mathbb{U}$ allows us to re-express the product in the important second form given in the equation and shows that the product is still in $\mathbb{U}$. It also clarifies the oddity that pictures of a torus seem to imply different roles to the two circles, each one going around the hole in a very different way. The form of the 2D pure waves shows that the two components are really treated identically.

This new simple wave is a complex-valued function that arises as the "product" (in the sense that the torus is the product of two circles) of two 1D simple waves. Since k_1 and k_2 are integers, it is clearly doubly-periodic with period 1 (due to the 2π factors) in each of its two coordinates. These new pure waves are of the form

$$\mathbb{R} \times \mathbb{R} \longrightarrow \text{torus} \longrightarrow \text{complex numbers}\,,$$

and they offer a decoupling of meaning that was not apparent before. In (11.5) the pure waves are essentially defined on the circle $\mathbb{U}$ and take values $\mathbf{e}(kt)$ on the same circle, though we interpret the latter as being complex numbers, so the circle plays two different roles. Now we have pure waves defined on the torus $\mathbb{T}$, but they still have values in $\mathbb{U}$ which are complex numbers. These new doubly-periodic waves with a frequency of k_1 around one circle and k_2 around the other will be seen to be the waves that stand as the basis of doubly-periodic functions. Figure 11.20 illustrates what these pure waves look

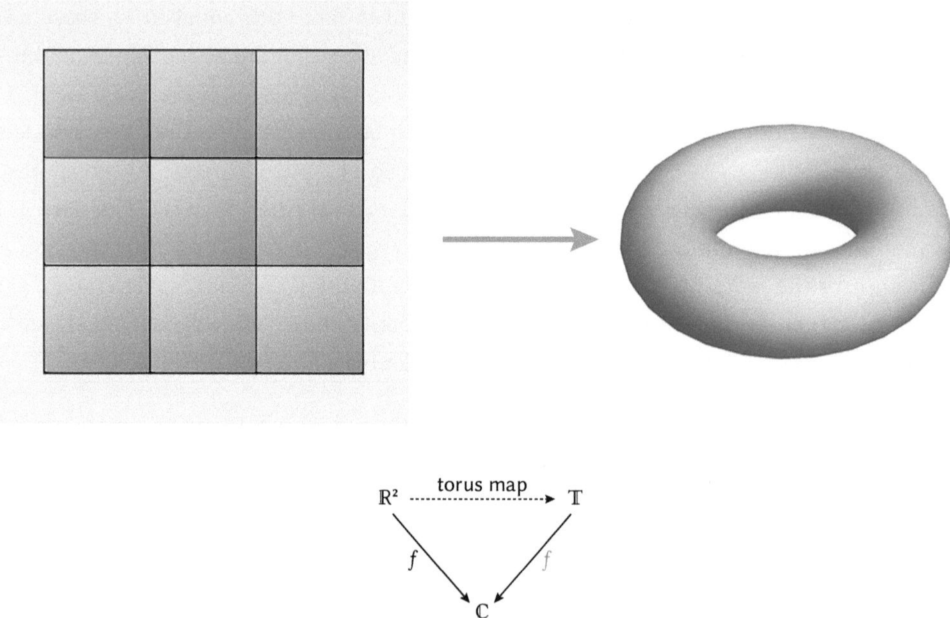

Figure 11.19: The figure visually suggests why doubly-periodic functions can be considered as functions defined on a torus, and vice-versa. On the left there is a representation of a doubly-periodic function f. Think of it, for instance, as recording the varying grey density (ignoring the black lines that are drawn simply to emphasize the block that is repeating). This varying density is completely determined by any one of the fundamental regions shown. Wrapping opposite sides of a fundamental region turns that region into a torus and allows us to think of wrapping the plane around the torus. Since each fundamental region wraps in exactly same way we can think of this wrapping as a mapping, indicated here as the "torus map" of the plane $\mathbb{R}^2$ around the torus $\mathbb{T}$. With this we can reimagine f as a function based on the torus rather than the plane. In the reverse direction, any function from the torus can be viewed as a doubly-periodic function on the plane simply by preceding it by the torus mapping.

like—very much like what we would expect a plane wave to be! But we need to explain the notation a bit.

Writing down coordinates like (k_1, k_2) in the two-dimensional setting, and later on (k_1, k_2, k_3) as we shall do in the three dimensions, is messy and not especially enlightening once we know what they mean. So mathematical notation has evolved to simply express such vectors by using a bold-face version of the same letter, in this case $\mathbf{k}$. Thus $\mathbf{k}$ will stand for some general vector of the form (k_1, k_2) or (k_1, k_2, k_3), according to the context. It could stand for (k_1, k_2, k_3, k_4), or for even higher-dimensional entities if that were our setting. A similar convention applies to (x_1, x_2), or to (x_1, x_2, x_3) in three-dimensional space: we simply use bold-face $\mathbf{x}$, with the context telling us the dimension. Another common notation is to write $\vec{x}$ instead of $\mathbf{x}$.

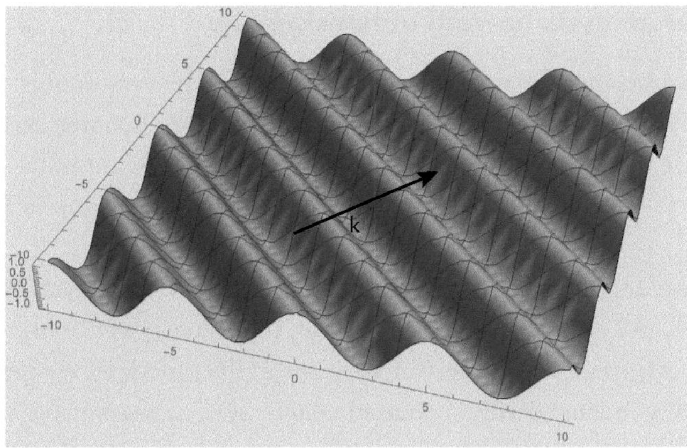

Figure 11.20: The real part of the plane wave $\mathbf{e}(2\pi\mathbf{k}.\mathbf{x})$ drawn as a function of $\mathbf{x} = (x_1, x_2)$. The wave "moves" in the direction of the "wave vector" $\mathbf{k}$ in the sense that the wave crests and troughs are at right-angles to $\mathbf{k}$ and are traversed in that direction for increasing values of x_1 and x_2. The imaginary part is similar.

Taking up this new notation, the next convention is to write $\mathbf{k}.\mathbf{x}$ for $k_1x_1 + k_2x_2$. This also applies in the 3D case where $\mathbf{k}.\mathbf{x}$ stands for $k_1x_1+k_2x_2+k_3x_3$, and similarly for higher dimensions. This is completely standard notation and it is called the *dot product* because of the dot used in the notation.[16]

There is more about it in §12.3.2 and Endnote 5. Since the dot notation is dimension-free in appearance, we will often use it in 1D too, where $k.x$ means kx. In this notation the pure wave (11.19) becomes the far simpler looking

$$\mathbf{x} \mapsto \mathbf{e}(2\pi\mathbf{k}.\mathbf{x}). \tag{11.20}$$

In this context $\mathbf{k}$ is called a *wave vector*, see Fig.11.20, as it completely specifies the wave. Following our earlier notation we shall also denote this pure wave by $w_{\mathbf{k}}$.

With this notation, we can express what we may guess to be the most general doubly-periodic function or observable (periodic with period 1 in two independent directions at right-angles to each other):

$$f(\mathbf{x}) = \sum_{\mathbf{k}\in\mathbb{Z}^2} c_{\mathbf{k}}\, \mathbf{e}(2\pi\mathbf{k}.\mathbf{x}), \tag{11.21}$$

where the complex-valued coefficients $c_{\mathbf{k}}$ are now indexed by wave vectors $\mathbf{k}$.

Comparing this with (11.13), which is the 1D version, we can see how the notation has been extended so that we recognize the very same form as we had before. The sum now runs over all integer pairs $\mathbf{k} = (k_1, k_2)$. Even the convergence condition looks the same, namely that $\sum_{\mathbf{k}\in\mathbb{Z}^2} |c_{\mathbf{k}}|^2$ is finite.

11.7.3 Fourier analysis beyond dimension one

We might guess from what we have seen in the 1D case that every doubly-periodic function is necessarily of this form. This is true under the same hypotheses as before, see §11.10.4, and once again, given such a function, the way forward is to determine the Fourier coefficients $c_{\mathbf{k}}$, and then see that they really do reproduce our function f through equation (11.21).

Fortunately we don't need any new ideas to do this: it works exactly the same way by using averages. We can get the constant term, the value $c_{\mathbf{0}}$ as the *average value* of f just as before. This time we mean the average value of the function over a range of values of x_1 and x_2 as they independently increase in value by 1, or equivalently, thinking of f as being defined on a torus, the average of f over the torus. With this observation we can use the same trick as before, obtaining the coefficient $c_{\mathbf{k}}$ by shifting the terms of f around so that the $c_{\mathbf{k}}$ term lies at the constant position.

$$c_{\mathbf{k}} = \mathrm{av}(\mathbf{e}(-2\pi\mathbf{k}.\mathbf{x})\, f(\mathbf{x})) = \mathrm{av}(\overline{\mathbf{e}(2\pi\mathbf{k}.\mathbf{x})}\, f(\mathbf{x}))\,.$$

Remember that the averages here are understood to be taken over the set of values that $\overline{\mathbf{e}(2\pi\mathbf{k}.\mathbf{x})}\, f(\mathbf{x})$ takes as $\mathbf{x}$ traverses the entire torus, see (11.14) and the cautionary note of §11.6.2. Here, as before, we are counting on intuition to fill in what averages are. In §11.9 we define the concept more rigorously.

This process works. We can indeed express any doubly-periodic function as a Fourier series of the form appearing in equation (11.21). Fourier analysis works in any dimension, even ones that we cannot picture, just by increasing the number of coordinates.

11.8 Eigenspaces and eigenfunctions

A primary interest in pattern is the search for invariants, phenomena that arise out of the dynamics that don't change. These are often considered to comprise the real essence of the pattern. Linearization, enlarging our view from a basic state space to spaces of functions defined on that state space, offers new forms of invariance, but not perhaps ones that come immediately to mind. Functions that do not change under the dynamics are functions that are constant on trajectories. These can be important—the conservation of energy appears in this form—but they can tell us little about aspects of change that are entailed by the dynamics. A more general feature is to consider *subspaces* of functions or observables that are invariant, in the sense that it is *as subspaces* that they do not change with the dynamics even though the functions within them may. This is an idea that we have seen before: a *set* may be invariant under transformation or change of some form, even though the individual elements are not. A circle, as a set of points, is invariant under rotation, but the points all move. Here we deal with subspaces.

Going back to our study of the doubly-periodic functions we know from equation (11.21) that they are expressible in terms of the pure waves

$$w_{\mathbf{k}} : \mathbf{x} \mapsto \mathbf{e}(2\pi\mathbf{k}.\mathbf{x}), \tag{11.22}$$

defined by the wave vectors $\mathbf{k} = (k_1, k_2) \in \mathbb{Z}^2$.

Although we worked with pure waves because they are the most obvious 2D waves we can think of, it seems amazing that they turn out somehow to be "so right". In fact there is another, and deeper, reason why they are so important. They are invariants, or better said, each one defines a one-dimensional space $\mathbb{C}w_{\mathbf{k}}$ which is invariant under the translation action of $\mathbb{R}$ on $\mathbb{T}$.

Consider the way in which we treat $\mathbb{T}$ as the state space with the dynamics given by the natural action of $\mathbb{R}^2$ on $\mathbb{T}$ as seen in (11.18). We have

$$\begin{aligned}(\mathbf{u} \triangleright w_{\mathbf{k}})(\mathbf{x}) &= w_{\mathbf{k}}(-\mathbf{u}+\mathbf{x}) = \mathbf{e}(2\pi\mathbf{k}.(-\mathbf{u}+\mathbf{x})) \\ &= \mathbf{e}(2\pi\mathbf{k}.(-\mathbf{u}))\mathbf{e}(2\pi\mathbf{k}.\mathbf{x}) = \mathbf{e}(-2\pi\mathbf{k}.\mathbf{u})w_{\mathbf{k}}(\mathbf{x}).\end{aligned} \tag{11.23}$$

In short

$$\mathbf{u} \triangleright w_{\mathbf{k}} = \mathbf{e}(-2\pi\mathbf{k}.\mathbf{u})w_{\mathbf{k}} = \overline{\mathbf{e}(2\pi\mathbf{k}.\mathbf{u})}w_{\mathbf{k}} \in \mathbb{C}w_{\mathbf{k}}.$$

In words, the dynamical action of each $\mathbf{u} \in \mathbb{R}$ on $w_{\mathbf{k}}$ is to take it to a scalar multiple $a(\mathbf{u})w_{\mathbf{k}}$ of itself. Here this scalar is the complex number $\mathbf{e}(-2\pi\mathbf{u}.\mathbf{k}) \in \mathbb{C}$. Thus every translation $\mathbf{u} \in \mathbb{R}^2$ takes the one-dimensional space $\mathbb{C}w_{\mathbf{k}}$ into itself: the subspace $\mathbb{C}w_{\mathbf{k}}$ itself is invariant under the action of $\mathbb{R}^2$, i.e. under the dynamics of the system.

A function f that has the property that the action of the dynamics simply alters it by scaling it by a complex number, or equivalently the dynamics leaves $\mathbb{C}f$ invariant, *is an invariant* of the system. These invariants are called *eigenfunctions*. So each $w_{\mathbf{k}}$ is an eigenfunction and the corresponding invariant subspace $\mathbb{C}w_{\mathbf{k}}$ is called an *eigenspace*. The scalars $a(\mathbf{u})$ are called *eigenvalues*.

This idea of an eigenfunction is a purely dynamical property and it immediately raises the question of exactly what are the eigenfunctions for the torus dynamical system? It has an amazing answer: save for the property that we can multiply by nonzero scalars, the pure waves $w_{\mathbf{k}}$ are the *only* eigenfunctions. Although we developed things from the pure waves because of their simplicity, it turns out that they are intrinsically special: they are dynamical invariants. Looking back we can see that the Fourier analysis says that these dynamical invariants form a basis through which all the functions on $\mathbb{T}$ can be expressed, (11.21). Parallel results for periodic functions apply in all dimensions.

There is an important insight here. Even if we had not been lucky enough to start with what later turn out to be invariants of the dynamics they would appear anyway if we were to search for the eigenfunctions of the system—its invariants in the space of observables. In fact we have the remarkable conclusion that the space of observables

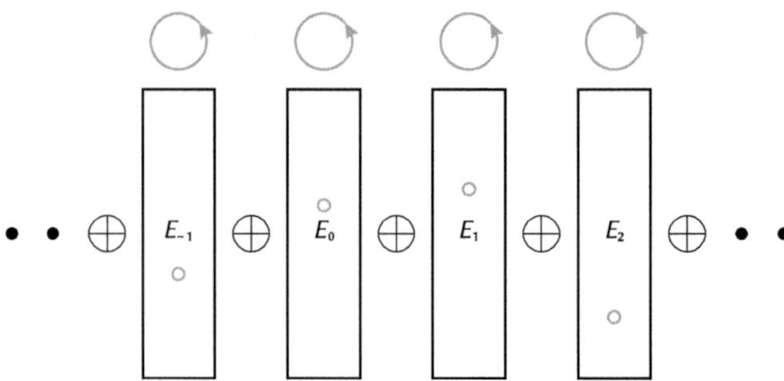

Figure 11.21: This is a schematic interpretation of what the eigenspace decomposition of $L^2(\mathbb{T})$ relative to the action of time looks like. Each box represents the (complex) 1-dimensional eigenspace $E_{\mathbf{k}}$ generated by the pure wave $w_{\mathbf{k}}$. The arrowed circles are intended to show the defining property of these eigenspaces, namely that each is invariant under the dynamics of translation. Each doubly-periodic function is expressible as a sum, each of whose components is in one of these eigenspaces. The circle dots are meant to suggest the choosing of coefficients $c_{\mathbf{k}}$, one in each component eigenspace. $L^2(\mathbb{T})$ is the entirety of functions that arise from all possible choices of $c_{\mathbf{k}}$, subject only to the convergence condition of square summability: $\sum_{\mathbf{k}} |c_{\mathbf{k}}|^2 < \infty$.

is decomposable into subspaces that are invariant under the dynamics. This is not just a feature of the systems modeling periodicity. It is a general feature that applies quite generally to dynamical processes and in particular to pattern systems. Generally it goes under the name of *spectral theory*. Finding and describing these invariants is a key part of it.

There is another important insight here. We set up the theory of waves in terms of complex numbers because of the beautiful way they allow us to encompass the idea of cyclical motion. This could all have been done without the complex numbers. However, when it comes to the invariants of the observables, we see that the eigenvalues are of the form $\mathbf{e}(-2\pi\mathbf{u}.\mathbf{k})$, which *are* complex numbers. This is absolutely characteristic of the study of eigenfunctions: corresponding eigenvalues will often be complex numbers, not real numbers. The complex numbers are indispensable in spectral theory.

11.8.1 An example of eigenfunctions

Consider for instance the simple rotation R of 90° on a circle. Evidently R^4 is the identity transformation, i.e. the transformation that does nothing. Not surprisingly it is related directly to the equation $x^4 = 1$. But this equation has only two real solutions, namely ± 1, neither of which captures the fact that the rotation is through $1/4$ of a circle. The full set of solutions of the equation are $\{\pm 1, \pm \mathrm{i}\}$, and indeed it is the latter two that are truly involved with the eigenfunctions associated with the 4-fold nature of the rotation. In fact we know that R is multiplication by $\mathbf{e}(2\pi/4)$ on the unit circle $\mathbb{U}$, and $\mathbf{e}(2\pi/4) = \mathrm{i}$.

To continue this example, instead of dealing with all of $\mathbb{U}$, let's reduce things to just four points on the unit circle, $X := \{1, \mathrm{i}, -1, -\mathrm{i}\}$. They make the four vertices of a square, which, to keep things clearer in what follows, we will also designate by $\{x_1, x_2, x_3, x_4\}$. The rotational symmetry group of this square is generated by our rotation R, which amounts to

$$x_1 \to x_2 \to x_3 \to x_4 \to x_1 .$$

Any complex-valued function $f : X \longrightarrow \mathbb{C}$ can be expressed as a 4-tuple (c_1, c_2, c_3, c_4) of complex numbers describing what happens to each of the four vertices. From this we can see that the entire space $\mathscr{F}$ of functions is just four-dimensional over $\mathbb{C}$ (since it requires exactly four complex numbers to define each possible function).

Now we want to look for eigenfunctions with respect to the dynamics of the rotation R. This amounts simply to rotating the four numbers defining the function—see §11.10.2 for more on this. One such is $f_1 := (1, 1, 1, 1)$ which is invariant under rotation. Its eigenvalue is 1. Another is $f_2 := (1, -1, 1, -1)$, which rotation changes to $(-1, 1, -1, 1) = -(1, -1, 1, -1)$. Its eigenvalue is -1. Next comes the more interesting $f_3 := (1, -\mathrm{i}, -1, \mathrm{i})$. Here is how it works: $Rf_3(x) = f_3(R^{-1}x)$, so we are looking at the four numbers $\{f_3(x_4), f_3(x_1), f_3(x_2), f_3(x_3)\}$. This is the function denoted by $(i, 1, -\mathrm{i}, -1)$. But this is $i(1, -\mathrm{i}, -1, \mathrm{i})$, so $Rf_3 = \mathrm{i}f_3$. Thus f_3 is an eigenfunction with eigenvalue i. Finally we have $f_4 := (1, \mathrm{i}, -1, -\mathrm{i})$ which rotates to $(-\mathrm{i}, 1, \mathrm{i}, -1) = -\mathrm{i}((1, \mathrm{i}, -1, -\mathrm{i})$ with eigenvalue $-\mathrm{i}$.

The four functions defined by f_1, f_2, f_3, f_4 serve as a basis for $\mathscr{F}$ in the sense that every function is uniquely a linear combination of these four. Thus we have decomposed $\mathscr{F}$ as the sum of four eigenspaces:

$$\mathscr{F} = \mathbb{C}f_1 + \mathbb{C}f_2 + \mathbb{C}f_3 + \mathbb{C}f_4 .$$

Without the complex numbers we could not arrive at such a decomposition.

The study of periodicity that we have introduced here has led quite naturally to the idea of spectral invariants in the form of eigenfunctions and eigenvalues. What is important for our conception of pattern systems in general is that it applies to them too. In the next section we fill in the mathematical blanks that show how this happens. This can be omitted on first reading, but ultimately it is useful in better understanding the ideas behind quantum mechanics.

The material of §11.9 and §11.10 fills in technical details around the definitions of integration and the dynamics of observables and measures. It is important for completeness, but not necessary to the overall understanding of what follows.

11.9 From states to functions: how do we go further?

Our ideas about patterning are based on the ideas around a state space, events, dynamics, and the quantification of these. When we bring in observables, which are by definition complex-valued functions on the state space, we move into a more nuanced setting. Although no single observable can be truly representative of an entire pattern system, when we consider its full space of observables we can see the system in ways that both include and go beyond what we can see otherwise. Our attention expands from dynamics applied to states to dynamics applied to observables. There is a need to expand the idea of a pattern system to include this new feature. The aim of this section is to see how it is done.

Instead of a state space, which is, when it comes down to it, just a *set*, we are interested in a set of observables, or functions. defined on that state space. The important additional feature is that this set of observables is closed under addition and scalar multiplication by complex numbers. In this way it has its own *algebraic structure.* As we have seen, such sets are called *linear spaces.*. For instance, the set of all periodic functions defined on $\mathbb{R}$ with period 1 is a linear space because the sum of any two of them is also a periodic function and scaling such a function again results in another. That's what we mean by being *closed under addition and scalar multiplication.* The set of complex-valued functions on the unit circle $\mathbb{U}$ is also linear space, in fact it is essentially the same as the space of functions with period 1. Similarly we have the linear spaces of doubly or triply periodic functions (and there are many more: spaces of continuous functions, differentiable functions, rapidly decreasing functions, L2-functions, measurable functions, etc.).

Our conceptual ideas around pattern systems ought to extend to fully include observables, but this raises two basic questions: how does the dynamics and how do the processes of quantification through measures extend to observables? This is important both to our present story about waves and later to our discussion of quantum theory, which is founded on ideas around linearity in precisely the sense that we speak of it here. Fortunately, once one has grasped the methodology, the process is rather straightforward. The key connecting idea is *mesa functions.*

11.9.1 Simple functions

Suppose that we have some pattern system $\mathcal{X} = (X, S, \mathscr{E}, \mathbf{m})$, with state space X, event space $\mathscr{E}$, measure $\mathbf{m}$, and some set S of operators of change. Turning our attention to observables we want to consider some linear space $\mathscr{F}$ of observables or complex-valued functions $f : X \longrightarrow \mathbb{C}$. Again, to say that it is a *linear space* is to require that addition of functions of $\mathscr{F}$ and scalar multiples of them are still in $\mathscr{F}$. A natural place to begin is with the basic subsets E that are the events that make up $\mathscr{E}$. From the point of view of pattern the subsets of $\mathscr{E}$ represent a level of distinction that we can make within the state

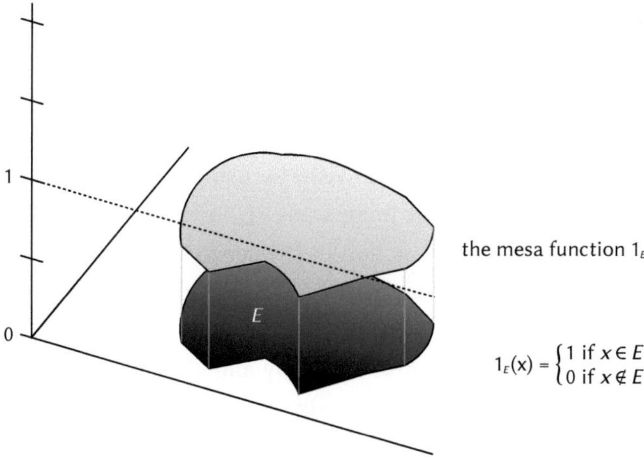

Figure 11.22: The figure shows the plot of the mesa function $\mathbf{1}_E$ (with its "shadow" below) defined by a region E in the plane. The mesa function simply answers the question "is x in the set E"?: 1 if yes and 0 if no. For each event E there is a corresponding mesa function $\mathbf{1}_E$.

space. If $E \in \mathscr{E}$ then the very simplest function that describes this event is the function $\mathbf{1}_E$ that is defined by

$$\mathbf{1}_E(x) := \begin{cases} 1, & \text{if } \ x \in E, \\ 0, & \text{if } \ x \notin E. \end{cases} \tag{11.24}$$

It takes the value 1 if the state x is in E and the value 0 otherwise. This type of function is called a *mesa function* since its plot reminds us of the mesas that are so common in the southwest USA—flat table-like protrusions of land arising out of the desert floor, see Fig.11.22. We can think of a mesa function as being the functional equivalent of the event that describes it.

Once we have defined these mesa functions we can combine them to define the so-called *simple functions* by taking *linear combinations*:

$$f = c_1 \mathbf{1}_{E_1} + c_2 \mathbf{1}_{E_2} + \cdots + c_n \mathbf{1}_{E_n} = \sum_{k=1}^{n} c_k \mathbf{1}_{E_k}, \tag{11.25}$$

where $E_1, E_2, \ldots, E_n$ are events and $c_1, c_2, \ldots, c_n$ are complex numbers. These events need not be disjoint, but if they are then $f(x) = c_k$ if $x \in E_k$ for some k and is zero everywhere else. If the events are not disjoint then $f(x)$ will be the sum of those c_k for which $x \in E_k$, and zero if it is in none of them.

It is rather clear that such functions (which are observables in our terminology) are more nuanced expressions than just events alone, for we are now weighting events by complex values and looking at linear combinations that we can form in infinitely many ways. The way in which we have defined them guarantees that the set of simple functions is indeed a *linear space* of functions.

Still, simple functions are rather chunky things, rather like the slices of wood used to shape contours in a crude 3D-representation of a hilly landscape. But just as we could approximate a circular disk with rectangles and so reach its area as a limiting process by making the rectangles vanishingly small, so we can approximate a huge variety of functions, as smooth or as jagged as we wish, by using simple functions. A computer monitor can seemingly represent any imaginable 2D landscape, but what is displayed is really no more than just a simple function in the sense that we have been speaking: a matrix of very small square pixels each of which is assigned a certain amount of luminous intensity. The mathematical version is the same but it does not have to limit the degree of refinement to the level of pixelation. By allowing unlimited pixelation, that is by allowing limits of simple functions approximations, we can arrive at a vast variety of possible functions, much as we can get the area of a circle as the limit of approximations by rectangles. This set of functions is a linear space and is referred to as the space of $\mathscr{E}$-measurable functions. It is much larger than the space of simple functions. The importance of the "simple" functions is that they are indeed simple and serve as stepping stones by approximation. We say more about this in §11.10.1 below.

11.10 Integration: measures lift to functions

Measures are mathematical entities that assign numbers to sets. What makes them so important is that they are also able to assign numbers to functions, and this is the basis of what we normally call integration.

We will continue to assume that we are dealing with a pattern system $\mathcal{X} = (X, S, \mathscr{E}, \mathbf{m})$ in which $\mathbf{m}$ is a measure that assigns values to the events $E \in \mathscr{E}$. The idea behind extending $\mathbf{m}$ to functions is rather transparent: start with the mesa functions and simply define

$$\mathbf{m}(\mathbf{1}_E) := \mathbf{m}(E)\,. \tag{11.26}$$

Next, extend this to simple functions by linearity: for f defined in (11.25) we define

$$\mathbf{m}(f) = \mathbf{m}(\sum_{k=1}^{n} c_k \, \mathbf{1}_{E_k}) := \sum_{k=1}^{n} c_k \, \mathbf{m}(\mathbf{1}_{E_k}) = \sum_{k=1}^{n} c_k \, \mathbf{m}(E_k)\,. \tag{11.27}$$

First $\mathbf{m}$ gives the numerical value $\mathbf{m}(E)$ to the mesa function defined by the event E, and then it extends to simple functions by simple linearity. Obviously not all functions are as simple as simple functions, but before we go on we should introduce some notation.

The process that we have just defined, which allows us to extend the idea of a measure on events to a measure on functions or observables, is called *integration*. This is integration in the same sense that integration is spoken of in calculus (not the sense in which we speak of the integration of pattern systems). There is a historical notation for this that dates back to Gottfried Wilhelm Leibniz (1646–1716), who along with Newton was the co-inventor of the processes of differentiation and integration in calculus, and it is still in common use. It will probably look more familiar:

$$\mathbf{m}(f) = \int_X f\, d\mathbf{m}\,. \tag{11.28}$$

The $\int$ sign is an extended letter "S" that stands for summing, X is the region over which we are integrating, f is the function being integrated, and $\mathbf{m}$ is the measure that we are using. The little "d" has little meaning in this context. It is a throwback to the time when the sum was thought of as constituted of "infinitesimally small" increments (very small "deltas") of the variable of the function.[17] For us it is just a notifier that says what follows is the applicable measure. This is all pure notation, and a very useful one, but it does not add any new information.

So integration is really a process of "taking the measure" of the function. The integration learned in a standard calculus course is what is called *Riemann integration.* It is different in the way in which it is derived and is not as powerful. It leads to *identical answers* in ordinary classical situations where the underlying space is ordinary Euclidean space (the line, the plane, 3D-space, etc.), though it is not really adequate for Fourier analysis or for probability theory. In calculus courses much time is usually spent going into algorithms for actually computing the numerical values of integrals. In measure theory courses much time is spent on the intricacies of convergence of limits. Here, we are concerned neither with evaluation nor technical details, but rather with the conception of how measure transforms into integration. The measure that is involved in ordinary calculus is the Lebesgue measure on the line (length), and similarly in higher dimensions it is the Lebesgue measures of area and volume. However, Lebesgue's theory goes far beyond these situations, for wherever there is a measure, no matter how strange, there is a corresponding notion of integration.

11.10.1 Extension from simple functions to arbitrary functions

Having defined integration for simple functions, we next see how it extends to all $\mathscr{E}$-measurable functions. A function $f : X \longrightarrow \mathbb{C}$ is called a *positive function* if it is *real-valued* and $f(x) \geq 0$ for all $x \in X$. If g is another positive function on X and $g(x) \leq f(x)$ for all $x \in X$ we write $g \leq f$.

The idea behind extending integration beyond simple functions begins with the idea that we should be able to approximate any positive function f as closely as we wish by using simple functions, approximating from below:

$$g = \sum_{k=1}^{n} c_k \, \mathbf{1}_{E_k}$$

where the c_k are non-negative real numbers and $g \leq f$, Fig.11.23. For simple functions we know how to measure them and we would expect this measure to be less than or equal

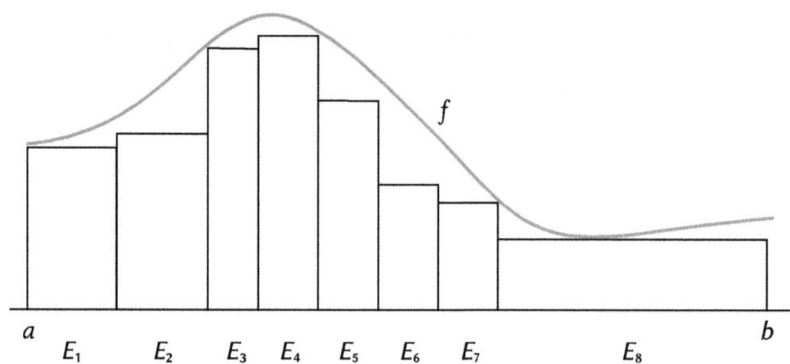

Figure 11.23: Approximating a positive function f by a simple function on the interval $[a, b]$ on the real line. If c_k is the height of the box over E_k then the simple function in question is $\sum_{k=1}^{8} c_k \mathbf{1}_{E_k}$. Its integral is obtained as the areas of the boxes, i.e. $\sum_{k=1}^{8} c_k \,\ell(E_k)$, where ℓ is the measure of length. There is no need for it to be a particularly good approximation. The integral of f over the interval $[a, b]$ is the supremum (the least upper bound) of the numerical values of these sums, taken over *all possible* simple functions (which may be composed from trillions of mesa functions).

to any putative value of $\mathbf{m}(f)$:

$$\mathbf{m}(g) = \sum_{k=1}^{n} c_k \,\mathbf{m}(E_k) \leq \mathbf{m}(f)\,. \tag{11.29}$$

This leads us to define $\mathbf{m}(f)$ to be the *least* real number for which the inequality in (11.29) holds for *all* simple functions g for which $g \leq f$. In a single mathematical statement

$$\mathbf{m}(f) \;:=\; \sup\{\mathbf{m}(g) \;:\; g \text{ is a positive simple function and } g \leq f\}\,. \tag{11.30}$$

The standard mathematical notation sup (standing for *supremum*) is read as the *least upper bound.* We want the smallest number b which bounds from above all the numbers $\mathbf{m}(g)$ arising from positive simple functions g for which $g \leq f$. Then we define $\mathbf{m}(f)$ to be this least upper bound: $\mathbf{m}(f) = b$. If there is no such number (because there is no upper bound to all these numbers $\mathbf{m}(g)$), then we write $\mathbf{m}(f) = \infty$.

This then is how we deal with positive functions: approximate by simple functions g from below and define $\mathbf{m}(f)$ as the smallest real number that bounds all the values $\mathbf{m}(g)$ that arise in this way.

This deals with positive functions. The process lifts to complex-valued functions by showing that they can always be written as a simple linear combination of four real-valued positive functions, and hence reduced to the case that we have just covered. The details are found in the Endnotes.[18]

This is how measures are extended from their set-theoretical origins to measures defined on functions. They are, in effect, integrals. This then is how the Lebesgue integral is defined.

In this discussion we have not stated any restrictions on the functions that we are speaking of. Actually there are restrictions, but they are quite mild. It is required that they be *measurable*, or more specifically *$\mathscr{E}$-measurable* functions, since $\mathscr{E}$ is the domain of the measure $\mathbf{m}$.

> A function $f : X \longrightarrow \mathbb{R}$ is measurable if and only if for all Borel sets B of $\mathbb{R}$, $f^{-1}(B)$ is an event of $\mathscr{X}$ (that is, $f^{-1}(B) \subset \mathscr{E}$). §6.6.4.

More simply, for f to be measurable it is sufficient that for all intervals $[a, b] \subset \mathbb{R}$ we have $f^{-1}([a, b]) \in \mathscr{E}$, since the closed intervals generate the sigma-algebra of Borel sets of $\mathbb{R}$. Remember that $f^{-1}(B)$ is by definition the set of all points $x \in \mathbb{R}$ for which $f(x) \in B$. We are asking that these sets be events. A complex-valued function is measurable if the real and imaginary parts of the function are $\mathbf{m}$-measurable.[19]

The definition we have given is an explanation of what $\mathbf{m}(f)$ means. That is we have conceptually defined what is called integration in calculus. It is something quite different to actually determine the numerical values of $\mathbf{m}(f)$. The real genius of Newton and Leibniz was to show how this can be done. However, at the level of our discussion this is immaterial—it is the idea that is important. We have learned how measures can measure functions.

In keeping with (11.28) we will also use the alternative notation $\mathbf{m}(f) = \int_X f \, d\mathbf{m}$ and call $\mathbf{m}(f)$ the integral of f over X with respect to the measure $\mathbf{m}$. If f is measurable, so is its absolute value $|f|$ (the function defined by $|f|(x) := |f(x)|$). The function f is said to be *integrable* with respect to $\mathbf{m}$ if

$$\int_X |f| \, d\mathbf{m} < \infty \,. \tag{11.31}$$

We will denote the set of all complex-valued measurable functions arising from $\mathscr{X}$ by $\mathscr{M}(\mathscr{X})$, $\mathscr{M}(X, \mathbf{m})$, over even $\mathscr{M}(X)$, according to what is clear from context. This is a linear space and we will refer to it as the *space of measurable functions of $\mathscr{X}$*.

11.10.2 Dynamics lifts to observables (functions)

We come now to the way in which the dynamics of the pattern system $\mathscr{X} = (X, S, \mathscr{E}, \mathbf{m})$ extends from change at the level of the state space to change at the level of observables (functions on X). Suppose that $f : X \longrightarrow \mathbb{C}$ is an observable, that is, a complex-valued function on X, and suppose that $s \in S$ is an *invertible* operator of change on X. It is natural that the changes that s institutes can be "seen" by the observable. Changes on X should be reflected in changes in what is observed. We have already seen this with waves, and it can be made to work in the same way in general: that is, define the action of s on the observable f to give rise to the new observable (function) defined by

$$(s \triangleright f)(x) = f(s^{-1} \triangleright x) \,. \tag{11.32}$$

This is the idea that we have seen before: transform the function by transforming its domain (the set on which it is defined) in the opposite direction. This makes sense as long as the operator s in question is invertible. In this chapter on waves the dynamics has been based on translation, that is transformations of the type $\mathbf{x} \mapsto \mathbf{u} + \mathbf{x}$, which is translation by the vector $\mathbf{u}$ and is invertible ($\mathbf{x} \mapsto -\mathbf{u} + \mathbf{x}$).

Returning to the question at hand and the pattern system $\mathscr{X} = (X, S, \mathscr{E}, \mathbf{m})$ consider how the idea of change looks at the level of mesa functions. Suppose that E is an event and $s \in S$ is an invertible operator. Then according to the definition

$$s \triangleright \mathbf{1}_E(x) = \mathbf{1}_E(s^{-1} \triangleright x).$$

But now we note that $\mathbf{1}_E(s^{-1} \triangleright x) = 1$ if and only if $s^{-1} \triangleright x \in E$, otherwise it is 0. In turn this says that $\mathbf{1}_E(s^{-1} \triangleright x) = 1$ if and only if $x \in s \triangleright E$. Thus

$$s \triangleright \mathbf{1}_E = \mathbf{1}_{s \triangleright E} \,. \tag{11.33}$$

This shows quite directly how the dynamics of the observables directly corresponds to the dynamics of states: the event E becomes $s \triangleright E$ and the mesa function $\mathbf{1}_E$ becomes $\mathbf{1}_{s \triangleright E}$.

From mesa functions we go on to simple functions by extending the operator by linearity: with f defined in (11.25),

$$\mathbf{s} \triangleright f = c_1 (s \triangleright \mathbf{1}_{E_1}) + c_2 (s \triangleright \mathbf{1}_{E_2}) + \cdots + c_n (s \triangleright \mathbf{1}_{E_n}) = \sum_{k=1}^{n} c_k \; \mathbf{1}_{(s \triangleright E_k)} \,. \tag{11.34}$$

From here, though we won't go through the details, we can extend the dynamics from these simple functions to arbitrary measurable functions through approximations, just as we extended measures.

Fig.11.24 illustrates what happens for a function on the real line where the dynamics is translation left or right along the line.

In defining how an operator s lifts from its effects on mesa functions to becoming an operator on the space of simple functions, we do it in a way that "respects" the algebraic structure of that space. That is, we require that s be a *linear operator*: for $f, g \in \mathscr{F}$ and $c \in \mathbb{C}$ then

$$s \triangleright (f + g) = s \triangleright f + \mathbf{s} \triangleright g\,;$$
$$s \triangleright (c f) = c\,(\mathbf{s} \triangleright f)\,.$$

This property extends quite naturally to arbitrary functions when they are approximated by simple functions (see equation (11.34)).

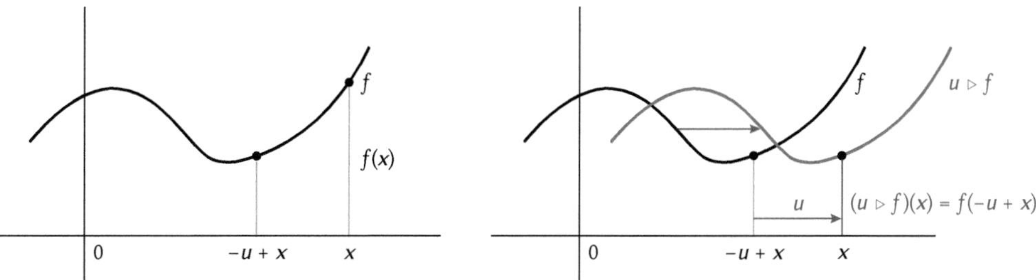

Figure 11.24: In this figure we show how the operation of translation along the x-axis affects a function or observable f. On the left we see a *plot* of a function f. On the right we also see the function $u \triangleright f$ (with $(u \triangleright f)(x) := f(-u + x)$) whose graph is that of f shifted to the right by u. So the operator u shifts the plot of f in the same direction as u—that is, to the right if $u > 0$.

11.10.3 Dynamics lifts to measures

We know that from the point of view of dynamical systems and pattern systems, the situation in which the measure $\mathbf{m}$ is invariant with respect to the dynamics is especially important. This suggests that the dynamics, which so far we have lifted to observables, may also apply to measures. This is indeed true. Given any measure $\mathbf{n}$ defined on the event space $\mathscr{E}$ we can define the transformed measure $s \triangleright \mathbf{n}$ by

$$(s \triangleright \mathbf{n})(E) := \mathbf{n}(s^{-1} \triangleright E). \tag{11.35}$$

Although we have been assuming here that the operators in question are invertible, in this present situation this is not necessary. We have already given meaning to what s^{-1} means when we are dealing with sets and it is our standard assumption that s^{-1} maps events to events (see §6.5.3, §6.5.6, and equation (6.2)).[20]

If s is invertible we can see that this definition matches the dynamics that we have placed on mesa functions, since we have

$$(s \triangleright \mathbf{m})(\mathbf{1}_E) = (s \triangleright \mathbf{m})(E) = \mathbf{m}(s^{-1} \triangleright E) = \mathbf{m}(\mathbf{1}_{s^{-1} \triangleright E}) = \mathbf{m}(s^{-1} \triangleright \mathbf{1}_E),$$

see (11.33) with s replaced by s^{-1}. This then extends to simple, and finally to all, measurable functions.

Put in this context, to say that $\mathbf{m}$ is an *invariant* of the dynamics is to say that $s \triangleright \mathbf{m} = \mathbf{m}$ for all operators $s \in S$. That would mean that for all events E,

$$\mathbf{m}(s^{-1} \triangleright E) = \mathbf{m}(E),$$

which is indeed how we defined invariance at (6.4). This in turn leads to the fact that if the measure $\mathbf{m}$ is invariant under the dynamics then the dynamics does not affect the measure of any observable:

$$\mathbf{m}(s^{-1} \triangleright f) = \mathbf{m}(f) \text{ , or in Leibniz notation,}$$

$$\int_X s^{-1} \triangleright f \, d\mathbf{m} = \int_X f \, d\mathbf{m} . \tag{11.36}$$

Averages

Up to now we have relied on intuition to talk about averages in determining Fourier coefficients. Thus in (11.16) we have described the Fourier coefficient belonging to the frequency k in a periodic function f on the circle $\mathbb{U}$ to be

$$c_k = \text{av}\left(\overline{\mathbf{e}(2\pi kt)}\, f(2\pi t)\right) .$$

Now we have defined integration of functions we can define these averages more precisely: they are integrals. This is understood to mean the average is being taken over the entire domain of the function. So far we have relied on intuition to understand what this *average* means. The true mathematical definition of this average is based on integration using the natural measure of arc-length on the circle $\mathbb{U}$. In more detail, let us denote arc length on the circle by α. Since the length of the circumference is 2π, we have $\alpha(\mathbb{U}) = 2\pi$. Being a measure on $\mathbb{U}$ it extends to have meaning as a measure on the periodic functions which are defined on $\mathbb{U}$. For any such function g we have its measure or integral, which we can write as

$$\alpha(g) \quad \text{or equivalently} \quad \int_{\mathbb{U}} g \, d\alpha .$$

The *average* (value) of g on $\mathbb{U}$ is then *defined* to be the value of this integral divided by the arc-length of the entire circle:

$$\text{av}(g) := \frac{1}{2\pi} \int_{\mathbb{U}} g \, d\alpha .$$

Quite generally, given a pattern system $(X, , S, \mathscr{E}, \mathbf{m})$ with $\mathbf{m}(X)$ finite, the *average* value of any function f on X is defined to be

$$\text{av}(f) = \mathbf{m}(f)/\mathbf{m}(X) = \frac{1}{\mathbf{m}(X)} \int_X f \, d\mathbf{m} .$$

If $\mathbf{m}$ is invariant with respect to some $s \in S$ (respectively all $s \in S$) then the average value of f on X is unaltered if we apply s to f (respectively any $s \in S$), as we see from (11.36).

11.10.4 The functions of Fourier analysis

With all this in hand we can return to an unfinished discussion of the true setting for Fourier analysis. Our basic conception has been that periodic functions can be expressed as sums (perhaps infinite in extent) of pure waves scaled by complex coefficients. The key process is that of determining those coefficients. But for what class of periodic functions is that really true?

The setting of Fourier analysis in 1D is complex-valued functions defined on the unit circle $\mathbb{U}$. We know that it can be viewed as the state space of a dynamical system on which $\mathbb{R}$ acts by rotations. The natural idea of an event is the set of all states in such-and-such part of the circle, in other words within some arc on the circle. Thus the event space for $\mathbb{U}$ is defined by agreeing that any closed arc on the circle is an event, and then taking the sigma-algebra of all subsets of $\mathbb{U}$ that arise from these (that is, whatever subsets of $\mathbb{U}$ arise through arbitrary finite and countable unions and intersections of these). The natural measure on $\mathbb{U}$, call it $\mathbf{a}$, is *arc length*, which assigns to each arc on the circle its length. With that we have defined a $\mathbb{U}$ as a measure space. And, we might note, arc length is not affected by rotation, so $\mathbf{a}$ is actually an invariant measure.

With this we can define the space of functions to which the Fourier analysis applies, that is, the functions $f : \mathbb{U} \longrightarrow \mathbb{C}$ which have spectral decompositions of the form shown in equation (11.13). They are those that have two properties:

(i) f is a measurable function with respect to $\mathbf{a}$ (see §11.10.1);

(ii) $\int_{\mathbb{U}} |f|^2 \, d\mathbf{a} < \infty$.

This space of functions (it is a linear space) goes by the mathematical name of the *space of L^2-functions*, $L^2(\mathbb{U}, \mathbf{a})$. We will see this type of space occurring again, since it is also at the heart of quantum theory. See ‡§19.4.1 for more detail.

The extension of this to doubly or triply periodic functions is the same, now replacing $\mathbb{U}$ by the corresponding torus and using the natural measures given by surface area or surface volume on these. Measurability and square integrability remain the required conditions. Then the Fourier analysis proceeds as we have described.

11.10.5 Spectral analysis

We can apply these ideas in the generality of arbitrary pattern systems $\mathcal{X} = (X, S, \mathcal{E}, \mathbf{m})$, where $\mathbf{m}$ is s-invariant for every $s \in S$. Along with this dynamical system we have its space of observables. The typical, and most useful, restriction on the observables is that they be $\mathbf{m}$-measurable and square integrable:

(i) f is a measurable function with respect to $\mathbf{m}$ (meaning measurable in the sense above);

(ii) $\int_X |f|^2 \, d\mathbf{m} < \infty$.

The functions satisfying this condition are called *square-integrable functions.* It is standard notation to denote the linear space that they compose by $L^2(\mathcal{X})$ (ell-two of $\mathcal{X}$).[21]

The second condition is equivalent to the statement that

$$\sum_{k=1}^{\infty} |c|^2 < \infty\,. \tag{11.37}$$

A *non-zero* function $f \in \mathcal{M}(X)$ is called an *eigenfunction* if the subspace $\mathbb{C}f$ that it generates is actually invariant under the action of the dynamics:

$$s \triangleright \mathbb{C}f \subset \mathbb{C}f \quad \text{for all } s \in S\,. \tag{11.38}$$

The space $\mathbb{C}f$ itself is called an *eigenspace*. To say that $\mathbb{C}f$ is invariant under s means that there is a scalar $a \in \mathbb{C}$ such that

$$s \triangleright f = af\,.$$

This scalar a is called an *eigenvalue* of the operator s. Different $s \in S$ will usually have eigenvalues. In the example above we saw that the operator of translation by $\mathbf{u}$ produces the eigenvalue $\mathbf{e}(-2\pi\mathbf{k}.\mathbf{u})$ when applied to the pure wave (and eigenfunction) $w_{\mathbf{k}}$. Different translations will produce different eigenvalues, though we might notice that, quite generally, they all belong to $\mathbb{U} \subset \mathbb{C}$. (The zero function, the unique function on X which takes on the value 0 for all $x \in X$, is *not* considered to be an eigenfunction because it is trivially invariant and does not generate a one-dimensional space: $\mathbb{C}\,0 = \{0\}$ is zero-dimensional.)

In these few lines lies one of the big steps in our development of pattern: how pattern can be viewed from the higher perspective of observables where invariance takes the form of eigenspaces, their eigenfunctions, and their eigenvalues. Finding eigenspaces and their associated eigenfunctions and eigenvalues goes under the general name of *spectral analysis*.[22]

11.11 Conclusion

The frame of reference for the chapter is waves. Since light is certainly a wave-like phenomenon and since it is the prime source of energy on which our planet depends for survival, we have emphasized just how much its actual physical nature has affected the development of our own physical organs, both their powers and their limitations.

The mathematical side begins with the simple idea of a wave, the cycling up and down motion of a simple sine-wave in time, and its strong connection to circular motion. Concentrating on these simple periodic wave forms we are led to the conclusion that we are dealing with a rather fundamental dynamical system: a circle and rotation around that circle. The simplest waves are pure waves of just one frequency. An arbitrary wave is a mixture of such waves composed of numerous frequencies and intensities, or amplitudes. In this chapter we have restricted our attention to waves that are based on some fundamental frequency, much as the sound and timbre of a musical instrument often arise out of a variety of harmonics based on one underlying base frequency. Such waves are *periodic*, repeating at the frequency of the fundamental, and they can be considered as functions defined on a circle.

Interest then is in the relationship of a wave to its underlying frequencies and their magnitudes. This goes by the name of *Fourier analysis*, and in more general terms, *spectral analysis*. As we will see in the subsequent chapters, these ideas are foundational in areas wherever there is repetition, and in the end they also lie at the foundations of quantum mechanics.

Waves are by nature continuous—they represent a continuous repetitive flow of intensity. Yet a simple wave can be described by a single number, its frequency. Fourier analysis connects these two aspects. In the case of a periodic wave, which we think of as a function on a circle, the effect is to express it as a collection of numbers which are its component frequencies and their intensities. This can be reversed so that such a collection of numbers is interpreted as defining a continuous periodic wave. This duality manifests a remarkable relationship between the continuous and the discrete. The continuous is the circle with its dynamics of rotation: the discrete appears in the integers which label frequencies.

The study of periodicity leads rather naturally to another part of mathematics, the complex numbers. The complex numbers are an extension of the real numbers. Thought of in geometrical terms, where the real numbers are represented by a number *line*, the complex numbers are viewed as the points of a *plane*, with the real numbers seen now as comprising just the x-axis of the plane and the new "pure imaginary" numbers comprising the y-axis. Each complex number is then a combination of a real and an imaginary part. They gain their algebraic structure by defining an addition and a multiplication that extend their counterparts in the real numbers. Addition is just what one would expect, adding the real and imaginary components. The multiplication is based on addition of angles, and it is here that the stellar role is played by the unit circle $\mathbb{U}$, the circle of radius 1 centered on the origin. This circle consists of all the points $\mathbf{e}(t) := \cos(t) + \mathrm{i}\sin(t)$, Fig.11.6, and it is a *group* under this multiplication. Since periodic functions can be viewed as functions defined on this circle, and it incorporates rotation through its algebraic structure of multiplication, it becomes the natural way in which to study periodicity.

What ultimately lies behind the Fourier analysis of periodicity is the group-theoretic duality that links the circle $\mathbb{U}$ and the integers $\mathbb{Z}$: the continuous wave has its internal frequencies, each with their own intensities or magnitudes. The duality is expressed through the Fourier transform, §11.6.2.

Up to this point, wave has been taken to mean one-dimensional wave. Waves, like ocean waves, are two-dimensional forms of periodicity. Repetitive patterning in two or three independent directions, though now spatial rather than temporal, is omnipresent in Nature and human culture. What can be simpler than a distinctive block that is then used repetitively to fill in a flat expanse or to fill in space? This is the basis of the vast field of crystallography, and is to be found everywhere in human cultures where there are floors to tile, brick walls to erect, and pyramids to amass.

Fortunately the basic idea of higher dimensional periodic waves and multiple component frequencies is relatively simple once the notation is properly extended. Whereas in the one-dimensional case periodic functions can be considered as functions defined on a circle, in two dimensions doubly-periodic observables can be reinterpreted as functions on a torus, which is the formal product of two circles and informally the surface of a donut. In the same way the triply-periodic functions that arise naturally in crystallography can be expressed in terms of a three-dimensional torus. It may be hard to visualize the product of three circles, but mathematically the formalism extends effortlessly, simply by adding in one more coordinate.

The main message is that the Fourier analysis proceeds just as before, with virtually no new ideas required. Again we find a continuous-discrete duality in play, this time with the torus related to pairs of integers, $\mathbb{Z}^2$, giving the frequencies in the two independent directions of periodicity. In the case of crystals, this becomes $\mathbb{Z}^3$.

Throughout all of this, the dynamics, though certainly implicit, is not explicitly accounted for. The last part of the chapter remedies this and shows clearly how the dynamics is not just there but actually can be seen as the basis of what is going on. It is here that we see the beginnings of a new idea. We are used to dynamics at the level of a state space and events that articulate distinctions among states and their various itineraries. However, the dynamics automatically lifts from change at the level of the state space to change at the level of the space of observables (complex-valued functions defined on the state space). The reason is simple: a function f on the state space describes it in reference to the way we articulate and label the individual states. If we think of the dynamics as transforming states to other states, then we are shifting the frame of reference and so altering the function. Although we might say that we are looking at the same function, the dynamical action of translation, whether that be in 1D, 2D, or 3D, effects changes in its explicit description.

Once we see this, we lift our attention from dynamics on the state space to the corresponding dynamics on the linear space of observables. The result leads to a very different interpretation of the pure waves, and one of the key new ideas of the chapter. Now pure waves are seen to be eigenfunctions of the dynamics. Each pure wave gives rise to a one-dimensional subspace of the observables which is *invariant* under the dynamics—a so-called *eigenspace*. The pure waves are responsible for all the eigenspaces. There are no others. This reveals the Fourier decomposition as a manifestation of the dynamics.

Here lies a deeper message about understanding pattern in a way that shifts attention away from a state system and its event space and towards the observables of that pattern system. In doing so we are led to look not just at one observable at a time, but rather at entire sets of observables, whether they be periodic functions, doubly-periodic functions, functions defined on a torus, measurable functions, and so on. Crucially, these sets of observables are more than just sets: they have an algebraic structure because we can add them and scale or multiply them by complex numbers. They are *linear spaces.* The dynamics of the state space is still there, but it now appears at the level of these spaces of observables where it offers insights into pattern that are not otherwise possible.

Turning attention from the state space and its dynamics to spaces of observables and how those dynamics play out there is hugely relevant. The real way we know the world is by observing it. As societies develop they tend to make observations at increasingly detailed levels that look increasingly mathematical, attaching specific values to various observed events and looking for patterns in what they find. The records of the ancient astronomers are just like this. The passage of development is then from observations to the formulation of dynamical models that can explain them and predict their future values. In reality this is all we can ever do in explaining Nature. Whether our methods are overtly mathematical or not, we observe, we build conceptual models, observe more, correct those models, observe more, and replace old models with new ones that fit the data better. Sometimes several models combine into one that then brings deeper meaning to both. The realization of the vastness of the spectrum of light is the famous example.

The *perception* of change requires "something" relative to which it can be perceived—something that doesn't change, or at least changes in some different way. The history of thought reveals our endless human quest to find a totally stationary "something" beyond it all, a something that we can truly rely on, relative to which all change might be expressed. Asian thought, notably Taoist/Buddhist thought, teaches that it does not exist, except as the great totality of all being. And of that nothing can be said!

Paradoxically what we have learned from science is that the invariance is not to be sought in things, but in the laws or principles by which change takes place. The speed of light, the measure of total energy, Maxwell's equations which define the dynamics of electro-magnetic fields—it is in these that we find invariance. Hydrogen atoms may be little dynamos of activity, but all hydrogen atoms manifest the same patterning and that is what makes them hydrogen atoms. Physics is based on the assumption that there are invariances, and that it is through them that we can understand and predict the unfolding of change.

In the unfolding of the physical world the material aspects are in continual change, but we presume that the processual aspects themselves, the principles of the system, remain unchanged. With all the uncertainties around dark matter, dark energy, the nature of time, and the non-locality of space, we see that we still have much to learn about these principles.

CHAPTER 12

Return

12.1 The nature of return

Return is the movement of Way
and yielding is the method of Way
All beneath heaven's ten thousand things
are born of Presence
and Presence is born of Absence.

Daodejing, Ch. 40, adapted from [81]

Just forty years before the authors were born the main consensus of physics was that there were no such things as atoms. Now we all know differently. The entire material world, from fleas to giant suns, is a shimmering dance of atoms and atomic particles. Yet these atoms turned out not to be quite the indivisible units of matter that we might have expected. "Particles" is not the right word, for the atomic world is not about things but about processes. What we can experience and measure are manifestations in form and structure that are dynamic in origin. In other words, at its core, the material world is not "material" as we had imagined. Rather it is systemic patterning. In mutual resonance, atoms can arrange themselves into forms without number and even the periodic table itself is a manifestation of pattern. Neither divisible into categories of particles/waves nor into categories of discrete/continuous, the atomic world defies expression in ordinary words. The only way we know of articulating this world, at least from a scientific perspective, is through the language of mathematics. More specifically it is through the language we have been developing around system dynamics and patterning.

This chapter and the two following it are an exploration of the story of atomism and its remarkable interpretation within the language of pattern and patterning in quantum mechanics. However, there is an important part of this story which touches on another central idea in pattern: *return*.

The movement of Tao is return. So says Ch. 40 of the *Daodejing*. Return is also a phenomenon that arises quite naturally out of the theory of dynamical systems, and we have already encountered it: the Poincaré recursion theorem of §7.9, put loosely, says that we can expect events (not states!) to recur . Choose any event of non-zero probability and any state in that event: with probabilistic certainty the trajectory of that state will return to this event.[1]

We might be tempted to use the word repetition here, which is indeed a very natural and common sense interpretation of pattern. However, repetition can carry a sense of regularity, which is not intended. The Poincaré theorem actually gives no indication of how long we may have to wait for an expected repetition, how different states in the same event may behave, or how often or how regular the repetitions may be. We prefer to speak of *return*.

The Taoist view is based on the idea that things, living and otherwise, arise out of the cosmic process, and ultimately return to it—a sense of arising out of, and returning to, an everywhere potent and embracing emptiness. Life forms arise, go through processes of youth, adulthood, and old age, finally to return to their atomic constituents. Nowadays we know that even the great suns are not immortal and their dissolution usually means the distribution of the higher atoms of the periodic table along with the dissolution of their accompanying planets. And we also know that those great clouds of atoms can gather again to form new suns and planets to orbit them. Were it not for these arising and returning cosmic processes, life would not be possible.

Coming down from these celestial heights, we can see on a more mundane level that what lies behind this is, *recognition*—re-cognition, the cognition of something known before. In one way or another this ability, conscious or not, is fundamental to all forms of life, and it always demands a certain assessment of sameness: when is something the same as before? Recognition is a highly fluid process, but it occurs at all levels, even at the bio-molecular level where receptors recognize and accept other molecules on the basis of shape and fit. It even goes to the atomic level, for the outcome of the interaction of a photon with an atom is a question of fit. Certainly at the human level we have seen how the act of recognition can vary according to the circumstances of the moment and our own intentions.

Put into a narrower context, consider the question of spatial long-range order. This term is not especially well defined in mathematics, but we have a good intuitive sense that coherence across a spatial region can bring it into a unity. Crystals, tiling patterns on floors, and wallpaper patterns are the most obvious examples, where the long-range order is so imposing that just seeing one local patch suffices to know the pattern in its entirety. This is repetition in its strongest form—periodic—and the focus of Ch.11.

More complicated, but still certainly within the realm of spatial long-range order, are the revolving positions of the Earth and Moon around the Sun. Already we saw in

Ch.11 how this form of repetition is not periodic, because the cycles involved do not have commensurate periods. But in this chapter we will see that there is a concept of *almost-periodicity*, which accounts for the remarkable coherence across time and space in these planetary movements. This is what underlies the Meton cycle of the Moon's cycles that we discussed in §2.2.

Even these seem very special when compared with the types of long-range order that we effortlessly place on the world. For instance, flying across Canada one may see the Rocky Mountains, the endless prairies, the ancient rock and lakes and coniferous forests of the Canadian shield, the arctic tundra, and so on. Going on foot through any of these one has a continuing sense of return. Everywhere you look, every step you take, is both different and the same. Even if it were years ago, there is still an instant sense of return upon seeing them again. Each zone, which may be spread out for hundreds of kilometers, is an example of a coherent ecological zone with its own vegetation and wildlife, its own sense of belonging to a common sameness. Mathematical equivalents into which these forms of long-range order can be teased are largely unexplored, but we will get some insight into them by looking at the expansive field of *point processes*.

How mathematics has developed around the key concept of long-range order and its accompanying idea of return are the basis for this chapter. This very same mathematics was involved in finally convincing scientists that the atomic world is real, and it remains a fundamental tool in crystallography and beyond.

12.2 Atoms

It seems difficult to believe now, but in 1900 there was no consensus among physicists for an atomic theory of matter. It was not because there was no idea that this was a possible hypothesis. Chemists and mineralogists interested in crystals were believers far earlier.[2] Observing the fixed and simple ratios in which compounds react together and the ways in which crystals cleaved along planes not only suggested the reality of atoms, but also suggested that they could quite naturally form themselves into regular lattices. Even a complete classification of crystal lattices had been accomplished through the work of Bravais in the 1850s and the collaborative efforts of Federov and Schönflies in the late nineteenth century. Boltzmann, whom we met in §4.4.5, and J. Willard Gibbs in the United States had already put forward a kinetic theory of atoms to explain the second law of thermodynamics as a statistical law rather than an absolute law. We have seen how statistical laws can bring about almost perfect certainty.

However, there were strongly opposed views. Ernst Mach (he of the Mach numbers related to the speed of sound and a highly influential Austrian physicist) and the physical chemist Wilhelm Ostwald were among those who chose to deny their existence. They argued that the laws of thermodynamics need not be based on a mechanics which dictated the existence of invisible atoms in motion. Ostwald in particular advocated the view that thermodynamics dealt only with energy and how it is transformed in the everyday world (www.aps.org/publications/apsnews/200502/history.cfm). Mach's opposition was based on a strong empirical standpoint that what could not be detected physically could not be accepted as part of physics. This reached an unfortunate extreme, when Mach is quoted as saying in 1900 that "If the belief in the existence of atoms is so essential to you, then I secede from the physical mindset. I do not want to be a physicist any more and I renounce any scientific appreciation". Such was the attitude which was so devastating to Boltzmann's life. Even Max Planck initially opposed the idea, though eventually he realized that Boltzmann must be right. It is to Planck that we owe the statement "A new scientific truth does not become accepted by convincing its opponents, but rather by waiting until they have all passed away and a new generation is growing up from the very beginning with this new idea".

As it turns out Mach, with his strong empirical stand, deeply influenced Einstein, and it is rather ironic that perhaps it was this that led Einstein to give the first empirical evidence of atoms. This was his Brownian motion paper of 1905 that we saw in Ch.5. In turn Einstein passed this empirical stance on to Pauli, and Pauli then to Heisenberg. By the time of Heisenberg in 1925 atoms were empirically known to exist, but the situation in atomic physics was one of total bewilderment. It was Heisenberg's deliberate separation of what was *observed* from what might be a *formalism that could describe it* that was foremost on his mind in making his great breakthrough into what we now call quantum theory [137].

Feelings and emotions are bound deeply into the human fabric. Our quest, our passion, for understanding takes place over a background that depends on a multitude of personal and cultural biases that are woven, often without our awareness, into our consciousness.

12.2.1 Epicurean thought

Of course the idea that there ought to be some fundamental (and indivisible) particles that made up the material world is an old one. In classical Greece Democritus was an early advocate of this theory and he is usually attributed as being the father of atomism. His theory was taken over (without much credit) by Epicurus (341–270 BCE), who greatly expanded it and used it as the basis of perhaps the first fully materialistic and non-teleological philosophy of the Western world. For Epicurus the universe is composed of atoms and empty space. Atoms move and arrange themselves into patterns that make up material things. Just as the finite set of letters of the alphabet can be arranged to form countless numbers of words, so too a finite number of distinct atomic types can arrange to form the multiplicity of forms we see around us. Epicurus even concluded that the movements of atoms have a built-in aspect of randomness, the famous (or infamous) *swerve* of Epicurean physics. There must be many worlds, many other living forms, infinitely many novel arrangements which atoms can form. The body and mind are material in origin, and the latter cannot survive without the former. Death is the dissolution of form [69].

Epicurus' primary aim, though, was not a theory of the world, but a theory of what human life involves and how best to live. Man's greatest need is *ataraxia*, a word meaning equanimity—freedom from pain and suffering. To him it was ataraxia that was the basis of pleasure, not the indulgence of the senses. Death, he thought, was nothing to fear, for when we are alive it is not there and when we are dead there is nothing to experience it. Like many of us, thousands of years later, he did not think that cultural norms and conventions of morality were written in the stars or by the hands of gods, but rather were of human origin.

Underpinning this naturalistic view of things Epicurus posited a theory of science based on this atomic picture of things, which he believed could account for all phenomena. As with most of Greek theoretical science, his ideas were based on thought and observation but lacked the role of experiment. Atoms were assumed to be indivisible, which turned out not to be the case, the behavior of particles at the nanoscale turned out to be more bizarre and unbelievable than he or anyone could have imagined, and Epicurean science contains a great deal of what we now know is nonsense. But it was based on some very rational ideas—the senses are our surest way to experience the world, the world is a very physical place, its existence is not teleological (i.e. not purpose or goal oriented), and it is in a constant state of change and return. It has a modern sound to it. What is missing from this is any conception of electro-magnetism or any other fields of force, the wholeness and coherence of the Nature, and what we might call the *systems idea* involving randomness tempered by implicit principles (the laws of physics) which underlie the evolution of the emergent properties we experience.

Epicurus was a prolific writer, but little of what he wrote has survived. However, one notable work did survive: *De Rerum Natura* (On The Nature of Things) written by the Epicurean Titus Lucretius Carus (c. 60 BCE) [108] some two hundred years after the death of Epicurus. Assuredly it is the only full-length science text written entirely in dactylic hexameter, and it has been seen, quite universally, as a masterpiece of Latin literature. Were it not for its somewhat miraculous re-discovery in 1417 it too may be have been lost, for it certainly had its enemies. Following the ideas of Epicurus it presented a purely materialistic picture of the world with no metaphysical spiritual underpinning and no guiding theological principles [69, 114].

In the centuries since its rediscovery Lucretius has been deeply influential in the development of Western thought, especially in the ideas around the material nature of the world. Copernicus, Galileo, Kepler, Bacon, Descartes, Hume, Montaigne, Newton, Erasmus, Darwin, Jefferson, Thoreau, Einstein, Santayana, Calvino, are just a few of the famous names that come to mind [65]. *The Nature of Things* is still a remarkable book to read. True, much of its science is plain wrong (perhaps much of our present science may be wrong too) but it can be very persuasive, and its unrelenting materialistic viewpoint has parallels with much of contemporary scientific thinking. Above all Epicurus and Lucretius believed that to the extent that we truly understand the world, we can achieve contentment in this life, finite though it is. One need not doubt either mind or the human spirit, but for Epicurism (and as often perceived by modern neuroscience too) those are emergent manifestations of what in reality are the dynamical processes—vastly structured, but purely atomic, constitutions of senses, body, and brain existing within a vast network of other beings, living and inanimate. Undoubtedly there remain many questions that we cannot truly answer, that perhaps are even unanswerable. In any case, it is clear that a great proportion of the human population does not accept such conceptual views of the world.[3]

> A novel topic is struggling strenuously to fall upon your ears, a novel aspect of nature to reveal itself. But nothing is so simple that it is not rather difficult to believe at first, and likewise nothing is so great or so wonderful that our wonder at it does not diminish little by little. First and foremost, consider the pure splendor of the sky and all within its confines—the random-roaming stars, the moon, and the sun radiant with dazzling light. Suppose that all these marvels were now revealed to mortals for the first time and were suddenly and unexpectedly thrust before their eyes, what more wonderful spectacle than this could be imagined, what spectacle that people would be less prepared to conceive as credible, if they had not witnessed it? None in my opinion, so marvelous would this sight have been. As it is, however, the spectacle has so satiated us that it has palled, and no one thinks it worth gazing up at the lambent precincts of the sky. So no longer let dismay at mere novelty cause you to reject a theory from your mind, but rather weigh it with penetrative judgement; then, if you consider it to be true, concede victory or, if it is false, equip yourself to fight it. For, since the totality of space out beyond the ramparts of our world is infinite, my mind seeks the explanation of what exists in those boundless tracts which the intelligence is eager to probe and into which the mind can freely and spontaneously project itself in flight [108, Lucretius, 2.925].

All this aside, it was the case that as late as the early 1900s, some 2200 years after Epicurus, the situation was still unresolved. Atoms are too small to see. It was not a question of insufficiently powerful microscopes. It is the nature of the visible spectrum. For atoms to be discernible they have to be separated by at least half a wave length of the source of electromagnetic waves (photons) illuminating them. The wavelengths of visible light are in the range 400–700 nanometers, while the radius of a typical atom is on the order of one thousandth of this. Effectively visible light cannot distinguish things this small. We might understand this by an analogy. A small stick standing in the ripples of a pond will leave its impression on the waves that pass through it. The same stick standing in high ocean surf makes no visible impression on its waves.

Two separate advances in the first part of the twentieth century, both of them about patterning, but otherwise quite different, strongly supported the atomic theory. The first was the 1905 paper of Einstein on Brownian motion, which we have already seen in §5.3. The second, in 1912, was the discovery of X-ray diffraction of crystals, which offered a completely new way to understand the long-range order of materials and offered striking visual evidence of the atomic lattice structure of crystals. This is the subject we come to next.

12.2.2 Diffraction

Diffraction gratings are optical grids of finely spaced periodically repeated lines, ridges, or rulings on a planar surface. By 1899 diffraction gratings with 4700 lines per millimeter were possible. This brings them into the region of the wave lengths of visible light. Light diffracts through the gratings and the path differences, depending on the path of light through closely spaced grids, lead to patterns of interference and so diffraction patterns. In polychromatic light, different wavelengths interfere differently with each other and this results in clear separation of colors, Fig.12.1. This effect is familiar in the colorful spectra on CD and DVD disks where the track spacings form such a grid.

In late 1911 Max von Laue was talking with Paul Peter Ewald who was writing a Ph.D. dissertation on diffraction. Laue had been studying the physics of crystals and this conversation got him thinking about whether crystals, that he conceived to be periodic arrays of atoms, might also diffract if one used X-rays, which have much shorter wavelengths than visible light. Photographic film can be made sensitive to X-rays, so the X-ray effects could be made visible.

This hunch was right. Fig.12.2 shows diffraction from a crystal of copper sulfate. The discrete nature of the images was of course highly suggestive of the discrete nature of the structure it was revealing, but initially it was considered important because it revealed the wave-like properties of X-rays. Remember that this is still before the age of quantum mechanics when the wave-particle duality of photons was finally unraveled. Laue received the Nobel Prize for Physics in 1914. It was the father and son team of William Henry Bragg and William Lawrence Bragg who made the mathematical connection between image and crystal structure (for which they received the Nobel Prize for Physics in 1915) and subsequently went on to make huge advances in revealing explicit crystal

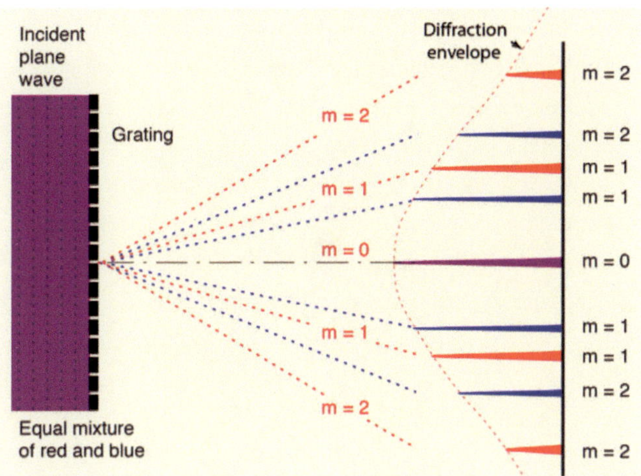

Figure 12.1: A diffraction grating affects different wavelengths (colors) of light in different ways. The figure illustrates the distinct bandings of red and blue as a mix of red and blue light passes through such a grating. The values of m refer to the repetitions of the same color at different angles. With permission, R. Nave, http://hyperphysics.phy-astr.gsu.edu/hbase/phyopt/grating.html\protect\protect\leavevmode@ifvmode\kern+.1667em\relax.

structure through diffraction. At least half a dozen Nobel Prizes have been awarded directly for the discovery and development of techniques around diffraction, and far more when it comes to instances where diffraction played a crucial role (for example in revealing the spiral structure of DNA). Later, electrons and neutrons, both of which have wave-like as well as particle attributes, were used to create diffraction images (and more Nobel Prizes!). As it turns out, almost all atoms and molecules can be teased into crystal structures and so made into excellent subjects for diffraction. Diffraction has become a completely standard tool in determining the atomic structure of complex molecules – hemoglobin, penicillin, vitamin B12, DNA, and some 90,000 more.

Still, the nature of diffraction images is initially confusing. The dots are not the atoms! Each dot indicates periodic repetition in the direction indicated by the position of the dot relative to the center of the diffraction pattern. Waves reinforce in some directions and cancel in others. The size of a dot is an indication of the intensity of radiation received at that point, which in turn depends on the coherence of the waves in that direction. We are not looking at a direct image of the crystal but rather at an image of the combined effects of interference that it has on X-rays that pass through it. In fact it is a Fourier transform, and as we now know, the Fourier transform tends to interchange long-range global structure and local structure. This is why diffraction is such a powerful tool in understanding long-range aspects of crystallography. But the effect is even more subtle and less transparent than it might first appear. It took many years and huge efforts to learn how to unravel the complexities of complex molecules using diffraction. It also

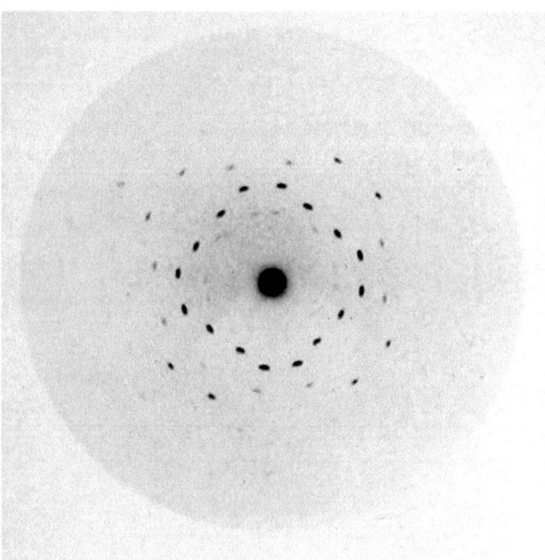

Figure 12.2: An example of X-ray diffraction from a crystal of copper sulfate taken from von Laue's 1912 paper. Von W. Friedrich, P. Knippinig, M. Laue, /en.wikipedia.org/wiki/X-ray_diffraction. Public Domain.

took many years to fully understand the mathematics which describes it. Many surprises continued to come.

Next we will explore how the dots arise, how they are directly related to repetition, and how the world of crystallography was shaken by more recent discoveries.

12.3 The mathematics of diffraction

12.3.1 The effects of diffraction

Diffraction has become a mainstay of modern materials science. Yet it is an effect whose origins are buried deeply within physical theories. It is not a question of the mathematics being difficult: it is the physics itself. Ultimately we are dealing with quantum mechanical effects and the manifestation as waves of what we usually think of as particles. Whether it is X-rays (photons), electrons, or neutrons, for all of these are able to produce diffraction, wave-particle duality dominates the world at the nano-scale of things. Thinking in terms of waves is intuitively the simplest. We will talk in terms of X-rays.

At the outset we have a collection of atoms. In terms of incoming X-rays, each atom is viewed as being a scatterer, that is, it will scatter the energy of any X-ray that interacts with it.[4] Each atom absorbs the X-ray and then emits it again, at the same wavelength, but now in the form of a spherical wave. For the purposes of illustration it is easier to stick to two dimensions and planar waves in which the X-rays will be emitted from each atom as a circular wave, Fig.12.4. The mathematics works out in the same way, and in crystals,

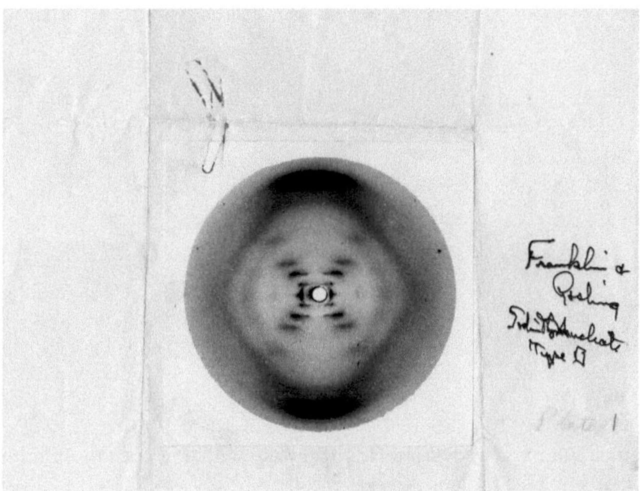

Figure 12.3: Photo 51: Rosalind Franklin's famous diffraction image of DNA that gave Watson and Crick the definitive insight into the spiral structure of DNA. The accompanying story of the failure to properly recognize her vital contribution to this momentous discovery is one of the sadder episodes in this remarkable field. Public Domain.

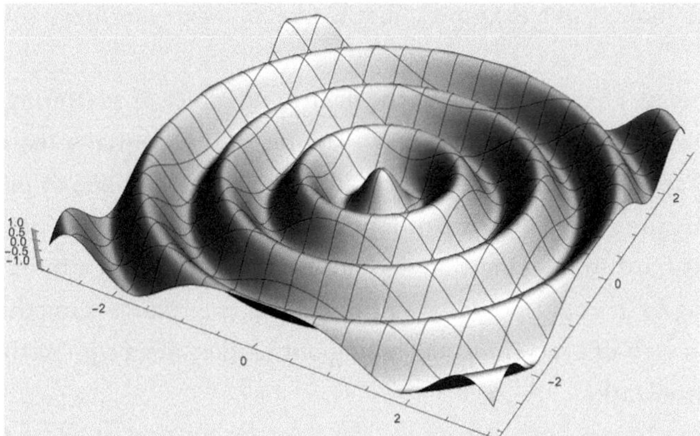

Figure 12.4: A circular wave. As we saw in Ch.11, the mathematical formulation of a wave is complex-valued. Here only the real part is shown.

where structural planes are dominant, the planar viewpoint is often an appropriate one anyway.

The effect of the scattering of an X-ray from two scatterers (atoms) is to produce a wave that combines them both, such as in Fig.12.5. The two waves "interfere" with each other, with their amplitudes complementing each other in some places and opposing each other in others.

It is not hard to imagine that the effect of the interference of scattered waves from a

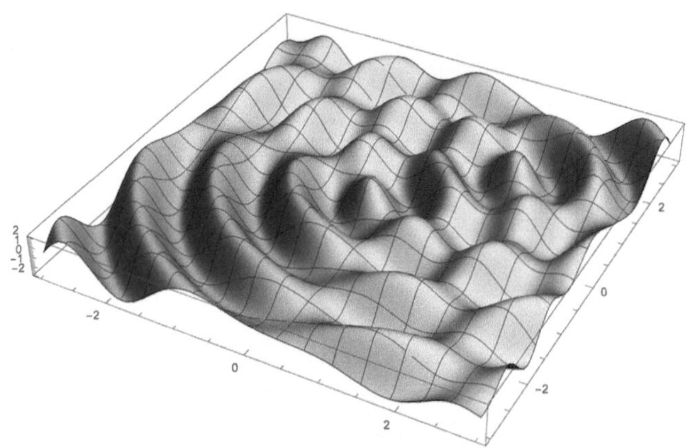

Figure 12.5: When circular waves overlap they interfere with each other. Again only the real component is shown.

large number of closely spaced points can be rather significant. Just how this plays out depends enormously on the arrangement of the atoms, which is precisely the reason that diffraction is useful to determining atomic structure. Before taking this up in more detail, it is interesting to look at two examples, just to see how extraordinary this interference effect can be.

In the first image, Fig.12.6, we show the scattering effect of scattering from a set of 400 randomly chosen points lying in a disk. The result is essentially featureless except for a single great spike in the middle. In the second, Fig.12.7, we see 81 points placed in a square grid. This time the scattering produces a remarkable grid-like pattern of spikes. In both cases it is hard to imagine that something of the order of a hundred waves, all of the form of Fig.12.4, could possibly so interact as to produce something as we see here. Flatness is a sign of everything cancelling out. Spikes are a sign of the waves' high degrees of synchronization.

It is interesting to see how this image develops by looking at the diffraction from square grids of $9, 25, 49, 81$ points in turn, Fig.12.8. The positions of the spikes don't change as we increase the number of points in the grid, and in all four cases the diffraction image of spikes extends indefinitely. What changes is that as the grid gets larger the spikes get higher and sharper. This is the indication that the pattern is more pronounced. If we were able to go to the limit, where the square grid went on forever, these spikes would be perfect, each with no width at all.

The salient observation is that the diffraction image is connected to the pattern of atoms that create it in a very conspicuous way. What is also apparent is that the diffraction image is related to the overall structure of the crystal, not so much the local detailed structure. That's what makes it so important. What is going on behind the scenes is what we want to explain.

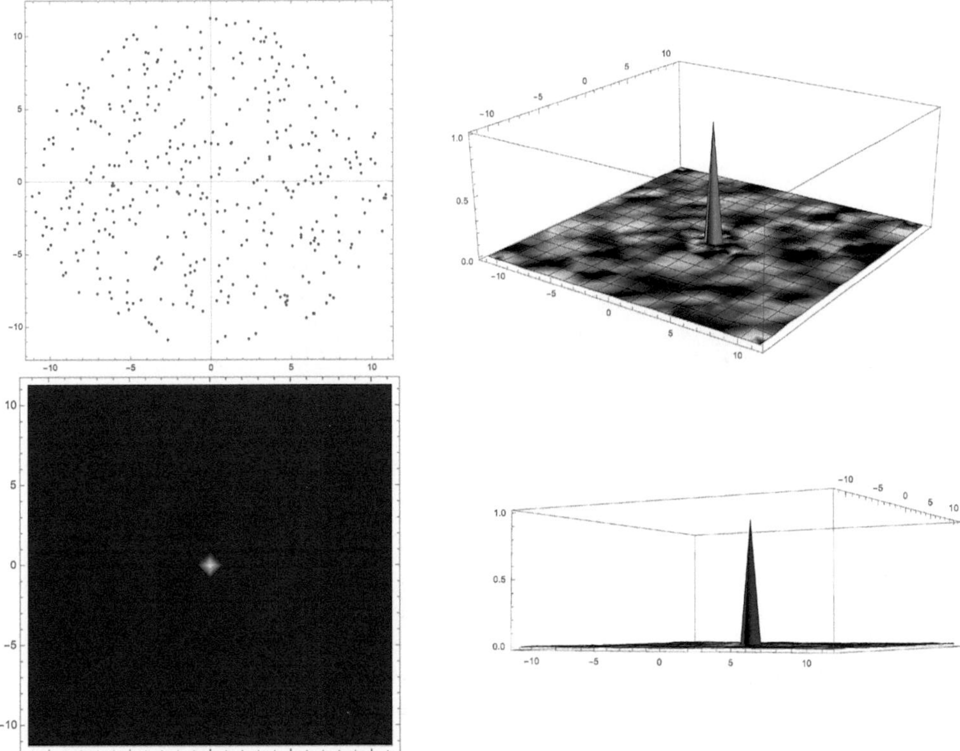

Figure 12.6: The figure shows a plot of 400 random points in a disk of radius just over 11 with average density equal to 1. The graph also shows a plot of the diffraction over a square of roughly the same size. The outcome is basically featureless except for a single central spike at 0.

12.3.2 The appearance of phase differences

When an X-ray interacts with an atom it is absorbed and then re-emitted as a spherical wave of the same frequency centered on the atom. When the same wave interacts with two different atoms the two spherical emissions will overlap each other creating a new wave that is a combination of both. It is how these combinations amplify or diminish each other that is the source of diffraction. The effect takes place in 3D but, as we have said, the 2D equivalent is just as relevant and easier to picture. Waves are spatial-temporal entities, but here we can ignore the temporal part. It is the relative positions of pairs of atoms that lie at the heart of the matter. As we know, the mathematical expression of a pure planar wave is of the form $w_{\mathbf{k}} = \mathbf{e}(2\pi(\mathbf{k}.\mathbf{x}))$, where $\mathbf{k}$ specifies both the direction of motion and the wavelength and $\mathbf{x}$ specifies extension in space (here two-dimensional, see (11.22)). Although this is a complex-valued function, in the end we will see that the total effect of diffraction is real-valued.

With this in mind, consider the scattering from two atoms, Fig.12.9. Two scatterers

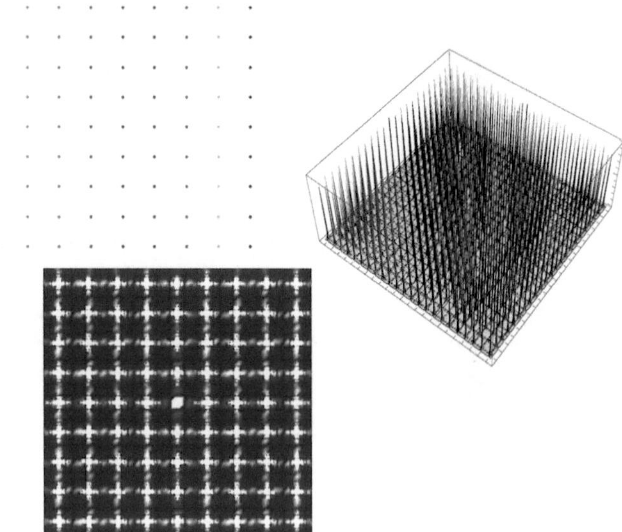

Figure 12.7: The figure shows a plot of 81 points in a square lattice grid and the resulting diffraction, both as a plot of the diffraction (right) and as a density plot (lower). The outcome is a series of spikes that look just like the central one and, as the density plot reveals, also laid out in a square gird. But, as the figure of spikes suggests, diffraction is not constrained to a square, but actually extends unbounded. The coherent patterning of the lattice grid of scatterers brings about consistency of phase differences, and this reveals itself with the intense peaks where these phases agree forming a secondary lattice, the *reciprocal lattice*.

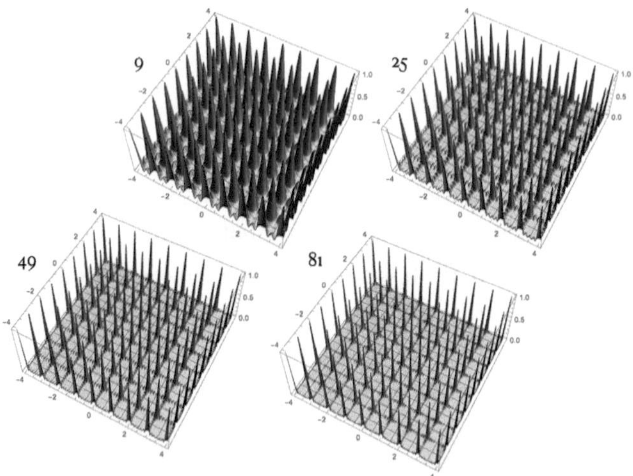

Figure 12.8: The figure shows diffraction from square grids with 9, 25, 49, 81 points. We have limited the graphics to an 8 by 8 square. The spike-like shapes extend in the same way over the entire plane even though we are scattering over only a finite number of points. As the grid is refined the spikes remain in the same places but become ever more spike-like.

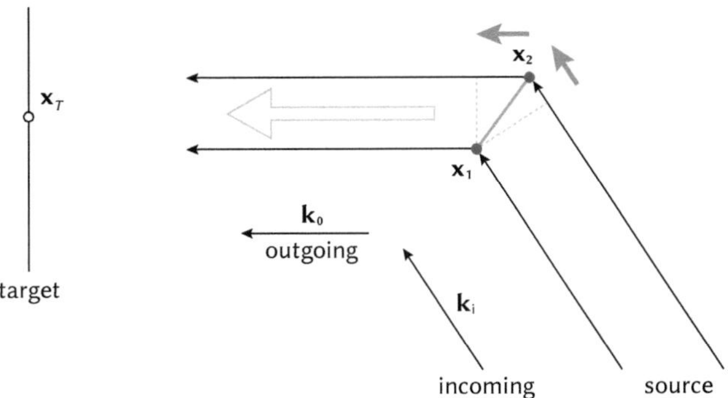

Figure 12.9: Two scatterers, $\mathbf{x}_1, \mathbf{x}_2$, shown in blue, are impinged by X-rays specified by $\mathbf{k}_i$, which gives both direction and wavelength from a source which we can take to be effectively infinitely far away. We consider their combined effect where they meet at some point on the target (say a photosensitive material) at some considerable distance in the outgoing direction $\mathbf{k}_o$. The change in direction, $\Delta\mathbf{k} = \mathbf{k}_o - \mathbf{k}_i$, produces a difference in the path lengths from the source to target due to the different positions of the two scatterers, indicated by the two thicker arrows. The effect is that the two outgoing waves differ in phase. The diffraction image arises out of the combined effects of these phase differences when taken over all pairs of atoms in some finite sample of material.

are interacting with an *incoming* wave whose direction and wavelength is $\mathbf{k}_i$. Each atom produces its own circular wave and we examine the *outgoing* wave fronts which have the same wavelength, but now in some particular direction $\mathbf{k}_o$. In experimental terms the incoming wave front is initiated so far away relative to the distance between the two atoms that it can be considered to appear flat.[5] The wavefront as it reaches the first atom $\mathbf{x}_1$ is indicated by a dotted line in the figure. Again, seen from the perspective of the target (say a photographic plate), which is distant from the perspective of atoms, the wavefront emanating from the atom $\mathbf{x}_1$ is also flat, as shown by the second dotted line.

What we can see is that from the perspective of $\mathbf{x}_1$ the incoming and outgoing wavefronts from the atom at $\mathbf{x}_2$ have to go further (the two arrowed segments of the outer path) before they line up again with the outgoing wave from $\mathbf{x}_1$. The effect is a phase difference, which means that the two waves are, in general, out of sync. It is these phase differences that account for the interference effects that we are trying to quantify.

This phase shift (which is a lag in the illustration we have given) depends only on the sum of these two arrowed segments, which in turn depends simply on the difference between the positions of $\mathbf{x}_1$ and $\mathbf{x}_2$, which is $\mathbf{r} := \mathbf{x}_2 - \mathbf{x}_1$, and the difference between the two directions $\Delta\mathbf{k} := \mathbf{k}_o - \mathbf{k}_i$. Each of these is a two-component vector: $\mathbf{r} = (r_1, r_2)$ and $\Delta\mathbf{k} = (k_1, k_2)$. The actual phase difference can be expressed succinctly by using the dot product that we introduced around equation (11.20): it is simply

$$\mathbf{r}.\Delta\mathbf{k} = -(r_1 k_1 + r_2 k_2) = -(\mathbf{x}_2 - \mathbf{x}_1).\Delta\mathbf{k}\,.$$

What is important here is just to note that we have a simple way to express what we need out of the rather complicated picture presented by Fig.12.9. The final analysis that leads to the diffraction is worked out in the next section. Skipping these details, the final result is found at equation (12.4).

12.3.3 From phase difference to diffraction

We finish the analysis by comparing the two paths after accounting for the path difference: that is where the large left-pointing arrow begins. Suppose the target point, way off to the left, is at $\mathbf{x}_T = \mathbf{x}_1 + \mathbf{k}$ (so $\mathbf{k}$ is some large multiple of $\mathbf{k}_o$). We put the T subscript to remind us this is a target point that is essentially the same for all the scattering that we are looking at, and that it is at a considerable distance relative to the size of the atomic sample that we are studying. This assumption allows us to assume that the parallel rays in the direction $\mathbf{k}_0$ from the two (and ultimately any other nearby scatterers) meet at the same target point, see Endnote 5. At the target we see that the waves there are $\mathbf{e}(2\pi\mathbf{x}_T.\mathbf{k})$ and $\mathbf{e}(2\pi(\mathbf{x}_T.\mathbf{k} - \mathbf{r}.\Delta\mathbf{k}))$. The total contribution of the two waves is their sum, which can be written as

$$\mathbf{e}(2\pi\mathbf{x}_T.\mathbf{k}) + \mathbf{e}(2\pi(\mathbf{x}_T.\mathbf{k} - \mathbf{r}.\Delta\mathbf{k})) = \mathbf{e}(2\pi\mathbf{x}_T.\mathbf{k})(1 + \mathbf{e}(-2\pi\mathbf{r}.\Delta\mathbf{k}))\,. \qquad (12.1)$$

We are interested in the relative difference that arises out of this difference of paths due to the change $\Delta\mathbf{k}$ in directions from source to atom to target, so the key part of this is $1 + \mathbf{e}(-2\pi(\mathbf{r}.\Delta\mathbf{k}))$, which we can write as

$$\mathbf{e}(2\pi(\mathbf{x}_1 - \mathbf{x}_1)).\Delta\mathbf{k} + \mathbf{e}(2\pi(\mathbf{x}_1 - \mathbf{x}_2).\Delta\mathbf{k})\,.$$

Writing the 1 in the form $\mathbf{e}(0) = \mathbf{e}(2\pi(\mathbf{x}_1 - \mathbf{x}_1))$ on the left seems strange, but the reason for doing so quickly becomes clear. What we have written so far is the part that depends on just *two atoms and as seen from the perspective of atom* $\mathbf{x}_1$.

Let us suppose now that we are looking at a sample that has a number of identical atoms located at $\mathbf{x}_1, \mathbf{x}_2, \mathbf{x}_3, \dots, \mathbf{x}_N$. Then again, seen from the point of view of atom $\mathbf{x}_1$, we would have

$$\begin{aligned}&\mathbf{e}(2\pi(\mathbf{x}_1 - \mathbf{x}_1).\Delta\mathbf{k}) + \mathbf{e}(2\pi(\mathbf{x}_1 - \mathbf{x}_2).\Delta\mathbf{k})\\ &\qquad + \mathbf{e}(2\pi(\mathbf{x}_1 - \mathbf{x}_3).\Delta\mathbf{k}) + \cdots + \mathbf{e}(2\pi(\mathbf{x}_1 - \mathbf{x}_N).\Delta\mathbf{k})\,.\end{aligned}$$

If instead we looked at it from the point of view of atom $\mathbf{x}_2$, it would read

$$\begin{aligned}&\mathbf{e}((2\pi(\mathbf{x}_2 - \mathbf{x}_1).\Delta\mathbf{k}) + \mathbf{e}((2\pi(\mathbf{x}_2 - \mathbf{x}_2).\Delta\mathbf{k})\\ &\qquad + \mathbf{e}(2\pi(\mathbf{x}_2 - \mathbf{x}_3).\Delta\mathbf{k}) + \cdots + \mathbf{e}(2\pi(\mathbf{x}_2 - \mathbf{x}_N).\Delta\mathbf{k})\,,\end{aligned}$$

and so on for $\mathbf{x}_3, \dots, \mathbf{x}_N$. Since no atom is more important than any other here, all of these scenarios must take place. The final result is extraordinarily simple, namely the sum of all possible $\mathbf{e}((\mathbf{x}_i - \mathbf{x}_j).\Delta\mathbf{k})$, where $1 \le i \le N$ and $1 \le j \le N$. This includes all the cases where $i = j$ in which case there is a term $\mathbf{e}(2\pi(\mathbf{x}_i - \mathbf{x}_i).\Delta\mathbf{k}) = 1$. This whole sum can also be written as the double sum

$$\sum_{i=1}^{N}\sum_{j=1}^{N} \mathbf{e}(2\pi(\mathbf{x}_i - \mathbf{x}_j).\Delta\mathbf{k})\,. \tag{12.2}$$

This then, finally, is the *diffraction* at $\mathbf{x}$.

This double sum can be rewritten in the form of a product:

$$\sum_{i=1}^{N}\sum_{j=1}^{N} \mathbf{e}(2\pi(\mathbf{x}_i - \mathbf{x}_j).\Delta\mathbf{k}) = \left(\sum_{i=1}^{N} \mathbf{e}(2\pi\mathbf{x}_i.\Delta\mathbf{k})\right)\left(\sum_{j=1}^{N} \mathbf{e}(-2\pi\mathbf{x}_j.\Delta\mathbf{k})\right). \tag{12.3}$$

This arises from the simple fact that $\mathbf{e}(2\pi(\mathbf{x}_i - \mathbf{x}_j).\Delta\mathbf{k}) = \mathbf{e}(2\pi\mathbf{x}_i.\Delta\mathbf{k})\,\mathbf{e}(-2\pi\mathbf{x}_j.\Delta\mathbf{k})$. Notice the sign change in the second factor. The target point is the same for all the atoms and so, looking at equation (12.1), it represents an overall common factor of $\mathbf{e}(\mathbf{x}_T.\Delta\mathbf{k})$ which does not affect the shape of the diffraction.

Usually the expression that we see is divided by the number of atoms N, see §12.3.4. The diffraction images shown in this section include this normalization.

Given that the atoms and their arrangement are fixed, the *diffraction is a function of* $\Delta\mathbf{k}$, the difference between the directions $\mathbf{k}_i$ (from the source) and $\mathbf{k}_o$ (to the target).

We can expect the diffraction to vary considerably as this difference is varied.

12.3.4 Immediate observations

What the diffraction looks like is revealed by letting the only variable in the expression for the diffraction, namely $\Delta\mathbf{k}$, vary. Since $\Delta\mathbf{k}$ is the change in directions between the incoming and outgoing directions of radiation, it is essentially an angle. In the actual measurement of diffraction in a physical experiment this angle can be varied, so in effect it traces out a circle (or a sphere in 3D—the *Ewald sphere*). Mathematically we are not constrained by such equipment and $\Delta\mathbf{k}$ can not only be varied in angle, but also in length. This allows us to freely determine which values of $\Delta\mathbf{k}$ might be of significance to a material in hand. In doing this $\Delta\mathbf{k}$ loses its meaning as the difference between source and target vectors and we might just as well simplify the notation and simply use $\mathbf{k}$.

If we look at equation (12.3) a little more carefully we can note a clear symmetry between $2\pi\mathbf{x}.\Delta\mathbf{k}$ and $-2\pi\mathbf{x}.\Delta\mathbf{k}$. This has two very visible consequences. The first is that the diffraction is the same for $\Delta\mathbf{k}$ and $-\Delta\mathbf{k}$ so the diffraction images should always look centrally symmetric (the diffraction at points diametrically opposed with respect to the origin is the same).

Next, since it is always true that $\mathbf{e}(-t) = \overline{\mathbf{e}(t)}$ no matter what the value of t is, the second sum in the product shown in equation (12.3) is the complex-conjugate of the first. Since the product $z\bar{z}$ of any complex number is a real number greater than or equal to zero (it is the square length $|z|^2$ of z), we see that the diffraction is always real-valued and non-negative. Writing it in terms of the free two-component variable $\mathbf{k}$ we come to the important conclusion, which we can take as the *mathematical definition of diffraction*:

Mathematically, the diffraction arising from a set $\mathbf{x}_1, \mathbf{x}_2, \dots, \mathbf{x}_N$ of identical scatterers in the plane (or 3D-space) is the real-valued function

$$\frac{1}{N}\left|\sum_{j=1}^{N} \mathbf{e}(2\pi \mathbf{x}_j.\mathbf{k})\right|^2 \qquad (12.4)$$

of the variable $\mathbf{k} \in \mathbb{R}^2$ (or in $\mathbb{R}^3$ in the 3D case).

The appearance of the averaging factor $1/N$ is not overtly important for our discussion here, but from a physical point of view as more atoms are added we should expect that the local intensity of the effect must diminish if the energy involved remains the same. It represents more accurately what happens as N gets very large, and importantly from a mathematical point of view what happens as $N \to \infty$.

The final observation is that if we set $\mathbf{k} = 0$, then every single term $\mathbf{e}(2\pi \mathbf{x}_j.\mathbf{k}) = \mathbf{e}(0) = 1$, so we get that the diffraction at the origin is N. There is always a peak at 0. In fact it is not possible for the diffraction to ever exceed this. No value of $\mathbf{k}$ can make the sum larger than having all their summands equal to 1. This explains why in the situation of random points, where the randomness of the phase differences wipes out all sense of coherence, there is still a definite spike at zero. In summary:

The diffraction is always real, non-negative, and centrally symmetric with a spike at 0.

12.3.5 Why lattices are special

The underlying feature of diffraction is the patterning arising from the interference effects of waves scattering from an arrangement of atoms. We have already seen in Fig.12.6 and Fig.12.8 just how powerfully the actual arrangement affects the outcome.

What is it about a lattice of scatterers that makes the diffraction look like a lattice of spikes? The explanation of this is both simple and revealing. Suppose that the scatterers are in a lattice arrangement, which for simplicity we will assume to be part of the square lattice $\mathbb{Z}^2$. Then each lattice point has coordinates $\mathbf{x}_j = (u_j, v_j)$, where u_j and v_j are

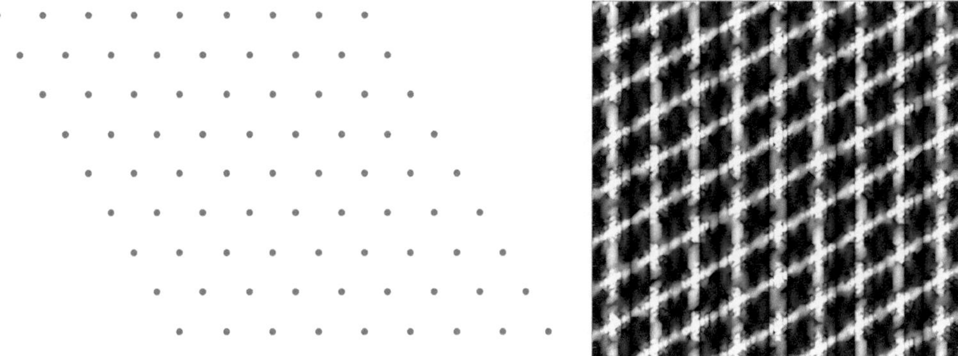

Figure 12.10: The figure shows a partial plot of 81 points from a hexagonal lattice grid. Out of the resulting diffraction, shown here as a density plot, emerges the reciprocal, or dual, lattice, which is also a hexagonal lattice, though at a different orientation.

integers. Now looking at (12.4) we can see at once that for any choice of $\mathbf{k} = (k_1, k_2)$, *where k_1 and k_2 are also integers,* $\mathbf{x}_j.\mathbf{k} = u_j k_1 + v_j k_2$ is an integer too. These resulting integers may all be different, but no matter what integers they are, in the formula for diffraction we are looking at terms of the form

$$\mathbf{e}(2\pi \times \text{ integer}) = 1\,,$$

so all the terms in the diffraction $\mathbf{e}(2\pi\mathbf{k}.\mathbf{x}_j) = 1$. This just goes back to the old fact that in going around the circle an integral number of times we end up where we started. Consequently each summand that we see in the formula for the diffraction is actually a 1, just as it was when $\mathbf{k} = 0$. In other words, we find a spike at each $\mathbf{k} \in \mathbb{Z}^2$. The consistency in the phases manifested by the lattice of atomic scatterers is due to the fact that for all $\mathbf{x} \in \mathbb{Z}^2$ (the lattice in physical space) and for any $\mathbf{k} \in \mathbb{Z}^2$ (in the so-called *reciprocal space*), $\mathbf{x}.\mathbf{k}$ is an integer.

So the spatial and orientational coherence leads to wave coherence and constructive interference (or interaction) of the waves at these points. At other points this constructive interference does not take place and the overall interference is destructive, to the point of reducing the total wave to almost nothing. The resulting pattern is that of sharp spikes at the points of the reciprocal lattice.

The situation just indicated is genuine enough, but it does hide one essential feature—namely that the lattice L of the scatterers and the lattice L^* of the points of constructive interference (the *reciprocal lattice*) are not necessarily the same. For more details see the Endnotes.[6] Fig.12.10 is an example when L is a hexagonal lattice.

The spike-like appearances of the diffraction that we see here appear in physical experiments as the result of accumulation of photons on a target. Typically such targets have been photographic materials which register the locations of significant accumulation as circular dots, their radii reflecting the various intensities. To remove confusion about whether we are dealing with one or many photons in the scattering process, it may help to know that experiments in quantum mechanics have repeatedly shown a remarkable phenomenon. Photons can be emitted one-by-one from the source. The outcome of each single photon will be the arrival of a scattered photon at some point on the photographic plate. Where it arrives is random, but random subject to a probability law that very highly favors the places where the intensity will ultimately be greatest. As photon after photon passes through the configuration of atoms, the diffraction image that we expect emerges. What is so strange about the process is that each source photon must really "experience" the entire collection of atoms and their arrangement, just as we have described above. It emerges as a scattered wave, but at the target has to reveal itself at a single point. The arrangement of the atoms determines the probability law that over the course of a large number of events (incoming photons) reveals itself as the diffraction pattern. This is one of the wonders of the famous wave-particle duality. We delve into this further in Ch.13.

12.3.6 The making of the crystallographic paradigm

Crystals were first characterized as such by the fact that their atoms are arranged periodically, based on lattice symmetry. There is an arrangement of atoms, visualized as being in a fundamental cell, and this fundamental cell repeats indefinitely in three spatial directions to fill out a part of space, Fig.12.11. This is the simplest orderly way we can imagine to create extended three-dimensional collections of atoms, and it appears everywhere in Nature. Most molecules, including such monsters as DNA and hemoglobin, can be crystallized; that is to say collections of them can be brought into lattice arrays, where they can be analyzed by methods of diffraction. This remains a crucial experimental way in which the shapes of bio-molecules are determined.

As we have just seen, the diffraction from crystals, being based on scatterers located on a lattice, appears as just a set of point-like spots located at the points of the reciprocal lattice. If, as is quite typical, there are atoms of different scattering strengths in the fundamental cell then the resulting point-like spots may not all have the same intensity, but generally there is still overall lattice symmetry.

In principle a perfect crystal produces perfect points and nothing else. This is called *pure point diffraction*, and the bright spots are called *Bragg peaks*. This is the idealization that would occur if the crystal were perfect and infinite in extent. In reality the Bragg peaks are distinct, but on close examination they are seen as peaks rather than points and there is usually some sort of diffuse background due to random movement of the atoms, the finiteness of the material, and so on. On photographic materials the intensity of the peaks appears as somewhat diffuse anyway: the more intensity the wider the spot.

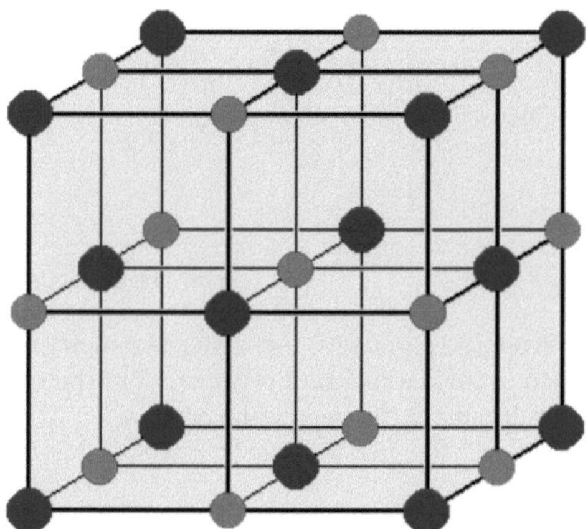

Figure 12.11: The figure is a schematic image of a single fundamental cell of the face-centered cubic lattice of NaCl (ordinary salt). Sodium atoms are indicated by black balls and chlorine by grey. The eight smaller cubes appearing here do not all have the same (translational) atomic arrangements, so none of them can serve as a fundamental cell. As it is pictured there are 14 chlorine atoms and only 13 sodium atoms. In fact they occur equally in the full crystal. If it were pictured with the grey balls at the corners, it would look the other way around.

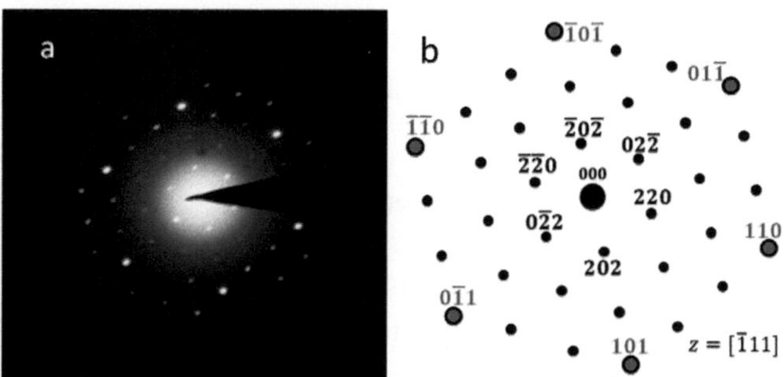

Figure 12.12: Spot diffraction pattern of quenched and tempered steel. The form of the indexing is the standard way of assigning triples of integer coordinates to the Bragg peaks, reflecting the lattice structure of the diffraction. Open access, IntechOpen, intechopen.com/books/modern-electron-microscopy-in-physical-and-life-sciences/electron-diffraction .

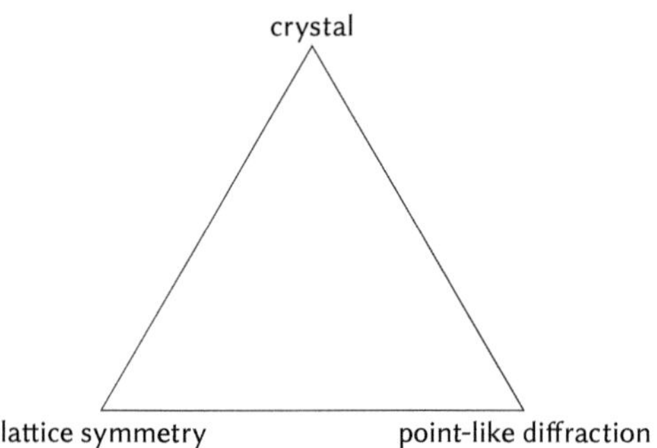

Figure 12.13: The crystal-diffraction paradigm: the lattice symmetry of crystals and the point-like diffraction that they produce are just different expressions of the very same thing. This paradigm collapsed in the last decade of the twentieth century.

Theoretically in the case of crystals all the peaks have the same intensity, but the realities of the situation usually make the spot at 0 the most intense of all. See Fig.12.12.

Not surprisingly over time a sort of paradigm emerged—here we call it the *paradigm of crystal-diffraction* —that suggested that in the material world crystals, lattice symmetry, and pure point diffraction are indicative of the same phenomenon, Fig.12.13. This paradigm was deeply entrenched in the minds of virtually all crystallographers, but it turned out to be false! The story behind this, which we tell in ‡§18.2, is quite marvelous, interweaving history, logic, imagination, Turing's halting problem, and contention. For a more detailed mathematical discussion of this see [115].

12.3.7 The inverse phase problem with diffraction

There is one drawback to all of this, and it is the source of one of the fundamental difficulties that confronts anyone working with diffraction. According to our definition (12.4), the diffraction is based on

$$z := \sum_{i=1}^{N} \mathbf{e}(2\pi \mathbf{x}_i.\mathbf{k}),$$

in which we can think of a z as a function of the variable $\mathbf{k}$. If this were truly the diffraction it would be a straightforward matter to use the Fourier transform to extract the individual terms and determine the locations $\mathbf{x}_i$ of the atoms. Unfortunately the diffraction is not this sum, but rather its absolute square, which is a real number. There is no way to know what z is from only knowing $|z|^2$. It is not the square root that we would require that is a problem— we can easily compute $|z|$. The problem is that z is a complex number and is really $z = r\mathbf{e}(t)$ for some real number t. But $|z| = r$, so the *phase information*, the value of t, is completely missing. The diffraction, beautiful as it is, does not contain all the information needed to reconstruct exactly what produced it. It was Linus Pauling who first pointed out that there are even different crystal structures that can produce the same diffraction! The reconstruction of the missing phase information has to be based on other knowledge about the atoms involved, their relative masses and scattering strengths, and their potential locations. One major advance was the realization by Lindo Patterson that equation (12.2) could be used to at least extract information about all the inter-atomic directions and spacings $\mathbf{x}_i - \mathbf{x}_j$.

Especially in the first part of the twentieth century, without the raw computing power of modern computers, it was both an extremely difficult and tedious task to extract the physical structure of crystals from their diffraction. Nonetheless the incredible work was done, including determining the atomic structures of a great array of materials such as benzene, hemoglobin, insulin, vitamin B12, and, perhaps most famously, DNA. One can only admire these people— the father and son team William Henry and William Lawrence Bragg, Dorothy Hodgkin, Lindo Patterson, Linus Pauling, J. D. Bernal, Kathleen Lonsdale, Robert Robinson, Peter Debye, Roslyn Franklin, and many others—who devoted so much time and skill to these daunting studies.

12.4 Long-range order

12.4.1 Introduction

Long-range order refers to the apparent coherence that we experience across separations of distance and time in recognizing continuing or reappearing elements of sameness. In its simplest form, long-range order is just plain repetition. Crystals are the prime examples. This space-filling form of order, simple repetition of a block of atoms so as to fill out space, is manifested universally over the material world. Of course this "simple repetition" is a balancing act of countless electrons and inter-atomic forces, but the fact remains that most molecular structures can be crystallized and these crystal forms represent stable energy configurations that can be maintained for significant periods of time.

The level of precision of this repetition in crystals can be exquisite. Ghirardi describes the existence of 20 cm-long crystals that have an accuracy of less than a billionth of a centimeter end to end [63]. He goes on to describe an amazing experiment of neutron diffraction in such a crystal. Imagine the 20 cm-long crystal tilted slightly with respect to the horizontal, so one end is lower than the other. The lower end is closer to the center of the Earth than the upper end, and so the gravitational force of the Earth is larger at the lower end (minutely so) than at the upper. This produces a distortion in the crystal lattice and this distortion can actually be observed by the neutron diffraction. The hidden forms of precision that lie everywhere around and within us are staggering.

This is the long-range order of crystals, but long-range order goes far beyond such seemingly sterile forms of repetition. Think of dynamical systems that depend on chance.

There we have also seen recurrent order, though the underlying repetitivity is of a very different nature. It is now expressed in the form of a probability law and the law of large numbers which governs expectation, not deterministic reappearance. Yet in its own way it is just as powerful a form of long-range order as anything else. Weaker still seems the Poincaré recursion theorem of §7.9 that shows that recurrence is a feature almost built into the concept of pattern.

Between these extremes, and no doubt even beyond them, the whole question of what long-range order means remains largely uncharted. Perhaps to understand it fully would be to understand the Cosmos in its entirety.

12.4.2 Almost-periodicity

In §11.6.1 we raised the distinction between combining pure waves which have a common period and pure waves that don't. This is the very question that we raised in §2.2—the difference between commensurate versus incommensurate cycles. The case of *commensurate* cycles reduces to the problem of functions on a circle, which we have treated in some depth. Now we return to the incommensurate case, and a good place to start is the type of example that we saw in Ch.2 at Fig.2.7. Let's start with two simple waves, one with period equal to 1 and the other with period equal to the golden ratio φ, and add them together,

$$f(t) := \sin(2\pi t) + \sin(2\pi t/\varphi)\,. \qquad (12.5)$$

We know that the pair 1 and φ are incommensurate (they have no common measure) and so these two waves have no common period. As a result their sum cannot be periodic. Yet, as Fig.12.14 shows, it certainly is something remarkably close to periodic. Here we see that 13 is an *almost-period* in the sense that $|f(t) - f(t+13)|$ is *uniformly* small, i.e. bounded by a single small number, no matter what the value of t. This is the sense that Harold Bohr had in mind when he defined almost-periodicity. In fact 13 is by no means the only almost-period that we can find in this example. There are many more, and there are much better ones than 13. But they are not obtained, as one might first think—simple combinations like $26 = 13 + 13$ are actually worse than 13. We saw this before in §2.2. We noted that 365 days is a reasonable approximation for the length of a year, but $2 \times 365 = 730$ days is not as good an approximation for two years, and take a few hundred of these 365-day years and the seasons are completely out of line with the calendar. In our present example 21 is a better almost-period than 13. In our golden ratio example we can actually have the uniform matching to be as close as we like, but the closer the match the larger the almost-periods are. So there are almost-periods p of different "goodness", depending on how uniformly small we want $|f(t) - f(t+p)|$ to be.

Why does this happen? The answer is more straightforward than one might expect. Any irrational number can be approximated by rational numbers as closely as

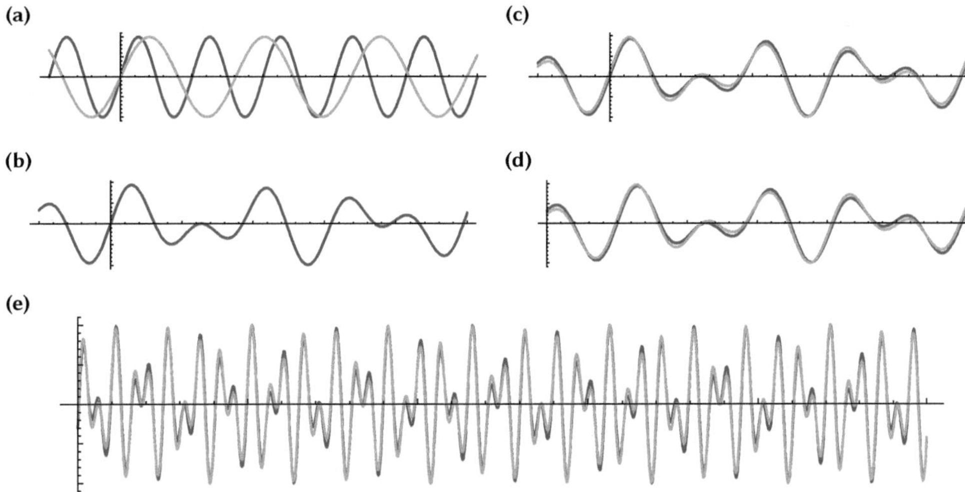

Figure 12.14: (a) shows plots of $\sin(2\pi t)$ (black) and $\sin((2\pi/\varphi)t)$ (grey), which are periodic with periods 1 and φ respectively. (b) is a plot of the sum of these two: $f(t) := \sin(2\pi t) + \sin((2\pi/\varphi)t)$. This new function is totally aperiodic. It has no periods at all, but it does have "almost-periods". The two plots (c) and (d) on the right-hand side show the overlaid plots of $f(t)$ and $f(t + 13)$ over two ranges, first of $[-1, 5]$, second over a range of $[88, 94]$. Finally at the bottom we see the plots of $f(t)$ and $f(t + 13)$ over the extended range $[50, 100]$. The degree to which they match is amazing. This matching is retained everywhere. It is certain that 13 is not a genuine period of this function, but it is an *almost-period.*

we wish. In the case of the golden ratio, the ratios of successive Fibonacci numbers, $1, 1, 2, 3, 5, 8, 13, 34, 55, \ldots$ are good approximants that get better and better the further we go.[7] Underlying this is the fact that

$$\frac{1}{1} \quad \frac{2}{1} \quad \frac{3}{2} \quad \frac{5}{3} \quad \frac{8}{5} \quad \frac{13}{8} \quad \frac{21}{13} \quad \frac{34}{21} \quad \cdots \longrightarrow \varphi = 1.6180339887\ldots .$$

We can see why these ratios are "almost" periods by comparing $f(t)$ and $f(t+13)$ using (12.5). The first term $\sin(2\pi t)$ doesn't change if we replace t by $t + 13$, because whole numbers are periods of $\sin(2\pi t)$. For the second term, $\sin(2\pi t/\phi)$, by comparing the arguments we see that

$$\frac{t+13}{\varphi} - \frac{t}{\varphi} = \frac{13}{\varphi} \sim \frac{13}{13/8} = 8\,,$$

which is close to an integer: $13/\varphi = 8.03444\ldots$. Similarly, $55/\varphi = 33.9919\ldots$ and this leads to

$$\begin{aligned} |f(t) - f(t+13)| &< 0.22 \ \text{ for all } t, \text{ and,} \\ |f(t) - f(t+55)| &< 0.052 \ \text{ for all } t\,. \end{aligned} \qquad (12.6)$$

Almost-periodic order is a relaxation of the exactness of periodic order, but still retains a very significant sense of repetition.[8] What we have just seen is actually an example of what is a very general phenomenon, which was first established by Harald Bohr:[9]

Bohr's theorem: any linear combination of pure waves, i.e. any expression of the form

$$c_1\mathbf{e}(2\pi r_1 t) + c_2\mathbf{e}(2\pi r_2 t) + \cdots + c_n\mathbf{e}(2\pi r_n t)\,, \tag{12.7}$$

where the r_j are any real numbers and the c_j are any real or complex numbers, is an almost-periodic function of the variable t in exactly the sense that we have been discussing. Furthermore, any almost-periodic function can be written in this way, or as a *limit* of such as $n \to \infty$.

So we are combining periodic functions, but with no restriction at all on their periods $1/r_j$, and the result is always almost-periodic. Conversely every almost-periodic function arises in this way.

We have stated this in a one-dimensional setting, but it applies in higher dimensional settings too, where the r_j are replaced by points $\mathbf{r}_j$ in some higher dimensional space, the variable t by a variable vector $\mathbf{k}$ from the same space, and the terms $\mathbf{e}(2\pi r_j t)$ are replaced by $\mathbf{e}(2\pi\, \mathbf{r}_j.\mathbf{k})$, where the dot product appears again.

Hence, to the extent that its various periods remain unchanging, the Earth-Moon-Sun system is guaranteed to be almost-periodic. That it should have such a good almost-period as the Meton cycle, which is really relatively small, is perhaps a stroke of good fortune.

Now, thinking about Bohr's theorem, an interesting idea emerges. Almost periodic functions, as expressed in (12.7), are related to the set of values r_j that produce them. Think of these r_j as points on the real line. Then we find a link. If we go back to (12.2), which is the diffraction of a finite set of scatterers (atoms) we see that, according to Bohr, it is an almost-periodic function in the variable $\Delta\mathbf{k}$. So the diffraction of a finite set of scatterers is an almost-periodic function. But to actually get Bragg peaks out of the diffraction needs more than this. What is needed is that all, or at least significant numbers, of the waves $\mathbf{e}(2\pi r_j t)$ come very closely into sync so as to produce Bragg peaks. Unless there is some coherence to their arrangement this almost-periodicity on its own is not enough to produce the phenomenon of pure point diffraction. But it is a link.

What sort of arrangements of atoms or scatterers will lead to the strong correlations required for Bragg peaks?

12.4.3 The golden mean tiling and its diffraction

We are going to give a famous and very revealing example of a point set on the line that is almost-periodic but definitely not periodic (it has no translational symmetries), yet still produces pure point diffraction. One remarkable thing about it is that it is essentially self-generating. Fig.12.15 explains how the point set comes about.

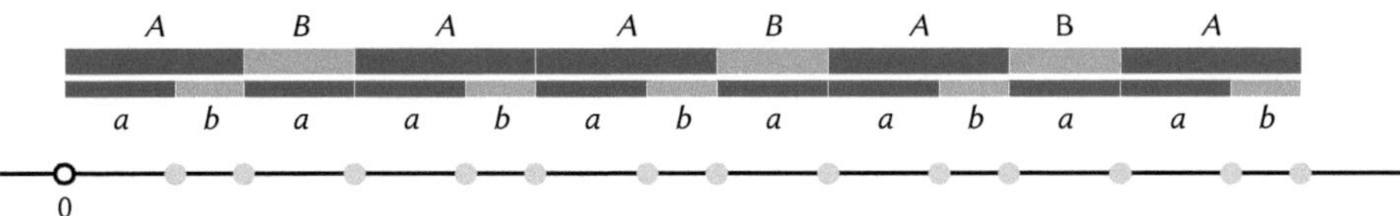

Figure 12.15: On the positive real line, shown at the bottom, we see a series of points (indicated by grey dots). Their positions mark the right-hand ends of the black a and grey b intervals that we can think of as tiling the entire real line to the right of 0. The lengths of these are in the golden ratio $\varphi = (1+\sqrt{5})/2$: black tiles have length φ and grey ones length 1. The layout of the intervals, and hence the dots, is determined by a curious rule. Above the a-b tiling is a derived tiling with tiles A and B that produces exactly the same sequence of letters as the original: *abaababa*... versus $ABAABABA\dots$. The relationship between the two tilings is clear: each tile A comes from an adjacent pair *ab* and the tiles B come from the remaining tiles of type a. The length of an A tile is $\varphi+1=\varphi^2$ and that of a B tile is φ: they are versions of a and b tiles scaled up by a factor of φ. Knowing the rule that connects them it is easy to continue the two tilings as far as one wishes. For instance, the next large tile is an A because the sequence up to *abaababa* continues as *abaababaab*, and that adjoins the pair *ab*. The whole thing can be derived from the initial pair *ab*, each of the two sequences forwarding the other. The overall effect is that in scaling up by a factor of φ we get the very same sequence: the sequence is *self-similar*.

The tiling

The construction of this tiling, which we call the *golden mean tiling*, is based again on using self-similarity in the ratio $\varphi : 1$. This self-similarity is evident in the figure. The golden ratio is defined by the equation $\varphi^2 = \varphi + 1$, or equivalently $(\varphi+1) : \varphi = \varphi : 1$. This actually guarantees that the tiling does not repeat—there is no point in the emergent sequence of a and b tiles where the thing just starts all over again, repeating exactly everything that went before. Looking at the figure, we can see why. Any period T of the smaller tiles would infer another one of period φT for the larger tiles. But these larger tiles are composed of the smaller tiles, so that periodicity of the larger tiles would infer the periodicity of smaller tiles, but with a period of φT. But the incommensurability of φ and 1 makes this pair of periods mutually incompatible. This is a well-known method of producing aperiodic tilings.

There may be no periods, but nonetheless the amount of repetitivity is incredible. No matter what finite substring of consecutive letters we choose, no matter how long, that same substring reappears infinitely often and even with bounded distances between repetition. Again it is easy to see why. The whole picture that we see emerges from the initial pair *ab*. Moreover, wherever this pair reappears, out of that pair the entire sequence will be born again to be seen later along in the tiling. Everything born is born again! Infinite repetition, but it is not periodic.

What we are seeing is a geometric form of almost periodicity: the sequence we have constructed is almost-periodic in almost any sense one might take for this term.

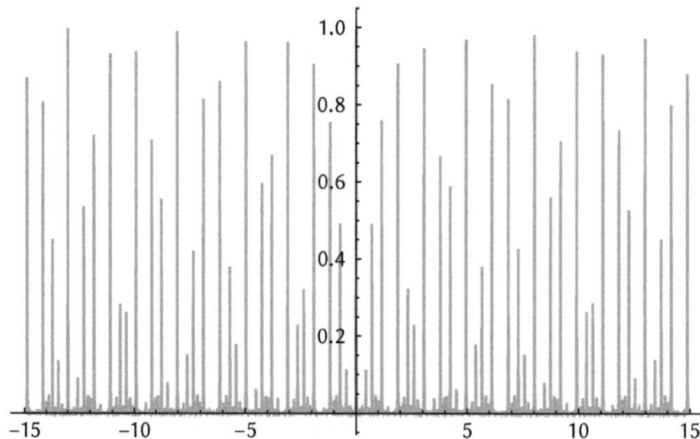

Figure 12.16: Here is the diffraction from the 55 points of the Fibonacci substitution sequence shown in Fig.12.15. What happens here is due to reinforcement of waves at particular points and cancelling out of those waves everywhere else. This is not misleading: the diffraction image from the entire substitution sequence (either one-sided or two-sided) is actually perfectly pure point. Neither is the figure misleading in suggesting that there is a considerable variety in the intensities. Each peak indicates the frequency of the repetition in steps of length equal to its position. The strongest peak is at 0, as expected. The diffraction is shown between −15 and 15, but it looks much the same however far we go out in either direction.

The diffraction

We now move from tiles to point sets. Suppose that we take the set of points that we have created in Fig.12.15 to be a set of atomic scatterers on the line, and think about their diffraction. Fig.12.16 shows the diffraction from the first 55 points from the tiling. Evidently there are strong Bragg peaks, but this time they are neither all equal in intensity, nor are they all nicely spaced out, as we would get from a crystal. What is important, however, is that what we see here it a genuine indication of what happens if we take the diffraction of the entire tiling sequence. There are infinitely many Bragg peaks, a countable number of them, and apart from these point-like peaks the rest of the diffraction is zero. This is *pure point diffraction* from a point set that has *no translational symmetries.* It is not a crystal in the usual sense of the word.

The Fibonacci tiling itself is not visually striking, but it does make some of the ideas behind aperiodic tilings apparent. Fig.12.17 shows the beautiful *shield tiling* of Franz Gähler. If we think of the vertices of the tiles as atomic scatterers, then we can ask about the diffraction. Fig.12.18 shows what emerges—a magnificent pure point image. Notice remarkably that it appears (and actually is) 12-fold symmetric, yet the original tiling has no 12-fold symmetry! The importance of diffraction is that it sees what it is often hard for us to see—global structure as opposed to local structure. Diffraction depends on long-range repetition, and as far as almost periodicity is concerned the shield tiling implicitly involves 12-fold rotational symmetry, even if the actual tiling itself does not have such a

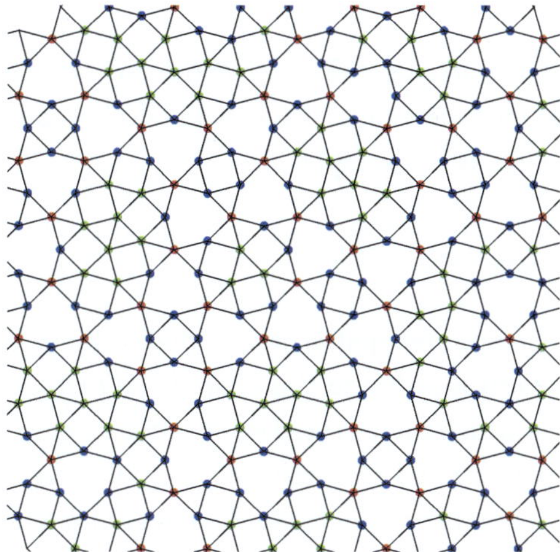

Figure 12.17: Shown here is *shield tiling*, so called because the tiles are triangles, squares, and the distinctive shields. In this particular representation we have colored the vertices of the tiles (red, green, blue, black) according to the four types of surrounding tiles that they can have. For instance, the blue vertices are surrounded by two triangles, a square, and a shield in a distinctive arrangement or pattern. We will call this tiling $\mathcal{S}$, though most often we are interested its set of vertices, which clearly define the tiling itself. For more details on the shield tiling see [9].

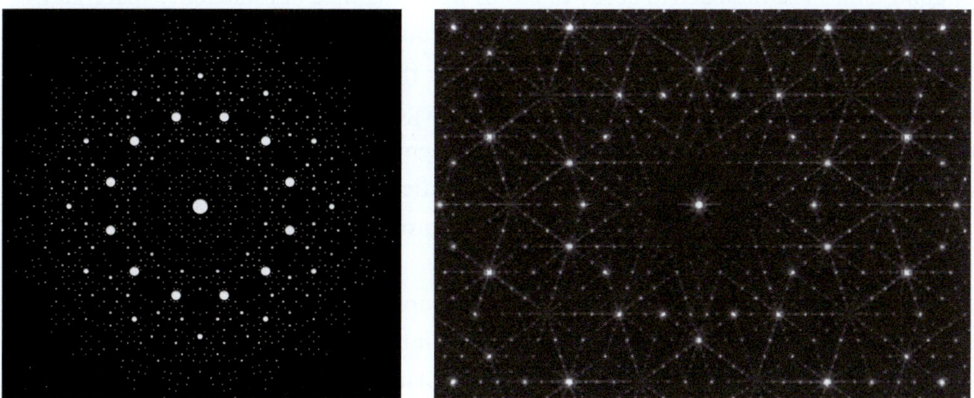

Figure 12.18: Diffraction from two famous tilings. On the left is the diffraction of the (any) shield tiling. On the right the diffraction of a rhombic Penrose tiling (see Fig.18.7), Dirk Frettlöh, tilings.math.uni-bielefeld.de.

symmetry. We will return to what this really means.

The pure point diffraction from aperiodic sets that we have just seen was supposed to be impossible in the world of crystallography. But it happened, and aperiodic crystals (quasicrystals) really exist, see ‡§18.2. It shows again how difficult it is to avoid making unexamined assumptions, or equivalently, how easy it is to make assumptions of which we are not aware.

At this point rather than pursuing diffraction in the context of tilings it is more revealing to first look at the far more general and widely applicable context of point processes, returning to the shield tiling as an example.

12.5 Point processes

Long-range order, or coherence over spatial or temporal distances, is a nebulous concept. As we pointed out in the introduction to this chapter, large geographical areas provide similar ecological zones in which there is very evident consistency. While detail by detail things are the result of chance and circumstance, overall there is a definite sameness. Here, we find ourselves at another one of those sameness-difference situations. But even as we write it down, we can see a clue: details may be based on chance, but laws of probability can bring coherent shape to randomness. This is the principle that underlies an important subdomain of stochastics: *point processes.*

A *stochastic (or random) point process* is a system in which points get randomly distributed in space. The initial investigations were point processes on the line, notably in a temporal sense. Certain events are happening and they happen at random moments in ongoing time. The radioactive decay of atoms from a sample of radioactive material will happen at random times, buses running along a route may arrive at a given bus stop at random times (even if there is a schedule), traffic flows may come to a halt at random moments of time, queues ebb and flow at cash registers, exceptional temperatures occur randomly, mortality tables are built on the seeming randomness of death in a large population. There may or may not be causal relations that lie behind these versions of randomness, but the effects are those of random processes and they are usefully treated as such.

Point processes also appear in higher dimensions. In ecological studies the distribution of trees, animal groupings, or forest fires fall into the notion of random distribution of points in the plane. Hurricanes hit cities and islands with both randomness and predictability, earthquakes strike at different places and times. Firing arrows at a target, or looking at data scattered across a graph are in the domain of planar point processes, while the distributions of atoms in a gas, apples on a tree, or air traffic around an airport lie in the three-dimensional domain. As we will see below, even very regular structures like crystals or almost-periodic tilings can be cast into the setting of point processes. Admittedly we are talking here only about distributions of points, but even this short list of examples makes it abundantly clear that point processes offer a formal setting in which we can model many natural situations. The important thing is that they are deeply connected with our subject of patterning.

Although there is no inherent limitation on dimension, we will restrict ourselves to point processes in the plane. The objects of interest are points sets $\mathscr{R}$ in the plane, which we can think of as the positions of trees, the originating centers of forest fires or lightening strikes, and so on. Think of $\mathscr{R}$ as standing for randomness. From the mathematical point of view, since we are interested in long-range order, it is easiest not to impose any restriction on the physical extent of the sets of points we want to consider. We will assume that they may be unbounded in extent and infinite in number. However, we do impose the condition that only finitely many points of such a point set may lie in any *bounded* region. We call such sets *locally bounded*. Any locally bounded set is either finite or *countable*. This inherent limitation of a locally bounded set $\mathscr{R}$ being at most countable can be seen by covering the plane by a tiling of equal sized squares. By assumption each of these squares contains at most a finite number of points, so the total number of points is a countable union of finite sets, hence itself countable, see § 7.2.2. However, unlike the situation with tilings that we have been looking at up to now, points of a locally bounded set can end up arbitrarily close to one another, and there can be vast numbers of points in some regions and none in others. We are interested in randomness.

At the outset we make no further restrictions. We begin with the set $\mathscr{L}$ of all possible locally bounded point sets $\mathscr{R}$ in the plane. This is going to be the context, the state space, of the dynamical system we are going to create. As we go on we will shape this set by a probability law with the aim of understanding the resulting point sets from the point of view of long-range order. There are two sorts of effects that we can think of in terms of long-range order. The first is whether there are some sort of expectations that we can place on the numbers or pattern shapes of points that we see when we choose a patch of the plane at random. The second is what we can expect to learn about a patch of points of $\mathscr{R}$ that is distant from some patch of points of $\mathscr{R}$ that we actually know everything about. In effect this means comparing point sets $\mathscr{R}$ with translated versions $x + \mathscr{R}$ of themselves, as we have been doing with almost-periodic point sets.

The vast collection $\mathscr{L}$ of all possible locally bounded point sets is translation invariant: if $\mathscr{R}$ is locally bounded, so is any translate $\mathbf{x}+\mathscr{R}$ of it, and with this we have the makings of a dynamical system

$$\mathbb{R}^2 \times \mathscr{L} \longrightarrow \mathscr{L}\,.$$

To shape this into a pattern system we need to create an event space and a measure. The event space arises out of a very simple idea: counting. For each bounded region A in the plane and each point set $\mathscr{R} \in \mathscr{L}$, we can count how many points of $\mathscr{R}$ actually lie in A. Every $\mathscr{R} \in \mathscr{L}$ is locally bounded so it has to have some *finite* number k of points in A. We can classify the point sets of $\mathscr{L}$ by the number of points that they have in A. This leads us to define

$N(A,k)$ is the set of all point sets in $\mathscr{L}$ which have exactly k points in A.

This number k can be $0, 1, 2, 3, \ldots$, and it leads to a natural partition of $\mathscr{L}$:

$$\mathscr{L} = N(A, 0) \cup N(A, 1) \cup N(A, 2) \cup N(A, 3) \cup \cdots .$$

This is the origin of our event space. Each $N(A, k)$ is a *subset* of $\mathscr{L}$ and is taken to be an event. If we select any bounded region A in the plane and randomly take a locally bounded point set then that set will have some number of points in k, and that is an "event", both in colloquial and mathematical senses.

We take the event space $\mathscr{E}$ to be the smallest Boolean subalgebra of subsets of $\mathscr{L}$ which contains all the sets $N(A, k)$ where A runs over all the (Lebesgue measurable) *bounded subsets* of the plane, see §6.5.1. In plainer words, we are taking all the subsets of $\mathscr{L}$ that arise out of finite or countable unions and intersections of sets of the form $N(A, k)$. This is really about the simplest thing we can say about a point set in the plane: the number of points that it has in the various bounded regions of the plane that we choose to take. This then is the event space.

In principle it is an easy step to impose a probability measure on this. We just take any measure $\mathbf{p}$ on $\mathscr{E}$ that has total measure equal to 1 (remember that, by definition, measures are defined on event spaces, §6.6.2). We can interpret this measure as shaping the system $\mathscr{L}$, for it declares which outcomes (events) are likely and which are not:

> the probability that a randomly chosen locally bounded point set has k points in the bounded region A is $\mathbf{p}(N(A, k))$.

As usual, the probability measure is not saying *how* we choose sets randomly, but rather it is stating what the expectations of various patterns (e.g. being in $N(A, k)$) are.

It is not immediately obvious what the scope of this definition is, but we will see that it is considerable. However, we can get one idea of what we might mean by coherence or long-range order:

> A point process $(\mathscr{L}, \mathbb{R}^2, \mathscr{L}, \mathbf{p})$ is *stationary* if $\mathbf{p}$ is invariant under translation.

This is a familiar idea, and its meaning here is that the probability of k points falling in the bounded region A is not changed if A is translated to somewhere else: for all $A \in \mathscr{L}$ and for all $\mathbf{x} \in \mathbb{R}^2$, $\mathbf{p}(N(A, k)) = \mathbf{p}(N(\mathbf{x} + A, k))$. It's far weaker than saying they have the same number of points: it is only a statement about probability. But what it does say is that the way in which chance distributes points in the plane is the same everywhere, and hence is a form of long-range order or, as we call it, *coherence*. Point processes are not necessarily assumed to be stationary, but here we will restrict ourselves to such.

12.5.1 Poisson processes

Poisson processes are the historical ancestors of all random point processes, producing point sets that seem to match our instinctive idea of what a random set of points might

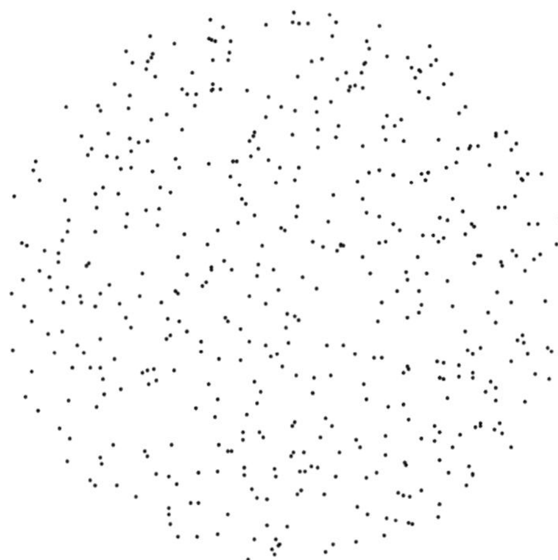

Figure 12.19: This is the typical appearance of what we can expect to see when points in a circular disk are distributed according to the standard Poisson distribution in two dimensions.

look like, see Fig.12.19 and also the point set that appears earlier in Fig.12.6. The idea behind the Poisson distributions might be best imagined as falling rain:

(i) the number of points falling in a region A of the plane is, on average, proportional to the area of A;

(ii) what happens in any particular region A is independent of what happens in any other region B that is disjoint from A.

There is only one prescribed variable, the density. The density is the expected number of points per unit area and is preassigned. This then results in explicit formulae for the probabilities of events, which naturally enough depends on area.[10]

Poisson processes in 1D were among the earliest stochastic processes studied. The regions A there are typically intervals on the real line, with the formulas for $N(A, k)$ being based on Lebesgue measure on the line, i.e. length. Such a process can be thought of as taking place in time, imagined as a machine that moves along the line dropping points at random. What is important is that after dropping a point the process works just like it is starting all over again. Think of a bus route where buses run at a particular frequency (so many buses per hour, on average) but there is no schedule and buses arrive at random times. Most of us have had the experience of the Poisson effect when schedules go awry, and after waiting fifteen minutes for a bus on a route where there is supposed to be a bus every ten minutes we find that two buses arrive within a few seconds of each other.

The one-dimensional situation is commonly cast in terms of an ordered system of events (for instance time, in our example). In higher dimensions, where there is no natural ordering available, the theory is cast in the form we have presented it, where we think in terms of entire point sets rather than point sets built up over time. We have seen this issue in the context of infinite zero/one strings: are they built up in a sequential way or are we simply presented with infinite strings? We have often found the latter more useful.

It is a characteristic of Poisson distributions that they allow clumping: points can be very close together, with no limit as to how close. Such distributions do not match the distributions where some sort of minimal separation is relevant—bushes competing for water in barren ground or hummingbirds competing for territory. In atomic physics atoms take up space and so are spatially separated and *fermions* like electrons are bound by the *Pauli exclusion principle*: so that no two can simultaneously be in the same state in a quantum system. The quantum formalism of this leads to *determinantal distributions.* (The unusual looking word "determinantal" arises from its connection with determinants of matrices.) Such point processes, which have applications well beyond quantum physics, make clumping unlikely, though not absolutely forbidding it. Fig.12.20 shows representative samples where the difference in "clumping" between a Poisson and a determinantal process is rather evident.

We should make it very clear, though, that we are dealing with random processes here and any particular outcomes, such as we show here, do not necessarily reflect the expected overall outcomes of their respective laws. In principle a stationary point process can produce any arrangement of any finite number of points in any given region. It is the *probabilities* of occurrence of the various finite arrangements of points that are important. The probabilities don't force the situation but they can, and do, very significantly influence the shapes of the outcomes that we should expect in any particular trial, even to the point of making some configurations of points essentially impossible, and other essentially inevitable.

12.5.2 Tilings as dynamical systems

It might seem strange that there is anything "dynamic" about tilings, but in fact they can very usefully be put into the context of point processes, the points in this case being the vertices of the tiles that constitute the tiling. We want to look at this a little more carefully, for it shows the enormous capacity of point processes to express pattern and gives us more insight into long-range order. For explicitness we will use the shield tiling as our model, but thinking of it now as a point set determined by its vertices. For simplicity we will simply talk of shield tilings and allow context to determine whether we are talking about tiles or vertices of tiles.

The first step is to collect together the "family" of shield tilings. What should we mean by a shield tiling anyway? We have given examples of parts of a shield tiling and then said that what we see just continues looking essentially the same forever. Putting aside the

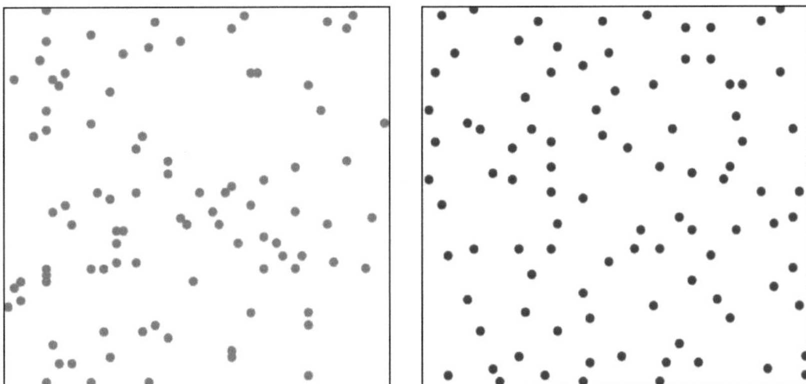

Figure 12.20: On the left a sample from a typical Poisson distribution. On the right a sample from a typical determinantal distribution. Poisson processes are more prone to clumping (points very close together) than determinantal point processes.

question of actually producing a shield tiling, we zero-in on an important fact: there is no unique shield tiling. Of course if we have one and translate it around we get seemingly different shield tilings, since there is no translational symmetry in a shield tiling. But it is not just that. There are infinitely many (in fact uncountably many) different shield tilings that lie in the same overall orientation as the one we have started from, and this is true even if we were to consider translations of a tiling to be the same tiling. What makes them all shield tilings is they are *locally indistinguishable.*

Two point sets from $\mathscr{L}$ (or two tilings) are said to be *locally indistinguishable* (LI) if any finite patch (no matter how large) of points (or tiles) that we see in one we can also find somewhere in the other. By this we mean by sliding the entire point sets around by translations, the two patches can be aligned perfectly. The definition is self-explanatory: two tilings are LI if we cannot tell them apart in any local way. No matter what size of disk D one takes, and no matter where one puts it down in one of the point sets, the very pattern of points and tiles that one sees within it can be found by putting the disk D down somewhere in the other tiling.

As the state space X for our new dynamics of the shield tiling we take the *LI-class* of one shield tiling $\mathscr{S}$—that is, the entire set of all tilings that are locally indistinguishable from $\mathscr{S}$. Certainly any tiling is locally indistinguishable from any of its translates, so it is immediate we have a translational action

$$\mathbb{R}^2 \times X \longrightarrow X .$$

In particular X contains $\mathscr{S}$ and all its translates $\mathbf{x} + \mathscr{S}$. But not all tilings locally indistinguishable from $\mathscr{S}$ are translates of $\mathscr{S}$. However, they are all shield tilings. This space X is called the *dynamical hull* of the shield tiling.

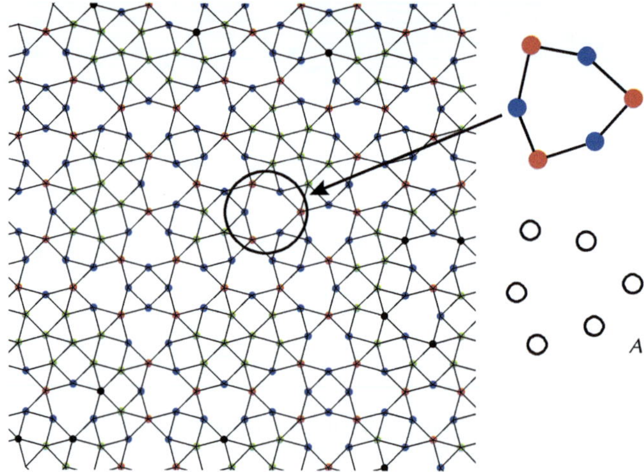

Figure 12.21: A single shield is being represented by a region A consisting of six small disks.

This gives us the dynamical setting and it looks very much like the setup for a point process. The difference is that a point process uses the set $\mathscr{L}$ of *all* the locally finite point sets as its state space, and is based on randomness, whereas our shield tiling dynamical system X consists only of the vertex sets of shield tilings and there is no randomness involved. Even so, the shield tiling system can be seen as a stochastic point process! Initially it seems hard to imagine how a stochastic point process could manage to produce just the vertex sets of shield tilings and with probability 1, nothing but the vertex sets of shield tilings, using only the counting functions $N(A,k)$. However, it is not as hard as it seems. The idea is to find a probability measure on $\mathscr{L}$ that gives measure zero to the totality of all point sets that we don't want to have occur.

Consider the fact that there is a minimum to how close two different vertices of a shield tiling can be. Suppose that we scale our shield tilings to make this minimum equal to 1, so any two points of any shield tiling are always at least 1 unit away from one another. That means that any disk A of radius less than one half, no matter where it is put down, can contain at most one point of any shield tiling. So simply insisting that $\mathbf{p}(N(A,k)) = 0$ for all $k > 1$ for all such disks will effectively (that is, up to probability zero events) eliminate any point sets in $\mathscr{L}$ that have points less than distance 1 apart.

More constraining, let's consider the shape of the tiles, the shields, for instance. Shields come in twelve different orientations in any shield tiling. Take the shield shape in one of its orientations. This time take A to be a fixed region composed of six minute disks, one centered at each of the six vertices of this shield. The more minute the disks the more A becomes the exemplar of the shield pattern, see Fig.12.21. Now the probability $\mathbf{p}(N(A,6)$ (i.e. that a randomly chosen shield tiling will have six points in this particular region A is not very large, but it is certainly positive since we are dealing with all the shield tilings in the shield hull and many of them will fit this particular region A. But both $\mathbf{p}(N(A,5)) = 0$ and $\mathbf{p}(N(A,7)) = 0$, for if five vertices line up so will the sixth, and there is no way to get seven points of a shield tiling into this set A.

We see that by using differently shaped bounded regions, we can, in principle, specify the probabilities associated with any events $N(A,k)$ by simply imposing the frequencies of these events as we see them in some shield tiling $\mathscr{R}$: shields, triangles, squares, and local configurations of points in general. Using the frequencies of occurrence of larger and larger finite patterns and making the little disks ever more fine, we construct a probability measure $\mathbf{p}$ on $\mathscr{L}$ that is consistent (since it is derived from what happens in $\mathscr{R}$) and when used as the basis of randomness can only produce point sets that look like shield tilings. This needs to be taken to the limit (vanishingly small disk sizes) to arrive at the limiting measure that produces only shield tilings.

As it happens there are explicit ways to construct shield tilings and write down the probabilities of the various events, but what we have said is quite generally applicable. It is the *frequencies of occurrence of the point patterns* that appear in shield tilings that makes the law of the process, the measure $\mathbf{p}$. Once the law $\mathbf{p}$ has been established then, with probability one, any point measure randomly chosen from the system $(\mathscr{L}, \mathbb{R}^2, \mathscr{E}, \mathbf{p})$ will have a point set that is the vertex set of a shield tiling. What's more, $\mathbf{p}$ is invariant under 12-fold rotation. The invariance that is not even possible in any particular shield tiling is nonetheless implicit within it, and emerges explicitly in the pattern system that embodies their entirety.

Admittedly there is a gap here: we will indeed obtain only shield tilings in this way, but we derived $\mathbf{p}$ on the basis of one shield tiling $\mathscr{R}$. Would every shield tiling lead to the same probability measure on the event space? The answer is yes, see §12.5.3 below.

Poisson processes, determinantal processes, and shield-tiling processes, and countless others, are all stochastic point processes and they all use exactly the same state space and the same translational dynamics:

$$\mathscr{X} = (\mathscr{L}, \mathbb{R}^2, \mathscr{E}, \mathbf{p})\,. \tag{12.8}$$

Each is uniquely shaped by its law or probability measure—we might even say its *underlying principle.* It was out of this type of thinking that our initial insight into the relationship between pattern and dynamical systems arose.

From the point of view of a Poisson point process, the occurrence of a shield tiling is a probability zero event. From the point view of a shield tiling point process the probability of a distribution satisfying the Poisson conditions is zero. Both systems have the same entire collection of locally finite subsets of the plane, and the same translational dynamics, yet the two different measures "see" very different things. There is an important idea bound in this. It is not just the way things are, but the way we choose to see them that is important. The discussion of the Cantor set in §12.9 below is simpler and offers a more direct insight into this sort of phenomenon—two measures that each fail to see what the other can see.

There are always benefits to being able to understand very different things from a common perspective. The subject of point processes encompasses a huge variety of pattern systems based on the distribution of points in the plane (and more generally in Euclidean space of any dimension). As such it provides an important setting for extending our conceptions of long-range order.

12.5.3 Ergodicity

Recall that a dynamical system is ergodic if the state space cannot be divided into two invariant parts of positive measure, §8.5.1. In the present context a point process is ergodic if the space $\mathscr{L}$ cannot be broken into two translation-invariant parts, both with positive measure with respect to $\mathbf{p}$. *All of our examples presented here are ergodic.* This brings us back to the Birkhoff ergodic theorem of §8.5.3. Although we have only discussed the Birkhoff ergodic theorem in the context of a single discrete operator, the same sort of conclusions can be drawn here, where the dynamics is based on $\mathbb{R}^2$.[11] If the system is ergodic then any trajectory in $\mathscr{L}$ almost surely reveals the same statistics as the entire ensemble. Since a trajectory simply means translating a point set $\mathscr{R}$ in all possible ways, or equivalently looking at all parts of the point set, this means that with almost surety any randomly chosen point set of the point process really is representative of the statistics of the entire ensemble of point sets in $\mathscr{L}$. This then is a strengthening of long-range order. With this idea in place we can return to the question of almost periodicity.

Almost periodicity in point processes

As we have seen before, §11.10.5, the formalism of the stationary point process (12.8) will lead us to the space of observables $L^2(\mathscr{L}, \mathbf{p})$ (the space of all square-integrable complex-valued functions on $\mathscr{L}$). See also ‡§19.4.1. We have seen the importance of this in periodic processes, where this linear space of functions decomposes into one-dimensional eigenspaces. These eigenspaces are invariant under the action of translation and display the spectrum of simple periodic frequencies which underlie the dynamics. Here we found that the space is cut up into one-dimensional spaces which are not further decomposable, a situation in which the spectral decomposition is called *atomic*.

It is not always the case that $L^2(\mathscr{L}, \mathbf{p})$ has such an atomic decomposition—in fact quite the opposite: *that is a marker of almost periodicity.* There is a very clear and clean result that summarizes how much of what we have been talking about fits into the overall picture of point processes. To state it we need to assume a bit more about the nature of random point sets from our point process: there is a *minimum separation* between distinct points in the sets that arise from the process. We will assume that distinct points of our point sets are always separated by at least some minimal distance.[12] This is sometimes called a *hard core* condition. Neither Poisson nor determinantal processes satisfy this condition, but it is a natural one to assume in the study of physical arrangements of atoms or living entities where physical conditions effectively separate individuals from one another.

Theorem 12.5.1 *Let* $(\mathscr{L}, \mathbb{R}^2, \mathscr{E}, \mathbf{p})$ *be a stationary planar point process which is hard core and ergodic. Then the following three conditions are equivalent:*

(i) the point sets in $\mathscr{L}$ *are pure point diffractive (all with the same diffraction), almost surely;*

(ii) the spectrum of $L^2(\mathscr{L}, \mathbf{p})$ *is atomic;*

(iii) the functions $f \in L^2(\mathscr{L}, \mathbf{p})$ *are all almost-periodic.*

In this result we see three very different features brought together: almost periodicity, pure pointedness of the diffraction, and the atomic nature of the spectrum. Each is equivalent to the others: if one is true so are the other two. Thus pure point diffraction is deeply related to *almost-periodicity*, not periodicity that crystallographers had long supposed. The condition is a condition of long-range order and, as we had expected, it is deeply related to the law $\mathbf{p}$ of the process. We have stated it in terms of the two-dimensional context of this discussion. The theorem is independent of dimension: it works for point processes in which the point sets are on the line, in 3D-space, or indeed, in any dimension.[13]

12.6 Some facts about stationary point processes

All three of the point processes that we have discussed are stationary. In that sense they each express a form of spatial coherence. The probabilistic feature of *stationarity* says that no matter what bounded region A of space we consider, the probabilities $\mathbf{p}(N(\mathbf{x} + A, k))$ do not depend on the translation $\mathbf{x}$. Although in any particular point set R drawn from a point process the number of points lying in A and in $\mathbf{x} + A$ will most likely be different, still the *expected values* of these numbers are the same. This means that our expectation of pattern is the same everywhere, which is certainly one sense in which we can interpret long-range order. If "return" is interpreted as seeing again what we have previously seen then this is return, for probabilistically whatever local patterning we see we can expect to see elsewhere, whether that be understood in temporal or spatial terms.

The intensity of a point process

Continuing with this, given any bounded region A in the plane there we can consider the average number of points, or expected number, of points that will fall in A as we go over the entire set of all point sets R in $\mathscr{L}$. Specifically, the expected number of points is

$$0\,\mathbf{p}(N(A,0)) + 1\,\mathbf{p}(N(A,1)) + 2\,\mathbf{p}(N(A,2)) + 3\,\mathbf{p}(N(A,3)) + \cdots .$$

Since we can apply this idea to every region A in the plane, this naturally gives rise to a measure I on the plane for which $I(A)$ is this expected number of points in A. This measure is called the *intensity* measure of the process. Now if we are assuming that the process is stationary, this intensity does not change under translations, in other words $I(\mathbf{x} + A) = I(A)$ for all $\mathbf{x} \in \mathbb{R}^2$. However, as we have mentioned before, see §6.6.4, the only measures on $\mathbb{R}^2$ that are invariant under all translations are the Lebesgue measure ℓ of area, and its multiples. Thus we must have

$$I(A) = D\ell$$

for some positive number D. This number D is called the *intensity of the process*. D is the expected number of points per unit area of point sets drawn from $\mathscr{L}$.

Mixing

As we have just seen, to have a complete decomposition of $L^2(\mathscr{L}, \mathbf{p})$ into one-dimensional spaces, that is, for the spectrum to be atomic, is quite special. While there is always one eigenspace in $L^2(\mathscr{L}, \mathbf{p})$, namely the subspace of constant functions on $\mathscr{L}$, i.e. $\mathbb{C}\,\mathbf{1}_{\mathscr{L}}$, since the constant function $\mathbf{1}_{\mathscr{L}}$ is unchanged by translation, it is quite possible that there are *no other one-dimensional eigenspaces.* The Poisson and all determinantal point processes belong to this opposite extreme. Such processes are what is called *mixing*, and actually they are really quite typical of point processes.

Mixing is best approached from a physical interpretation which gives a deeper insight into the concept of long-range order. Suppose we have an event E of $\mathscr{E}$ and a point set S which is in E. We might ask to what extent we can predict that some translate $\mathbf{x} + S$ is also in E. For example we might have the event $S \in N(A, 1)$, meaning that S has one point in the bounded region A of the plane. We might then ask what we can infer about $\mathbf{x} + S$ belonging to the same event, i.e. $N(A, 1)$. To say that $\mathbf{x} + S \in N(A, 1)$ is the same as to say that $S \in N(-\mathbf{x} + A)$, reflecting, as we have seen before, that the action of dynamics on a state space implies dynamics on the event space. With this in mind, the particular question we have asked is how $N(A, 1)$ and $N(-\mathbf{x} + A, 1)$ are related, if at all. Does the fact that S has exactly one point in A have any implication that it also has exactly one point in $-\mathbf{x}+A$? This is a question of long-range order, which we can state more generally as asking how the behavior in one region of a point set relates to its behavior in another. It all depends on the point process, which in turn depends on the law of the process. A point process is *mixing* if in the long run we can infer nothing.

Formally, a stationary point process $(\mathscr{L}, \mathbb{R}^2, \mathscr{E}, \mathbf{p})$ is said to be *mixing* if for every pair of events E and F in $\mathscr{E}$, and for all sequences of $\mathbf{x}_k$ of translations where $|x_k| \to \infty$ as k

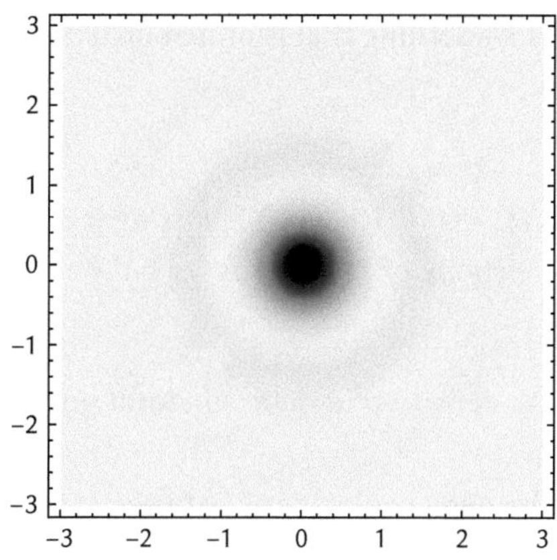
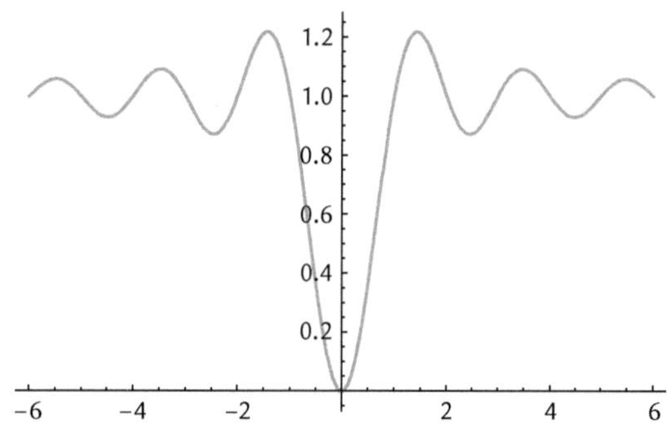

Figure 12.22: Two plots showing the rotationally symmetrical distribution of probability of points of a determinantal point process of intensity equal to 1, *given that there is a point at the origin.* The point process illustrated here is a determinantal sine process with the distribution of probability density around one point situated at the origin given by $\left(1 - \left(\frac{\sin(\pi|x|)}{\pi|x|}\right)^2\right) \ell$, where ℓ is the Lebesgue measure of area [11]. The curve on the right is the radial profile of density function f seen on the left. It shows a strong "repulsion" from the point located at the center (points are unlikely to be close to each other) that leads to a ring of increased probability density slightly greater than 1 (which we might expect since the intensity is 1), and then a leveling as we move further out. The value it levels towards is the intensity of the process, namely 1. The meaning of this can be interpreted by looking at the left-hand plot: given any region A in the plane, the expected number of points in this region is the ratio of the accumulated density over A divided by the area of A. The flatness outside the immediate vicinity of the origin indicates that the distribution of the second point resembles what would happen in a Poisson process, with this ratio approaching the intensity of the process (number of points per unit area). This shows the declining influence of the point located at the origin, and hence, overall, the lack of long-range influence of any local arrangements of the points. This is a sign of mixing.

increases, we have

$$\lim_{k\to\infty} \mathbf{p}(E \cap (-\mathbf{x}_k + F)) = \mathbf{p}(E)\,\mathbf{p}(-\mathbf{x}_k + F) = \mathbf{p}(E)\mathbf{p}(F)\,.$$

The far right-hand side is the condition of probabilistic independence of events E and F. The second equality is just a consequence of the stationarity of the process: translation does not alter probability. The first equality is the key. It says that as the distance $|\mathbf{x}_k|$ of the translation $\mathbf{x}_k$ gets very large the events E and $-\mathbf{x}_k + F$ become essentially independent. This even applies when $F = E$.

Think in terms of a liquid. It is quite likely that in the local spatial arrangement of atoms, neighboring atoms influence each other. In that case we would expect to make some statements about the local distribution of atoms around a particular atom. But as distance increases, this influence decreases, until it is negligible.

Mixing implies ergodicity (not obvious), which means that we expect the same statistics on average, no matter where we look in the space. We will still see the return of point pattern that we have seen before, but that return is not something that is probabilistically dependent on what we see in a particular region now.

Now mixing is directly associated to the spectrum.

Theorem 12.6.1 *An ergodic stationary point process* $\mathcal{X} = (\mathcal{L}, \mathbb{R}^2, \mathcal{E}, \mathbf{p})$ *is mixing if and only if the one-dimensional eigenspace of constant functions of* $L(\mathcal{L}, \mathbf{p})$ *is its only one-dimensional eigenspace .*

Fig.12.22 offers a visual insight into the probability density of a determinantal point process.

Taken with Thm. 12.5.1, the two extremes show the deep connections between long-range order and the spectrum of $L^2(X, \mathbf{p})$. The negative result of the mixing theorem does not mean that spectral analysis has no value in mixing point processes. Far from it. It means that the spectrum is far more subtle. It is a highly developed part of mathematics.

12.7 Curie's principle

Looking at the shield tiling, Fig.12.17, one feels that there is something decidedly 12-fold rotationally symmetric about it. But no shield tiling has 12-fold symmetry. It's not even possible to get tiles of this tiling to wrap with 12-fold symmetry around a point. Nonetheless, its hull is perfectly 12-fold symmetric!

How can this be? Fig.12.23 suggests why. In §12.5.2 we have looked at the shield tiling from a dynamical point of view within the context of point processes. Here we find the dynamical hull, which integrates into a single entity the entire range of locally indistinguishable shield tilings. The hull is perfectly 12-fold symmetric, and so too is the measure $\mathbf{m}$ since the measure is based on local pattern frequencies and it is no surprise that the rotated versions have just the same frequencies of finite patterns as the original. In the end it is this measure that is crucial to the diffraction, and this explains its rotational symmetry.

What we see here is an example of a collection of ideas that go by the name *Curie's principle.* Pierre Curie, he of the famous Curies of radio-activity, wrote a seminal paper in 1894 on the question of symmetry and its appearance in cause and effect relationships. His statement is that the symmetries of the causes are to be found in the effects. However, the manifestation of the effects can lack some aspects of the symmetry of the cause. The idea is important, though it is hard to formulate precisely what Curie meant and harder still to state as a universal principle (see [31] for a recent philosophical discussion). However, as it appears here, it is clear. The "cause" is the measure, the law of the process, that shapes the outcome of the process. The individual "effects" may not fully or explicitly express the symmetry, but their totality does.

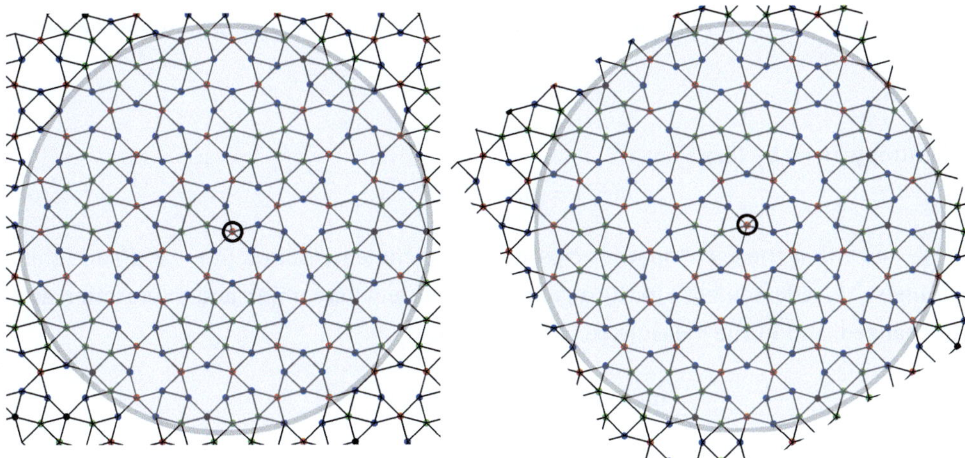

Figure 12.23: Rotating a shield tiling from $\mathcal{S}$ through 30° (one twelfth of a circle) yields a tiling locally indistinguishable from it, and hence also in $\mathcal{S}$. Here we have rotated the patch of Fig.12.17 through 30° around the the circled red point. Ignore the square frame, which is not part of the tiling, and look within the circled region. The two circular patches are not the same, but taken in their entirety one can surmise that the two tilings are locally indistinguishable and hence both members of the same hull. The hull $\mathcal{S}$ has 12-fold rotational symmetry.

This is another example of a hidden pattern that is revealed only when taken in a larger context. All shield tilings have exactly the same diffraction, see Fig.12.18, revealing the 12-fold symmetry in all its beauty. Locally we cannot see the symmetry, but the diffraction is a global property, and does. Perhaps in our present context we could paraphrase Curie's principle to say that the inherent invariances of a pattern system may be missing in its emergent effects.

12.7.1 Laws or freedoms?

It has been pointed out that when it comes to laws (physical or societal), the laws that tell you what must happen or what you must do are worse than the laws that tell you what can't happen or what you must not do, the reason being simply that whatever is not forbidden is allowed and so offers more room for creativity and more freedom. It has been suggested that, at least in particle physics, perhaps the laws forbidding (e.g. the Pauli exclusion principle) are the more important ones. Generally we find that Nature seems to explore whatever is not forbidden. This question of whether things are defined directly or indirectly by negation might be seen also as an example of the Curie principle.

In a pattern system $(X, s, \mathscr{P}, \mathscr{E}, \mathbf{m})$ based on an initial partition $\mathscr{P}$, the resulting event space is the smallest collection of subsets of X containing $\mathscr{P}$ and closed under unions, intersections, complementation, and the transformations deriving from the dynamics. In

§8.3.3 we have seen that the evolution of events leads to the increasingly complex events

$$\mathcal{P}^{(1)}, \mathcal{P}^{(2)}, \mathcal{P}^{(3)}, \dots, \tag{12.9}$$

and entropy given by (8.9):

$$h(\mathcal{P}) = \lim_{k\to\infty} \frac{1}{k} H(\mathcal{P}^{(k)}). \tag{12.10}$$

We know that the event space is closed under complementation ($E \mapsto X\backslash E$) and this "symmetry" is reflected both in the algebraic structure where union and intersection are interchanged, and in the measure $\mathbf{m}$ where we have

$$\mathbf{m}(E) + \mathbf{m}(X\backslash E) = \mathbf{m}(X) = 1\,.$$

To make this symmetry look clearer, write E' for the complement $X\backslash E$, so then $E'' = E$. Then for each event

$$P_0 \cap s^{-1} \triangleright P_0 \cap \cdots \cap s^{-k} \triangleright P_k\,, \tag{12.11}$$

which represents the decreasing set of events coursing over a transition of events through k iterations of the dynamical operator s, we have its complement

$$P_0' \cup s^{-1} \triangleright P_1' \cup \cdots \cup s^{-k} \triangleright P_k' \tag{12.12}$$

representing the corresponding increasing set of negations that imply it. The set $\mathcal{P}^{(k)}$, which is the union over all events (12.11), is then matched with the set $\mathcal{P}^{(k)'}$, which is the intersection of all their negations (12.12).[14]

If we wish then, we can interpret the events of (12.11) as representing the freedoms available once all the restrictions of (12.12) have been imposed. Here existence is implied by the nature of absence. In our case, we are beyond the simple dichotomies of Boolean algebras since everything is moderated by the measure, but since the measure matches the complementation, it fits into the same picture. Increasing tendency (larger measure) in one corresponds with decreasing tendency (smaller measure) in the other.

The idea of looking at complements of events as having the same information but in a very different form is in fact the basis of a mathematical duality theory put forward by Marshall Stone (1903–1989) that matches the algebraic structure of the event space with the topological structure of what are called Stone spaces. This has the effect of putting action (presence or absence of events) prior to the spaces on which these actions take place [133].

Although it may seem remote, both in Daoist and Buddhist thought, *negation* is a primary concept and is often seen as the creative side, or even source, of existence. This is the content of the Ch. 2 from the *Daodejing* (already quoted in §4.1.2) pointing out that concepts always arise in dual pairs:

> There can be no existence without non-existence
> No difficult without easy
> No long without short
> No high without low, etc. [193, Wilson]

Distinguishing invokes duality, but it is a duality that frames a whole, just as the famous taiji symbol (Fig.4.1). Returning to our event space, complementation maps $\mathscr{E}$ into itself, *interchanging* the algebraic structure. There is an overall symmetry that reflects the fact that events cannot be framed without their negations. The complements of events are just as good a way of describing the Boolean structure as the events themselves. In this way we might envision the unfolding of events as more like the exploration of whatever is not forbidden.

12.8 Pattern out of nothing: morphogenesis

> One of the true mysteries in biology is the emergence of a complex organism from a single fertilised egg cell. This initial cell is spatially more or less homogeneous; it certainly cannot be held to contain a blueprint of all the structures to be formed in the sequel. This is obvious from simple observations: for example, the cleavage of a sea urchin egg may result in two complete embryos, or a single fertilised human egg may develop into a pair of identical twins. Thus, self-organisation must be at work, rather than a rigid execution of a predefined programme.[8, Ellen Baake].

Among the most intriguing and mysterious features of Nature is its ability to create differentiated form where there is apparently none. Cell division with its subsequent differentiation in the development of a single cell to a complete animal seems miraculous. How does one cell divide over and over again to produce the trillions of cells that are an entire living creature with all its multitude of different parts and functions? Any part of polyphydra is capable of regenerating a complete animal. Even the far simpler repetitive patterns in the fur and feathers of animals and birds are amazing in their diversity and their ability to maintain themselves and grow along with the growth of the animal. This sort of robust yet flexible spatial patterning surely cannot be directly programmed into the surface skin of such animals. What sort of process can lie behind this self-organization that produces distinctive differentiated pattern where none previously existed?

To some extent we can already imagine how answers to this question may go. We know that pattern systems can produce distinctive outcomes through their dynamical evolution. The outcome of the dynamics of a system may be vastly different than anything we might anticipate. The explicit ingredients to such a pattern system may be very

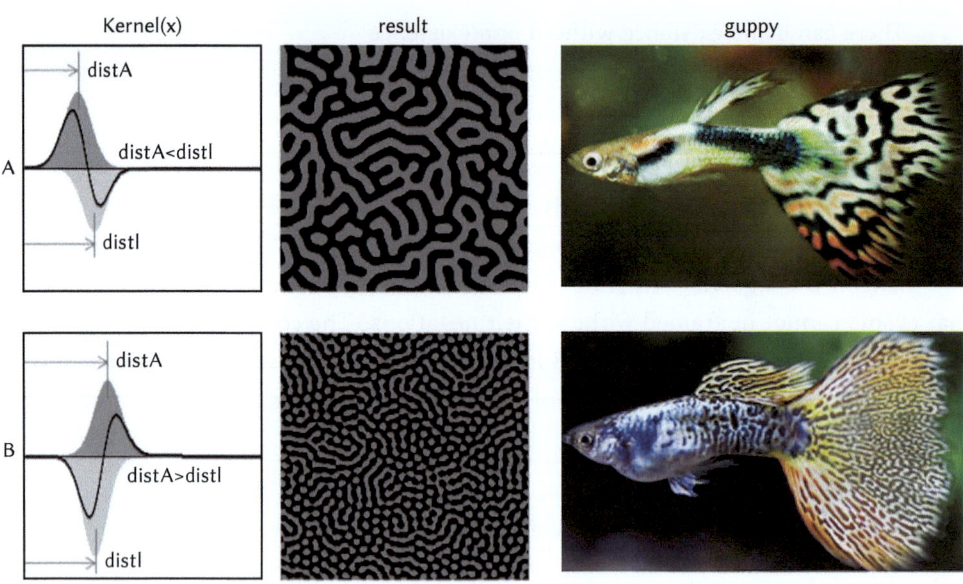

Figure 12.24: Shown are two patterns arising out of a random pattern through the diffusion process described in the text. The patterns are compared with the actual patterning in the tails of guppies. The left side of the figure shows the combinations of two Gaussian curves that give rise to these patterns. Although every run from a random starting position ends up with a different pattern, the patterns arising from a given pair of Gaussians have recognizably similar overall appearances, see Fig.12.27 [100]. CC BY 4.0 Deed Attribution 4.0 International.

hard to find, but we know in general terms how this apparent self-organization can happen. Nature's self-made tapestry is indeed wondrous, and though we should never lose that sense of wonder, it is not beyond the possibility of observing, modeling, and further understanding.

One form of emergent patterning that has been understood arises through processes of diffusion. Its initiator was Alan Turing, the same Turing so famous for cracking the Enigma code in WWII and laying down the formalization of computation. Unfortunately his death in 1954 ended his work in this direction very prematurely, and it took many years before his seminal paper of 1952 [173] was fully appreciated.

Here, rather than treat Turing's work directly, we briefly discuss some more recent work by Shigeru Kondo, on modeling the generation of the visible characteristic patterns that we see on animals. This is based on the same overall idea put forward by Turing, but is simpler to explain. It operates as a form of point process [100, Kondo]. Fig.12.24 shows the actual patterning in two fish and beside it comparable results from a pattern system model that is in effect a stochastic point process.

The process can be described as one of local activation and long-range inhibition. Imagine a material cellular surface which is initially undifferentiated and in which each cell produces two types of enzymes that dissipate to neighboring cells. The first promotes a certain activity of the cell, say pigment production, and the other inhibits it. The first acts at once and dissipates quickly. The other dissipates more slowly and acts more slowly. The effect is that from the point of view of one cell there is local promotion of activity, but further away there is inhibition of this activity. Fig.12.25 shows what is intended.

Now we have to imagine what happens if every cell is participating in this process, each promoting activity in its local area, but inhibiting it further away. The activity/inhibition profile is placed on every cell and the *accumulated* interplay of promotion and inhibition of activity from it and all its surrounding neighbors determines whether or not that cell retains or changes its pigmentation. In principle the initial state of the cellular surface may be quite homogeneous, but in reality any randomness in the capacities of the various cells to produce enzymes, which assuredly will be the case, will be enough to initiate change that spreads in time to produce very striking global effects. The process is governed by two circular Gaussian distributions which parameterize the activation and inhibition parts of the process. The global effect produces distinctive patterns according to the specific values of the parameters involved in defining these two Gaussians and their separation. Although the process is stochastic, so the explicit results will differ from case to case, the outcomes are quite consistent in the overall appearance. Fig.12.24 and Fig.12.27 are examples of such patterns compared with actual patterns in guppies. Setting aside details, the idea of the mathematics is fairly straightforward and can be viewed as an example of a point process. Our canvas is a pixelated square, say of size $n \times n$ with n^2 square pixels each of which is weighted with a value between 0 and 1. We consider it to be an $n \times n$ matrix $M = (M_{ij})$, where $0 \leq M_{ij} \leq 1$ for all i, j. Here we have taken $n = 100$. The *state space* is the set $\mathscr{M}$ of all such matrices.

The effect of the dynamical action to an element M of $\mathscr{M}$ is to alter its matrix entries by placing the profile kernel (which in our case is 13×13, see Fig.12.25) at each pixel square in turn, multiplying the kernel values with the values in the corresponding pixels that they overlay, and adding them all together. This process is called *convolution* (see Endnote 15 for more explanation). There are some considerations for what happens at the edges of our canvas, but that is rather unimportant here. The overall tendency is to retain the sign of a point locally and to reverse it further away. But every point is affected both by its nearest neighbors and points which are further away. After we have done this at *all* the points, we add up the results and trim them back into the range $[0, 1]$. We can think of the trimming as representing the idea that pigment can range from none to a certain maximum, here designated by the range 0 to 1. Thus after convolution the matrix of values may have changed, but it is still in $\mathscr{M}$ again.

This then is the process: convolve, trim back, and then repeat. The repeating process can be thought of as the on-going distribution of pigment activity. If n is very large, then

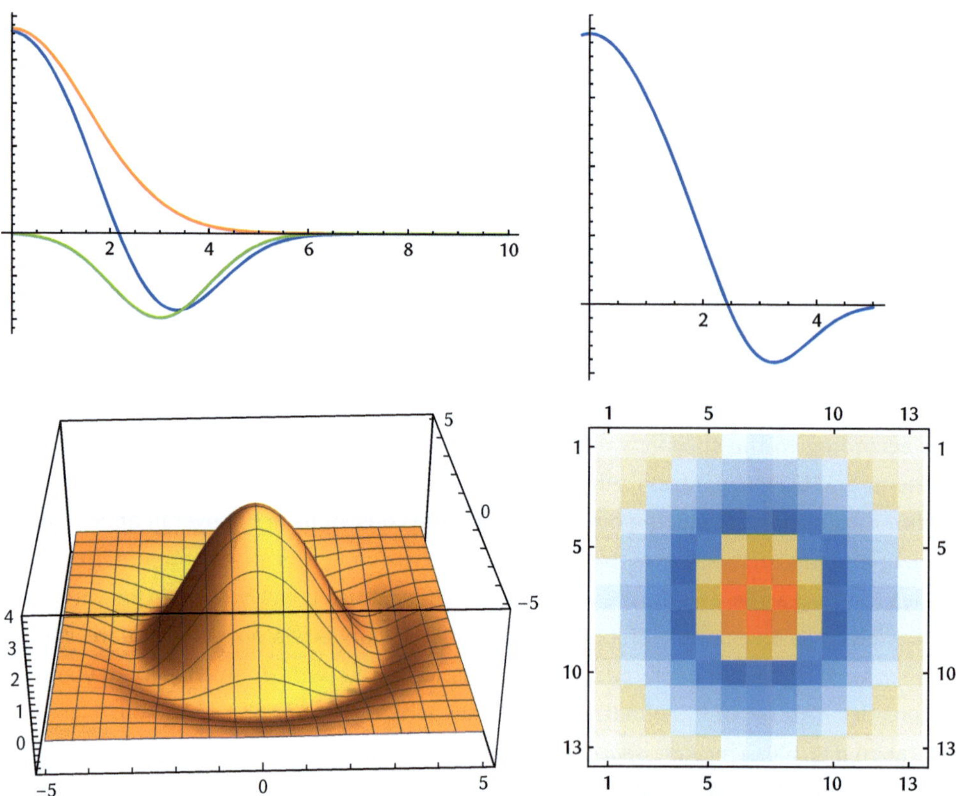

Figure 12.25: The distribution of pigment activation/inhibition is based on the difference (top left and right, black) between two Gaussian curves, one positive and one negative (both shown in grey), which are displaced from one another. We have shown only the profile of the distribution—the process diffuses outwards in all directions (bottom left). The accumulated effect is initially positive, but it decreases outwards and then becomes negative. At even further distances the effect has largely disappeared. The final picture (bottom right) is a pixellated version of the first image, made into a 13 × 13 grid with the numerical values of the activation being replaced by grey-shading. This grid is referred to as the *kernel* and it is the mathematical object used in the calculations.

Figure 12.26: Five consecutive steps, left to right, of evolution of a random pattern in a 100×100 pixellated square, starting from a random pixellation. Notice the stabilization of the image.

Figure 12.27: Five runs, each of five steps of the convolution process, showing the patterns emerging from five different random pixellations. The same and not the same.

in effect we are looking at a point process, just as we have been describing above.[15]

The process begins from a randomly chosen state in $\mathscr{M}$. Then the effect is generally for the process dynamics to rapidly stabilize, though certainly not always to the same state. In Fig.12.27 the successive effects starting from five different random matrices are shown. Even with totally different random initial pixel states, a strong patterning will occur, and although the outcomes are different, they are all distinctively similar. Further, if part of the resulting pattern is obliterated or randomized, the process will self-repair leading to a new final state that is still recognizably in the same family.

The dynamics tends towards producing stable states. In fact in our examples the stable states seem to consist almost entirely of zeros and ones—the two extremes of the interval $[0, 1]$. As Fig.12.27 suggests, there are many stable states, and the dynamics is far from ergodic: individual trajectories do not explore the entire state space.

Although we are used to pattern emerging out of the context of an evolving pattern system, and this certainly exemplifies it, the effects here and the simplicity of getting them are striking. The parameters of the situation are the exact specification of the two Gaussians, their mean and variance, and also the distance between their centers. Although we have not done so here, it is possible to make some directions more favorable for diffusion than others and produce definite features of alignment.

The principle of an interplay between activation and inhibition seems to be a common source of the distinguishing and patterning that we see everywhere. Neurons activated by the senses often suppress the activation of alternative receptors, thus enhancing and distinguishing the effect. When some muscles constrict, others that oppose them must be inhibited from acting. In the present case we have a model in which activation at a point has a local effect that promotes pigmentation nearby while actively suppressing it further away. The effect is distinctive patchwork areas of pigmentation.

12.9 Cantor dust

12.9.1 Self-similarity

Up to now we have considered spatial repetition in the form of translational or rotational transformations that match up parts of space. This is repetition where size does not change.

The Fibonacci tiling of Fig.12.15 suggests another form of repetition where it is the scale that changes. The construction of the Fibonacci sequence and the tiling is based on self-similarity where the scale is increasing. The initial tiling with a and b tiles is replaced (or we should perhaps say reconsidered) as a tiling with A and B tiles, which are versions of a and b scaled up by a factor of the golden ratio φ. It is immediately tempting to repeat this, and replace the A and B tiles with scaled up tiles $\mathbf{A} = AB$ and $\mathbf{B} = A$, where again the scaling is by a factor of φ, and indeed this works, and there is no need to stop there. We could do this forever. We know that the Fibonacci chain is not periodic (although it is repetitive), but it does have perfect repetition if we allow for changing scale.

What we see in the Fibonacci case is self-similarity going up in scale. There is also self-similarity going down in scale. There is one example that we all know extremely well, based totally on self-similarity, that goes in both directions, scaling upwards and downwards. This is the decimal system. Think of a unit interval on the real line, e.g. $[0, 1]$. The interval from $[0, 10]$ consists of ten of these unit intervals one after another. Call it a *decade*. The interval from $[0, 100]$ consists of ten of these decades, call it a *centade*. And so it goes on to with ten centades making up the interval $[0, 1000]$, and so on. This is self-similarity in increasing scale.

It works the other way, when we go into decimals. The interval $[0, 1]$ is made up of ten intervals of length $1/10$, and each of those is made up of ten intervals of length $1/100$, and so on. Our entire decimal system is based on this lovely idea of self-similarity. Of course 10 is not special. What we have shown is base 10, but it can be done with any integer greater than 1, most famously base 2, which we have used so often in this book.

The ancient mathematicians and astronomers of Sumerian and Babylonian times (c.a 3000 BCE) used base 60 [175]. This system was adopted by Greek astronomers, notably Ptolemy (ca. 150 CE) who based much of his work on their detailed records of the changing heavens. Some of that influence still lingers with us today—our minutes and seconds in the division of time and our use of degrees, minutes, and seconds for angles.[16]

Self-similarity also lies at the heart of the fascinating world of fractals. Fractals get their name because they embody fractional dimensions, but the usual way we see them is in some form of self-similarity with ever decreasing scale. We readily appreciate this form of repetition because it appears in all its glory in bounded regions that we can grasp at a single glance.

The fractal world is well represented in popular culture, especially in the famous Man-

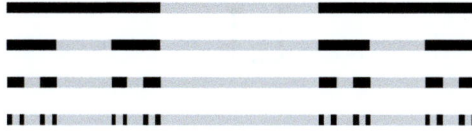

Figure 12.28: Four iterations of the process of removing middle thirds. The Cantor set C is whatever is left over after repeating the process forever.

delbrot set, and there are numerous excellent and beautifully illustrated mathematical presentations. Rather than adding to what is already so well described, here we are going to treat this fractal world with one famous example. Our example is in a one-dimensional setting and derives its interest from its simply defined but highly unintuitive nature—more like dust than anything substantial—and from its stature as the prime exemplar of the nature of self-similarity. It also introduces a type of measure that we have not yet seen, the *singular continuous* measure. Finally, it fits naturally into our idea of pattern systems, and remarkably makes its appearance in many unexpected ways in all sorts of pattern systems, especially tiling dynamics.

12.9.2 The Cantor set

Constructing the Cantor set

The *Cantor set* C is a subset of the unit interval $[0, 1]$ in the real line. It is easy to get the general idea of C. Begin with the interval $[0, 1] = \{x \in \mathbb{R} : 0 \leq x \leq 1\}$. From this interval remove the middle one-third. This leaves two pieces. From these remove their middle thirds. This leaves four pieces. Repeat, and from these four remove their middle thirds, see Fig.12.28. Do this forever. What is left over is the Cantor set, C. Not surprisingly, there isn't much. Surprisingly there is more than one might first think.

To make this look like a dynamical system, we need somewhere to put all the points we are removing. So, to the interval $[0, 1]$ we adjoin another point, which we will think of as some sort of trash can. We denote it by $\sqcup$ which looks appropriate. So we define $X := [0, 1] \cup \{\sqcup\}$. The union is disjoint. If it feels more comfortable one could replace $\sqcup$ by some number, say 2. But its value has no relevance to what is going to take place, so we prefer to keep it non-numerical. The Cantor set can now be seen as an emergent pattern inside a dynamical system based on X.

The dynamics just formalizes the simple mantra: throw away middle thirds. Throughout it is important to remain aware of the difference in meaning between sets like $[2/3, 1]$, which is the set of all the numbers between $2/3$ and 1 *including* the end points $2/3$ and 1, and sets like $(2/3, 1)$, which is the set of all the numbers between $2/3$ and 1 *excluding* the end points $2/3$ and 1. These are *closed intervals* and *open intervals* respectively. The "throwing away" is throwing away *open* intervals, e.g. the first one is $(1/3, 2/3)$.

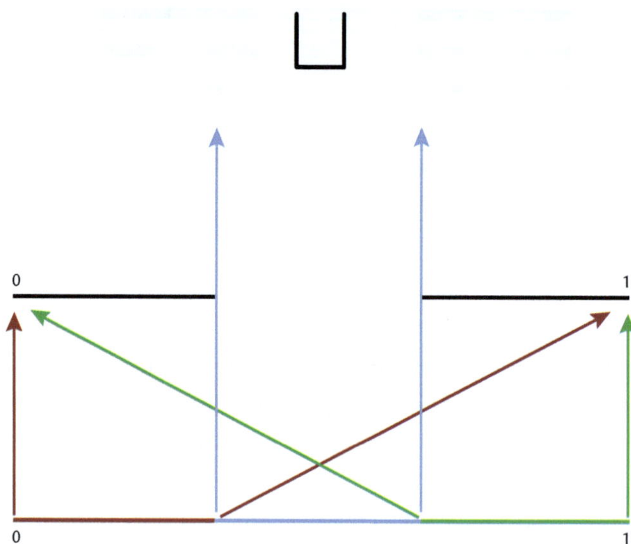

Figure 12.29: A diagrammatic representation of the dynamics of stretching and throwing away. The left third of the unit interval is stretched out to fill the entire interval. Similarly for the right third. The middle third is taken to the trash can $\sqcup$.

Define two stretching operations, one on the subinterval $[0, 1/3]$ and one on $[2/3, 1]$:

- on $[0, 1/3]$ the stretch is to multiply by 3: $x \mapsto 3x$;
- on $[2/3, 1]$ the stretch is to multiply by 3 and then shift the result back into $[0, 1]$: $x \mapsto 3x - 2$,

see Fig.12.29. Let's simply call the application of these two operations, "stretching by a factor of 3". Each of them stretches out an interval of length $1/3$ so that it exactly covers the unit interval $[0, 1]$.

Now we can define the dynamics on X. There is a single operator s. Its action on any point $x \in X$ is

- if $x \in (1/3, 2/3)$ then $s \triangleright x := \sqcup$ (throw away the middle third);
- if $x \in [0, 1/3]$ or $x \in [2/3, 0]$ then $s \triangleright x$ is the result of stretching x by a factor of 3;
- $s \triangleright \sqcup := \sqcup$.

With this we have the dynamical system (X, s). The first operation throws away the middle third, the second prepares the system for throwing away the next two middle thirds by stretching each of the two remaining pieces back over the interval $[0, 1]$. The third just says that what's in the trash stays in the trash.

The first application of s throws away the entire interval $(1/3, 2/3)$. Applying it twice further throws away $(1/9, 2/9)$ and $(7/9, 8/9)$, since the first application of s stretches each of them into $(1/3, 2/3)$. Applying s three times also throws away the further four intervals $(1/27, 2/27)$, $(4/27, 5/27)$, $(22/27, 23/27)$, and $(25/27, 26/27)$, and so on.

The *Cantor set* C is the set of points of X that survive the throwing away process, that is, $x \in C$ *if and only if its trajectory never gets to* $\sqcup$. Of course, once a trajectory hits $\sqcup$ it will never get out since $s \cdot \sqcup = \sqcup$. Whatever C is, it is clear that if $x \in C$ then also $s \triangleright x \in C$, by the very definition of C. So we have a *sub-dynamical system* using the action of s restricted to C:

$$s : C \longrightarrow C .$$

Perhaps surprisingly, $s \triangleright C = C$. Were it not so, some elements of C would be trashable after the application of s, which would mean they were never in C in the first place. Below we will see this in a far more explicit form.

Our impression that not much survives is, in some sense, very real. The total length of all the intervals that will eventually be thrown away can be worked out by noting the lengths of the initial few of these that we gave above, and extrapolating:

$$\frac{1}{3} + 2(\frac{1}{9}) + 4(\frac{1}{27}) + \cdots = \sum_{k=1}^{\infty} \frac{2^{k-1}}{3^k} = \frac{1}{3} \sum_{k=0}^{\infty} \frac{2^k}{3^k} = \frac{1}{3}\left(\frac{1}{1-\frac{2}{3}}\right) = 1 , \qquad (12.13)$$

where we just use the standard formula for adding the terms of an infinite geometric progression.[17]

Since $[0, 1]$ itself has total length 1, it looks like there is nothing left. In terms of Lebesgue measure, the measure of length on the line, C has measure 0.

But C is far from being empty! Certainly there are a few points like $0, 1, 1/3, 2/3$ which we can see will survive the discarding middle intervals process, so they are in C. What is amazing, though, is that C is actually an uncountable set, and it is not hard to see why. Think of expanding real numbers in the ternary system, that is, base 3. This is the base-three version of the self-similarity system that we discussed above. It uses just the three digits $0, 1, 2$. Thus, for instance, in the ternary system $0.1202 = 1/3 + 2/9 + 0/27 + 2/81$. Every real number x between 0 and 1 has an expansion $x = 0.x_1x_2x_3\ldots$ where the $x_k \in \{0, 1, 2\}$, and conversely every such ternary expansion represents a number in $[0, 1]$. Note that $0.2222222\cdots = 1$, for the same reason that in the decimal system $0.999999\cdots = 1$. This is what equation (12.13) says.

Now comes the nice part. We claim that any number $x \in [0, 1]$ which has a ternary expansion in which the digit 1 *never* appears is actually in C. To see why notice that the first part of the definition of s throws away all and only numbers in the middle third, and these are all of the form $0.1 * * * * * \ldots$. Now if we look at the stretching, it comes in two forms: if $x = 0.0x_2x_3x_4\ldots$ then $x \in [0, 1/3]$ and stretching multiplies it by 3: $3x = 0.x_2x_3x_4\ldots$, the left shift! Similarly, for the points in $[2/3, 1]$, all of which are of the form $x = 0.2x_2x_3x_4\ldots$, stretching gives $2.x_2x_3x_4\cdots - 2 = 0.x_2x_3x_4\ldots$. Again the effect is the left shift.

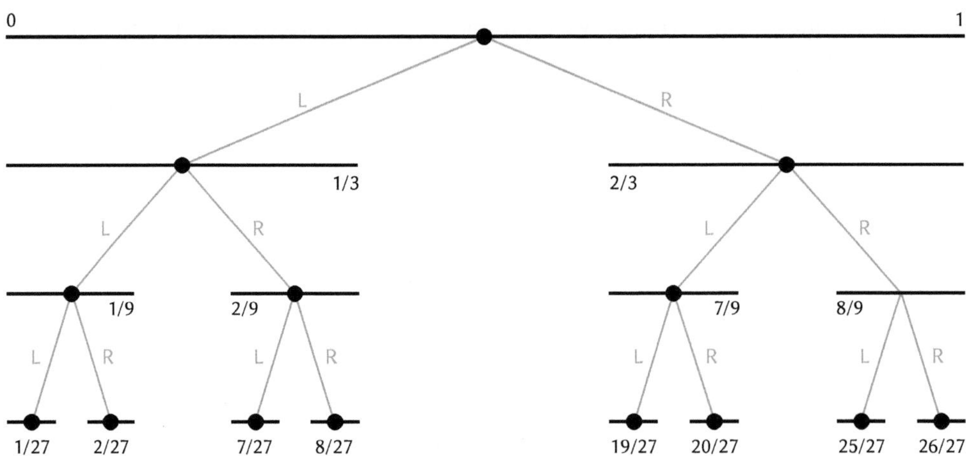

Figure 12.30: The first few steps in removing middle thirds to obtain a Cantor set. The process of choosing left (L) or right (R) intervals at each step gives rise to the binary tree, shown with grey edges. The fractions indicate the positions of the ends of the intervals that emerge out of the removal process.

So the stretching amounts to the left shift in both cases. If there are no 1s in the expansion, then left shifting of x can never produce a 1, no matter how many times we apply s, so s will never send it to the trash. So we see again that C is a dynamical system in its own right and now we see further that it consists *exactly* of the set of points that can be written in ternary form with no digits 1 appearing. We call this the Cantor (dynamical) system, and we now have a sense of what the Cantor set contains—all numbers in $[0, 1]$ whose ternary expansion contains no 1s.

Thinking in terms of choices required to make up a point lying in C we get another picture of the Cantor set. In creating a point $x = 0.x_1x_2x_3\ldots$ of C we have to make one of two choices at each step: 0 or 2. We can imagine each such point as describing a path in a binary tree: for each x_k write L if it is a 0 and R if it is a 2, and treat the resulting sequence as a series of instructions for walking down the tree, taking the left (L) or the right (R) branch at each node, Fig.12.30. The Cantor set is then pictured as the set of all left-right paths through a binary tree, each of its points represented by one of these infinite paths.

12.9.3 Conjugacy

Looking back at Fig.4.4 gives an interesting insight. There we viewed the shift space on two symbols $0, 1$ as a binary tree in terms of which each state of the shift space is seen as a path through the tree. The Cantor system is structured in just the same way. In fact we just saw that the action of s is in effect the left shift when the elements of the Cantor set are written in their $0/2$-ternary expansions.

Formalizing this relationship, it says that there is a natural one-to-one correspondence between the shift system $\{0,1\}^\infty$ and the Cantor system which takes each sequence of zeros and ones into a ternary number in which the zeros remain zeros, but the ones become twos: thus, for instance,

$$010011011\ldots \overset{\psi}{\leftrightarrow} 0.020022022\ldots .$$

We could write the Cantor set as $\{0,2\}^\infty$, which shows that in effect the two dynamical systems are the same. As long we ignore any numerical concepts and just treat $0,1,2$ as symbols and nothing else, then one is the shift system on the alphabet $\{0,1\}$ and the other on the alphabet $\{0,2\}$. As dynamical systems, $\{0,1\}^\infty$ and C are the same, they are *conjugate*, see §7.5.4.

Abstractly then, the Cantor set and the full shift are very much the same thing. In particular, since the state space of the full shift is uncountable (§7.4), so too is the Cantor set! Since there is a stationary (invariant) measure on the shift system—remember the standard probability measure that gives 0 and 1 equal probability—there must be a similar one on the Cantor set. It is only when we represent them in terms of the real numbers, interpreting strings as binary or ternary expansions of numbers, that they look so remarkably different.

On the one hand the $\{0,1\}$-shift maps *onto* the interval $[0,1]$ by this interpretation, and it has Lebesgue measure (length) equal to 1. On the other hand the $\{0,2\}$-shift maps onto a seemingly very sparse subset of $[0,1]$ that has zero Lebesgue measure. Such are the intriguing puzzles of around infinite sets![18]

12.9.4 A singular continuous measure

The Cantor set is "dust" because seen as part of the real line its Lebesgue measure is zero. Even the tiniest intervals in the real line have positive measure, but the Cantor set does not contain any intervals at all. Evanescent as it appears, it still exists, and it has a perfectly good stationary measure (which we will denote by κ) to go along with it, and it is a measure that completely turns the tables around. The entire set of points we *threw out* to create C, a set of points that had total Lebesgue measure equal to 1, actually has zero measure with respect to κ, while it is C that has measure 1.

This measure is most easily seen by simply matching it to the standard probability measure on the $0,1$-shift system, where the measure can be interpreted in terms of the probabilistic outcomes of coin tosses. In that picture the dictionary sets $[0]$ and $[1]$ each have measure $1/2$, the dictionary sets $[00],[01],[10],[11]$ each have measure $1/4$, and so on. We can just copy this over to the Cantor shift $0,2$-shift system and get a corresponding measure that we call κ.

What corresponds to the dictionary set $[0]$ of the $\{0,1\}$-shift is the corresponding dictionary set $[0]$ consisting of all $\{0,2\}$-strings that start with 0. Similarly, corresponding to $[1]$ we have $[2]$, consisting of all strings that start with 2. Each of these sets gets κ-measure equal to $1/2$. Likewise, each of the four dictionary sets $[00], [02], [20], [22]$ has κ-measure $1/4$.

A more visual way to see this is to imagine walking a path down the Cantor tree of Fig.12.30, starting from the top. At each step choose to walk left (L) or right (R) with equal probability $1/2$. After some steps one ends up in some interval. For instance if we take the steps LRR we end up in the interval $[8/27, 9/27]$. The probability we will end up there is $1/8$ since we took three steps and made three choices, each with probability $1/2$. Measure κ gives $1/8$ to that part of C that lies in $[8/27, 9/27]$. This same idea applies throughout. In this interpretation, $\kappa(I)$ for any interval I of $[0,1]$ is the probability that in a random walk of the tree we end up in the interval I. Thus, for instance κ gives the value 0 to the entire interval $(1/3, 2/3)$ since no walk will ever end up at a point of this interval. Similarly it gives measure zero to all the discarded middle thirds. It also gives measure zero to individual points. Finally we also define $\kappa(\sqcup) = 0$, the trash counts for nothing. In this way κ is defined on all of X, and in particular on $[0,1]$. We have

$$\kappa(X \setminus C)) = 0$$
$$\kappa(C) = \kappa([0,1]) = 1\,.$$

The mystery to this is where exactly does κ pick up any positive measure at all?

We can think of κ as a measure on (X, s) or on the subsystem (C, s), whichever is more convenient. We prefer C because it is a complete dynamical system in its own right and the trash $\sqcup$ was just a convenience to see how C arises, and in any case $\kappa(\sqcup) = 0$. The measure κ is stationary for the dynamics of (C, s), just as the standard probability measure on the standard shift system is stationary. This means that for any (measurable) set A in C, $\kappa(s^{-1} \triangleright A) = \kappa(A)$.

Once we have mapped the two systems into the real line by interpreting the strings as binary and ternary expansions, the standard measure of the $\{0,1\}$-system is Lebesgue measure (length) and the Cantor set looks like dust. Yet, from the point of view of κ this "dust" is richly structured. The measure κ is invariant with respect to the stretching operator s, Lebesgue measure is not.

Lebesgue measure is the only (up to an overall scaling factor) measure on the real line that is stationary under the translation action of $\mathbb{R}$. Translation reveals self-similarity, but it is self-similarity in which there is no change of scale. The Cantor set C is entirely located in the unit interval $[0,1]$, and is obviously not invariant under translation. However, it is invariant under the stretching operator s, and its measure κ is self-similar with respect to this operation. It is the only stationary probability measure on (C, s).

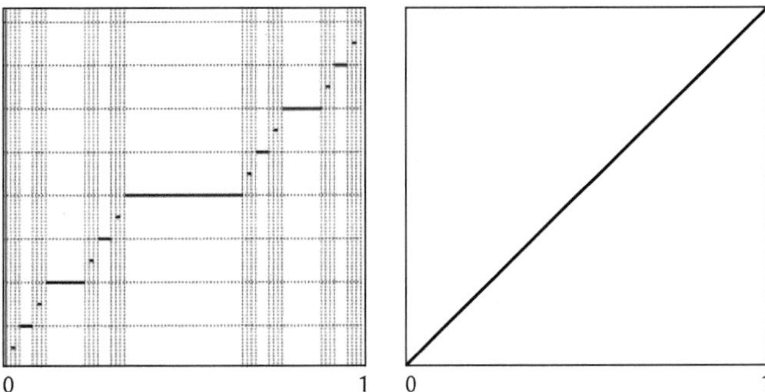

Figure 12.31: The devil's staircase. Comparison of the cumulative distribution functions of the Cantor measure κ (left) and the Lebesgue measure (right) on the interval $[0, 1]$. For the Cantor set only part of the distribution can be shown, but this is enough to show how the process continues in a self-similar type of way as middle thirds are removed. The Cantor set measure κ accumulates measure only *off* the middle thirds, so it looks constant almost everywhere—the horizontal lines in the graph. What's more, these horizontal lines eat up all of the interval $[0, 1]$ as seen by usual measure: the devil's staircase accumulates measure only through the points of the Cantor set, of which there are uncountably many but utterly disjointed. The Lebesgue accumulates measure smoothly and uniformly. Both measures give measure 0 to individual points, but they accumulate measure in totally different ways. The devil's staircase provides an example of an increasing function that is differentiable almost everywhere with 0 derivative. For more on this see [143].

The two measures do not actually see each other, each one assigning measure 0 to a set to which the other gives measure 1. One way to compare the two measures side by side is to look at the graphs of their *cumulative distribution functions.* These arise by looking at the total probabilities for the interval $[0, x]$ as x runs from 0 up to 1. The total probability of the set $[0, 1]$ is 1 in either case, so both functions end up at 1. But the way they get there looks vastly different. For Lebesgue measure the measure of $[0, x]$ is just the length of this interval, which is x. Its graph as x varies is a straight line. The graph of $\kappa([0, x])$ is far stranger, Fig.12.31. People love to invoke the devil when anything is wildly unusual and presents lots of conceptual difficulties, and the cumulative distribution function of κ is no exception: it is sometimes called the *devil's staircase.*

The measure κ is different from anything we have seen before. It is not like a point measure because it gives value 0 to every single point, whereas a point measure gets all of its positive values from its values on particular points. Neither is it remotely like measures of length, area, or volume that we have seen. Measures like κ, which give measure 0 to all points and which give all their positive measure to sets that have Lebesgue measure zero, are called *singular continuous measures.* They seem in some sense "orthogonal" to Lebesgue measures.[19]

Cantor sets along with their singularly continuous measures, bizarre as they might first seem, are in fact a mainstay of iterative dynamics. The zero-one shift system is a

Cantor set and, as we have seen, it is one of the most pervasive forms of discrete dynamics. Cantor sets represent a very distinct form of patterning that lies at the basis of discrete dynamics.[20]

12.10 Conclusion

Return. In the epigraph to this chapter we see that the *Daodejing* says that return is the movement of Dao. The enigmatic nature of the *Daodejing* makes it unwise to speculate too fully on what might be implied by the word "return" here, but the general idea is clear: we experience the processes of Dao as an endless emergence and dissipation of form, the ten thousand things arising into being, having their moments of existence, and then vanishing. The two key words here are 有 (you) and 無 (wu), which we see translated as *presence* and *absence*, and they bring us back to the very opening of the *Daodejing*. There it is said that distinctions are marked by words, and without distinctions we remain with the whole. Seen in this way, presence and absence are about distinction and lack of distinction.

This is not a question of things emerging from and returning to emptiness as though emptiness was some transcendent store-house of possibilities. Often we can watch clouds forming or clouds disappearing right before our eyes. We know that what has changed is not the material essence, but the form of appearances. The atoms and molecules that make up the air have not emerged or disappeared. It is the same with us as human beings. We die and the form is not what it was. The atoms that make up our body do not vanish in burial or cremation but will re-emerge in new forms, or even as transplanted organs. And beyond that pure material physicality, the effects of our lives, the ideas and human interactions that we have generated, are still present and may go on to influence the world for centuries. In terms of the Hua-yen philosophy, the entirety is a completeness—nothing to add, nothing to take away—and no aspect of it, however small, can avoid being essential to this completeness. It is form that changes.

Put in terms of pattern and patterning, absence and presence are about events. We experience the world in terms of what we distinguish. Where we don't distinguish there is absence: where we do there is presence. These two depend on each other. Change means that events change. If the focus of discrimination (the event space) changes there are new presences and new absences. As humans we have all the receptors and biological capacities to encourage us to connect with each other and form social groupings. We make extensive and nuanced use of these to bring about an eventful life. But to a mosquito a human being amounts to responses to exhaled carbon dioxide, sweat, and body heat, for each of which it has appropriate sensors in its antennae and its legs. Its event space, what is present and what is absent, is remote from ours, but just as real.[21]

Our own version of return is the return of process to prior events. The idea that return implies both distinction and recognition of prior events is the explicit content of one of the most basic outcomes of pattern systems that we have seen: the Poincaré recurrence theorem. But that theorem in itself does not give full flavor to more fundamental questions we might ask about the long-range order and coherence that are so important to our lives. Neither does it suggest any way in which to organize or classify the vast variety of forms recurrence can take. At the one end we have seen the perfect regularity of a periodic cycle and at the other end the randomness of a Poisson process.

What is common throughout is the possibility of framing change in the form of pattern systems. When all is said and done this in itself may be what we mean by *coherence*—that there is a systemic interpretation in which events fit together as a dynamical whole. Crucial to this interpretation is the *law* of the process. The dynamics is one thing but the frequencies in which events happen or can be expected to happen is another. The law, the measure, is what determines these and so determines what shape the dynamics will have and what will emerge from it as it evolves. We most often look for invariant measures, for those are the most compatible with dynamical process itself and bring coherence and repose to underlying change.

This chapter begins with the mildest deviation from periodic order, namely the *almost-periodic order* that arises when repetition meets incompatible repetition. This produces a very specific type of return which is characterized by return with bounded waiting time or return within bounded spatial difference. Those bounds depend on what it is one is looking for in terms of return, but each type has its own bounds for guaranteed repetition.

Periodic return means guaranteed return at specific temporal or spatial intervals while the Poincaré theorem is essentially guaranteed return at the level of events, but with no hints on when or where. Almost periodicity is a form of long-range order somewhere in between. If we think about natural processes that generate forms on the basis of the local environment, it is not surprising that in a relatively homogenous environment, similar forms would appear. Then we would perhaps be able to predict patterning far away from where we are, either in time or space, on the basis of what we can see around us. This is one interpretation of what we mean by long-range order. On this definition, almost-periodic order is a very strong form of long-range order, and it often reveals itself in a characteristic and beautiful way, namely the Bragg peaks of diffraction.

A running theme of this chapter is the story of diffraction, because it is a technique that produces images based on the *internal* 3D-structure of materials, particularly ones in crystal form. Initially diffraction was considered important in showing that X-rays are a part of the great electro-magnetic spectrum. At the same time it gave strong evidence of the reality of the atomic world. Pioneered in the early part of the twentieth century it soon became (and still remains!) a major technique in determining the internal arrangements of atoms in complex organic and inorganic substances. The most striking part of this is that crystalline structures produce diffraction images that are dominated by dot-like concentrations of intensity, the Bragg peaks—so much so that Bragg peaks became the distinguishing feature of crystals. When there is nothing else but Bragg peaks, this is called *pure-point diffraction.*

As it turned out, much to everyone's surprise, Bragg peaks are an indicator not of periodicity but of *almost periodicity.* This issue came to the fore through the discovery of almost periodic (but not periodic) tilings, which have pure point diffraction, and then somewhat later the discovery by D. Shechtman of phases of Al-Mn alloys in which there was an essentially pure point diffraction but with clear 5-fold symmetry that we have seen is impossible in 2D- and 3D-lattice structures. The story of pure point diffraction and the struggles in the scientific community to accept and understand what it really means is told in the Endnotes. Perhaps as much as anything it is a story of how beautifully mathematics, physics, and chemistry can inform each other, with twists and turns through aperiodic tilings, the halting problem, and the introduction of dynamical systems.

The idea of return is put into a much larger context when we turn to random (or stochastic) point processes—the laying down of points in space. Whether sets of points fall in 1D-, 2D-, or 3D-spaces, whether they be deterministic or random, well organized or not, well separated or not, symmetrically distributed or not, or whether they display deep correlations across space and time or not, they all fit under one roof. In the end it is the vast diversity of probability measures and their role within these processes as *systems* that brings forward the concept of coherence.

We have begun with the most famous of all point processes, the Poisson processes. These are quite the opposite of long-range order in the sense of predicting repetition of form on local observation. The very definition says that the distributions of points in separated regions do not affect each other at all, they are statistically independent. Instead the commonality, the coherence, is built around the idea of a *common density.* It is assumed that on average the number of points falling into any region (say of the plane, in the 2D case) is proportional to its area, the constant of proportionality being the *intensity*. There are no other assumptions. There are many point processes that randomly distribute points with no perceptible repeating patterns, the Poisson process being the grandfather of them all. It is often taken as the default random distribution of points. Still, it does not fit well with modeling the distribution of atoms of a gas for which there is a natural avoidance of very close clustering. Determinantal processes are random but make point clusterings less likely, and produce typical point patterns that are recognizably different from those of Poisson processes, see Fig.12.20.

What is surprising at first is that even very well-defined tilings, like the shield tiling or even periodic tilings, can be viewed as random point processes. This becomes possible as soon as one takes not just one tiling, but an entire family of tilings that are locally indistinguishable from each other under translations. Then the stochastic aspect arises by thinking about the frequencies of finite point patterns found in randomly chosen point sets according to the law of the process.

Since point processes are so flexible and appear so frequently both in the natural world and in our complex societal world, they offer a deeper perspective on what we really mean by *coherence*. What is common to all of these processes is that they are pattern *systems*. Each belongs to a system that is dynamical in nature and molded by some statistical law connecting events into a context of change. The parts fit together. Stationarity, ergodicity, mixing, pure pointedness, symmetry, and similarity (some of which are discussed in the Endnotes) are all properties associated with the law of the process.

Although we have only touched briefly on it, one of the most interesting ideas that arises in the metaphor of the self-made tapestry is *morphogenesis*. How totally amazing it is (and one can see even as far back as ancient Greek thought how deeply puzzling it was seen to be) that well-defined pattern can emerge from no discernible pattern at all; and can do so with the ever-differing but nonetheless ever-consistent likenesses that we see in the fur of animals, the scales of fish, and the feathers of birds. These are some of the amazing evolutionary possibilities of pattern systems. We have much to learn, but as the simple point process that we discuss shows, we can expect the same fundamental ideas of emergence of pattern out of difference and change.

Self-similarity is familiar to everyone in the form of fractals and their amazing graphical images (frequently lavishly displayed on T-shirts). Since it is well explored in the public eye, we have not dwelt on it in depth, but it is important to realize that their essence is dynamical. We have taken a look at the famous Cantor set, which can be considered to be the progenitor of this whole field. Apart from its unexpectedly close connection to the standard shift system, the Cantor set reveals a form of invariant measure that is very different from any other we have seen. It is a *singular continuous measure*. Here it plays a role totally complementary to that of the usual Lebesgue measure of length on the interval $[0, 1]$, each effectively ignoring the other. Singular continuous measures are often considered to be bizarre and difficult to understand. But what we see here is a measure that is compatible with self-similarity on a decreasing scale, while Lebesgue measure is compatible only with the self-similarity of translation (but no change of scale). Each has its domain. Importantly, both appear quite naturally in our deepening understanding of Nature.

Though we have mentioned it many times, it bears repeating again: what is most impressive about the whole scientific story is that all its various disciplines and subdisciplines tell a unified story about the world. We see this as one of the most significant revelations of our own lives, lived largely over the latter half of the twentieth century and now well into the twenty-first. The great tragedy is that we know full-well that we (humanity) exist as a small part of a vast interconnected web of being, but cannot discipline ourselves to live within the bounds that this implies.

Chapter 13

Quantum

13.1 Introducing a strange world

Up to now our examples of physical patterning and pattern systems have all been rooted in the world of classical physics. At the scale of atoms and molecules, classical physics breaks down and it is through quantum physics that we describe what we observe.

The mathematical system that models quantum physics, commonly called *quantum mechanics* or simply *quantum theory*, was developed in the first part of the twentieth century, and only with enormous intellectual struggles. It is unintuitive, mathematically very different from classical physics, and is riddled with all sorts of philosophical issues, going as far as the nature of human consciousness. Yet it is astoundingly accurate, predictive, and explanatory. Our modern technical world simply could not exist without our understanding of it, and neither would our contemporary understanding of Nature. No doubt this will be even more so in the future. All our present understanding of matter is based on ideas of quantum theory.

Now, the quantum world has to be as dynamical as anything else in our Cosmos, so quantum systems ought also to fit into our scheme of pattern systems. In this chapter we shall show how this comes about. However, the mathematics involved in quantum theory is different than that appearing in classical physics, and we should begin by stressing six points:

(i) Observables: In classical physics the observable features of a dynamical system are functions defined on its state space. We think of observing a system as measuring it, or more precisely assigning values to its states in some way that represents some physical quality. Measurement of some particular kind when applied to all the possible states of the system produces a function on the state space, the value of the function at any particular state being the outcome of the measurement or observation of that state. It is an implicit feature of this approach that the observation has no effect on the states. If we are observing the dynamics at a particular state, the observation of it (the value of the function in question at that state) makes no difference to the state, or at least represents the true particulars of that state as we measure it. The state and its transitioning in time are effectively independent of the act of observation. In quantum theory this no longer holds. The act of observation can cause a change of state and the observation itself only represents the changed state, not the one that pre-existed the observation. There is an important point here that we will continue to make: in terms of pattern systems, we can think of observation as the interaction of two pattern systems, the observer and the observed. The observer and the observed cannot be isolated from one another. They must physically interact, and at the quantum level the effects of interaction are inescapable.

(ii) Quantization: The aspect of quantum theory that gives it its name is that many (though not all) observable features of Nature at the quantum level are *quantized*: the observed values do not assume a continuum of values, but only either a finite or countable set of values. For instance, the energy levels of an electron around a nucleus of an atom can only take a discrete set of specific values, and no others. Such an electron can switch to a higher energy level in that discrete set of values, say by absorbing a photon, and then switch back to its previous level, or even to a different level, by emitting a photon. But the energies of those photons will be exactly those that match those changes of energy. Light emitting diodes work by emitting light as electrons change energy levels, and their constant color is a result of the quantization. Strangely the electron does not take any intermediary values as it switches energy levels. It is truly quantized. The failure of classical theories to explain the quantization or the explicit energy levels involved was a central impetus in the creation of quantum theory.[1]

(iii) Wave-particle duality: At the quantum level the distinction between point-like particles and wave-like behavior breaks down. Physicists argued for centuries about whether light was wave-like or particle-like, for it can quite convincingly show both types of effects. It turns out it is neither (or if you want, both). Photons, and other entities like electrons, neutrons, protons, and neutrinos, display wave-like and particle-like features according to the situation.

(iv) Indeterminism: There is the well-known fact that the outcome of an observation at the quantum level generally has a probabilistic note to it. One observation will lead to one single value, but observation of the same state under identical conditions may lead to a different result. So quantized or not, there are probability distributions associated with observables. Thus determinism is apparently lost and randomness, or chance, enters into the foundations of quantum theory. This is one of the most profound (and also most disquieting) ideas of the theory, for it runs counter to some of our deepest intuitions.

(v) Uncertainty: probabilistic indeterminism enters in a quantifiable way. Between some pairs of observables (notably position and momentum) this takes an explicit form in which the more certain one is about the numerical value of one variable the less certain one can be about the value of the second.

(vi) Dynamics: Quantum systems are dynamical systems, and their evolution can be thought of as continuous and deterministic but punctuated by discontinuities associated with change of state. Energy is an observable, and it supplies the impetus for change described by the famous Schrödinger equation (see §13.3 and (13.8)). The action of time is determined by the energy of the system (which itself can vary in time).

Is it strange that the quantum world conforms so poorly to our own intuitive views of the world? As macroscopic individuals with senses that discern only macroscopic distinctions, perhaps we should not be too surprised that our understanding of the physical world, derived from experience, culture, and genetics, does not match what we find at the atomic level. Even so, quantum theory is more than just unintuitive. It seems to defy our normal ideas of logical reasoning. Were it not for its enormous success in matching and explaining what is observed and its amazing predictive power it is hard to imagine how this bizarre form of patterning could ever have arisen in the human mind and how it could ever have been taken seriously. Certainly in the history of philosophy, no matter where, and even in the history of science itself, there were no premonitions of the quantum world that we have come to accept as reality.

Not surprisingly all these features of quantum theory have caused huge difficulties in physics. This is not because there is any doubt about what happens under repeated experimental conditions—these facts are used every day in the technology of lasers, electron microscopes, diodes, LEDs, MRI, and so on— but because they are so contrary to our intuitions. This is especially true of the probabilistic nature of observables. The primary objection is not that this probabilistic nature is observed, for we all know that many things appear to happen by chance and using probabilistic methods is the best way to deal with them. But it is our human nature to put chance down to our ignorance or incapacity to understand all the forces at work: we tend to believe that Nature is deterministic and effects are always due to underlying cause.

In the philosophical arguments around quantum theory this "underlying cause " goes under the name of *hidden variables*, the idea being that there are other variables, presently unknown to us, that would bring back causality if we could find them. Under this line of thinking quantum theory, as we now have it, is *incomplete*. Einstein was the most famous proponent of this camp, and he argued vigorously for the incompleteness of quantum theory. However, the standard interpretation of quantum theory today, the *Copenhagen interpretation*, that dates back to the founding fathers of the subject, and notably to Niels Bohr (the brother of Harald Bohr of almost periodicity that we have already discussed), is that quantum theory, as we have it, *is* complete. There is evidence, from many angles and many experiments, that quantum theory is complete and randomness is a true feature of what we can observe in Nature. We will come back to this with a discussion of the EPR paradox and Bell's inequalities in Ch.14.

Having said this, one of the things that one doesn't so often hear about is that there is perfectly deterministic temporal dynamical evolution in the mathematical description of any undisturbed quantum theoretical system. *It is in the act of interaction that determinism can be lost.* And any observation of a quantum system constitutes an interaction. What underlies the quantum system, and what one can actually observe about it, are not the same thing. The realization of this is one of the achievements of Heisenberg, but again it raises philosophical questions.

Another point that we should make is that up to now we have taken mathematical models as ways to express, better articulate, or predict processes about which we already have reasonable intuitions. We are not usually thinking that the model is in some way our ultimate description of what we are observing. However, this is the case in quantum theory. The assumption is that the mathematical model is the *only way* we have to really articulate what is going on, and intuitions taken directly from the macro-world that we live in are riddled with contradictions if we apply them at the quantum level. In this respect, the mathematics becomes the defining model of quantum theory, and as such we must become deeply aware of the ultimate human-ness of our understanding of Nature and its amazing degree of abstraction. It is an articulation of the underlying principles of Nature in abstract mathematical terms which are ultimately products of our own minds. This is the origin of Hawking's description of quantum theory as "the dreams that stuff is made of". This is in no way to deny the experiential reality of the world, but only to emphasize that we can only articulate it in human terms.

Popularizing quantum theory is difficult because of the great number of mathematical ideas involved and the fact that the mathematical description is really the only accurate articulation of it that we have. Verbal descriptions with no mathematical content can often make things look more mysterious and inaccessible than they really are, and can even lead quite notable figures wildly astray.[2].

Fortunately we have already seen many of the mathematical ideas involved, so we can give a picture of how these curious features of quantum theory fit into a coherent scheme. Of course our purpose is far from making us into quantum physicists or working through lots of quantum theoretical examples—which indeed can be dauntingly difficult—but rather to show the mathematical setting of it, and how it looks from a dynamical pattern point of view. In doing so, many of the ideas of quantum theory become clearer because in spite of the abstraction of the setting we can see how all they fit together and how the strange results of quantum theory are expressed and even appear right out of the formalism. Absorbing the formalism, even in the rather limited depth that we can take it here, does have its challenges. But the rewards make it worth the trouble. In the end we will find that we are still in the world of pattern as we have been developing it: the same axioms that have motivated our entire discussion are still here. The same ideas, but a very different setting.

Two final remarks. There are two fundamental invariants of Nature that arose to prominence in post-nineteenth century physics. The first is the speed of light ¢, which is a central feature of relativity theories. The second is *Planck's constant h* (more often appearing in its so-called *reduced form* $\hbar = h/2\pi$), which is equally central to quantum theory. The speed of light is huge compared to human notions of speed, and h is minuscule compared to human scales of perception.[3] We seem to embrace both scales at once. Classical physics is sometimes described as the situation in which h is set to zero and ¢ is set to infinity, which at the level of normal human experience is the appearance of things. Many formulas of quantum theory and relativity become the formulas of classical mechanics when this is done. This is why classical physics works so well.

What we are describing in this chapter should be seen as a short introduction to the theory of quantum processes, set at the level that does not include relativistic effects. This works fine as long as the particles or entities involved are not moving close to the speed of light or do not experience relativistic changes in mass. For instance, in lighter atoms relativistic effects are not so important, and quantum theory as we describe it here is appropriate. For heavier atoms relativistic effects become more important. The glittering color of gold and its imperviousness to tarnishing, for example, are known to be relativistic effects.[4]

13.2 What the mathematics looks like

> For the sake of persons of different types of mind, scientific truth should be presented in different forms and should be regarded as equally scientific whether it appears in the robust form and vivid coloring of a physical illustration or in the tenuity and paleness of a symbolic expression.
>
> James Clerk Maxwell, [70]

The great question facing physics in 1925 was the structure of atoms, notably the simplest of all atoms, the hydrogen atom. The issue had to do with explaining the behavior of the electron that somehow circled the single proton that was the nucleus. The initial idea had been that the atom was like a little planetary system with the positively charged nucleus attracting the negatively charged electron that circled around it. Such a model was described by Rutherford, but it suffered from an irredeemable problem: classical physics applied to such an orbiting electron showed that it would radiate energy and the orbit would collapse in an infinitesimal moment of time, something of the order of 10^{-11} seconds. Furthermore the possible energies of the electron of a hydrogen atom were observed to be quantized: they only existed at certain energy levels. As we have seen, electrons gain or loose energy in the form of photons, but only of very specific energies. No physical theory of the time could formalize what was going on. It needed a new way of thinking.

In 1925 the 23-year-old Werner Heisenberg published the first mathematical formalism for quantum theory. As Rovelli says, "In your twenties, you can dream freely" [137]. His paper originates out of his thinking about the observed quantum effects in the hydrogen atom. The paper was far from what was to evolve into the fully fledged theoretical foundation for quantum theory, but he had made a great conceptual breakthrough that opened the gates. He begins his paper stating that its purpose is "to establish a basis for theoretical quantum theory, founded exclusively on relationships between quantities which, in principle, are observable" [101, p.177]. This was to be a returning motif in his thinking.

Heisenberg's approach was very much oriented towards the discreteness of quantum effects, something that was an anathema to many physicists, and very much depended on the mathematics of matrices which was not part of the physicist's mathematical repertoire at the time. Thus, when very shortly thereafter Schrödinger introduced an alternative approach to quantum theory based on waves (which were continuous), there was a considerable sigh of relief from some quarters. Still, the discrete/continuous conundrum was to badger physicists for years, and in particular Heisenberg and Schrödinger held firmly to their initial positions. Actually their two theories are mathematically equivalent. Eventually Bohr came to hold the position that the two pictures were the two sides of one coin and that these two aspects of reality are complementary. That remains the present view. If we look back at our long discussion of waves starting at §11.6.3 we can see the discreteness and continuous rubbing against each other. Especially in the case of periodic waves where we have a complete description in each form, the two are related by the Fourier transform. These same features are relevant here.

The version of quantum theory that we present is non-relativistic quantum theory, the physics of processes that take place at speeds much less than the speed of light. It starts by looking like a wave theory, which in a sense it is. The discreteness arises once we start to look at what is observable. It retains an essential feature of Heisenberg's original thinking that concentrates ultimately on what is observable, and not on any supposed physical features that in principle cannot be observed. This was particularly relevant for the supposed orbits of the electron of a hydrogen atom, which are not observable.

What are observable are the energy levels of the electron and its general proximity to the nucleus, but not any precise simultaneous description of where it is and how it is moving at any moment. Nowadays we don't think in terms of an electron being in an orbit, but rather use the word *orbital* and think of its position in terms of a cloud of probability. The metaphor of a cloud may evoke vagueness, but in fact within the appropriate context, we will see that there is nothing vague about it at all.

Underlying any quantum system is a part of physical space, which we will take as the familiar Euclidean space. To keep things as simple as possible, we will usually take our space to be one-dimensional and bounded—an interval on the real line, a circle, or even a finite set of points on the real line. We will denote it by X, and generally take it to be an interval, say $[0, 1]$, on the real line. This restriction to 1D in no way bypasses the central issues of quantum theory. It just avoids having to worry about unbounded domains or the extra coordinates, dot products, and other complications of higher dimensions, all of which are subsidiary to our goal here. We will take up the 3D version when we come to § 14.2.

Difficulty

What is hard and what is easy in mathematics depends enormously on what a person has seen and understood before. What follows is certainly challenging in a number of ways. First of all there is the strangeness of what is involved, then the level of abstraction, and finally the actual mathematical notation and the process of absorbing what it means. As an encouragement to readers with less of a mathematical background it may help to keep in mind that the conceptual ideas in this mathematics have never been easy for anyone. What they do do is to give an introduction to some of the most profound insights into the nature of things that human beings have ever created. We have put some of the details into the Endnotes, but have tried hard to retain enough of the actual mathematical notation—the language of quantum mechanics—so as to articulate the underlying ideas and see how they naturally arise. As we have already pointed out, but it bears repeating, the mathematics of quantum theory takes on the unfamiliar role of being just about the only way we have to express what it is about. The observable outcomes of quantum theory are aligned neither with our sense of intuition nor with how we normally imagine that things can be explained. In the end the articulation and the explanation are embedded in the mathematics itself.

Since any philosophical discussion of quantum theory (and this is a field that naturally produces a *lot* of philosophical dispute!) ultimately comes down to the mathematics, the more one can see how the mathematics works, the better one is to understand what it says and what it does not say. That puts a lot of weight onto the founding assumptions of quantum theory, which is why we will finally present it in such terms. What is important to the book as a whole is that the quantum world can, presumably, be taken as one of our ultimate sources of pattern and patterning. The fact that in spite of its remarkable re-envisioning of all the parts it still fits into the very same framework that we have been developing all along, is important to our story.

As far as the conceptual requirements of this discussion, the basic ideas have already been laid down, particularly in Ch.11 on waves. The mathematical ideas that we develop here do involve a little bit of calculus, but only the conceptual ideas of differentiation and integration. An excellent book written for the layperson, and one that would make a good companion to this chapter, is Ghirardi's *Sneaking a look at God's cards* [63]. The title especially seems to fit how we may feel when we look at quantum theory.

13.2.1 Waves

The basic entities in quantum systems are complex-valued functions on set X. Here, as we have suggested, we will think of X as being some interval in the real line, say $[0, 1]$. We will speak of the complex-valued functions as being waves, for waves are things we have talked about in §11.3 and can intuitively imagine, and they turn out to be appropriate metaphors for the theory. However, do notice that from the very outset the waves are *complex-valued* functions.

Thinking of quantum theory as a wave theory, we can frame it within the mathematical ideas that we have already developed. These waves can be added and subtracted, scaled in amplitude, shifted in phase (by scaling by complex numbers on the unit circle), and undergo destructive and constructive interference. As such the set of waves under consideration forms a *linear space*, just as we have described before. This linear space is the *state space* of our quantum system and the individual waves are the *quantum states*. So we have the first big difference: the states of our quantum system are waves, not just elements of a set. Observables are going to be something quite different than we have formerly seen.

Loosely speaking the waves of quantum theory "represent" physical entities and carry probabilistic information about their position and movement. Typically, especially at the outset, these entities represent particles (electrons, photons), so we will refer to them as particles. This ambiguity between the concepts of particle and wave lies at the heart of quantum theory. As waves they are both spatial and temporal objects. Spatially they carry information about the probability distribution of the particle (which can be very localized in X, making it more "point-like") while temporally they take into account the movement of the particle. The famous uncertainty between position and momentum will be spelled out in §13.6.

Notationally we will designate waves using symbols like $\psi = \psi(x, t) = \psi_t(x)$. So for a fixed value t of time, the wave takes a (complex) value $\psi(x, t)$ at every $x \in X$, which *represents* the distribution of the particle on the line *in terms of probability*. As t changes the particle may move, and so also its distribution in space may change. It is sometimes easiest to think of $\psi(x,t)$ defined on the entire real line but always equal to zero outside the region X.

So at the outset the system is set up to deal with the probabilistic nature of quantum theory and its dynamical content. However, there is something very unusual about it. As we have already noted, the wave ψ is actually a complex-valued object—$\psi(x,t)$ is in general not a real number but a complex number—just as in our previous discussion of waves. However, ψ itself is *not* literally a probability wave: it is the square of its magnitude or absolute value, $|\psi|^2$, that is the probability wave. Thus the *probability distribution* of the particle at time t is actually given by $|\psi(x,t)|^2$, not $\psi(x,t)$ or $|\psi(x,t)|$, see Fig.13.1. Since the absolute square $|z|^2$ of a complex number z is $\bar{z}z$ which is always a non-negative real number, so too are the values of

$$\overline{\psi(x,t)}\psi(x,t) = |\psi(x,t)|^2 .$$

Just like other probability distributions that we have discussed earlier, for example the normal distribution, $|\psi(x,t)|^2$ is not itself a probability. Rather, if Δx is a tiny interval centered on x then the probability of the particle being in this interval is very close to being $|\psi(x,t)|^2\,\Delta x$. The integral is the sum over such intervals over the entire domain, in the limit as $\Delta x \to 0$. For any interval $[a,b] \subset X$ the probability that the state ψ will be localized to the interval $[a,b]$ at time t is

$$\int_a^b |\psi(x,t)|^2 dx .$$

Since it is supposed to be a probability distribution, we must have the total probability (the probability that the particle is somewhere) equal to 1:

$$\langle\psi|\psi\rangle := \int_X |\psi(x,t)|^2 dx = 1 . \tag{13.1}$$

The part to look at is the integral (this is integration with respect to the usual Lebesgue measure of length—see §11.10). On the left we have taken the opportunity to introduce the inner product angle-bracket notation. For now take it simply as notation. In §13.3.3 we will develop its meaning and come to see that it is a mainstay of quantum theory.

The quantity $\langle\psi|\psi\rangle$, which we can define for *any* wave by the integral of (13.1) is non-negative. Its square root (≥ 0) is denoted by $||\psi||$, so we have

$$||\psi||^2 = \langle\psi|\psi\rangle \geq 0 . \tag{13.2}$$

$||\psi||$ is called the *norm* of ψ and we can think of it as being an indication of the overall "size" of ψ seen from an integrative point of view. A state ψ is called *normed* if $||\psi|| = 1$,

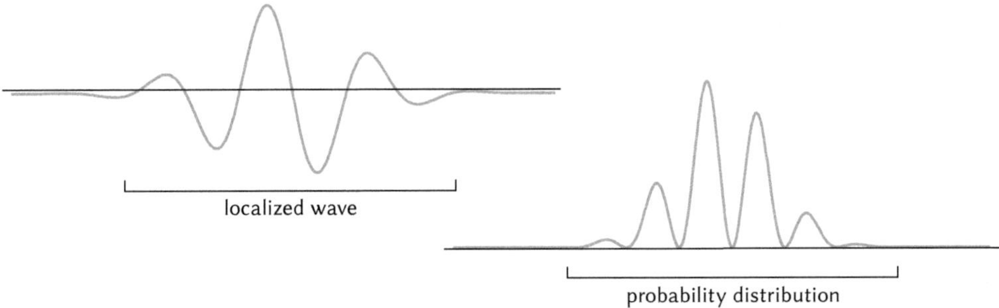

Figure 13.1: The idea of a wave packet (left) and its absolute square (right). The wave packet is a wave that is localized in position. Generally speaking waves are complex-valued, but for the sake of the picture we have taken a real-valued example (the probability wave *is* real-valued). When wave packets are properly normalized (scaled) their absolute squares are probability densities and the total area under their plots is equal to 1. Such waves are called *normed waves.*

which is equivalent to the probability condition of (13.1). Note that $||\psi||$ is a non-negative *number.* It is *not* the same as $|\psi|$, which is the non-negative *function* whose value at any $x \in X$ is $|\psi(x)|$.

Conceptually it is important to keep in mind that the waves ψ of quantum theory are not themselves probability waves—they are complex-valued functions. The explicit probability distribution of a normed wave or state is carried in its *absolute square* $|\psi|^2$, which is real-valued and non-negative. Although the complex-valued waves can interact, diffract, and enhance or inhibit each other, *no direct physical meaning* is given to them. We will come back to this issue of their physical existence later.

The basic objects of the subject are spatial-temporal waves ψ. Initially we don't worry about time, so we can imagine just a wave caught at one moment of time. Our principal interest is then spatial waves $\psi = \psi(x)$, which, for the moment we will simply call the *states* of the system. We then call the entire *linear space* of these spatial waves (which admits both addition of waves and the scaling of waves by complex-valued constants) the *state space* of the system and denote it by $\mathscr{H}$ (the notation derived from its interpretation as a Hilbert space, see §9 in the Endnotes). When we include time (which is energy based), it will appear in the form of dynamical action on these spatial states, and then states $\psi(x)$ will appear in their spatial-temporal forms $\psi(x, t)$.

Even with just these words we can see how far we have come in our search to articulate what pattern means. We still have a state space and we still have a time action upon it. We still have states that we can think of as single elements in the state space. But here these states are complex-valued waves that extend across our entire physical space of interest. These waves intrinsically carry information about probability density and also information about relative phases, one to another. Both of thees are vital to the outcomes of their interactions.

13.2.2 Observables

We have an interpretation for $|\psi|^2$, but the physics has nothing to say about the "material" properties of ψ itself. However, like waves that we have seen before, these quantum theoretical waves can interact with each other and even interact with themselves, so that like waves they can both reinforce or cancel each other, just as intersecting ocean waves create swells and eddies. In this way they create patterns of probability, which in turn determine the probabilistic outcomes of observations.

Observation or measurement of a system involves interaction. Dirac [46] put it in this way:

> ...it becomes important to remember that science is concerned only with observable things and that we can observe an object only by letting it interact with some outside influence. An act of observation is thus necessarily accompanied by some disturbance of the object observed.

The mathematical formalism of "observing" runs like this: for each classical observable (say position, momentum, energy) there is a corresponding quantum observable. These quantum observables are capable of producing a certain set of numerical outcomes. Some observables have only a finite number or countable set of outcomes (e.g. the energy of the electron of a hydrogen atom), and some can have a continuous range of values (e.g. the position of a particle in space, or on the real line). The mathematical ideas are similar for each, but conceptually it is simplest to stick to the finite and countable cases, which we do here as much as is possible.[5]

An observable A has a set of potential numerical outcomes $a_1, a_2, a_3, \ldots$ (we don't specify whether this is a finite or a countable list), each of which is a *real number*. Each of these outcomes corresponds to a particular state $\alpha_1, \alpha_2, \alpha_3, \ldots$ of $\mathscr{H}$. These are called *eigenstates* and $a_1, a_{2,3}, \ldots$ are called *quantum numbers.*. When the observation corresponding to A is actually made on a normed state ψ, one of the values a_j will be the outcome. Which one is is a matter of chance.

The probability that the outcome is a_j is symbolically written as $|\langle \alpha_j \mid \psi \rangle|^2$. This suggests mathematical content, but for the moment we can again take this as pure notation. Soon we will give it specific meaning, §13.3.3. The really remarkable fact is that not only is this the probability of the outcome being a_j, but the *actual state* of the system after the observation is α_j. So no matter what ψ was before, its state after observation is one of the states α_j. Which one is a matter of chance, the probabilities being given by the values $|\langle \alpha_j \mid \psi \rangle|^2$.

In this symbolism we must have

$$\langle \psi \mid \psi \rangle = |\langle \alpha_1 \mid \psi \rangle|^2 + |\langle \alpha_2 \mid \psi \rangle|^2 + |\langle \alpha_3 \mid \psi \rangle|^2 + \cdots = 1\,, \tag{13.3}$$

which summarizes the fact that it is certain (probability equal to one) that one of the potential outcomes will be observed. This aligns with and expands (13.1). The meaning of the left-hand term is that the total area of the absolute square of ψ is equal to 1, the requirement of a normed wave.

This then is the famous indeterminacy of quantum theory. Underlying it is a crucial idea that distinguishes quantum theory from classical theories. The state ψ, which is a wave, is seen by the observable A as some combination of other waves $\alpha_1, \alpha_2, \ldots$ with certain probabilities involved. We will see, in fact, that ψ is a *linear combination* of these waves:

$$\psi = c_1\alpha_1 + c_2\alpha_2 + c_3\alpha_3 + \cdots + c_n\alpha_n\,, \tag{13.4}$$

where the $c_j \in \mathbb{C}$ (or more generally a countable sum of such waves).

The ideas that waves (states!) can be expressed as linear combinations of other waves (states) is called the *principle of superposition*. This underlying linearity is a specific and crucial feature of quantum theory. Dirac explicitly says that in his development of relativistic quantum mechanics he was guided by insisting that this linearity be preserved.

We will see very shortly that the temporal action in quantum theory is deterministic, *up until the moment that an observation is made.* It is the action of observation that introduces probabilistic outcomes. After the observation (or as we prefer to say, after some interaction with some other quantum system), the state of the system is one of the states α_j, as we have noted. This is often referred to as the *collapse of the wave function*, but this is misleading because it suggests that the outcome is not a wave function. It is—it's just a different one, α_j, and the system will then continue its time evolution from this new state.

If one immediately repeats the same observation, the answer will most likely come out to be the same value a_j and the state will remain as α_j. But in time α_j will have evolved to another state, and this state will be an evolving superposition (linear combination) of its possible eigenstates $\alpha_1, \alpha_2, \ldots$. In making subsequent observations with the observable A, we are back into probabilistic outcomes again.

One of the effects of this situation is that observation destroys the present state ψ and replaces it by some other state α_j. This means that the observable is an operator of change in its own right. Strangely, and very differently than what happens in a classical situation, the observation gives virtually no information about ψ! It tells us only that $|\langle\alpha_j\,|\,\psi\rangle|^2 > 0$, i.e. that the state α_j must have had some chance of being the outcome. Furthermore it is incorrect to think that because a_j was observed ψ must have been in the state α_j all along. This is definitely not the case. The state ψ encompasses the entire suite of its potential outcomes until the observation is actually made. However, if we are able to set up the situation so that we can repeatedly observe the same state ψ, then we can make the same observation again and again, each time getting one of the potential outcomes α_j, and so by the law of large numbers infer the probabilities a_j of the outcomes.

From our perspective in pattern theory we can think of observation as an *interaction* between two pattern systems, the system being observed and the other called "the observer". The observer is not just a blank. The act of observation is, in effect, an engagement between dynamical processes. The act of observation is an interaction, and this interaction cannot be made without some physical exchange between the two pattern systems. At the macro-levels of human scales such interactions can often be made so carefully that the effect of the interaction is insignificant. However, at the quantum level this is no longer possible. This is not taken to mean that if we were more careful we would not have these issues. Quantum theory takes the view that this uncertainty of outcome is an absolute effect of observation and it cannot be overcome. Preferably we might say that the uncertainty is an absolute outcome of the interaction of two systems. We will see how this occurs in §13.3.

As an aside, this should raise some thoughts about the common terminology of even using the word "observables". It is natural enough in the lab where equipment is set up very specifically to test quantum theoretical ideas. But presumably quantum interactions take place all the time at the atomic level and do not require human beings to make them happen. To label all such interactions as observations is to present a distorted picture of what happens. We will come back to this, because it is relevant to our idea that pattern systems form larger integrated pattern systems when they interact.

All of this gives some idea about how observables enter into the quantum theoretical picture and what they do. But we are still some way from actually defining what observables are in mathematical terms. That is still to come.

13.3 Indeterminacy

This issue of the nature of observation emerges in a striking way in the Heisenberg uncertainty principle. This involves particular pairs of observables, and claims that the more accurately one knows one of them the less accurately one can know the other. This goes all the way to the extremes where to know one exactly is to know nothing about the other. This is not a matter of inaccuracy of observation. The two famous examples are: position-momentum and time-energy.

Indeterminacy is a consequence of the way in which observables work. If we want to make observations using observables A and B, we know that after observing a state ψ with A it will change state to some eigenstate α_j of A. So when we make the next observation B we are not looking at the same state ψ that we began with, but rather at whatever α_j transpired. Of course if we do it the other way around and start with observation B on ψ then the state will change to some eigenstate β_k of B. So again, when we next come to make the observation A we are not looking at ψ, but at β_k. This leads to the algebraic statement that $B \circ A$ (A followed by B) may not be the same as $A \circ B$ (B

followed by A). Some observables commute ($B \circ A = A \circ B$) and some do not.

The formalism of quantum theory is able to show that the dispersion of values of two observables around their mean values has a definite relationship in the form of an inequality. In the case of position and momentum, the two observables denoted by $\mathbf{X}$ and $\mathbf{P}$ do not commute, and the result reads

$$\Delta_\psi \mathbf{X}\, \Delta_\psi \mathbf{P} \geq \hbar/2 . \tag{13.5}$$

Here $\Delta_\psi \mathbf{X}$ and $\Delta_\psi \mathbf{P}$ are the numerical quantities representing uncertainties. The important point is the inequality. If one of the uncertainties is close to 0 then the other has to be relatively large so that the value of their product is at least $\hbar/2$. Here $\hbar = h/2\pi$, the reduced form of Planck's constant, which itself is incredibly small. That is why we never see these uncertainties at the human scale of things. But at the atomic level these uncertainties are very real, and are actively exploited in electronics. We will derive this particular *uncertainty principle* in §13.6.

13.3.1 Quantum dynamics

So far we have been talking about physical states being waves, but of course lurking in this is the famous wave-particle duality. Particles (photons, electrons, neutrons, indeed all matter) are supposed to have a wave-like and particle-like nature. Thus particles are considered to be waves, which, in our formalism, refer to states in our space of waves $\mathscr{H}$. But particles should also be particle-like and so be localized in space. Thus there is the notion of a *wave packet*, which is a state ψ, but one in which there is a small local region of the physical space $\mathbb{R}$ in which most of the wave is situated. We need to formalize what that means more precisely, but intuitively the values of $\psi(x)$ essentially become negligible outside of this local region, see Fig.13.1, with the implication that the probability of their occurrence is negligible. At the quantum scale, the scale of atomic particles, the spread of the wave-like component is significant.

Now, since we are thinking in terms of particles that move, we have to bring in the temporal side of the picture. Evidently quantum systems evolve in time and we ought to be able to talk about the resulting dynamics. The dynamics of quantum theory has to have the same sort of framework that we have developed for dynamical systems. That means we are looking for an action of time on the state space:

$$\begin{aligned} \mathbb{R} \times \mathscr{H} &\longrightarrow \mathscr{H} \\ (t, \psi) &\mapsto t \triangleright \psi = T(t)(\psi) , \end{aligned}$$

where time is expressed as usual as a real-valued parameter t. We are going to discard the usual $\triangleright$ notation here because we will need to go into the action of time in some detail. So we write $T(t)$ to denote the operator that moves states according to time. In this notation, if ψ is the state of the system at time 0, say, then $T(t)(\psi)$ will be the state to which ψ has evolved at time t. As usual, $T(0)(\psi) = \psi$, so $T(0)$ is the identity operator. Common notation is to write ψ_t for $T(t)(\psi)$ and to write $\psi_t(x)$ or $\psi(x,t)$ for the value of the state $(T(t)(\psi))(x)$ at $x \in \mathbb{R}$.

The dynamics we are talking about are the dynamics of the quantum system left alone to evolve as it does. Any act of observation is likely to change the state of the system in some abrupt way, and at that point the system's time evolution is interrupted and will continue from this new state. So observation will make discrete and non-deterministic jumps. But left alone the quantum system actually does evolve *deterministically*, and that evolution is taken to be smooth and continuous.

If $T(t)$ takes normed states to normed states, as it should, then it must preserve the norms of the waves of $\mathscr{H}$. It also should preserve the superposition, or linear combination, of waves, so we expect that any linear combination $c\psi + d\phi$ of waves ψ and ϕ would transform as $cT(t)\psi + dT(t)\phi = c\psi_t + d\phi_t$ in time. This just says that $T(t)$ is a *linear operator* for each time t. In keeping with the probability condition (13.1) that all normed states must satisfy, and since the action of time $T(t)$ takes normed states to normed states, it is required that

$$\langle T(t)\psi | T(t)\psi\rangle = \langle \psi|\psi\rangle \,. \tag{13.6}$$

This condition goes under the suggestive name of the *unitary requirement*, and operators that satisfy it are called *unitary operators*. This strong condition actually leads to the common form of the Schrödinger equation (13.8), as we shall now see.

We are talking about the evolution of the system in time, but it is energy that is the source of change, its effect being manifested moment by moment. So energy is responsible for the instantaneous rate of change of each state ψ in the system and $T(t)\psi$ is the accumulated effect of change over a course of time. Schrödinger's equation is about the instantaneous rate of change of states, and hence about energy. Since we start with the time operator T, we want to derive a statement about the energy from its incremental changes.

For a fixed ψ we compare $T(t + \Delta t)(\psi)$ and $T(t)\psi$, over a very short period of time Δt (not the same concept as in Δ_ψ of uncertainty). The rate of change of ψ_t is then given by

$$\begin{aligned}\frac{\psi_{t+\Delta t} - \psi_t}{\Delta t} &= \frac{T(t+\Delta t)(\psi) - T(t)(\psi)}{\Delta t} = \left(\frac{T(t+\Delta t) - T(t)}{\Delta t}\right)(\psi) \\ &= \left(\frac{T(\Delta t) - T(0)}{\Delta t}\right) T(t)(\psi) = \left(\frac{T(\Delta t) - \mathbf{1}}{\Delta t}\right)(\psi_t)\,.\end{aligned}$$

The last step used the fact that $T(t + \Delta t) = T(t)T(\Delta t) = T(\Delta t)T(t)$ and $T(0) = \mathbf{1}$ (the identity operator). Now letting $\Delta t \to 0$ we arrive on the left side with the *derivative* $d\psi_t/dt$ of ψ with respect to time and on the right with the new linear operator

$$H^* = \lim_{\Delta t\to 0} \frac{T(\Delta t) - T(0)}{\Delta t}\,, \tag{13.7}$$

and so the equation

$$\frac{d\psi_t}{dt} = H^*\psi_t, \quad \text{or, putting in the spatial variable } x, \quad \frac{d\psi_t(x)}{dt} = H^*\psi_t(x).$$

In effect this defines the *energy operator* $H := \mathrm{i}\hbar H^*$ and the Schrödinger equation

$$\mathrm{i}\hbar\frac{d\psi_t(x)}{dt} = H\psi_t(x). \tag{13.8}$$

Here i is the usual square root of -1 and $\hbar$ is the reduced Planck's constant. The reason for bringing these unlikely looking constants here—that is, multiplying the equation on both sides by $\mathrm{i}\hbar$—will be explained in §13.3.2 below. Energy H is actually an *observable* in the technical sense that we will come to in §13.4.1. The letter H honors William Hamilton (1805–1865) who long prior to the quantum age had given an elegant formulation of classical dynamics which was later taken up in quantum mechanics. We will see his name arise again in Ch.19. The minor complications of these last few steps are worth it from the insight that they give us into the connection between energy, change, and time that are made quite visible. Energy as the physical source of change is not something that we have discussed much so far, but here the connection is manifest. Does time produce change or does change produce time? They are bound together with the operating principle behind them being energy.

13.3.2 Wave packets

Since the basic entities of quantum theory are waves, we should go back and look at our basic description of waves in §11.4.2. The starting point was the simple circular function

$$t \mapsto \mathbf{e}(2\pi t).$$

Here t was taken as a real variable and the resulting function simply described a point moving around the unit circle in the complex plane, with period 1. We begin with simple waves and use them to derive more complicated waves, notably ones that are localized in space, by taking linear combinations of simple waves, a so-called *superposition* of waves, see Fig.13.3.

The waves of quantum theory are made up in the very same way with the very same entities as the waves that we studied before. However, here the waves depend on two real parameters, x for position (on the line in our case) and t for time.

The mathematical description of the simplest wave function for a particle moving along a line has the basic form

$$\begin{aligned}\psi_t(x) = \psi(x,t) &= \mathbf{e}(2\pi(-ft + kx)) \\ &= \mathbf{e}(2\pi(-ft))\,\mathbf{e}(2\pi kx)).\end{aligned} \tag{13.9}$$

Here we use physicists' standard conventions, where the functions are given Greek letters like ψ, ϕ and the symbol f is reserved for frequency. (In the physics literature this will appear as $\psi(x,t) = e^{2\pi\,\mathrm{i}(-ft+kx)}$.)

With x fixed, the meaning of f is the *time frequency* of the wave (f cycles/second). With t fixed, k is the reciprocal of the *spatial wave length*, $k = 1/\lambda$, see Fig.11.9, since every time x increases by λ, kx increases by 1 and $\mathbf{e}(2\pi kx))$ goes through one cycle.

Viewing the wave moving along the real line left to right, if time is held constant increasing x means looking at the profile of the wave further to the right, so the profile seems to move left. If instead position is held constant increasing time means seeing the profile move to the right (we see earlier parts of the profile appearing). If frequency and time are matched so that $-ft + kx = 0$ then

$$\frac{x}{t} = \frac{f}{k} = f\lambda,$$

which says that the rate of linear progression of the wave is f cycles per unit of time.

In the second line of (13.9) we have used the general fact $\mathbf{e}(u+v) = \mathbf{e}(u)\mathbf{e}(v)$ to split the wave as a product of its time and space parts. This separation into temporal and spatial components can be quite useful in breaking a quantum theoretical problem into two stages, dealing with them one at at time.

Putting the two together, in one time cycle of the wave, the wave will move through one wavelength, so the wave is moving (to the right) at a velocity of $f\lambda$. We can follow the peak of the wave by setting $x = f\lambda t$ (velocity × time). Since $\lambda = 1/k$ we can rewrite this as $xk = ft$. Then

$$\psi(x,t) = \mathbf{e}(2\pi(-ft + kx)) = \mathbf{e}(0) = 1,$$

so this peak value of 1 is seen to follow x as it moves to the right at velocity $f\lambda$, just as we saw above. This is how we can correlate the symbolism with the idea of a moving wave.

Two earlier results from theoretical physics enter here. The wavelength is also a measure of the *momentum* of the particle, which is

$$p = h/\lambda = hk, \tag{13.10}$$

where again h is Planck's constant.[6] This is the famous formula of de Broglie in which he hypothesized the wave-like nature of particles. He writes,

> The fundamental idea of my [1924] thesis was the following: The fact that, following Einstein's introduction of photons in light waves, one knew that light contains particles which are concentrations of energy incorporated into the wave, suggests that all particles, like the electron, must be transported by a wave into which it is incorporated. ...My essential idea was to extend to all particles the coexistence of waves and particles discovered by Einstein

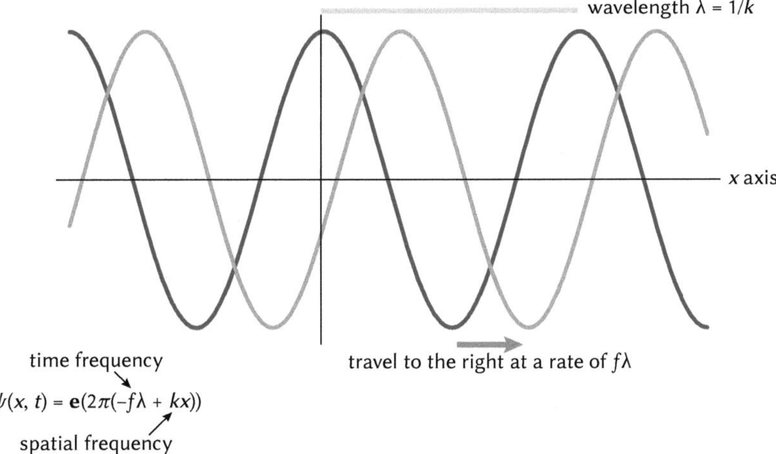

Figure 13.2: We show the real part of the moving wave ψ given by $\psi(x,t) = \mathbf{e}(2\pi(-ft + kx)) = \mathbf{e}(2\pi(-ft))\,\mathbf{e}(2\pi kx)$. That is, we are showing $\cos(2\pi(-ft + kx))$. Here x is the line coordinate and t is time. The dark curve shows the form of the wave when $t = 0$. The time frequency of the wave (cycles/sec) is f and the spatial frequency (cycles/meter) is k. The lighter curve suggests the movement of the wave in time. Looking at the peak (which occurs when $-ft + kx = 0$) we see how its position, the coordinate x, increases as t increases.

> in 1905 in the case of light and photons. ...With every particle of matter with mass m and velocity v a real wave must be associated.

Along with this we have the energy of the particle,

$$E = hf\,.$$

This is the famous Planck–Einstein formula for a wave-like particle with frequency f, which dates from 1900 and 1905 respectively.

Using a little bit of calculus, this provides an opportunity to see how this comes out from the theory we have so far. Energy is what provides the change of a wave packet in time. To see the action of energy precisely we need to look at the instantaneous rate of change of the wave at each moment. Going back to our discussion of energy, we can plug ψ of (13.9) into the Schrödinger equation (13.8) and see what happens. From calculus we get that

$$\frac{d\psi}{dt} = \frac{d\,\mathbf{e}(-2\pi(ft - kx))}{dt} = (-2\pi\,\mathrm{i}f)\mathbf{e}(-2\pi\,(ft - kx)) = (-2\pi\,\mathrm{i}f)\psi\,,$$

so we have

$$H\psi = \mathrm{i}\hbar H^*\psi = \frac{\mathrm{i}h}{2\pi}H^*\psi = \frac{\mathrm{i}h}{2\pi}\frac{d\psi}{dt} = \frac{\mathrm{i}h}{2\pi}(-2\pi\mathrm{i}f)\psi = hf\psi = E\psi\,.$$

For a more detailed explanation of this see the Endnotes.[7] This reveals that the energy operator H, which we derived from looking at the natural way that time should act, really is connected to energy in the anticipated way. It also shows something else that connects us back to what we have seen before in §11.8, and notably equation (11.23). We see here that ψ is an *eigenfunction* of the observable H and its *eigenvalue* is the energy E, precisely as it should be. E is a number, H is a linear operator. This little calculation also shows the reason for the strange looking constants involved with the definition of H: they are there to make sure that this last equation holds.

This description of a particle based on (13.9) clearly gives the wave structure, but equally clearly it is not localized in the sense of Fig.13.1. Nor does it have the probability property of (13.1). In reality things are finite, and so too must be the waves representing particles. Particles are usually envisioned as being in the form of wave packets localized in space and frequencies, see Fig.13.3. The caption to this figure shows how this issue is broached. It shows how a pure wave can be shaped by a Gaussian profile into a localized waveform. Before this shaping the wave is a single frequency, but the localization process leads to a distribution of frequencies. The tighter the localization, the greater the distribution of frequencies. In the figure we see the Fourier transform of the shaped wave (which we know reveals its underlying frequencies) has a continuous distribution of frequencies, so the wave packet is a complex superposition of pure waves. In addition the figure shows how the wave packet can be recovered from its Fourier transform. Here there is also a suggestion of the mathematical origin of the uncertainty principle. The width of the Gaussian profile is in reciprocal relation to the width of its Fourier transform, so that as one expands the other contracts.

13.3.3 The inner product

It is time to fill in the definition of the angle bracket $\langle \mid \rangle$ that is so prevalent in quantum theory. It is called the *inner product*, taking any two waves and producing out of them a complex number.

The inner product of two waves $\psi, \phi \in \mathscr{H}$ is defined by

$$\langle\psi|\phi\rangle := \int_X \overline{\psi(x)}\phi(x)dx\,. \tag{13.11}$$

To explain this it is easiest to go back to the dot product, which we have seen in §11.7.2 and is also discussed in the Endnotes of §12.3.2. The dot product of two finite vectors $\mathbf{z} = (z_1, z_2, \dots, z_n)$ and $\mathbf{w} = (w_1, w_2, \dots, w_n)$ is

$$\mathbf{z}.\mathbf{w} = z_1\, w_1 + z_2\, w_2 + \cdots + z_n\, w_n \tag{13.12}$$

if all the entries are real numbers. If now $\mathbf{z}$ and $\mathbf{w}$ are complex vectors then the corresponding "product" is defined by

$$\langle \mathbf{z} \,|\, \mathbf{w}\rangle := \overline{z_1}\, w_1 + \overline{z_2}\, w_2 + \cdots + \overline{z_n}\, w_n\,. \tag{13.13}$$

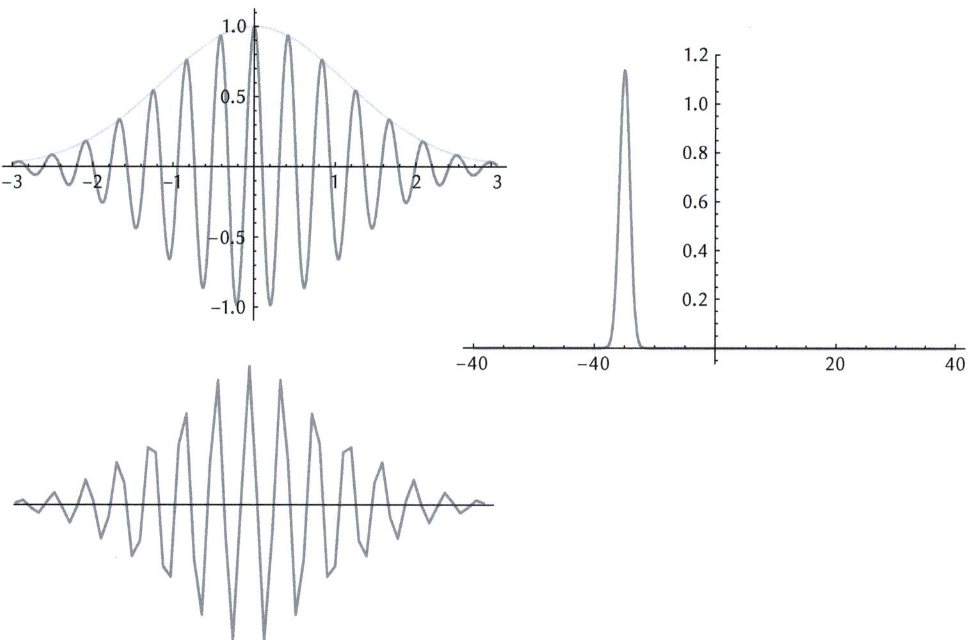

Figure 13.3: Top left, a simple wave is shaped by a Gaussian profile into a spatially concentrated wave packet. The resulting wave packet has a complicated spectrum made up of a superposition of simple waves which can be derived out of its Fourier transform, shown on the right. The Fourier transform is a smooth curve indicating that there is a continuum of frequencies involved in it, clearly concentrated into a small region. That the Fourier transform indeed carries the same information as the initial wave packet is seen here by using a discrete inverse Fourier transform based on sampling the Fourier transform at 60 places (frequencies): the result is an approximate form of the original wave packet. Since it is based on a finite sample of the Fourier transform and is a straight line plot between data points, it looks rather rectilinear. Nonetheless, it is a genuine approximation and could be improved to any desirable precision by using more sample points. It shows how a wave packet can be construed as a superposition (linear combination) of simple waves. The area under the absolute square of the wave should equal 1 for this to be a normed wave.

There is an example of this in §13.5.2. When we are dealing with waves ψ, ϕ that are continuous functions of position x, the finite sums of (13.13) are replaced by the continuous summing of integration, see §11.10. We don't need to know how to compute these integrals, so just treat them as continuous sums, of if you prerfer, averages.

However, "why $\overline{\psi(x)}$"? If we were dealing only with real-valued functions then the complex conjugation would not do anything and we would have just a continuous version of the dot product. But with complex variables the complex conjugation becomes important. We can get insight into why it appears by considering a complex number z. Whereas we can get the absolute value of a real number x by taking the square root of x^2, to get the absolute value of the complex number z (that is, r if $z = r\,\mathbf{e}(t)$) we must take the square root of $\bar{z}z$. Let's call this last the *absolute square.* Taking the square root of just z^2, though it exists, is $\pm z = \pm r\,\mathbf{e}(t/2)$, which clearly is not a real number in general. It is the absolute square that we need. So it makes sense that for a complex wave ψ, the real number that measures its size would be the square root of the integral (continuous sum) of $\overline{\psi(x)}\psi(x)$. In fact that is how we defined what it means for ψ to be a probability distribution in (13.1).

Given a pair of complex numbers z and w, the absolute square of their sum is

$$\overline{(z+w)}\,(z+w) = \bar{z}\,z + \bar{z}\,w + \bar{w}\,z + \bar{w}\,w\,,$$

where we have expanded the product by assuming linearity. Apart from the absolute squares of z and w we also see the terms $\bar{z}w$ and $\bar{w}z$ (which are conjugate to each other). Thus we can see that the definition of the inner product is a natural outcome of attending to linearity.

The wave form of this is

$$\langle\psi+\phi|\psi+\phi\rangle = \langle\psi|\psi\rangle + \langle\psi|\phi\rangle + \langle\phi|\psi\rangle + \langle\phi|\phi\rangle\,,$$

and

$$\langle\psi|\phi\rangle = \overline{\langle\phi|\psi\rangle}\,. \tag{13.14}$$

We say that ψ and ϕ are *orthogonal* if $\langle\psi|\phi\rangle = 0$ (and hence also $\langle\phi|\psi\rangle = 0$). If ϕ and ψ are orthogonal then

$$\langle\phi+\psi|\phi+\psi\rangle = \langle\phi|\phi\rangle + \langle\psi|\psi\rangle\,,$$

and conversely this equation implies that ϕ and ψ are orthogonal.[8]

The inner product is genuinely linear in its second variable and what is called *conjugate linear* in its first variable:

$$\begin{aligned}\langle\psi|a\phi_1 + b\phi_2\rangle &= a\langle\psi|\phi_1\rangle + b\langle\psi|\phi_2\rangle\,,\\ \langle a\psi_1 + b\psi_2|\phi\rangle &= \bar{a}\langle\psi_1|\phi\rangle + \bar{b}\langle\psi_2|\phi\rangle\,.\end{aligned} \tag{13.15}$$

Summing up we can say that what we have is a linear space $\mathscr{H}$ consisting of all the complex-valued functions on the interval X and the inner product defined in (13.11) defined by integration over X.[9]

An important inequality, which is used in the proof of the uncertainty principle, is the *Cauchy–Schwarz inequality* which says that for any $\psi, \phi \in \mathscr{H}$,

$$|\langle\phi|\psi\rangle| \leq \|\phi\|\,\|\psi\|\,. \tag{13.16}$$

See the Endnotes for a short proof.[10]

The *norm* of ψ, see (13.2), is $\|\psi\| = \langle\psi|\psi\rangle^{1/2}$. *Normed waves* are those whose norm is equal to 1. Any non-zero wave ψ can be *normalized* by scaling it by $1/\|\psi\|$.

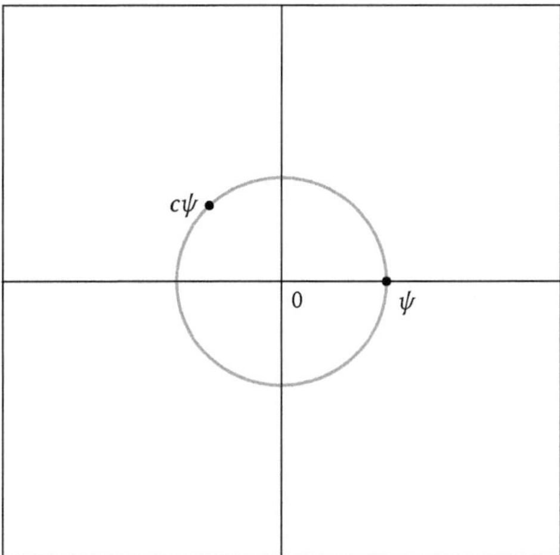

Figure 13.4: If ψ is a normed vector then the subspace that it generates is $\mathbb{C}\psi = \{c\psi : c \in \mathbb{C}\}$. The figure represents the one-dimensional complex subspace of $\mathcal{H}$ generated by ψ. It is a plane, a copy of the complex plane, so it looks two-dimensional as a real space, which we have represented by the square. The circle within it is of radius 1 about 0 and all the points on it are expressible as $c\psi$ where $|c| = 1$. All of these are normed vectors, $\langle c\psi \mid c\psi \rangle = 1$, and all are actually considered to *represent* the same physical state.

This brings us to a crucial concept in quantum theory. If ψ is a normed wave and $c \in \mathbb{U}$ (i.e. c is a complex number on the unit circle, or equivalently $\bar{c}c = 1$) then $c\psi$ is also normed:

$$\langle c\psi | c\psi \rangle = \int_X \overline{c\psi}(x) c\psi(x) dx = \bar{c}c \int_X \overline{\psi}(x)\psi(x) dx = \langle \psi | \psi \rangle \,.$$

> The formalism of quantum theory says that as waves or states, ψ and $c\psi$ (where $c \in \mathbb{U}$) are different states of $\mathcal{H}$ if $c \neq 1$, but from the *physical point of view*, they *represent* the very same *physical state.*

Fig.13.4 gives a more visual explanation.

This does not mean that the coefficients c_j that appear in the expression of a quantum wave, for instance in (13.4) below, can be *individually* changed by phase factors—their relative phase factors are crucial. However, if they are *all* changed by the same phase factor c that would change ψ to $c\psi$ and all the coefficients to cc_j, and this is considered to contain exactly the same physical information as ψ. This is subtle because we are dealing both with what happens at the level of the waves themselves as mathematical objects, and also what happens at the level of observation as physical objects. This is another instance of the difference between the observable and the actual underlying formalism. We will see how this plays out in what follows.

13.3.4 The physicality of waves?

This ends our initiation into the world of quantum theory. But there is one final comment we should make before we go into a more detailed version of the theory. Schrödinger (1887–1961) could not accept the fact that there are discontinuous quantum jumps in the energy of an electron with no intermediate energy values. He felt that his wave theory overcame this by using continuous waves where the quantization came out of the nature of observations. Heisenberg, though he admitted the power of the wave approach, felt that the world was ultimately discrete. If quantum jumps without intermediate states seem bizarre, Heisenberg was quick to point out that Schrödinger's waves actually had no underlying physical meaning. Schrödinger tried hard to overcome this objection, but never could.

The problem is this. The basic states of a quantum system are waves ψ. We have already seen that these waves are the basis of the theory, and it is their interference, self-interference, and entanglements that make the quantum world work. But what are these waves in reality? Although $|\psi|^2$ has a genuine physical interpretation as a wave of probability distribution, no physical meaning can be attached to ψ itself. All that can be observed comes from observables, and although the resulting values of these depend on ψ, they are not ψ itself. Knowing the values $|\psi(x)|^2$ as x ranges over the space X only determines the value of $\psi(x)$ up to a phase factor: if $c \in \mathbb{U}$ then $|c\psi(x)| = |\psi(x)|$, so the phase factor is left unresolved. This is the same issue that arose in our discussion of diffraction. Thus underlying quantum theory, which is the theoretical basis of our present understanding of the physical world, are non-material entities that can only be interpreted through observables. Ultimately they are evanescent, perhaps conceivable only in the form of mental symbols. We are left looking at a form of emptiness. Yet immaterial as these waves may be, they still represent tendencies that transform dynamically in time and space, and interact in ways that can be precisely formulated and lead to predictable physical events.

Thus we come back to Heisenberg's point of departure: the necessity to deal *exclusively* with relationships between quantities which are, in principle, observable. Philosophically this is a positivist position. What we can know about reality is, ultimately, what we can derive from our senses. Needless to say, this is a wonderful playground for further philosophical musings. It points to the ultimate failure of words and concepts to capture the actual essence of Nature. We can measure, we can predict, and using quantum theory we can integrate huge ranges of the world that we observe into coherent systems of thought with amazing explanatory powers. Yet, in the end, as has been pointed out so often in the history of thought, there always remains something unsaid, or even unsayable, a Heraclitean *logos*.

Philosophically the observed and the observer become a larger pattern system, an integration of two pattern systems. As we have seen in §8.4, an integrated system is not simply the product of its two parts: it is what we might conceive of as a *gestalt*, a form other than either of the parts that make it. It is an intertwining of events. And this points to what we have already suggested is misleading terminology. In using the words "observer" and "observation", we seem to place the context into that of a lab with experimental apparatus and human minds probing the outcomes of experiments. But seen from the point of view of pattern systems this is the interaction of two pattern systems, one of them being the observer who has framed the experiment. But do quantum effects require observers as such? If a photon passes through a camera lens it may be construed as a wave, and when it ultimately interacts with a sensor behind the lens there is an interaction that may produce an incident photon in one pixel on the sensor. As such the quantum wave that represents the sensor has altered. A digital image caught by a camera is nothing more than an interpretation of the counts that each pixel of the sensor has received. However, the quantum effect is in the context of light, lens, and sensor, and their combined quantum states. Stated this way, it is the quantum state of the combined lens-sensor that is changing according to the photons that interact with them. The state of the photon has not suddenly been reduced to something that is no longer a quantum "wave". Its form is now merged as part of a larger system which is still interpretable in the usual quantum theoretical terms. We do not take the position that the individual mind of the observer is the ultimate expression of reality, though it might be his or her ultimate realization of reality.

13.4 The nature of observables

Now that we have introduced the basic formalism around quantum mechanics, we can understand better what observables are about. We do this in two steps: the first says what the effect of an observable is, the second says what an observable actually is, as a mathematical entity.

After observation with an observable A, any state (wave) $\psi \in \mathscr{H}$ will appear in one of a possible collection of states α_j, $j = 1, 2, 3, \dots$ (which are other states in $\mathscr{H}$), and produce with it a corresponding numerical observation a_j (which is a *real* number) with some particular probability.

The formalism says that ψ (and this applies to any state in $\mathscr{H}$) is uniquely expressible as a superposition of these states α_j, in the sense that ψ is uniquely expressible :

$$\psi = c_1\alpha_1 + c_2\alpha_2 + c_3\alpha_3 + \cdots . \tag{13.17}$$

The $c_1, c_2, \dots$ are complex numbers, the $\alpha_1, \alpha_2, \dots$ are waves. If ψ is a *normed* wave these coefficients actually encapsulate the probabilities, namely

> the probability that the observable A will result in the numerical value a_j is $|c_j|^2$, and in that case the state is changed from ψ to the state α_j.

We will then find that

$$1 = \langle\psi|\psi\rangle = |c_1|^2 + |c_2|^2 + |c_3|^2 + \cdots , \tag{13.18}$$

which says that the probability of the outcome of observation being one of the states is 1. (This actually says that the states α_j are mutually orthogonal, something we will explain below.)

The nature of the observable A can now be explained. First of all, it is a *linear operator* on $\mathscr{H}$, meaning that A is a mapping

$$A : \mathscr{H} \longrightarrow \mathscr{H} .$$

It takes states to states, or in more intuitive language, waves to waves, and its effect on any linear combination of states (13.4) is

$$A(\psi) = c_1 A(\alpha_1) + c_2 A(\alpha_2) + c_3 A(\alpha_3) + \cdots . \tag{13.19}$$

This is the linearity part. Second, it acts in a very special way on the waves α_j:

$$A(\alpha_j) = a_j \alpha_j . \tag{13.20}$$

This is where the *numerical* values of measurement, a_j, come in. These are real (as opposed to complex) numbers. If we go back to ideas that arose in §11.8 we see that what we are saying here is that the waves α_j are eigenfunctions for the linear operator A and the a_j, which are real numbers, are the corresponding eigenvalues.

In this context these eigenfunctions are called *eigenstates*, since they are still states in $\mathscr{H}$. The set of real numbers $\{a_1, a_2, a_3, \dots\}$ is called the *spectrum* of A. These are the complete set of outcomes (measurements) of the observable A. It is customary to simplify notation and write $A\psi$ instead of $A(\psi)$ or $A \triangleright \psi$, and we will do that from this point on.

So the effect of A on ψ is a combination of linearity of (13.19) and the important equation (13.20):

$$A\psi = c_1 a_1 \alpha_1 + c_2 a_2 \alpha_2 + c_3 a_3 \alpha_3 + \cdots . \tag{13.21}$$

Following the representation shown in Fig.11.21, we have Fig.13.5 showing how all these parts now fit together. The extraordinary interpretation of equations (13.17) and (13.21) shows the remarkable compactness of the mathematical language of quantum theory.

What is important when considering the observable A is the set of eigenstates and eigenvalues of this as an operator on $\mathscr{H}$. There are many possible observables, and, as we shall explain below, each is an operator on $\mathscr{H}$ and each comes with its own spectrum. If $\mathscr{H}$ is finite dimensional, we find that A has a finite set of eigenstates that serve as a "basis"

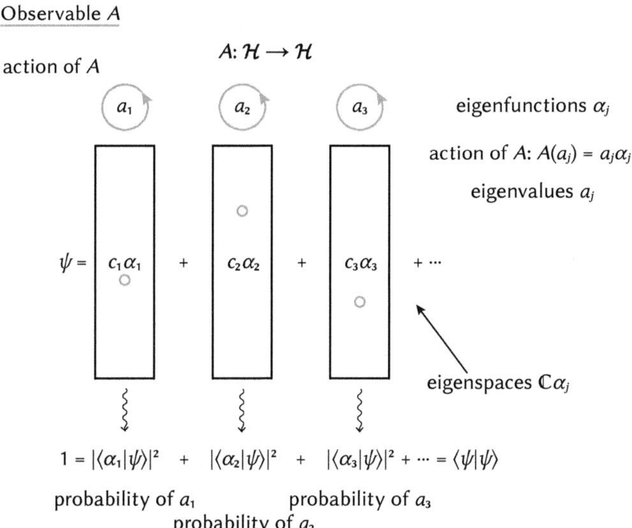

Figure 13.5: A schematic interpretation of how observables work (compare with Fig.11.21). Every observable has its own eigenspaces $\mathbb{C}\alpha_j$, and any ψ can be written as a linear combination (superposition) of these α_j. The corresponding eigenvalues a_j are what arise out of observation, along with a change of state from ψ to α_j. Which eigenstate/eigenvalue emerges is a question of probability. Specifically the pair α_j, a_j will be the result of the observable with probability equal to $|c_j|^2 = |\langle\alpha_j|\psi\rangle|^2$.

for the space $\mathscr{H}$—a framework in terms of which *every* state can be uniquely expressed as a linear superposition. (There are parallel, but mathematically more difficult, results in infinite dimensional situations that need not concern us here. As an aside, the theory of quantum computing/information is finite dimensional.)

Different operators produce different frameworks for the same state space $\mathscr{H}$, though always with the same number or cardinality of eigenstates. We can think of these just as we do of coordinate systems. We can coordinatize the plane with the usual x and y orthogonal axes. But any orthogonal pair of lines through the origin can be taken as axes and produce its own framework for coordinatizing the plane. What happens in $\mathscr{H}$ is the same principle. The number of states that make up such a framework is what we call the *dimension* of $\mathscr{H}$. Below, in the example of polarization in §13.5, we will look in detail at a quantum theoretical system in which the dimension is two.

13.4.1 Observables in more detail

We have seen how an observable A looks from the point of view of its action. But this is not really a definition of what an observable is. We are defining A in terms of what is observed—the states α_j, the numerical values a_j that are measured, and the probabilities of their occurrence. But it should be the other way around. We should say what it means to be an observable and then see all these details emerge from that.

13.4.2 Orthonormality

The very first thing we can do with the inner product is to define what it means to be an observable. Any operator A on $\mathscr{H}$ can be looked at in terms of how it relates to the inner product.

An operator A, that is to say a linear transformation $A : \mathscr{H} \longrightarrow \mathscr{H}$, is called *self-adjoint* or *Hermitian* if for all waves $\psi, \phi \in \mathscr{H}$,

$$\langle A\psi|\phi\rangle = \langle\psi|A\phi\rangle . \tag{13.22}$$

What this actually implies is not clear at the outset, but it is certainly not a property shared by most linear operators on $\mathscr{H}$.

Observable: By definition, an *observable* is a self-adjoint operator on $\mathscr{H}$.

As an example, the energy operator H is an observable.[11]

Self-adjoint and Hermitian are more mathematical terms—"observables" sound more physical, but they are the same. A crucial fact (that is not easy to show in a few lines) is that every self-adjoint operator (observable) does what we have seen above: it breaks the space $\mathscr{H}$ into eigenspaces $\mathbb{C}\alpha_j$ where $A\alpha_j = a_j\alpha_j$ for some real number a_j. This is not something that every operator on $\mathscr{H}$ will do. Since these eigenspaces are subspaces, we can always rescale α_j so that it is a normed vector, and we shall always suppose that this is done. Thus, any observable leads to the decomposition of $\mathscr{H}$ of the type we began with in §13.4.[12] This then is the formalism that underlies quantum theory. Fig.13.5 is a summary.

Along with each observable we have a splitting of $\mathscr{H}$ into eigenspaces. But each observable produces its own splitting and in general different observables produce different splittings . It is this fact that leads to the issue of non-commutativity that we will come to with the uncertainty inequalities.

The very first thing we can see using our definition of self-adjointness is that the eigenvalues are all *real numbers* (not simply complex). If $A\alpha_j = a_j\alpha_j$ then

$$\langle A\alpha_j|\alpha_j\rangle = \langle a_j\alpha_j|\alpha_j\rangle = \overline{a_j}\langle\alpha_j|\alpha_j\rangle, \quad \text{whereas}$$
$$\langle\alpha_j|A\alpha_j\rangle = \langle\alpha_j|a_j\alpha_j\rangle = a_j\langle\alpha_j|\alpha_j\rangle .$$

But these are supposed to be equal. The only way this can happen is if $\overline{a_j} = a_j$, which is the same as saying that it is a real number.

As we have said, two waves are said to be *orthogonal* if their inner product is zero. In general two subspaces of $\mathscr{H}$ are said to be orthogonal if each wave in one is orthogonal to each wave in the other. A simple consequence of our definition of self-adjointness is that the different eigenspaces of an observable A are mutually orthogonal. This is also straightforward to see: if α_j and α_k are two eigenvectors of A with *different* eigenvalues a_j and a_k then

$$\begin{aligned}\langle A\alpha_j|\alpha_k\rangle &= \langle a_j\alpha_j|\alpha_k\rangle = \overline{a_j}\langle\alpha_j|\alpha_k\rangle = a_j\langle\alpha_j|\alpha_k\rangle, \quad \text{since } a_j \text{ is real, whereas}\\ \langle\alpha_j|A\alpha_k\rangle &= \langle\alpha_j|a_k\alpha_j\rangle = a_k\langle\alpha_j|\alpha_k\rangle\,.\end{aligned}$$

Since $a_j \neq a_k$ the only way these can be equal is if $\langle\alpha_j|\alpha_k\rangle = 0$. Thus α_j and α_k are orthogonal.

Assuming that all the eigenfunctions α_j are normed and all their eigenvalues are different, we can summarize inner product relations between these various eigenstates as

$$\langle\alpha_j|\alpha_k\rangle = \begin{cases}1 & \text{if } j = k\,,\\ 0 & \text{if } j \neq k\,.\end{cases} \tag{13.23}$$

The set of eigenstates $\{\alpha_j\}$ form an *ortho-normal basis* of $\mathscr{H}$ (they are all of norm 1 and they are all mutually orthogonal).

If ψ is of the form of equation (13.17), we find that

$$\begin{aligned}\langle\alpha_j|\psi\rangle &= \int_X \overline{\alpha_j(x)}\,\psi(x)\,dx = \langle\alpha_j|\sum_k c_k\alpha_k\rangle\\ &= \sum_k\langle\alpha_j|c_k\alpha_k\rangle = \sum_k c_k\langle\alpha_j|\alpha_k\rangle = c_j\,.\end{aligned}$$

Thus the coefficients c_j can be extracted out of ψ by the same sort of process that appears in 11.6.2.

This gives meaning to what we wrote in (13.3): the absolute squares of the c_j are the probabilities:

$$|c_j|^2 = |\langle\alpha_j|\psi\rangle|^2\,. \tag{13.24}$$

In keeping with the assumption that the observable aspects of a state are unchanged by an overall phase factor, note that none of this changes if we replace *all* the α_j by $c\alpha_j$ for some $c \in \mathbb{U}$. The inner product changes by an overall factor of $\overline{c}c = 1$, in other words, it does not change.

13.5 Polarization

13.5.1 A familiar phenomenon

One of the simplest examples of how the quantum language works is the familiar phenomenon of polarization. Light reflected off roads or other horizontal surfaces is polarized so that the wave forms themselves tend to be oriented in the horizontal direction.

Polarizing sunglasses are designed to transmit photons polarized in the vertical direction and to be opaque to ones in the horizontal direction. The result is that road glare is largely eliminated. Sunglasses are polarizing filters.

Perhaps not surprisingly, the phenomenon of polarization is utilized in the biological world. Various species of ants use the position of the Sun in navigating direct homeward journeys to their nests, in spite of the complicated meandering exploratory routes that they used to go out. But several species of desert ant (*Cataglyphis bicolor*) can sense polarization patterns in the sky to establish the approximate plane of the solar meridian and so navigate stretches of featureless terrain [84, p. 367].

In spite of its overall familiarity, polarization produces some surprising effects. It is actually a quantum theoretical process, and since its formalization is relatively simple, working through it gives a taste for how it works, as well as explaining the strange behavior of polarization.

The light emerging from a (perfect linear) polarizing filter is polarized in one direction. Let us take such a filter and choose its orientation to be *vertical.* Now suppose that we put a second polarizing filter behind it, in the sense that the light from the first filter now passes through this second one. If the orientation of this second filter is also vertical then it has no effect and transmits all the polarized light coming to it from the first filer. If, however, the orientation of the second filter is horizontal, at right-angles to the first, then none of the polarized light from the first filter will pass through it (at least in principle).

What happens at other angles is that a proportion of the incoming light is transmitted through the filter. Specifically the proportion is $\cos^2(\theta)$, see Fig.13.6. As θ increases (or decreases) from 0 (vertical) to $\pm 90^\circ$, the proportion of light transmitted decreases from all to none. (In this section we have preferred to use the more familiar degrees rather than radians.) This is Malus' law, a fact that has been known for ages (E. L. Malus, 1775–1812). Central to our discussion will be $\theta = 45^\circ$, for which we have $\cos^2(45) = 1/2$. Half the photons will pass through this filter.

This does not sound overly surprising. But consider the following. Suppose the second filter is set horizontally ($\theta = 90^\circ$). No light passes through both filters. Now suppose we put a third polarizing filter *between* the first and second filter and this new filter is set at 45°. Amazingly now, $1/4$ of the light gets through all three filters. This is hard to fathom. Adding more obscuration would seem unlikely to improve the transmission. The underlying reason this works is that *if a photon passes through a filter then it is polarized to the orientation of that filter.* There is a change of state. Passing through a filter constitutes an observation. Since the angle of the middle filter is at 45°, $1/2$ the photons will pass through it, and they will be polarized to the orientation of that filter. The photons that successfully get through this filter are now oriented at an angle of 45° to the horizontal filter, so again $1/2$ of them will successfully pass through it (and now be oriented horizontally). Thus $\frac{1}{2} \times \frac{1}{2} = \frac{1}{4}$ of the photons will pass through.

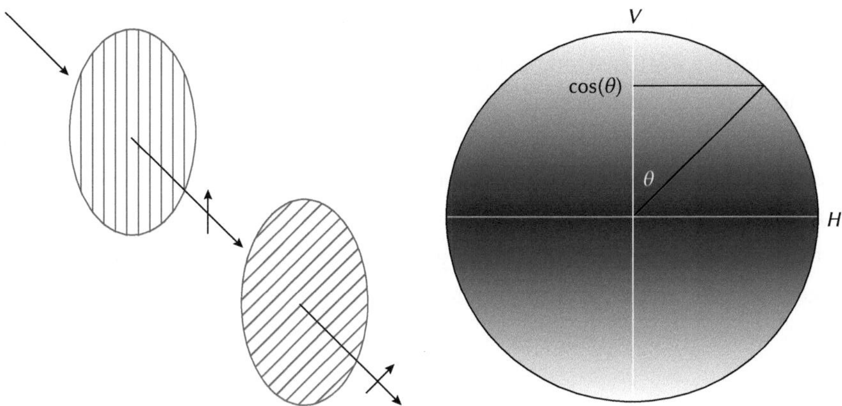

Figure 13.6: Unpolarized light is polarized vertically after passing through a vertical polarizing filter. A second filter, polarizing at an angle θ to the vertical, will only pass (vertically polarized photons) with probability $\cos^2(\theta)$. Those that do pass through emerge polarized at angle θ. If the second polarizer is set for horizontal polarization ($\theta = 90^\circ$) then the probability of a vertically polarized photon passing through it is $\cos^2(90) = 0$. The half-way point is at angle 45°, where $\cos^2(45) = 1/2$. The effect is indicated in the right-hand side of the figure. A circle of radius 1 is shown and the amount of vertically polarized light passing through a polarizing filter set to the angle θ is suggested by the shading.

A second experiment reveals the true quantum nature of what is going on. Light is made of photons, and if the intensity of photons passing through a polarizing filter is reduced enough, one can record what is happening one photon at a time. So suppose that the incoming photons are all polarized in the vertical direction and we put a second filter at the 45° orientation. We have seen that the intensity of the light is reduced by a half. But what happens at the level of a single photon? The photons that emerge are not altered in frequency (or energy), but only half of them make it through the filter. In fact the process is random. Whether a photon passes through the second filter is not predictable and is a matter of fifty-fifty chance.

This sort of phenomenon appears no matter what the angle θ is. Photon by photon the acceptance is a Bernoulli process with the chance of passing through the filter equal to $\cos^2(\theta)$. In view of the famous formula $\sin^2(\theta) + \cos^2(\theta) = 1$, the chance of not passing through the second filter is $1 - \cos^2(\theta) = \sin^2(\theta)$ (see the Endnote 8 in §11.5.1).

The effect of the two filters is to produce a random yes/no outcome, the relative probabilities being governed by the angle θ. There lies behind this a deep and fundamental question. What does this randomness mean in this context? Does it mean that these outcomes are the result of "God playing dice", as Einstein disparagingly put it, or is it just an indication of our lack of knowledge about other hidden, but quite deterministic, physics lurking in the background that we have yet to learn about?

The traditional Copenhagen interpretation is that quantum theory is complete—there are no hidden variables which will be able to restore determinism. Nature really does embody randomness in this strong sense. This question is basically irrelevant from the point of view of the engineer designing devices that are subject to quantum atomic and nanoscale effects. Quantum theory is astoundingly successful. But from a philosophical point of view it is profound. Are there truly random events in Nature? The completeness interpretation says yes. Over the years there has also been a steady trickle of work that seeks to test this randomness hypothesis, and all of it points to the correctness of the Copenhagen interpretation. We will discuss this later in the context of the EPR paradox, §14.5.

13.5.2 The formalism of polarization

To put this experimental evidence into a quantum theoretical space we need a state space and an inner product. Here, since there is no temporal aspect to consider and only one thing we are concerned with, namely the angle of polarization of two filers, things can be reduced to the simplicity of a two-dimensional state space

$$\mathcal{H} = \mathbb{C}\psi_V + \mathbb{C}\psi_H\,,$$

where the two states ψ_V and ψ_H represent photons oriented in vertical and horizontal polarizations. The meaning of this is that every state ψ of the system can be expressed uniquely as a *linear combination* (superposition) of these two states, i.e. in the form

$$\psi = c_V\psi_V + c_H\psi_H\,,$$

where c_V, c_H are complex numbers. If you wish to think in terms of functions or "waves" we can think of this as this as representing the function that takes value c_V on ψ_V and c_H on ψ_H. The inner product is *defined* to make ψ_V and ψ_H into an orthonormal basis:

$$\begin{aligned} \langle\psi_V|\psi_V\rangle = \langle\psi_H|\psi_H\rangle = 1\,, \\ \langle\psi_V|\psi_H\rangle = \langle\psi_H|\psi_V\rangle = 0\,. \end{aligned} \tag{13.25}$$

Our interest is in states of the form

$$\psi_\theta = \cos(\theta)\psi_V + \sin(\theta)\psi_H\,. \tag{13.26}$$

Each of these represents a photon polarized to the orientation θ. In what follows keep in mind that $\cos(\theta)$ and $\sin(\theta)$ are real numbers, which simplifies the algebra. Since $\cos^2(\theta) + \sin^2(\theta) = 1$, ψ_θ is normed, i.e. it has norm 1.[13] In this notation $\psi_V = \psi_0$ and $\psi_H = \psi_{90}$, but we will continue to use the V and H terminology. It is important to understand that equation (13.26) does not mean that ψ_θ is a mixture of the two photons ψ_V and ψ_H. We are talking about individual photons in the state ψ_θ.

At this point we think only in terms of one polarizing filter. Consider what happens to the photon ψ_θ, polarized at angle θ, when it is passed through a vertical polarizing filter. The vertical makes an angle θ with respect to the orientation of ψ_θ, so the probability that it will pass through it *and be observed as* ψ_V is $\cos^2(\theta)$. But note that since $\langle \psi_V \mid \psi_H \rangle = 0$,

$$\begin{aligned} |\langle \psi_V | \psi_\theta \rangle|^2 &= |\langle \psi_V | \cos(\theta)\psi_V + \sin(\theta)\psi_H \rangle|^2 \\ &= |\cos(\theta)\langle \psi_V | \psi_V \rangle|^2 = |\cos(\theta)|^2 = \cos^2(\theta), \end{aligned}$$

in agreement with (13.24). So the formalism works as it should in this case. In the same way, if ψ_θ were to pass through a horizontal filter, the probability that *it will be observed as* ψ_H (which is oriented at angle $90 - \theta$ to ψ) is $\cos^2(90 - \theta) = \sin^2(\theta)$, and again this is what the formalism gives:

$$|\langle \psi_H | \cos(\theta)\psi_V + \sin(\theta)\psi_H \rangle|^2 = \sin(\theta)^2 .$$

At this point we might clarify something that may be a bit puzzling. We have been talking about vertical and horizontal as though they were somehow distinguished by Nature itself, and in defining our state space $\mathscr{H}$ we essentially used them as determining the directions of the axes, much as we use x, y-axes in coordinatizing the plane. But these are relative terms, and it should not matter what pair of orthogonal orientations we use to set up the formalism. So suppose that instead we took two other axes K and $K^\perp$ oriented at angle θ to the vertical, and consider the two corresponding states (for convenience angles are in degrees).

$$\begin{aligned} \psi_K &= \psi_\theta = \cos(\theta)\psi_V + \sin(\theta)\psi_H , \qquad (13.27) \\ \psi_{K^\perp} &= \psi_{90+\theta} = \cos(\theta + 90)\psi_V + \sin(\theta + 90)\psi_H = -\sin(\theta)\psi_V + \cos(\theta)\psi_H . \end{aligned}$$

Now it is not hard to see that

$$\mathscr{H} = \mathbb{C}\psi_K + \mathbb{C}\psi_{K^\perp} ,$$

the simplest explanation being that just as K and $K^\perp$ are at angle θ to the vertical and horizontal, the vertical and horizontal are at angle $-\theta$ to K and $K^\perp$, so we should have

$$\begin{aligned} \psi_V &= \cos(-\theta)\psi_K + \sin(-\theta)\psi_{K^\perp} , \\ \psi_H &= \cos(-\theta + 90)\psi_K + \sin(-\theta + 90)\psi_{K^\perp} = -\sin(-\theta)\psi_K + \cos(-\theta)\psi_{K^\perp} . \end{aligned}$$

These indeed are solutions to the equations above.

The principle being evoked here shows the deeper meaning of the superposition principle. The state space $\mathscr{H}$ is a linear space and $\{\psi_V, \psi_H\}$ is a *basis* for it, in the sense that every state in $\mathscr{H}$ is uniquely expressible as a linear combination of these two—a *superposition* of these two. But these two states are not somehow distinguished over all others.

Any orthonormal pair of states $\{\psi_K, \psi_{K^\perp}\}$ would be just as good. As we have mentioned before, this inherent linearity is a crucial feature of quantum theory.

Observing the effect of a polarizing filter ought to be formalized into an observable, a linear operator on our state space $\mathcal{H}$. The observable for vertical polarization is the operator whose effect on states is

$$A_V : c_V\psi_V + c_H\psi_H \mapsto c_V\psi_V = 1c_V\,\psi_V + 0\,c_H\psi_H\,.$$

The second version shows that ψ_V and ψ_H are eigenvectors with the eigenvalues 1 and 0. Thus the effect of A_V on the state $\psi_\theta = \cos(\theta)\psi_V + \sin(\theta)\psi_H$ corresponding to polarization with orientation θ is

$$A_V(\cos(\theta)\psi_V + \sin(\theta)\psi_H) = \cos(\theta)\psi_V = \langle\psi_V|\psi\rangle\psi_V\,. \tag{13.28}$$

In particular the outcome is either change of state to ψ_V with probability $|c_V|^2$, or absorption of the photon into the filter. The last term shows that the effect of the observable A_V on any state ψ is simply

$$A_V : \psi \mapsto \langle\psi_V|\psi\rangle\psi_V\,.$$

This situation for the observable A_θ that applies to the observable for a polarized filter at angle θ is exactly parallel to this, with V replaced by θ.[14]

With this we can explain the three-filter paradox. We will assume that the first filter has been passed and we are looking at the state $\psi_V = \psi_0$. Now we want to see what happens with the subsequent filters at 45° and 90°. We know that this state will not pass directly through a horizontal filter:

$$A_{90}(\psi_0) = \langle\psi_0|\psi_{90}\rangle\psi_{90} = \langle\psi_V|\psi_H\rangle\psi_H = 0\,.$$

But now consider the two observables A_{45} followed by A_{90}:

$$\begin{aligned}\psi_V &\xrightarrow{A_{45}} \langle\psi_{45}|\psi_0\rangle\psi_{45} = \frac{1}{\sqrt{2}}\psi_{45}\\ &\xrightarrow{A_{90}} \langle\psi_{90}|\frac{1}{\sqrt{2}}\psi_{45}\rangle\psi_{90}\\ &= \frac{1}{\sqrt{2}}\langle\psi_{90}|\psi_{45}\rangle\psi_{90} = \frac{1}{\sqrt{2}}\frac{1}{\sqrt{2}}\psi_{90}\\ &= \frac{1}{2}\psi_{90}\,.\end{aligned}$$

This gives us the correct result that there is a probability of $(1/2)^2 = 1/4$ of a vertically polarized photon passing through the two filters, and those that do are polarized in the horizontal direction.

There is another feature here that is important. What we have done amounts to composing two operators A_{45} followed by A_{90}, that is $A_{90} \circ A_{45}$. The *order matters.* If the order were reversed we would have $A_{45} \circ A_{90}$, which would have taken the initial state ψ_V to zero. This importance of order is the subject of the next section.

13.6 The uncertainty relation

13.6.1 What the uncertainty relation says

We come now to one of the strangest aspects of quantum theory, the uncertainty relation, that states that some pairs of observables cannot both be known with perfect accuracy at the same time. This relationship is not a vague statement related to the difficulties of making two independent observations. It is a quantitative and provable fact that comes directly out of the formalism we have discussed. It goes so far as to say that if one were to know one of the variables with perfect accuracy, one could say nothing about the other variable at all.

The two most famous examples are position-momentum and time-energy, but it is best to start by treating this problem in a rather general way since it is really a straight mathematical consequence of what we have already learned.

Commutativity

Issues really begin with the problem of making two observations of two different types (e.g. the position of a particle and its momentum). We know making an observation will usually result in a change of state. The system might be in state ψ when we take the observation A, but after making it its state will have changed so some state α_j, which is one of the eigenstates of the operator representing that type of observation, and the probability of this occurrence is $|\langle\alpha_j|\psi\rangle|^2$. When we follow this immediately by observation B of a second kind we will get some state β_k, but its probability of occurrence is based on the state α_j and so is $|\langle\beta_k|\alpha_j\rangle|^2$. Thus

$$\psi \xrightarrow{\text{prob. } |\langle\alpha_j|\psi\rangle|^2} \alpha_j \xrightarrow{\text{prob. } |\langle\beta_k|\alpha_j\rangle|^2} \beta_k\,.$$

If we were to perform the two observations in the other order we would have

$$\psi \xrightarrow{\text{prob. } |\langle\beta_k|\psi\rangle|^2} \beta_k \xrightarrow{\text{prob. } |\langle\alpha_j|\beta_k\rangle|^2} \alpha_j\,.$$

So the order of the two observations is important!

In the world of classical physics, where observables are functions, the order is not important since the effect of observation is not supposed to change the state of the system: the product of two complex-valued functions $(fg)(x) = f(x)g(x)$ is the same as $(gf)(x) = g(x)f(x)$ for the simple reason that multiplication in $\mathbb{C}$ is commutative. In the quantum world observables are operators and, as we can see, order is important. In fact the only way that this problem gets avoided is if *the two operators have the same eigenstates*, which in turn is equivalent to saying that the two operators *commute*, that is to say, for all states ψ, $B(A(\psi)) = A(B(\psi))$. Another way to say that as operators they commute is to say that $B \circ A - A \circ B = 0$. The commutativity of pairs of operators becomes a fundamental issue.

Expectation

The outcome of making an observation on a state ψ with an observable A is probabilisitic, namely one of its eigenvalues a_j with probability $|\langle \alpha_j \,|\, \psi \rangle|^2$. A basic question is how these observations deviate from their mean or average, and one of the important quantifiers of this deviation is the variance. There is some discussion of this in Endnote 18, see (5.1). If r is a real-valued random variable then we write $\mathbb{E}[r]$ for its expected (or average) value. The *variance* of r is based on the expected amount that r deviates from its average value $\mathbb{E}[r]$. The deviation for a single trial is $|r - \mathbb{E}(r)|$, the size of the difference between the random variable r and its expected value. The *variance* is the *expected value* of the square of the deviation:

$$\mathbb{E}[(r - \mathbb{E}[r])^2] .$$

The non-negative square root of the variance is called the *standard deviation*, which we denote by $\Delta(r)$.[15] In this notation $\mathbb{E}[(r - \mathbb{E}[r])^2] = \Delta(r)^2$.

There are several appearances of the symbol Δ in this chapter and they come with different meanings. They all carry some connotation of difference, and they are all standard notation. This should not cause confusion since they appear in different contexts.

There is a standard calculation that makes it easier to calculate the variance:

$$\begin{aligned} \Delta(r)^2 &= \mathbb{E}[(r - \mathbb{E}[r])^2] = \mathbb{E}[r^2 - 2r\mathbb{E}[r] + \mathbb{E}[r]^2] \\ &= \mathbb{E}[r^2] - 2\mathbb{E}[r]\mathbb{E}[r] + \mathbb{E}[r]^2 \\ &= \mathbb{E}[r^2] - \mathbb{E}[r]^2 . \end{aligned} \tag{13.29}$$

Note here that $\mathbb{E}[r]$ itself is a number, not a random variable, so $\mathbb{E}[\mathbb{E}[r]] = \mathbb{E}[r]$ and $\mathbb{E}[2r\mathbb{E}[r]] = 2\mathbb{E}[r]\mathbb{E}[r]$.

In the physics world, the expected value of the observable A in relation to the normed state ψ has its own special notation $\langle A \rangle_\psi$ and there is a very tidy way in which to write it:

$$\langle A \rangle_\psi = \langle \psi \,|\, A\psi \rangle . \tag{13.30}$$

This is explained as follows: if $\psi = c_1\alpha_1 + c_2\alpha_2 + \cdots$, then $A\psi = c_1 a_1 \alpha_1 + c_2 a_2 \alpha_2 + \cdots$, and

$$\langle \psi \,|\, A\psi \rangle = \langle c_1\alpha_1 + c_2\alpha_2 + \cdots \,|\, c_1 a_1 \alpha_1 + c_2 a_2 \alpha_2 + \cdots \rangle = |c_1|^2 a_1 + |c_2|^2 a_2 + \cdots .$$

The last step uses the fact that the α_js form an orthonormal basis (13.23) and the fact that $\langle c_1\alpha_1 \,|\, c_1\alpha_1 \rangle = |c_1|^2$, etc. Looking at the right-hand side we see all the possible outcomes, each multiplied by $|c_j|^2$. Since $|c_j|^2$ is the probability of the value a_j being the outcome, this expression is the expected value of the observable A on the state ψ.

Using (13.29) and rewriting it in physics notation, the variance $(\Delta_\psi A)^2$ of the operator A applied to the state ψ is

$$(\Delta_\psi A)^2 = \langle A^2 \rangle_\psi - \langle A \rangle_\psi^2 . \tag{13.31}$$

This is a quantitative statement about the *spread* of the values of ψ around its expected value with respect to the observable A. The smaller it is the more the values are concentrated around their mean or expected value, and hence the greater certainty in what the value will be. The larger it is, the more diffuse these values, and the greater the uncertainty of what the values will be.

With this we can write down the basic step of the famous uncertainty relationship. We begin with two self-adjoint operators (observables) A, B. The result says that for any state ψ whatsoever,

$$\Delta_\psi A \, \Delta_\psi B \geq \frac{1}{2} |\langle \psi \,|\, (AB - BA)\psi \rangle| . \tag{13.32}$$

So it is an inequality based on the product of the standard deviations of the two observables relative to the state ψ. If A and B commute, that is $A \circ B - B \circ A = 0$, then the right-hand side of the inequality is zero, and it doesn't say anything since we already know that the left-hand side is non-negative. The importance of the relation is when A and B do *not* commute, and $(AB - BA)\psi \neq 0$. For then, if we really have a good idea of the expected value of the observable A on the state ψ (perhaps we have constrained the situation so there we can be pretty sure where a particle is located), then the expected *deviation* of ψ from its expected value ought to be small, so $\Delta_\psi A$ is small. However, the inequality tells us that the smaller $\Delta_\psi A$ is, the larger $\Delta_\psi B$ will have to compensate for it. In other words we cannot have such a good idea of what the outcome of observable B (say the momentum) will be. In fact, if we had really nailed down the value of A, so $\Delta_\psi A = 0$ then $\Delta_\psi B$ would be infinite, so we would know nothing about it at all! The opposite happens when the roles of A and B are interchanged. This is what the uncertainty relation tells us.

13.6.2 Why the uncertainty relation is true

We already know enough to prove this result, and we want to actually show it here because of its central importance. The result is slightly easier to show if we assume the expected values of A and B are zero: $\langle A \rangle_\psi = \langle B \rangle_\psi = 0$. (This can be arranged by replacing A by $A - \langle A \rangle_\psi I$ where I is the identity operator, in other words, subtracting $\langle A \rangle_\psi$ from A throughout, and similarly for B.) With this assumption and using the fact that A is self-adjoint, we get from (13.31) that

$$(\Delta_\psi A)^2 = \langle A^2 \psi \rangle_\psi = \langle \psi | A^2 \psi \rangle = \langle A\psi \,|\, A\psi \rangle = ||A\psi||^2 \quad \text{or, more simply,}$$
$$\Delta_\psi A = ||A\psi|| , \tag{13.33}$$

and similarly for B.

Now we are going to use six facts in a row:

- A and B are self-adjoint;
- the conjugate symmetry relation (13.14);
- for any complex number $z = a + ib$, $z - \overline{z} = a + ib - (a - ib) = 2ib = 2i \operatorname{Im} z$;
- for any complex number $z = a + ib$, $|\operatorname{Im} z| = |b| \leq (a^2 + b^2)^{1/2} = |z|$;
- the Cauchy–Schwarz inequality, see (13.16);
- equation (13.33).

We have

$$
\begin{aligned}
|\langle \psi \,|\, (AB - BA)\psi \rangle| &= |\langle \psi \,|\, AB\psi \rangle - \langle \psi \,|\, BA\psi \rangle| \\
&= |\langle A\psi \,|\, B\psi \rangle - \langle B\psi \,|\, A\psi \rangle| = |\langle A\psi \,|\, B\psi \rangle - \overline{\langle A\psi \,|\, B\psi \rangle}| \\
&= |2 \operatorname{Im}\langle A\psi \,|\, B\psi \rangle| \leq 2|\langle A\psi \,|\, B\psi \rangle| \\
&\leq 2||A\psi||\,||B\psi|| = 2\Delta_\psi A\, \Delta_\psi B\,.
\end{aligned}
$$

This is the mathematics underlying the uncertainty relation (13.32) that we were trying to show. It is also an impressive example of the power of the mathematical notation to pack so much into so few lines.

13.6.3 The uncertainty relation for position and momentum

The most famous uncertainty relation, the one that seems to be familiar to everyone who speaks of quantum mechanics, is that between position and momentum (though commonly spoken in terms of position and velocity. Here we explain what it actually says and how it arises out of the general uncertainty relation, leaving the more mathematical details to the Endnotes.

Position and momentum are observables represented by two Hermitian operators $\mathbf{X}$ and $\mathbf{P}$. The state space is the usual $\mathscr{H}$ of waves on an interval X on the real line. The observable for position, the *position operator* is defined by

$$
(\mathbf{X}\psi)(x) = x\psi(x)\,, \text{for all } x \in X, \psi \in \mathscr{H}\,.
$$

In other words, it is multiplication by x. Since $x \in X$ is real, we see that $\mathbf{X}$ is indeed an observable:

$$
\langle \mathbf{X}\phi|\psi\rangle = \int_X \overline{x\phi(x)}\,\psi(x)dx = \int_X \overline{\phi(x)}\,x\psi(x)dx = \langle \phi|\mathbf{X}\psi\rangle\,.
$$

The observable for momentum is

$$
\mathbf{P} = -i\hbar\frac{d}{dx}\,.
$$

Apart from the coefficient in front, this is differentiation with respect to x. The definition of this observable derives from intuition that we can gain from De Broglie's (13.10), stating that the momentum p of a photon with spatial wavelength λ is $p = h/\lambda$. Recall the mathematical description of such a photon in (13.9). Since we are only dealing with position we can ignore the time part and consider $\psi : \psi(x) = \mathbf{e}(2\pi x/\lambda)$, where $x \in X$. Then, see the Endnote 7 of §13.3.2,

$$\frac{d}{dx}\mathbf{e}(2\pi x/\lambda) = \frac{2\pi \mathrm{i}}{\lambda}\,\mathbf{e}(2\pi x/\lambda),$$

from which

$$\begin{aligned}\mathbf{P}\psi &= -\mathrm{i}\hbar\frac{d}{dx}\mathbf{e}(2\pi x/\lambda) = -\frac{\mathrm{i}h}{2\pi}\frac{d}{dx}\mathbf{e}(2\pi x/\lambda)\\ &= -\frac{\mathrm{i}h}{2\pi}\frac{2\pi\mathrm{i}}{\lambda}\mathbf{e}(2\pi x/\lambda) = (h/\lambda)\,\mathbf{e}(2\pi x/\lambda) = p\,\psi\,.\end{aligned}$$

This says that ψ is an eigenvector for $\mathbf{P}$ with the eigenvalue $p = h/\lambda$, which is the momentum.

With this we need only to work out $(\mathbf{XP} - \mathbf{PX})(\psi)$ and use (13.32) to arrive at the uncertainty relation

$$\Delta_\psi \mathbf{X}\, \Delta_\psi \mathbf{P} \geq \hbar/2\,. \tag{13.34}$$

We leave the details of this, including an explanation of why momentum is indeed an observable, to the Endnotes.[16]

There is another famous uncertainty relation that derives from the relationship between the time operator T and the energy observable H: the energy-time uncertainty relation. This is considerably more subtle than the position-momentum relation. In fact T is not even an observable. We refer to the Wikipedia article on this for more explanation [174].

13.7 Conclusion

Although the mathematics of quantum theory is still recognizably system and pattern based, with its state space, its operators of change, and its events through which change is manifested, it is radically different from the mathematics of "classical" pre-quantum times. Quantum theory was conceived in order to account for the bizarre outcomes of the early experiments at the atomic level. Now, over a hundred years later, as we have learned far more about the atomic and subatomic worlds, it has evolved into an incredibly accurate and predictive foundation of science and an everyday tool in the world of science and engineering. At some level all of this has been absorbed into our collective consciousness, though perhaps with little understanding of what it means. Ordinary life goes on.

However, quantum theory raises deep and difficult philosophical questions. Foremost is the question of chance. Quantum theory as we have presented it assumes that the events following an observation, or more generally following the action of an observable, are based on chance. The standard *Copenhagen interpretation* of quantum theory, championed by Niels Bohr and Werner Heisenberg, claims that this is not chance in the form of ignorance of hidden deeper causalities yet to be discovered, but is indeed pure random chance. This understanding deeply undercuts our usual ideas of cause and effect, and has proven deeply disturbing to many, Einstein in particular. We are by neurological design (Hebbian learning for instance) and by cultural norms, brought to think in terms of cause and effect. The strict principle of cause and effect runs deep in Western thought and its teleological assumptions make it difficult to accept any acausal theory of reality. However, no attempt to produce a hidden variables version of quantum theory has been convincing, and as we will see in the next chapter with Bell's inequalities, the evidence does not support there being one.

Pure randomness in Nature does not seem conducive to our underlying hopes that the Way is one of ever-continuing progress. Words like progress, or better/worse, are human constructions, relevant at a particular time and place and for a particular group of people, but relative in nature and local in scope. Not absolutes.

The positive side of this is one of freedom. With strict cause and effect out of the picture, we see that we are not just parts in a deterministic machine. Nature as we know it, is the outcome of both chance and law, and their interplay is one of endless creative possibilities and wonders. One has only to look at the fecundity of living and evolving forms that fill every environmental niche on Earth to see its amazing potential for creation and exploration.

We might add that even without the issue of chance the common ideas of cause and effect are surely naive. If I am in hospital because in crossing the road I was hit by a car, I could say that my carelessness and/or the driver's carelessness caused the accident. But why was I crossing the road, and why was that driver at that spot at that time. Then we go back to the cause of me crossing the road was that I needed a haircut, and the cause of the driver being distracted was thoughts of his mother who had just fallen and broken her hip. But why did she fall and why did I choose this day to go for a haircut? And so it goes until a vast array of circumstances going back to why the sperm that conceived me managed to beat out all the rest, and then into the deep past of my, and no doubt also your, ancestors.

We look for simple answers because life requires making decisions on the basis of incomplete information. But the reality we experience is not the outcome of simple sequential chains of events but rather a vast tree of past events that stretches into the timeless past. It is neither possible to take into full account what has shaped the present nor how that will shape the future. Standard interpretation of quantum theory says that in addition the present is shaped by continuing events of pure chance. Still, there are those underlying laws, and we have seen that even in systems that evolve on the basis of chance, those laws can lead to almost inevitable conclusions. A teleological progression shaped on the basis of human hopes and desires might be wishful thinking, but it is quite possible that certain evolutionary patterns of the Cosmos are inevitable.

The second deep issue arising from quantum theory is that of the so-called objective observer—the watcher who watches as an outside spectator with no influence on what is happening. The pairs of conjugate variables, where the knowledge of one diffuses knowledge of the other (the uncertainty principle), is one indication that the observer cannot record events without affecting the system. It is not just that we are too clumsy and have not refined our methods enough to avoid unwanted effects. Put in the language of pattern systems, any observation should be seen as an interaction, and any genuine interaction comes about with synthesis to some degree of one system with another. This presumably is not restricted to the laboratory and to the human experimenter. We can't imagine that this deep truth is only about a particular type of animal in a particular type of situation. Any model of the totality that sees all parts as players in the endless flow, as in the Hua-yen view, cannot have genuinely objective observers. In quantum theory we have seen that uncertainty is not something that was built in, but arises as an emergent property from the mathematics. This uncertainty has long been established as a verified fact and is fully utilized in modern day electronics.

The third deep issue lies in quantum entanglement and questions of locality. Locality is the idea that local events can only spread their effects across space via processes bounded by time (the speed of light). We know now that quantum effects are not localized in this sense. This issue is more clearly described in §14.5, which presents Bell's argument that locality in Einstein's sense would result in certain observable effects, certain inequalities. To Bell's surprise (he expected the opposite), it turned out that these inequalities failed under experiment. Recent experiments have shown that entangled particles, say photons, can be separated over distances measured in kilometers and yet retain instantaneous correlation with one another. Presumably the separation could be in light years. Evidently the Cosmos is deeply interconnected as an entirety in ways that we barely comprehend.

In these three ways we can see that quantum theory is more than just another way of looking at things and getting an "aha" moment. It is not something that was pre-imagined and then later revealed to be rather obvious, or something that we can look at and suddenly get a grasp on a new way of seeing. Even its greatest exponents admit that quantum theory defies the mind—what Stephen Hawking refered to as "the dreams that stuff is made of". It is so because at the atomic level Nature is revealed to be radically different from what we can actually experience. Quantum theory seems almost unimaginable both in what it says and in the astounding degree of accuracy to which it predicts. From any perspective, and particularly the perspective of Western philosophy, it can seem unnerving—worse than Darwinism in its dethroning of our treasured senses of transcendent importance and unending progress.

Still it contains within it an idea of great importance. We are not just some players put upon the stage of reality only to be removed after our part has been acted out. We are an intrinsic part of a living reality. We belong. This reality, however feebly we understand it, is who we are. Asian philosophies have traditionally been more organic and dynamic, carrying deep respect for the internal coherence and mutual inter-dependence of all aspects of reality. Dao is flow of emergence and return. Words and distinctions puncture reality into parts which, though necessary for life to survive, are relative in their nature and can obscure the underlying unity.[17] Buddhist thought treats ideas of mind and self in a similar way, stressing the complete interdependence pictured in Indra's net of jewels. Of course, there was nothing resembling quantum theory, and probably never could have been. These philosophies, especially those associated with the extended sphere of Chinese thought, have always been more about *how* to live rather than to explain *why* things are the way they are. Still, it is not hard to see that they easily align with what contemporary physics has to say, and can even lead to deeper insights of the mind-body-Cosmos trinity.

CHAPTER 14

Quantum patterns

Quantum theory merges the discrete with the continuous and matches exactness with chance and uncertainty. It sheds profound doubt on the cherished idea in science of the "independent observer" and strongly suggests that reality can never be reduced to completely observable entities. These deeply philosophical ideas are bound in a subtle, yet, when one looks back at it, rather concise and elegant mathematical framework. Quantum theory may not be an ultimate mathematical vehicle that humankind will develop in trying to understand Nature, but still, since its creation a century ago, it has continued to offer insights and exactitude without flaw. It is probably true, as is often said, that no one fully understands it. But we can most certainly say that about reality itself, and quantum theory has done its share to open our eyes to its profundity.

In this chapter we explore three additional features of quantum theory which show both its explanatory power in unfolding the patterning of the material world and its paradoxical power to lead us beyond what our senses can discern or our minds can fully imagine. These three are the periodic table, the strange property of spin, and the arguments that it raises against locality.

All three have important historical stories from which they originated, and all three depend on mathematics. Here we concentrate on the stories and setting the mathematical stage on which they have been articulated. There are extensive details in the Endnotes for those who would like to see more.

14.1 The periodic table

14.1.1 The amazing order of the elements

"If, in some cataclysm, all of scientific knowledge were to be destroyed, and only one sentence passed on to the next generation of creatures, what state-

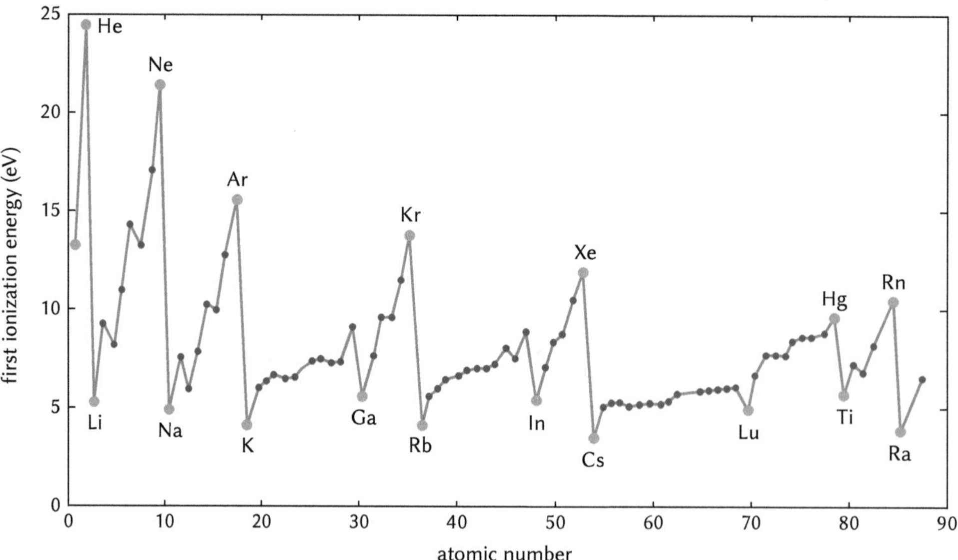

Figure 14.1: This graph shows what is called the first ionization energy for the elements when they are ordered according to atomic number. This is the amount of energy required to pull off one electron (the easiest one) from the atom. Ponor, CC BY-SA 3.0. sciencenotes.org/what-is-ionization-energy-definition-and-trend.

> ment would contain the most information in the fewest words? I believe it is the atomic hypothesis that all things are made of atoms—little particles that move around in perpetual motion, attracting each other when they are a little distance apart, but repelling upon being squeezed into one another. In that one sentence, you will see, there is an enormous amount of information about the world, if just a little imagination and thinking are applied."

This quotation of the famous physicist Richard Feynman leads off an online piece [89], where the author goes on to suggest that if one single *graph* were to be passed on he would choose Fig.14.1. This goes well beyond the statement that there are atoms. It shows that there are different types of atoms, they can be listed in some sort of order, we have information about their nature, and there is some sort of remarkable repetitive process at work.

The far more familiar and iconic periodic table, Fig.14.2, is another way to see this repetitivity, as its name states so explicitly. C. P. Snow wrote about his first experience with the periodic table:

> "For the first time I saw a medley of haphazard facts fall into line and order. All the jumbles and recipes, and hotchpotch of the inorganic chemistry of my boyhood seemed to fit themselves into the scheme before my eyes—as if

	1	2															3	4	5	6	7	8	9	10	11	12	13	14	15	16	17	18
1	1 H																															2 He
2	3 Li	4 Be																									5 B	6 C	7 N	8 O	9 F	10 Ne
3	11 Na	12 Mg																									13 Al	14 Si	15 P	16 S	17 Cl	18 Ar
4	19 K	20 Ca															21 Sc	22 Ti	23 V	24 Cr	25 Mn	26 Fe	27 Co	28 Ni	29 Cu	30 Zn	31 Ga	32 Ge	33 As	34 Se	35 Br	36 Kr
5	37 Rb	38 Sr															39 Y	40 Zr	41 Nb	42 Mo	43 Tc	44 Ru	45 Rh	46 Pd	47 Ag	48 Cd	49 In	50 Sn	51 Sb	52 Te	53 I	54 Xe
6	55 Cs	56 Ba	57 La	58 Ce	59 Pr	60 Nd	61 Pm	62 Sm	63 Eu	64 Gd	65 Tb	66 Dy	67 Ho	68 Er	69 Tm	70 Yb	71 Lu	72 Hf	73 Ta	74 W	75 Re	76 Os	77 Ir	78 Pt	79 Au	80 Hg	81 Ti	82 Pb	83 Bi	84 Po	85 At	86 Rn
7	87 Fr	88 Ra	89 Ac	90 Th	91 Pa	92 U	93 Np	94 Pu	95 Am	96 Cm	97 Bk	98 Cf	99 Es	100 Fm	101 Md	102 No	103 Lr	104 Rf	105 Db	106 Sg	107 Bh	108 Hs	109 Mt	110 Ds	111 Rg	112 Cn	113 Nh	114 Fl	115 Mc	116 Lv	117 Ts	118 Og

Figure 14.2: The periodic table. The elements are tabulated in the ascending order of their atomic numbers, the numbers of protons in the nucleus. There are lots of variations on the geometrical representation of the periodic table. This one, due to William Jensen (1986), is one of the most instructive and natural. It is not as convenient typographically as the more familiar squatter forms, but it more faithfully indicates how the electron shells are filled. The "periodicity" is indicated in the columns, which collect together elements with similar chemical properties (and also similar outer electron shells).

> one were standing beside a jungle and it suddenly transformed itself into a Dutch garden."

Such is the power of abstraction.

The periodic table appears on the background walls of schools and labs, on T-shirts, and on refrigerator magnets. So commonplace and familiar is it that we lose the immensity of what it actually says, for the periodic table purports to lay out a complete list of the different fundamental "building blocks" out of which our material world is made. All matter, as we commonly know it, is made up of these elements—atoms—vast assemblages of them. As far as we know, the entire material Universe is based on these very same elements, operating under the very same physical and chemical processes everywhere.[1] These elemental blocks of material, about 90 types in all, are used over and over again, assembled and reassembled into mountains and lakes, lichen, leeches, plankton, trees, mice, and our very own bodies and minds—"endless forms most beautiful and most wonderful", in Darwin's words. Equally amazing is that this list is a patterned list: there is a natural ordering to it and, as we shall see, there is a simple numerology that expresses the buildup of the internal structure and the repetition. At the very foundation of the Universe there is pattern, and it is deep.

Where do these atoms come from, what is their internal structure, how do there come to be these elements and no others, what sort of relationship do they have to one another, do we know them all or are there others? The Universe is a place of energy, change, pattern; what are the patterns underlying the elements out of which we, and everything else, are made?

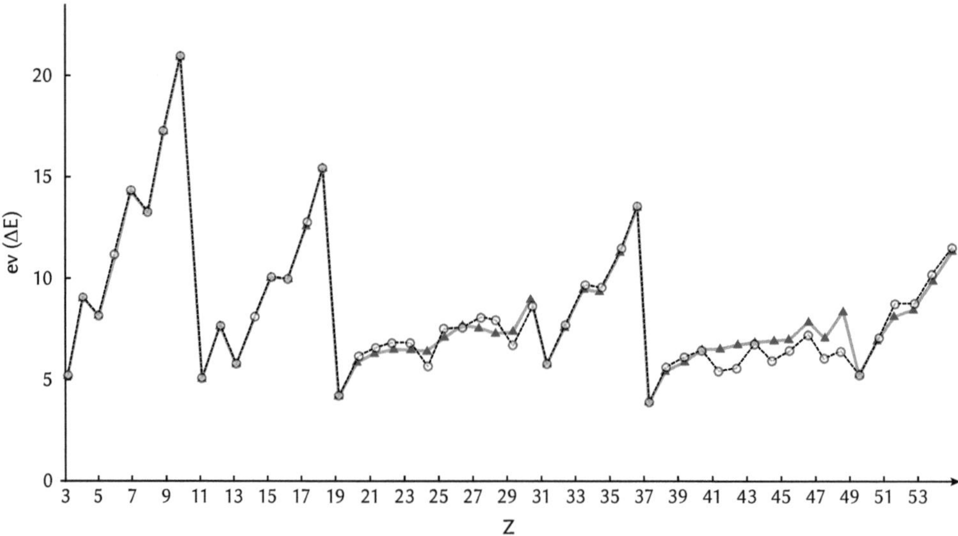

Figure 14.3: A comparison between the experimental and theoretical versions of the first ionization graph. The circular readings mark experimental results, the triangular readings the theoretical. E. Clementi, from [34], see also [141]. Courtesy E. Clementi and Springer.

Such answers as we have require stepping into the quantum world. We certainly have not unlocked all the secrets of Nature, nor are we ever likely to, but Fig.14.3, which compares the theoretical results on the so-called first ionization that follow from quantum theory with those derived from experiment, is very impressive. Quantum theory was not tweaked or especially formed to produce this pattern—it is not a theory specifically about the internal structure of atoms, but this is what it predicts. When we realize that on the right-hand side of this image, we are talking about atoms with over 50 electrons, each with its varied and interactive orbitals, and that classical physics cannot even explain the basic structure of the orbitals of the single electron of a hydrogen atom, we can appreciate its power.

Here we put forward a mathematical description of a simplified, but still remarkable, way of explaining the appearance and intrinsic meaning of the numbers that are used to classify the various elements. What is perhaps most interesting about this is that the results depend not only directly on quantum theory itself, but also deeply on symmetry, specifically the rotational symmetry of 3D-space, the orthogonal group SO(3). As much as anything else this story shows how symmetry has taken on such a remarkable role in the particle physics, and by extension in our theory of pattern.

The ancients, of course, were as keen to understand the nature of the physical world as we are, and naturally enough they asked questions similar to those that we have asked. They too speculated that there should be indivisible "elements" which in varying quantities could account for the variety of physical materials. In the Greek world they were earth, water, air, and fire, which seem to capture the basic ideas of solid-liquid-gas, and the mysterious energetic state-changing fire. As we have noted in §12.2.1 some Greek philosophers took an atomic model seriously, indeed called its basic constituents atoms, and constructed a far more elaborate theory of the physical world.[2] None of the Greek "elements" is an element in the modern sense, but they do show an attempt to classify the elements on the basis of their apparent physical properties, and that too is the theme that underlies the periodic table.

The scientific story of the periodic table is long and complex, and it is still a story in progress. The story so far really falls into two temporal ranges, and we can take 1900 as a rough dividing point. The periodic table was the inspiration of chemists, but the internal structure of the atom, which explains it, was the inspiration of physicists. For a compact and highly informative treatment we refer the reader to [141], from which we have drawn much of our historical material.

We will pick up the story in the early years of the nineteenth century when John Dalton, a young English schoolteacher/chemist of Quaker background introduced what was to be the beginnings of a new atomic theory. Dalton's ideas arose both from his studies on gases (the relationships between temperature, pressure, and volume) and the discoveries that gases always combined in simple numerical ratios of weight. This fact was also enunciated by Alexander von Humboldt and Joseph-Louis Gay-Lussac.

The main points of Dalton's atomic theory were:

- elements are made of extremely small particles called atoms;
- atoms of a given element are identical in size, mass, and other properties: atoms of different elements differ in size, mass, and other properties;
- atoms cannot be subdivided, created, or destroyed;
- atoms of different elements combine in simple whole-number ratios to form chemical compounds, and in chemical reactions, atoms are combined, separated, or rearranged.

Initial attempts to further these ideas were stymied by misconceptions that seem to us now as rather surprising. For instance, Dalton realized that water was made up of hydrogen and oxygen. His version of it was what we would write as HO, one atom of hydrogen and one atom of oxygen. But from their experiments Humboldt and Gay-Lussac deduced the formula that two volumes of hydrogen and one volume of oxygen made up two volumes of water vapor. This obviously didn't fit Dalton's model and it was hard to see how to fix it. This was solved by Amedeo Avogadro (1776–1856), who suggested that both oxygen and hydrogen existed naturally as diatomic molecules H_2 and O_2.[3] So it was two molecules of hydrogen and one molecule of oxygen that combined to form two molecules of water, H_2O [141]. Strangely, Dalton did not accept this theory

Table 14.1: Predicted and actual physical constants of germanium.

property	predicted	actual
relative atomic mass	72	72.3
specific gravity	5.5	5.47
atomic vol.	$13cm^3$	$13.2cm^3$
color	dark grey	greyish white
sp. grav. tetrachloride	1.9	1.887
boiling pt. of tetrachloride	100° C	86° C

since he firmly believed that like-atoms would repel each other and not form diatomic molecules.[4]

Understanding of the elements progressed further, when it was realized that certain triads of elements have very similar chemical properties and also a specific numerical relationship. For instance lithium, sodium, potassium, have atomic weights 7, 23, 39, and 23 is the average of 7 and 39. Pretty soon some twenty triads had been discovered, and chemists started to make tables of elements based on triads. It is an injustice to the many experimentalists who were involved in all this, that we skip forward to Dimitri Mendeleev who is considered the father of the periodic table. We know that it was on February 17, 1869, that Mendeleev first wrote down his table. It was not the first attempt, but he was the first to use such a table to make predictions of missing elements.

The definition at the time was that elements were those substances which could not be broken up further by chemical processes. Of course there was no prior way of knowing if there were elements that had not yet been discovered and there was no way of knowing if certain recalcitrant substances (notably argon) were elements or were just difficult to break up. Mendeleev's table was based on like chemical properties and increasing atomic weight. Finding missing slots in the table he made predictions, and these included what are now called gallium (discovered in 1875), scandium (1879), and germanium (1886).

Since germanium is a familiar name because of its many high-tech uses (infrared optics, wireless communication devices, solar panels, for instance), let's take it as an example of Mendeleev's predictions and compare those to what was found. Germanium was slotted to be element number 32 (Mendeleev called it eka-silicon because it was to be the next element in the column of the periodic table which began carbon-silicon-??. Indeed it is element 32 and does have similar chemical properties to carbon and silicon, e.g. it forms germanium-tetrachloride parallel to carbon-tetrachloride and silicon-tetrachloride. Table 14.1 compares the predictions for germanium and its actual properties [141]. This is an example of the power of pattern.[5] The pattern was not explained yet, but it was already predictive—the very same idea that all of us use all the time, but now used to predict new elements that up to that point had remained undiscovered.

Mendeleev's atomic predictions were not all correct, in fact only eight out of sixteen predictions that he made were correct. There remained doubters to the periodic table, and with good reasons. One was the discovery of argon. In spite of the fact that argon makes up almost 1% of our atmosphere (over twenty times the amount of CO_2 by volume), it was not separated as an isolated substance until 1894 (William Ramsey and Lord Raleigh). This was the first inert gas to be found, but there was no place in the periodic table into which to fit it. Still it resisted attempts to break it into simpler elements. It was not until 1900 that it was announced to be a new element (atomic number 18) and a whole new column, the so-called noble gases, all of which are inert, needed to be added to the table. This new column contained helium (2), neon (10), argon (18), krypton (36), xenon (54). Mendeleev's table was restored (and Mendeleev was delighted). We recall that helium had been discovered by spectral analysis of light from the Sun, but it was not isolated on Earth until 1895. These elements do not form other compounds under normal circumstances, hence they are inert.

Another problem with the periodic table was the ordering of elements. The obvious ordering is by atomic weight, which seems natural and is something that is feasible to determine. However, ordering by atomic weights leads to problems, one of particular difficulty being the pair of elements tellurium (52) and iodine (53). Mendeleev put these two elements in the correct order on the basis of their chemical properties, and assumed that the atomic weights that were known were in error. But the atomic weight of tellurium really is greater than that of iodine, and that created a mystery. Now we know that the ordering in the periodic table is based on the number of protons in the nucleus, which is the *atomic number*, and not the atomic weight (which depends also on the number of neutrons in the nucleus). But until the internal structure of atoms was known there was no way to explain the precise ordering of the elements in the periodic table.

This brings us finally to the arrival of physics on the scene. We have already noted the considerable opposition to the idea of atoms in the physics community, largely because of the enormous success of ideas based on continuity rather than discreteness. It was the chemists who had developed and embraced the idea of atoms, because their experiments so convincingly involved simple numerical relationships in the chemical interactions. The periodic table was their triumph. In the early 1900s, as it became clear for various reasons (of which we have seen Einstein's paper on Brownian motion and the diffraction experiments of von Laue), the atomic hypothesis became tenable and physics really entered the picture.

We have seen that Boltzmann and other researchers in statistical dynamics had already embraced the atomic hypothesis in the late nineteenth century. Also physicists had been studying electrons, and in 1896 J. J. Thompson and his coworkers discovered that electrons behaved like particles and were the carriers of negative charge. They also determined the mass of an electron to be about a thousandth of the mass of a hydrogen atom. Robert Milliken and Harvey Fletcher determined the charge on the electron, which is the fundamental unit of electrical negative charge. By 1914 Ernest Rutherford and others had conceived the first model of the atom as a dense positively charged nucleus surrounded by electrons, and it was at this point that the quantum world had to be directly confronted by physicists: even for the simplest atom, the hydrogen atom, the energy levels of its electron are discretized (or quantized, as it came to be known) and they cannot be reconciled at all with classical physics.

In 1913 Niels Bohr created the first quantized model of the hydrogen atom that attempted to reconcile the known discrete set of energy levels of its single electron. He also was the first to suggest that electrons fall into shells and that it was the outer shell that was implicit in the columns of the periodic table. This was the beginning, and the inspiration, for what was to follow—a full-blown theory that could accurately articulate what we now call the quantum world.

14.2 The structure of Atoms

The principles from which the periodic table arises are all pattern based. These involve both simple numbers (the discrete aspect) and complex geometrical structure (the continuous aspect) simultaneously, as well as many of the ideas that we have developed in this book. Although we cannot go into all the details, still it is revealing to see how the various features of pattern systems come together in this theory. Perhaps most interesting of all is to see how symmetry emerges as a powerful tool in understanding the intricacies of pattern in Nature.

The section starts with the numerology that reveals the basic structure of the periodic table. These numbers and the rules that interrelate them are just stated as though they appear by magic, but their origin can be seen by looking at the quantum theory for the simplest element, hydrogen. This makes up the second section, and it is here that symmetry arises as a key ingredient, first as the group of rotations SO(3) in 3D-space and second as the subgroup of rotations SO(2) around one axis. These are the groups SO(3) and SO(2) that we have introduced in Ch.9. Finally we come to an unexpected twist, one of those unsuspected secrets of Nature: there is a deeper level of symmetry beyond the rotation group SO(3), and this is the origin of spin. Most of the mathematical details appear in the Endnotes.

14.2.1 Four quantum numbers

Each element of the periodic table can be uniquely labelled by four numbers. These are:

- the *principal quantum number* n;
- the *angular momentum number* or *orbital shape quantum number* l;

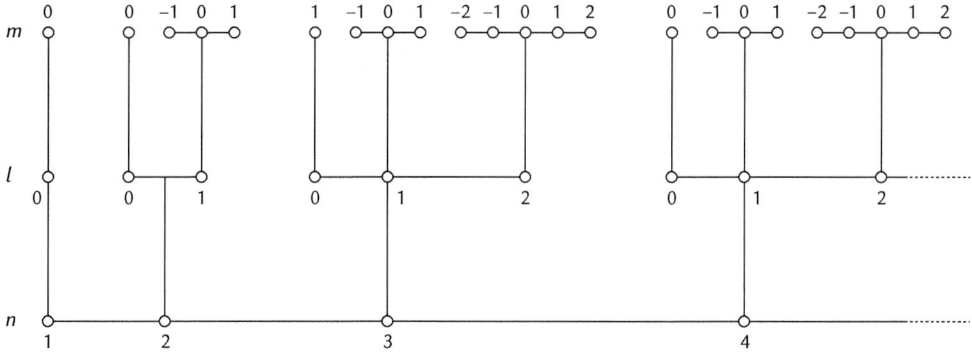

Figure 14.4: The figure shows which values of the quantum numbers $[n, l, m]$ are allowed according to the rules of (14.1). Here we see $[1, 0, 0]$, then $[2, 0, 0]$, $[2, 1, -1]$, $[2, 1, 0]$, $[2, 1, 1]$, and so on. The periodicity (repetition), from which the periodic table gets its name, is obvious. The number n is the shell number. There are n^2 possibilities for $[n, *, *]$, but when spin is included it doubles to get the maximum number of $2n^2$ electrons in the nth shell.

- the *magnetic quantum number* m;
- the *electron spin number* m_s.

The first three are whole numbers; the last, spin, of which we will have more to say later, takes just two values $1/2, -1/2$. The first three numbers are not arbitrary, but are inter-connected by the following somewhat strange looking rules:

$$\begin{aligned} &n = 1, 2, 3, \dots , \\ &l = 0, 1, \dots , n-1, \text{ altogether } n \text{ values}, \\ &m = l, l-1, \dots , 1, 0, -1, \dots , -(l-1), -l, \text{ altogether } 2l+1 \text{ values}. \end{aligned} \tag{14.1}$$

In words, the last of these says that m takes all integer values between l and $-l$. At the lowest values, $n = 1, 2$, these lists are interpreted as meaning that when $n = 1$ then l and m can only be 0, and when $n = 2$ then $l = 0$ with $m = 0$, or $l = 1$ with m equal to any of $1, 0, -1$. Fig.14.4 is a schematic interpretation of these rules. For each allowable choice of $[n, l, m]$ there is also the two-valued phenomenon of spin, so in the periodic table each triple $[n, l, m]$ appears twice, labelling two distinct elements. We will come to this in §14.3. These numbers are quantum numbers, in the sense that they arise out of a quantum mechanical interpretation of the atomic structure of atoms. We will eventually see how they all arise and why they are limited by the rules that we have just stated. But first some interpretation.

We know that the ordering of the elements in the periodic table is by their atomic number Z, and the atomic number of an element is by definition the number of protons in its nucleus. In an electrically neutral atom there are the same number of electrons as protons. In the earliest notions of twentieth-century atomic physics, the electrons were thought of as particles in orbit around the nucleus, much like planets around a star. This analogy was found to be incorrect, and we now think of the electrons as cloud-like distributions of probability around the nucleus. Each electron is in some particular quantum state, these states being states in the sense of the quantum theory that we have discussed above. A crucial fact (initially an assumption, and later understood as involving yet another fundamental aspect of symmetry) is the *Pauli exclusion principle*: *no two electrons in an atom can be in the same state.* As the atomic number increases so too does the number of electrons, and they fill in various *orbitals*, each of which is labelled by the three quantum numbers $[n, l, m]$. Finally, each orbital is divides into two states that are distinguished by another feature called *spin*. This is the fourth quantum number m_s.

The principal quantum number n is the primary descriptor of the energy of the orbitals that it labels, and these orbitals form one *shell*.

All told the total number of atoms in the nth shell, $[n, *, *, *]$ is $2n^2 = 2, 8, 18, 32, \dots$, numbers which are probably familiar from the periodic table and in any case are rather visible in Fig.14.2.[6]

The second quantum number l takes on values $0, 1, 2, 3, 4, \dots$ and is described as referring to *subshells* with the letter designations $s, p, d, f, g, \dots$.[7] Finally the third number m is the magnetic quantum number that further indexes electrons of each subshell.

Each allowable choice of $[n, l, m]$ of the first three quantum numbers describes one atomic *orbital*. As we have said, each orbital labels two successive elements of the periodic table, these being distinguished by the fourth quantum number, *spin*. The four numbers $[n, l, m, m_s]$ lie at the basis of the periodic table.

We can think of the ascending table of elements as the process of filling in the various shells, subshells, and orbitals. The way in which they fill is based on the overall principle of minimizing the total energy. It is known that the subshells fill in the order of increasing values of the sum $n + l$ of the first two quantum numbers, and within each value of $n + l$ the values go from lowest value of n upwards. This is illustrated in Fig.14.5. As it is there is no overall theoretical explanation for this rule, and it is not entirely consistent with the complexity of large atoms that occur later in the periodic table. The orbitals fill up in the order of increasing energy.

The overall label of an atom is the label of the last electron added.

14.2.2 The nature of the periodic table

The historical origin of the periodic table is the alignment of elements with similar chemical properties. We now know that from a chemical point of view it is electrons from incomplete shells that are the active ones. Completed shells play no role, and atoms whose

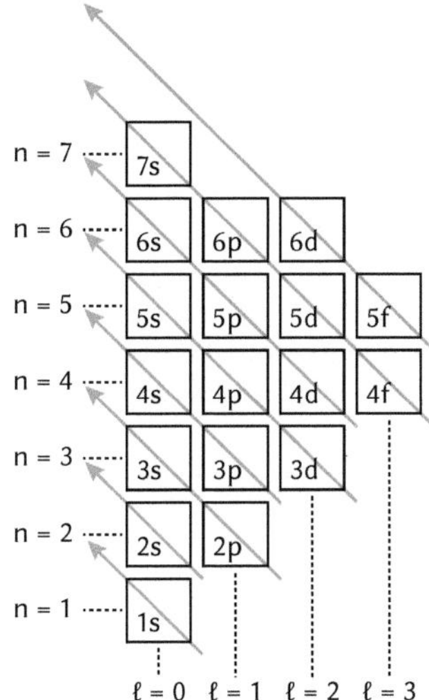

Figure 14.5: This figure shows the overall order in which the subshells are filled. This is called the $(n + l)$-rule. In Table 14.2 we see the transition from the end of the $2p$ subshell, through the $3s$, $3p$ subshells, finishing at the start of the $4s$ shell. At higher levels there are exceptions.

shells are all complete are inert, in the sense that they do not take part in chemical reactions: helium, neon, argon, krypton, ...

The columns of the periodic table list elements with similar incompleteness of their shells. For this reason it is common to describe elements in terms of the number of electrons in the most relevant subshell. So for example silicon—element number 14 with $n = 3$ and $l = 1$, and two electrons in the p subshell—is written as $3p^2$, standing for the more complete description of the arrangement of electrons $1s^2\, 2s^2\, 2p^6\, 3s^2\, 3p^2$. Table 14.2 shows how the third shell, $n = 3$, is filled out giving the electron structure of the corresponding elements. The underlying periodicity that gives the table its name is apparent. It should be noted that in spite of the fact that the electrons have different designations within the atomic structure and all are in different states, all electrons are indistinguishable as quantum particles.

14.2.3 Enter hydrogen

This then is the overall picture of the labelling scheme, but it leaves the origin of these numbers completely unexplained. The most direct intuition of these numbers is that they refer to vibratory states of the atom, stable "stationary" states which correspond to specific vibratory frequencies, much as we have seen in the much simpler form that we have suggested in Fig.2.7. Ultimately this is exactly what we will see. The periodic table reflects the way in which the electrons are accumulated into the atom as the number of protons in the nucleus increases. Since no two electrons can occupy the same state, i.e.

Table 14.2: Here we list the elements in order of atomic number from neon to potassium. These constitute the third shell ($n = 3$) sandwiched between the last element of the second shell ($n = 2$) and the first element of the fourth shell ($n = 4$). The notation clearly shows how the electrons enter and gradually fill first the s subshell and then the p subshell. To direct attention to where the change is taking place we have abbreviated the electron arrangement of the completed second shell by its last element, e.g. (Ne) instead of $1s^2 2s^2 2p^6$, and similarly for the last element of the third shell (Ar) instead of $1s^2 2s^2 2p^6 3s^2 3p^6$. The magnetic quantum numbers do not explicitly appear, but implicitly we see that for $n = 3, l = 1$, there are three values for m $(-1, 0, 1)$ and either one or two values of spin (for a total of 6 electrons) that gradually fill in the p subshell. All told there are $8 = 2 + 6$ elements in the third shell, and altogether $2 \times 3^2 = 18$ elements when the third shell is complete.

atomic number	symbol	element	electrons	n, l	shell
10	Ne	neon	$1s^2 2s^2 2p^6$	2, 1	2
11	Na	sodium	$1s^2 2s^2 2p^6 3s$ or or (Ne)$3s$	3, 0	3
12	Mg	magnesium	(Ne)$3s^2$	3, 0	3
13	Al	aluminium	(Ne)$3s^2 3p$	3, 1	3
14	Si	silicon	(Ne)$3s^2 3p^2$	3, 1	3
15	P	phosphorus	(Ne)$3s^2 3p^3$	3, 1	3
16	S	sulfur	(Ne)$3s^2 3p^4$	3, 1	3
17	Cl	chlorine	(Ne)$3s^2 3p^5$	3, 1	3
18	Ar	argon	(Ne)$3s^2 3p^6$	3, 1	3
19	K	potassium	(Ar) $4s$	4, 0	4

cannot have the same quantum numbers, as additional electrons are added, they acquire new vibratory states.

The simplest atom is the *hydrogen atom*, with a nucleus consisting of a single proton with a single electron orbital around it. Of course multi-electron atoms are far more complex and their entire electronic structures are relevant to their exact shape. However, as an approximation, it is possible to imagine a multi-electron atom as simply a hydrogen-like entity for which only the last electron added—the *radiant electron* as Tomogawa has called it—is the one whose state we are trying to describe. This reduces the study of multi-electron atoms to the study of one-electron atoms with a nucleus whose net charge is 1—a sort of variation on a hydrogen atom. We call such an atom a *hydrogen-like atom*. Somewhat amazingly the origin of the four quantum numbers can be derived from this very simplified point of view: it is a question of determining the entire set of states that such an electron can actually occupy.

The four numbers that designate the distinct atoms are quantum numbers, which have to appear from quantum observables. They arise from four different features. The first quantum number arises from quantized energy levels that the atom can assume. The second arises because the physics of the atom ought to be invariant with respect to its orientation in space. This hardly seems to be very restrictive, but it is, and it is here that the rotation group SO(3) of 3D-space becomes crucial. Although the symmetry

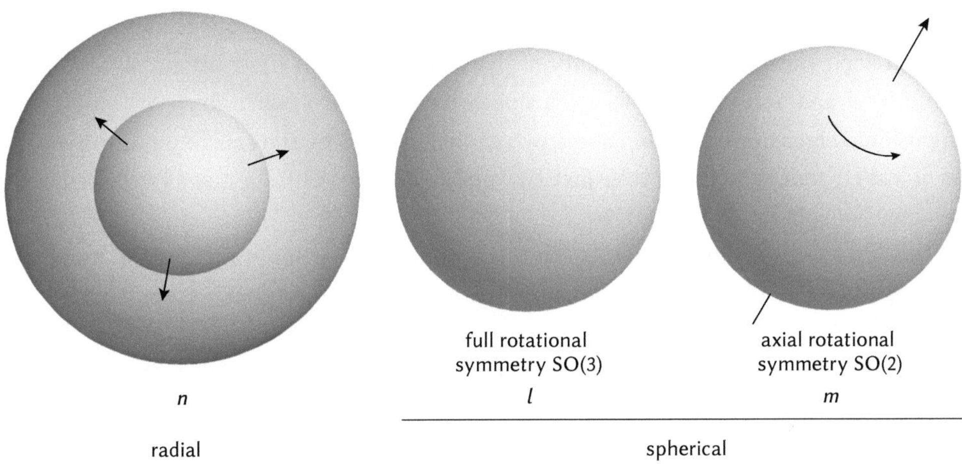

Figure 14.6: The geometric origins of the first three quantum numbers are radial expansion, which is an expression of the energy of the electron; the symmetry group of the sphere that expresses the invariance of the physics with respect to orientation in space; and the rotational symmetry that expresses the invariance around the axial magnetic field of the atom.

of SO(3) produces the quantum number l and corresponding eigenspaces, it does not produce *one-dimensional* eigenspaces which we need in order to get to the level of states.

The third quantum number arises because the atom is assumed to have an axis about which it is rotationally symmetric. This brings in the rotation group of 2D-space, SO(2), which actually splits the l-eigenspaces defined by l into "one-dimensional" spaces. See Fig.14.6.

The mathematics of all this begins with the same formalism that we have already developed, except now it is framed in terms of three-dimensional space. It starts with the hydrogen-like atom with its nucleus centered at the origin $(0, 0, 0)$ and the state space $\mathscr{H}$ consisting of all complex-valued functions $\psi : \mathbb{R}^3 \longrightarrow \mathbb{C}$. The atom is then described as being in some state ψ. The state is subject to the Schrödinger equation which defines the dynamics of the single electron. This equation can be split into a radial part related to the energy of the electron, from which the first quantum number arises, and a spherical part where the rotation groups SO(3) and SO(2) come in.

Most of the real mathematics involved with this classification of the first three atomic quantum numbers can be found in ‡Ch. 19. All of it is based on mathematical ideas that we have discussed so far. Though it is lengthy, it is very instructive to see how such results as these are derived. From the perspective of the book, it brings together many of the ideas we have been talking about, and in particular it sheds light on the profundity of symmetry as an embedded principle of Nature.

At this point, having arrived at one-dimensional eigenspaces corresponding to the different states and labelled by n, l, m, there would seem to be no further need, or even possibility, to go further. But there is. In total at this point we have accounted for n^2 states at each energy level. The reality of the situation is that there are possibilities for each orbital and actually $2n^2$ elements that arise from each shell. Each of these one-dimensional eigenspaces encompasses two forms, labelled by the fourth quantum number m_s called *spin*, which can take just two possible values ($\pm 1/2$). Usually these go by the names of spin up and spin down.

This last quantum number m_s is the most subtle, and in some ways is the most interesting of all since it arises from an entirely new intrinsic (and non-classical) magnetic property of the electron, *spin*. As we have suggested, it is a quantized two-valued *intrinsic* property of the electron, that is to say, every electron has spin. We will discuss the history of its discovery, its mathematical interpretation, and also the strange revelation that it brings about the symmetries of Nature.

14.3 Spin

14.3.1 What is spin?

It's amazing to think that there is a group of symmetries of which we are essentially totally unaware that lies at a foundational level in Nature. This group is deeply connected to the familiar rotation group SO(3), but in a very precise sense it is twice as large. Its importance lies in the uncovering of the last of the four quantum numbers that label the elements of the periodic table, and the possible electron states in a hydrogen atom.

At the start of §14.2.1 we introduced the four quantum numbers out of which the periodic table arises. The first three of these four numbers, $[n, l, m]$ can be explained in terms of the energy of the radiant electron and the invariance of the atom under rotations, see Fig.14.6. The actual details of this can be found in §19.2, with the rotation group SO(3) playing a fundamental role. What was unforeseen in the development of quantum theory is that there is actually a larger symmetry group involved.

At each stage of the process of going one step further in the $[n, l, m]$ series—what we have been calling "adding a radiant electron"—there are actually two electrons involved, adding one electron and then another with the opposite spin, though not necessarily one directly after the other. The electron filling for the consecutive atoms between neon and potassium is shown in Table. 14.3.

This cannot be explained by SO(3), but in fact, matching this doubling there is a larger group of symmetries that, to put it in anthropological terms, Nature and every electron "knows about" even if we don't—a larger group of symmetries that forms what is best described as a two-to-one covering of SO(3). In our ordinary experience this group seems entirely hidden from view, yet it is part of the underlying patterning of the world. We don't propose to develop the formalism of quantum mechanics that includes spin, but it

Table 14.3: An extension of Table 14.2 showing how electrons and spin evolve through the third shell of the periodic table, starting at the end of the second shell and ending at the beginning of the fourth.

a.n.	symbol	electrons	shell	spin
10	Ne	$1s^2 2s^2 2p^6$	2	⥮ · ⥮ · ⥮ · ⥮ · ⥮
11	Na	$1s^2 2s^2 2p^6 3s$ or (Ne)$3s$	3	⥮ · ⥮ · ⥮ · ⥮ · ⥮ · ↿
12	Mg	(Ne)$3s^2$	3	⥮ · ⥮ · ⥮ · ⥮ · ⥮ · ⥮
13	Al	(Ne)$3s^2 3p$	3	⥮ · ⥮ · ⥮ · ⥮ · ⥮ · ⥮ · ↿
14	Si	(Ne)$3s^2 3p^2$	3	⥮ · ⥮ · ⥮ · ⥮ · ⥮ · ⥮ · ↿ · ↿
15	P	(Ne)$3s^2 3p^3$	3	⥮ · ⥮ · ⥮ · ⥮ · ⥮ · ⥮ · ↿ · ↿ · ↿
16	S	(Ne)$3s^2 3p^4$	3	⥮ · ⥮ · ⥮ · ⥮ · ⥮ · ⥮ · ⥮ · ↿ · ↿
17	Cl	(Ne)$3s^2 3p^5$	3	⥮ · ⥮ · ⥮ · ⥮ · ⥮ · ⥮ · ⥮ · ⥮ · ↿
18	Ar	(Ne)$3s^2 3p^6$	3	⥮ · ⥮ · ⥮ · ⥮ · ⥮ · ⥮ · ⥮ · ⥮ · ⥮
19	K	(Ar) $4s$	4	⥮ · ⥮ · ⥮ · ⥮ · ⥮ · ⥮ · ⥮ · ⥮ · ⥮ · ↿

is enlightening to have a glimpse at this group which shows yet again how universally symmetry serves to conceptualize the processes of Nature.

As often happens in physics, the symptoms of a problem were evident long before any type of explanation. In 1892 Albert Michelson (1852–1931), famous for the Michelson–Morley experiment that showed that the speed of light is unchanged by the velocity of the frame of reference in which it is measured, made high-precision measurements of several lines in the hydrogen spectrum and concluded that they looked like they came from a double source. This was not initially considered to be a serious issue, but by 1915 there was consensus that this doubling into extremely close spectral pairs was a reality that could not easily be explained away. For instance, what we have called the 656 nm line in Fig.19.2 is actually two lines at around 656.3 nm separated by a gap of about 0.016 nm. The situation continued to bounce around during the time of the great intellectual struggles with quantum theory, until 1925 when two Dutch physicists, Samuel Goudsmit and George Uhlenbeck, suggested that the fine structure of the hydrogen spectrum could be explained by a new and intrinsic quality that electrons may have. It is called *spin*.

It is revealing to see what Samuel Goudsmit himself said about their discovery [124]:

> What the historians forget - and also the physicists - is that in the discoveries in physics chance, luck plays a very, very great role. Of course, we do not always recognize this. If someone is rich then he says "Yes, I have been clever, that is why I am rich"! And the same is being said of someone who does something in physics "yes, a really clever guy.....". Admittedly, there are cases like Heisenberg, Dirac and Einstein, there are some exceptions. But for most of us luck plays a very important role and that should not be forgotten.[8]

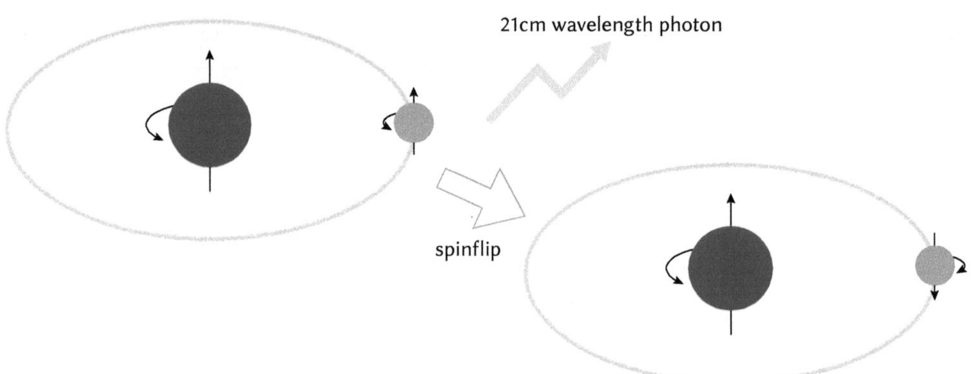

Figure 14.7: A schematic illustration of the meaning of spin flip. There is an energy drop associated with the transition from one state to the other, with the emission of a photon with wavelength of approximately 21cm (in the radio frequency range).

14.3.2 Spin in more detail

Spin is a magnetic phenomenon in the sense that it involves interaction with magnetic fields. Similar to the charge or the mass of the electron, spin is an *intrinsic quality* that is common to all electrons, not some extra that can or cannot be there. However, the word "spin" itself is somewhat misleading since it conjures up the electron spinning around some axis. As it is, when measured along any axis, spin can only take on one of two states—spin up and spin down—with associated numerical values $1/2$ and $-1/2$.

As is so typical for quantum mechanics, which value appears is probabilistic. If the direction of the axis (and magnetic field) of an atom is established, say in the positive z direction of 3D-coordinates (call this "up"), then the electron spin is either parallel (including direction) to this or "anti-parallel" to it, meaning that it is in the opposite direction. If a measurement for electron spin is made with a detector at some angle θ to this axis it is found that the probability of the spin being parallel (in the same direction) to the axis changes smoothly from 1 to 0 as θ ranges from 0 to π radians (0° to 180°) . At the same time the probability of it being anti-parallel increases from 0 to 1, the only outcomes being one or the other.

Actually this may be one time where the mathematical expression for the entangled up/down states may be clearer than words. Just as in the case of polarization we have two states, which we will designate $\mathbf{s}_\uparrow$ (spin up) and $\mathbf{s}_\downarrow$ (spin down), with the inner product relations

$$\langle \mathbf{s}_\uparrow \,|\, \mathbf{s}_\uparrow \rangle = 1 = \langle \mathbf{s}_\downarrow \,|\, \mathbf{s}_\downarrow \rangle, \tag{14.2}$$

$$\langle \mathbf{s}_\uparrow \,|\, \mathbf{s}_\downarrow \rangle = 0 = \langle \mathbf{s}_\downarrow \,|\, \mathbf{s}_\uparrow \rangle, . \tag{14.3}$$

A typical state in this system is

$$\psi_{\text{spin}} = c_1 s_\uparrow + c_2 s_\downarrow\,,$$

where c_1, c_2 are complex numbers with the usual probability condition $|c_1|^2 + |c_2|^2 = 1$, see (13.18). This gives the interpretation that $|c_1|^2$ is the probability that a test for spin in the vertical direction will be $s_\uparrow$, and $|c_2|^2$ the probability that it will be $s_\downarrow$.

The state representing spin has the form

$$\psi_{\text{spin}} = \mathbf{e}(-\theta/2)\cos(\theta/2)\mathbf{s}_\uparrow + \mathbf{e}(\theta/2)\sin(\theta/2)\mathbf{s}_\downarrow\,, \tag{14.4}$$

where θ is the angle of the detector to the axis of the atom. This is not as complicated as it looks. Ignoring the phase factors, we see the sine and cosine terms which are familiar from the polarization model. The two components of the entangled state are then further modified by two phase factors $\mathbf{e}(\pm\theta/2)$. Remembering the phase factors $\mathbf{e}(-\theta/2)$ and $\mathbf{e}(\theta/2)$ are on the unit circle $\mathbb{U}$, we see that

$$|\mathbf{e}(-\theta/2)\cos(\theta/2)|^2 + |\mathbf{e}(\theta/2)\sin(\theta/2)|^2 = |\cos(\theta/2)|^2 + |\sin(\theta/2)|^2 = 1\,.$$

This gives the probabilistic interpretation. As θ ranges from 0 to π, cosine and sine range from 1 and 0 to 0 and 1 respectively. The spin values associated with spin up and spin down are $1/2$ and $-1/2$.

What is particularly interesting is that the change is in terms of half-angles, not θ itself. This has a very peculiar effect. Look at how ψ_{spin} changes when we sweep a complete revolution, starting at any angle θ and increasing it to $\theta + 2\pi$. The value of $\mathbf{e}(\theta/2)$ changes to $\mathbf{e}((\theta + 2\pi)/2) = \mathbf{e}(\theta/2 + \pi) = -\mathbf{e}(\theta/2)$. Similarly $\mathbf{e}(-\theta/2)$ changes to $-\mathbf{e}(-\alpha/2)$. So a full revolution of θ changes ψ_{spin} to $-\psi_{\text{spin}}$.

Certainly ψ_{spin} is not the same state as $-\psi_{\text{spin}}$ but, as we know, it does represent the same *physical state*—-remember the state is not altered by multiplication by phase factors. However, the spin state ψ_{spin} is not transforming according to SO(3): a full 360° rotation should do nothing. So there is a difference between what is happening at the base level of states and what is happening at the levels that are observable. In our normal world, rotating something through 360° does not change its physical nature.

But to continue: if we now follow this first rotation by another 360° rotation, so that in total we pass through 720°, then $-\psi$ transforms to $-(-\psi) = \psi$, so we are now back where we started. So it takes *two* full 360° rotations to restore the quantum state to its initial state.

The strangeness of spin

Can this strange intrinsic property of spin make any difference that can be physically observed? Obviously not in any way that we are familiar with. But the answer is yes,

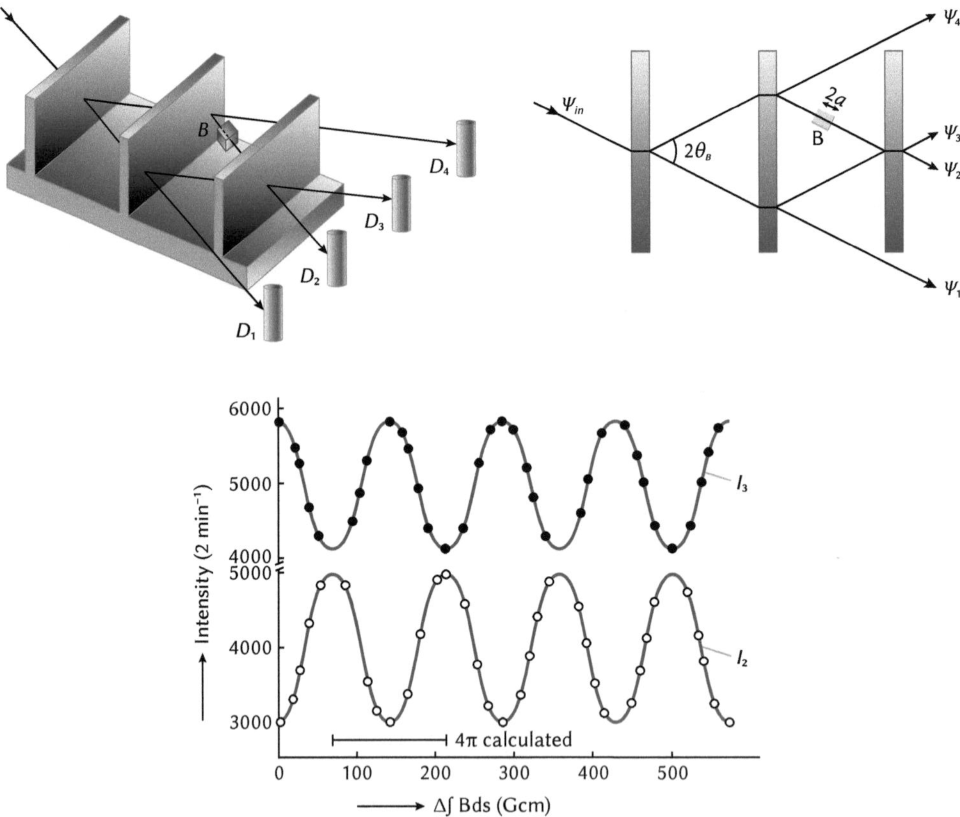

Figure 14.8: The figure shows the overall shape of the crystal and the complex split path that a neutron wave will take in passing through it. Notice the segment labelled B in which a magnetic field that can rotate the beam is located. The recombination of the two waves ψ_2 and ψ_3 after one of them has been rotated produces mutual interference that alters the probabilities that the neutron will be detected at detectors D_2 or D_3. These probabilities are shown on the vertical axis in the plot below, measured in terms of intensities of neutrons per unit time (totaling about 10, 000 every 2 minutes), as the shift of phase is increased along the x-axis. What is plain is that the effect is cyclical, but with a period of 4π, not 2π. The figures are taken from [21]. Courtesy Massimiliano Sassoli de Bianchi and Springer Publ.

and there are several famous experiments connected with it. We will look at one (which was performed by two separate groups independently) that is described in some detail in [21]. We begin by pointing out that electrons are not the only particles with spin. Neutrons have spin too, and these experiments were based on neutrons. In addition, neutrons have the same wave-particle duality as electrons. In the experiment a defect-free silicon crystal is cut to produce three planes of great perfection, see Fig.14.8. As the figure illustrates, a beam of neutrons incoming to the crystal splits twice to produce four descendant beams. Two of these (ψ_2 and ψ_3 in the figure) are allowed to recombine before exiting. A key feature is to have one of the two recombinant beams first pass through a local magnetic field, which causes phase shifting and hence mutual interference of the beams upon recombination. The rate of neutrons reaching the crystal is sufficiently low that it may be considered that there is only one neutron passing through the crystal at any given time, so the interference experienced by the two exiting beams ψ_2 and ψ_3 really is *self-interference* derived from a single wave. This interference is measured in terms of the probabilities that the neuron will be detected at detector D_2 or D_3. The results of these experiments, led by H. Rauch and S. Werner respectively, were reported in the mid-1970s, and show precisely the effect that we have been discussing. The key point is that the interference effect on the exiting ψ_2 and ψ_3 beams is periodic of period 4π, not 2π as one would expect, see Fig.14.8. This is explained in great detail in [21], which also has a discussion on the philosophical implications of these experiments.

The conclusion of all this is that spin does not transform under our familiar group of rotations SO(3). We expect a 360° rotation to leave the physics unchanged, and indeed in SO(3), a 360° rotation is the identity element! It leaves everything unchanged. But this same rotation somehow reverses up and down spin. Finding a mathematical basis for spin turned out to be hard. Heisenberg, Jordan, and Pauli are among the famous physicists who struggled with this, and it was Pauli who suggested that an extension of the state space beyond the set of functions on $\mathbb{R}^3$ was necessary—effectively an extension of $\mathbb{R}^3$ by the little two-dimensional spin system that we have described in (14.2). It was Paul Dirac who finally found the formalism that properly embraces it. In [75], Stephen Hawking writes about this paper:

> In 1928, a brilliant paper by Paul Dirac entitled *The quantum theory for the electron* presented a relativistic quantum theory and a relativistic replacement to the Schrödinger Equation which is now known as the Dirac Equation. Remarkably, Dirac's theory required the existence of anti-particles even though no such thing had yet been supposed. Thus he was able to predict the existence of the positron (which is the electron's anti-particle) before it has been experimentally detected! Equally astounding, Dirac was also able to show that by including relativity in quantum mechanics he could explain

the intrinsic angular momentum or 'spin' of electrons which had been an unsolved problems since its discovery.

Linearity and the probabilistic interpretation were crucial to Dirac's thinking. In passing, we point out that Hawking's book *The dreams that stuff is made of* [75] is a marvelous repository of many of the great papers that created quantum mechanics, including this famous paper of Dirac.

Underlying this phenomenon of spin there is a larger group, of which SO(3) is a sort of two-to-one shadow. This group is one of which we seemingly have no sensory awareness at all, but it is a feature of Nature, and it is the symmetry of this larger group that is relevant to spin.

The special unitary group SU(2)

The standard name for this group is the *special unitary group* and standard notation is SU(2). It is usually pronounced just as it looks: "ess-u-2". There is a unique "shadow" mapping

$$\rho : \mathrm{SU}(2) \longrightarrow \mathrm{SO}(3),$$

which is two-to-one in the sense that for each rotation $R \in \mathrm{SO}(3)$ there are two elements, $\tilde{R}$ and $-\tilde{R}$ that map to R. The mapping also respects the multiplication in the sense that

$$\text{if } \tilde{R} \mapsto R \text{ and } \tilde{S} \mapsto S, \text{ then } \tilde{R}\tilde{S} \mapsto RS\,.$$

This works no matter what $\pm$ signs are used with $\tilde{R}$ and $\tilde{S}$. This does actually say something! On the left side the group multiplication is in SU(2) and on the right side it is in SO(3). The shadow mapping respects the fact that both entities are groups, being compatible with their multiplication.

Of course there are lots of things missing in this very sketchy statement. Where does SU(2) come from, what do its elements look like, what is the group operation, and how does the shadow map arise? The simplest explanation, and one that does not involve any new physics, arises from yet another algebraic object, the four dimensional space $\mathbb{H}$ of *quaternions*. This might be considered as a four-dimensional analog of the complex numbers. Because it is interesting in its own right, and also gives a good idea of how SU(2) and SO(3) are related, we have included a section on the quaternions in ‡§19.1.

Wisps of the group SU(2) can be gleaned in a rather attractive and peculiar way. There is a famous body movement of hands, wrists, elbows, and shoulders where two full revolutions are required to restore the body to its starting state. A most beautiful form of it appears in the *Balinese legong dance* where it is performed with fans, see Fig.14.9. It is also known as *binasuan* (a Filipino dance). It is performed with fans, plates, cups, candles, or just bare hands in which the palms of both hands remain upwards while

Figure 14.9: These beautiful figure drawings illustrating the Balinese legong dance are those of Miguel Covarrubias. The key feature is the fan in the woman's right hand. Starting at the top upper right in the figure and following it across the lower line, it undergoes two complete revolutions, always same side face-up, returning to the original position—note that in the leftmost figure on the lower line the fan has made its first complete revolution. Originally published in *The Island of Bali* by Miguel Covarrubias [37, p. 208–209]. Reproduced by permission.

making two complete revolutions, the left palm rotating clockwise and the right palm counterclockwise. After one revolution the arms and wrists are highly rotated, but are restored to the customary position by continuing the rotation of the palms (in the same direction) for another full revolution. Altogether there are two complete revolutions of each palm: two revolutions to return to the original state. There are videos of this on the internet and it is easy to try it for yourself. Using one or both your hands, just trying to keep palms upwards the whole time. Don't hurt yourself! It is hard to keep the palms perfectly upwards facing the entire time, but some approximation of it is possible and shows the effect.

This effect can be interpreted directly as underlying spin, though it involves more mathematics than we can pursue here.

Spin is very peculiar. It doesn't correspond to anything we really know from normal experience, though it is possible that there are direct sensory consequences of spin. The quantum effects of spin may be directly observable in the visual field of some migratory birds, see §14.3.3 below. Spin transforms under a group that is larger than SO(3). It is unfamiliar, yet it does describe something that can be observed and measured, albeit so far only with very special technology. What is important for us here is that it explains the last quantum number and allows us to see that all four quantum numbers describing the hydrogen atom and the periodic table can be seen as aspects of symmetry. The symmetry is not just that of SO(3), but is that of SU(2).

14.3.3 The migration of birds and spin

It has long been established that migratory birds can sense the magnetic field and use it in navigating their long and amazing semi-annual flights. But until recently just how the magnetic field could lead to sensory perception has been a mystery. Now there are specific proposals of the biological processes that make this possible. At their root is a fundamental realization of how quantum mechanical effects can directly influence biological structures. No doubt there is much more still to be revealed on this, but at the heart of this particular story lies *spin* [144].

A single lone electron in the outer shell of an atom can, through a single quantum event, become entangled with the lone electron in an outer shell of another atom. This entanglement is in the form of coherent *spins* of the two electrons. These *radical pairs*, as they are called, can occur in biological systems. These entanglements are extremely brief, but experiments have shown that they last long enough for their parent atoms to show altered behavior under the influence of a magnetic field.

The particular effect observed is a change in autofluorescence [88]. Autofluorescence is the natural emission of light by biological structures when they have absorbed light. Certain blue light photoreceptors (cryptochromes) in living cells are capable of creating radical pairs that produce autofluorescence that is susceptible to an applied magnetic field. One particular instance is important. This is the flavin adenine dinucleotide (FAD). Flavoproteins utilize the unique and versatile structure of FAD to catalyze difficult redox reactions involved with the enzymatic processes of metabolism [55].

The details of this are evidently complex, but a visible effect is decrease and increase in autofluorescence under changes in the magnetic fields due to FAD in certain cells. This suggests that migratory birds may actually get direct visual clues to the Earth's magnetic field and their orientation with respect to it. This is something that will become clearer in time.

Evidently, hidden as it is, spin is not just some weird quantum mechanical effect that is necessary to formalize our understanding of the periodic table but otherwise irrelevant at the level of macroscopic creatures like birds and human beings. We know now that along with change there is invariance and this invariance can be carried in extraordinary forms of symmetry. We have had a taste of it and in ‡§19.2 there is more detail. But symmetry is pervasive and deep, especially in our understanding of the subatomic world. Tarasov and Wilczek [165, 190] offer very readable accounts of this.

The world is a unity, and no part of it is separable from any other. From the subatomic world through the atomic and molecular worlds, through the geological, biological, and mental worlds, and beyond our planet through the immensity of our Solar System, our Galaxy, and the Universe of which we are a part, there is a coherent and consistent wholeness. That is the great story that modern science has to tell.

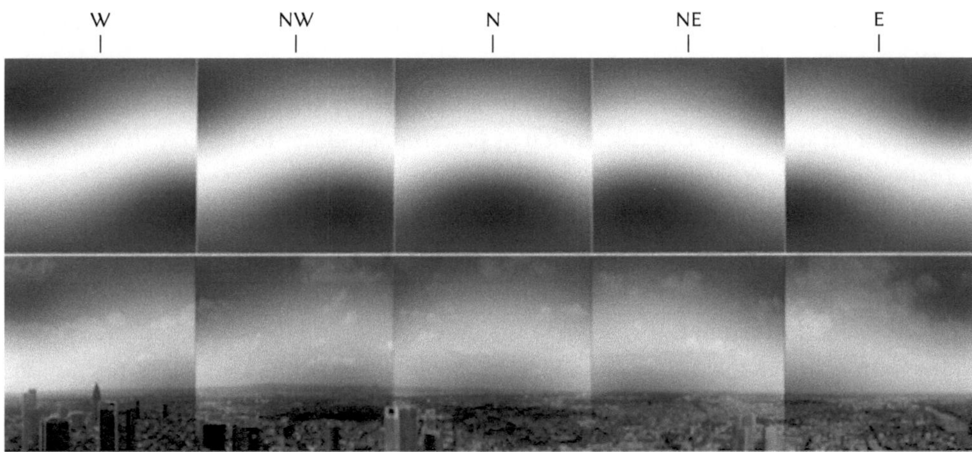

Figure 14.10: A suggestion as to how the magnetic field might become visible to a bird, with the visual field overlaid with darkening indicating various directions, west through north to east. This image was made with VMD and is owned by the Theoretical and Computational Biophysics Group, NIG Resource Center for Macromolecular Modeling and Visualization, at the Beckman Institute, University of Illinois at Urbana-Champaign. ks.uiuc.edu/Research/cryptochrome.

14.4 An appropriate language

What can we learn from this excursion into quantum theory and the periodic table? Perhaps overriding everything else is an appreciation for the extraordinary wonder of Nature, even at the smallest scales, and the great conceptual leaps necessary to make sense of it.

Language has arisen out of our interaction with the world as we can experience it, and so has come to articulate what the world is. But the quantum theory reveals a larger world, one in which common language fails. One can talk about quantum theory endlessly, but in the end the trail of words ends and what is left is the language and formalism of the mathematics. Galileo's famous dictum *"The laws of nature are written by the hand of God in the language of mathematics"* seems particularly appropriate here. Having some insight into this extended language seems like a prerequisite for talking intelligently about what is, after all, our own world. The objective of the last two chapters has been to offer some such insight.

For all its unusual aspects, the formalism of quantum theory is deeply system-based according with the same set of abstract principles that we have been developing all along, and pattern is what is physically manifested by the system. One of the purposes of discussing wave theory so extensively in the previous two chapters, apart from its intrinsic importance, has been to see the gradual conceptual evolution of ideas around dynamics, and hence around patterning. Most important in this development has been the shift in emphasis to linearization. We saw this shift first in Ch.11 and Ch.12, where the emphasis switched from states making up a set called the state space to observables which are functions on that state space and comprise the new state space. This new state space is a linear space, for observables can be added and scaled by real or complex numbers. It produces a far more nuanced description of pattern, one closer to the way in which we actually measure and study pattern.

Dynamics is always framed in terms of how states of the state space change, whether that be through time, spatial rotation and translation, or the shifting of symbolic sequences. Dynamics on the state space implies dynamics on the space of functions on that state space, which in this case were the observables, and dynamics appeared there in the form of linear operators on the space of observables. At the same time, the notion of pattern was refined from emphasis on invariance in the context of the partitioning of a state space to questions of invariant subspaces in the space of observables—the uncovering of pure periodic waves within the complex linear combinations of waves that make up the ones that we observe, for instance.

In quantum mechanics linearization becomes key, where it goes by the name of the *superposition principle*. The complex numbers are more than a convenience—they are an essential part of the formalism. There is a gestalt shift. There is still some underlying physical space, but the state space has become a linear space of complex-valued functions on that physical space. There are still observables, but these have become linear operators on this linear space. They themselves are dynamical operators, for they act as agents of change on the state space. An observable not only changes states, but also produces outcomes that are real numbers. However, these are found by searching for the invariant eigenspaces of the operator, the potential outcomes then being its eigenstates, and the resulting measurements their eigenvalues. Furthermore they do not appear deterministically, but rather probabilistically.

Most curious of all, the states ψ themselves are not physically observable. Only their effects which come in the form of probability waves $|\psi|^2$. This leads to the peculiar property that although a state ψ and a multiple $u\psi$ of it by a complex number u of absolute value 1 (an element of $\mathbb{U}$), which one can think of as a phase shift of ψ, are different states, they are considered to represent the *same physical state.*

14.5 EPR and Bell's theorem

This section revolves around a profound philosophical issue that arose in quantum theory, and to some extent still exists. Is quantum theory, as it is expressed presently, truly complete? Are there missing aspects to the theory that would replace its apparent basis in chance and restore causality? Are there rational causal explanations of its apparent "spooky" violation of locality?

Einstein was never completely happy with quantum theory, and we will look in some detail at one example that reveals where his uneasiness lay. Our main objective is to prove the Bell inequalities, which are consequences of an assumption of locality but are consistently violated by experimentation. Bell's ideas and variations of them are strong refutations of Einstein's ideas, and though significant philosophical positions still remain, the evidence against locality is convincing. In this section we want to offer some insight into what locality means and what it implies.

Our discussion is based significantly on the presentation of these ideas by Ghirardi [63], which goes through the historical, philosophical, and mathematical sides of this question in depth. Our setting is that of polarization. We begin with a single photon ψ which, according to our section on polarization, is expressible in the form

$$\psi = c_V \psi_V + c_H \psi_H \,,$$

where the coefficients, whose absolute squares are probabilities, satisfy $|c_V|^2 + |c_H|^2 = 1$. Recall that a polarization test done on ψ, say in the vertical direction, will produce YES with probability $|c_V|^2$ and NO with probability $|c_H|^2$. YES means the photon passes through the filter, thereupon being in the state ψ_V; NO means that it does not (and is absorbed into the filter). YES and NO would be interchanged if the test were for horizontal polarization.

The remarkable fact is that it is possible to create pairs of photons ψ_1 and ψ_2 that are so coupled that they respond *identically* to the vertical polarization experiment or to the horizontal polarization experiment. Furthermore they can be physically distanced from each other. To be clear, if ψ_1 passes the test YES for verticality (it passes through a vertical filter), then performing a test for vertical polarity for ψ_2 will also produce YES (it too will pass through a vertical filter). Similarly if ψ_2 passes vertical polarization, so does ψ_1. The same also holds for NO, which means that if one does not pass a vertical filter neither will the other. The two photons are said to be *entangled*. We will denote this entangled state by Ψ. We will abbreviate YES and NO to Y and N.

As far as notation goes we see that the outcome of a vertical polarization experiment is Y-Y or N-N. We express this in quantum theoretical terms by

$$\Psi = \frac{1}{\sqrt{2}} \psi_1^V \otimes \psi_2^V + \frac{1}{\sqrt{2}} \psi_1^H \otimes \psi_2^H \,. \tag{14.5}$$

We can parse this as follows. The first state is $\psi_1^V \otimes \psi_2^V$, which says that the first photon is in state V and the second state is also in state V, with the symbol $\otimes$ indicating that we are dealing with two separate photons. The second term has the other possibility, $\psi_1^H \otimes \psi_2^H$, where both photons are in the horizontal polarization. *These two states are the only two possibilities.* Both ψ_V and ψ_H are supposed to be normed, but we are dealing with their entangled combination Ψ, which also should be normed. Hence the coefficients $\frac{1}{\sqrt{2}}$, whose squares are $1/2$ indicating that each of the two possibilities of the outcome is equally probable and Ψ is a normed state.

It is a remarkable fact that although we have defined the entangled state Ψ in terms of two specific orthogonal directions, denoted as vertical and horizontal, it is actually independent of which pair of orthogonal directions we choose: if K and $K^{\perp}$ are another pair of perpendicular directions then

$$\Psi = \frac{1}{\sqrt{2}}\psi_1{}^K \otimes \psi_2{}^K + \frac{1}{\sqrt{2}}\psi_1{}^{K^{\perp}} \otimes \psi_2{}^{K^{\perp}}.$$

It is quite straightforward to see this, see the Endnotes.[9] In other words Ψ is *rotationally invariant.* The two directions indicated by V and H are not something intrinsic to it. We could have used the directions indicated by K and $K^{\perp}$ just as well and got exactly the same state.

There is no doubt that the situation that we have described is really quite strange. If one of the photons is tested then whatever the outcome the other will test the same way. Even though testing one produces a probabilistic outcome, the subsequent testing of the other is completely determined. The issue is this: the separation of the two photons can, in principle, be as large as we want. Tests on such entangled quantum particles have been made at distances over many kilometers. To make matters even stranger, if one particle is tested and the other one tested so soon after that nothing travelling at the speed of light could travel the distance between them, the results still match.

This outcome profoundly disturbed Einstein, and with two other physicists, Boris Podolski and Nathan Rosen, he wrote a famous paper—a sort of thought experiment, arguing that quantum theory was not complete [51]. This argument is called the *EPR paradox.* The subtleties behind this would take us too far from the subject of this book, however, the main argument revolves around the notion of *locality*. The EPR argument is based on

Einstein's locality: elements of physical reality objectively possessed by a system cannot be influenced instantaneously at a distance.

The conclusion of the EPR argument is that the apparent violation of locality that is witnessed by a pair of entangled photons is due to an incompleteness in the formalism of quantum theory. There must be hidden variables that are carrying information that determines the outcome of the measurement on the second photon once it is separated from the first.

What follows is an explanation of the famous Bell inequality and how it has come to be understood as a refutation of the hidden variable argument [17, 18].

14.5.1 Bell's argument

Bell's inequality is based on the following idea. Suppose that we have two observables A and B (we will think of them as observables arising from passing a photon through a polarizing filter) that are applied in two different regions of space, which we will denote by L and R (think of local and remote, or left and right, for instance). Suppose that among the outcomes of these observables we have a and b respectively with certain probabilities. There are three probabilities of which we take note:

$$p^L(A;a), \quad p^R(B;b), \quad \text{and} \quad p^{LR}(A,B;a,b)\,.$$

These stand for the probability of the observable A when applied in region L producing the value a, similarly with $p^R(B;b)$ when B is applied in the region R producing the value b, and finally the probability of the outcomes a and b when as a *combined system* the observations A at L followed by B at R are made. Einstein's locality says that since the elements of the system, the observables A and B, are at a distance they cannot influence each other simultaneously. The outcome of A cannot of itself affect the outcome at B. Bell's interpretation of this is that their probabilities must be independent. In other words,

Bell's locality: $p^{LR}(A,B;a,b) = p^L(A;a) \times p^R(B;b)$.

Quite independently of this, we actually know something about the probabilities arising from p^{LR}. We start by taking A as the observable $A_V = A_0$ that we introduced around equation (13.28) and B as the observable A_θ, which is filtration with a filter set to the angle θ with respect to the vertical. The values of the observation are then either Y(YES) or N(NO), once on the photon in region L where we test for vertical filtration A_V, and the other in the region R where we test for the filtration A_θ. To see how this works consider $p^{LR}(A, A_\theta, \mathrm{Y}, \mathrm{Y})$. The photon at L has probability of $1/2$ of passing through the vertical filter, and then it is in state ψ_V. The photon at R is now also in state ψ_V! Its probability of passing through the filter at angle θ is $\cos^2(\theta)$. We conclude that

$$p^{LR}(A, A_\theta, \mathrm{Y}, \mathrm{Y}) = \frac{1}{2}\cos^2(\theta)\,.$$

Similar reasoning gives rise to the following:

$$p^{LR}(A_V, A_\theta, \mathrm{Y}, \mathrm{Y}) = p^{LR}(A_V, A_\theta, \mathrm{N}, \mathrm{N}) = \frac{1}{2}\cos^2(\theta)\,, \qquad (14.6)$$
$$p^{LR}(A_V, A_\theta, \mathrm{Y}, \mathrm{N}) = p^{LR}(A_V, A_\theta, \mathrm{N}, \mathrm{Y}) = \frac{1}{2}\sin^2(\theta)\,.$$

For instance, if the photon in region L fails to pass the vertical filter, then we know for sure that the photon will fail it too. But that means that it cannot have any probability of passing through a vertical filter and so must be in the state ψ_H! Now the probability of ψ_H passing a filter set at angle θ to the vertical is $\sin^2(\theta)$.

Note that the sum of all four of these is $\cos^2(\theta) + \sin^2(\theta) = 1$, as it should be. In view of the rotational symmetry of the entangled pair, these values apply whenever the filters at L and at R are at angle θ to each other.

Now we come to Bell's argument. In this there will be four filter observables A, B, C, D which will be taken at various angles. At the moment we make no assumptions on these angles, but ultimately the angles we use will be

$$\theta = 0^\circ, 22.5^\circ, 45^\circ, 67.5^\circ .$$

We now define

$$\begin{aligned} E(A,B) = {} & p^{LR}(A,B,\mathrm{Y},\mathrm{Y}) - p^{LR}(A,B,\mathrm{Y},\mathrm{N}) \\ & - p^{LR}(A,B,\mathrm{N},\mathrm{Y}) + p^{LR}(A,B,\mathrm{N},\mathrm{N}) . \end{aligned} \tag{14.7}$$

This is just a number, adding and subtracting four probability values. There are parallel versions using other pairs of observables, e.g. $E(C, B)$, which we will use below.

Bell's locality assumption allows us to write this as a product:

$$E(A,B) = \left(p^{L}(A,\mathrm{Y}) - p^{L}(A,\mathrm{N})\right) \times \left(p^{R}(B,\mathrm{Y}) - p^{R}(B,\mathrm{N})\right) . \tag{14.8}$$

This is a crucial step. Whereas (14.6) is an outcome of our formulation of quantum theory, this factorization is based on the assumption of locality.

Since $p^L(A, \mathrm{Y}) - p^L(A, \mathrm{N})$ is the difference of two numbers which are between 0 and 1, their difference is between -1 and 1, and so

$$|p^{L}(A,\mathrm{Y}) - p^{L}(A,\mathrm{N})| \leq 1 . \tag{14.9}$$

The same goes for the other term.

Now Bell looks at the sum

$$|E(A,B) - E(A,D) + E(C,B) + E(C,D)| \tag{14.10}$$

(note the single minus sign) and proves that no matter what angles the four filters are set to, it is at most 2 in value:

$$|E(A,B) - E(A,D) + E(C,B) + E(C,D)| \leq 2 . \tag{14.11}$$

This is *Bell's inequality*.

Using the definition of $E(A, B)$ and the values given at equation (14.6), we have

$$\begin{aligned} E(A,B) &= \frac{1}{2}\cos^2(\theta) - \frac{1}{2}\sin^2(\theta) - \frac{1}{2}\sin^2(\theta) + \frac{1}{2}\cos^2(\theta) \\ &= \cos^2(\theta) - \sin^2(\theta) = \cos(2\theta) = \cos(2(\sphericalangle AB)) . \end{aligned} \tag{14.12}$$

Because we have various angles to consider, rather than give them all names we simply use the symbol $\sphericalangle AB$ to indicate the angle between these two polarizing filters. We have used the trigonometric identity of Endnote 8 of §11.

Using this and (14.9), we derive Bell's inequality in the Endnotes.[10] We encourage the reader to follow through the proof. It is definitely very clever, but it does not involve anything more than simple algebra. But the real importance of Bell's inequality is that it is NOT TRUE when the observables are those that we indicated above!

We can easily see this using the suggested values $\theta = 0°, 22.5°, 45°, 67.5°$ of the four filters A, B, C, D. The angles in question are

$$\sphericalangle AB, \sphericalangle AD, \sphericalangle CB, \sphericalangle CD, \text{ which are } 22.5°, 67.5°, 22.5°, 22.5°.$$

Plugging these into the cosine formula $\cos(2(\sphericalangle \quad))$ of (14.12), we get respectively

$$\frac{1}{\sqrt{2}}, \frac{-1}{\sqrt{2}}, \frac{1}{\sqrt{2}}, \frac{1}{\sqrt{2}},$$

so

$$E(A, B) - E(A, D) + E(C, B) + E(C, D) = 2\sqrt{2},$$

which is greater than 2!

Thus Bell's locality assumption, which is in effect Einstein's locality, is *incompatible* with what quantum theory predicts. Of course this is just theoretical, but experiments show that what quantum theory predicts is what happens, and this falsifying of the inequality has been carried out a great number of times. The inequality is violated in practice as well as in theory.

An important fact about this is that the argument completely skirts the issues around hidden variables. It does not prevent there being hidden variables involved. Just what variables may or may not be in play in the entanglement of the two photons is not the point. The point is that expressing Einstein's locality assumption as a result of the independence of two random variables already violates theory and experiment. That is why it is considered to show that no hidden variable theory is possible.

14.5.2 What are the conclusions?

Bell's inequality is a statement about probabilities, and as such it has to be tested in practice by repeated observations with the law of large numbers leading to conclusions about whether or not it is violated. In the situation which we describe, there are four filters. In each instance one has to decide such things as which of the four possibilities involved in the E functions one should choose first, and if that is randomized with what method of randomization. And having decided that, should the choice be made before or the photon pair is generated or "in flight". There are lots of potential loopholes in the argument. Initially the equipment for testing the passing or failing to pass was not suitably accurate to draw firm conclusions. It was the work of Alain Aspect in France in the 1980s that changed the level of precision. Over the course of the intervening years, many variants of Bell's inequality have been used, and loopholes have gradually been eliminated.

An excellent account of the philosophical issues around the implications of quantum theory and the history of the people, the ideas, and the experiments that surround them is to be found in *The Quantum Dissidents* [58]. There have always been those who have questioned the strangeness of quantum theory and have sought ways to explain its effects without recourse to the aspects of pure randomness that are part of the Copenhagen interpretation. There is good reason to do so! Quantum theory is a statement about the nature of reality, and it is so far removed from our intuitions about the nature of things that everything that it predicts needs to be analyzed in detail.

What is surprising is that within the physics community there has generally been very little interest in this type of pursuit, and dissenters have been generally discouraged from trying to put them to experimental tests. Quantum theory worked just fine—why waste time and money testing what was clearly a forgone conclusion? Still, there were dissenters and there were several attempts at hidden variable theories that would reproduce the experimental results of quantum theory, the most famous being that of David Bohm. The earlier experiments in the 1960s and 1970s to test Bell's inequality were, for varying reasons, performed by dissenters. They expected Bell's inequality to be satisfied in experiment and were generally disappointed that it failed to be so. Bell himself was one of them!

As the years passed and the technical potential for isolating and arranging things at the atomic level made spectacular advances, these questions started to be taken more seriously. Developments in quantum computing and quantum encryption, which depend on entanglement, have tested the ideas of non-locality entanglement in many ways. The present conclusion seems overwhelmingly to support quantum theory as it is formulated: it is complete in the sense that there is no evidence that hidden variables are needed to account for what has been observed. In the end we are left with the conclusion that the world of quantum theory really is as mysterious as it is cut out to be. The Nobel Prize in Physics 2022 was awarded jointly to Alain Aspect, John F. Clauser, and Anton Zeilinger "for experiments with entangled photons, establishing the violation of Bell inequalities and pioneering quantum information science" [121].

It should be remembered that what is being disputed is the idea of *local* causality. That does not deny what we might call a cosmic causality that arises as a functioning of the Universe in its interconnected entirety. It is notable that in mathematics the idea of randomness and of a random variable is defined in a way that quite clearly avoids dealing directly with what chance or randomness might really mean, see §5.1.6. It makes no statement about what chance is. It leaves quite open that chance is just an expression of the inevitable incompleteness of our knowledge of the entirety that lies behind all events, see [28, Afterword].

Recent advances in both physics and neuroscience suggest that the entirety in which we exist is a totally interconnected and interdependent unity. It is a unity in which perhaps neither law nor chance are true realities, but one in which our perceptions and conceptions of events *can be seen to be* outcomes of laws and chance. Even though it was not stated in these terms in classical Chinese thought, it is clearly in alignment with both Daoist and Mahayana Buddhist thought, which tended to move in the direction of the whole to the parts rather than the parts to the whole.

Beyond this, none of it denies the possibility that there will be other interpretations of quantum theory in the future. Physics is neither a fully unified theory nor a subject that claims to offer absolute truths. Understanding the nature of reality within our entwined relationship of human life and consciousness will undoubtedly remain a shifting target. Is one content with a theory that has components that seem only to have a mathematical or Platonic existence, or should one suppose that at the foundations there must be physical entities, even if it is not possible to observe them? Is it reasonable to suppose that things can be called physical entities if it is not possible to observe them?

14.6 Reality: what do we really know?

> There is no such thing as the pure *objective* observation. Your observation, to be interesting, i.e. to be significant, must be *subjective*. The sum of what the writer of whatever class has to report is simply some human experience, whether he be poet or philosopher or man of science.
>
> [170, Thoreau, May 6, 1854]

Thoreau lived long before the quantum age, but here he puts his finger on a crucial intuition: what we as human beings can express about the world is based on what we can experience. As Heisenberg put it, "What we observe is not nature itself, but nature exposed to our method of questioning." It is the world as seen from human perspectives. Niels Bohr's position on reality is summed up in his words "It's wrong to think that the task of physics is to find out how nature is. Physics concerns what we can say about nature." Yet Einstein thought otherwise: "What we call science, has the sole purpose of determining what is" [101, Ch. 11]. For some significant insight into this use of the word "is" see the Endnotes.[11]

At the start of the twentieth century the nature of reality, if not fully explained and still having a few paradoxical irregularities, was certainly understood by scientists as something that we could describe as an objective reality. Classical physics reigned supreme, and its underlying principle was mechanics. As Lord Kelvin (William Thompson), a leading physicist of the nineteenth century, perceived it

> ...before any considerable progress can be made in the philosophical study of nature a thorough knowledge of mechanical principles is absolutely necessary. It is on this account that mechanics is placed by universal consent at the head of the physical sciences [40, p.119].

Mechanical meant steam engines, electric motors, steel bridges, things meshing with things. This was the world seen as machines, objective entities that constituted reality and whose properties we could discuss as matters of fact.

This idea of mechanics as a practical tool at the human scale is important, but seeing it as an underlying feature of reality has proved untenable. Just some twenty years into the twentieth century classical physics had been turned upside-down and the old notions of reality had begun to unravel. Relativity theory (both special and general) had completely altered our understanding of space-time, and quantum theory was undoing both our long-held views about determinism and the role of cognition in our creation of "reality". Two of its greatest thinkers were in deep disagreement of the nature of things—a disagreement that they never reconciled.

These last two chapters have been an introduction to the extraordinary world of quantum theory. It is a mathematical model of ideas and their interpretation which have allowed us to articulate and predict our experiences with the atomic and subatomic worlds. From the point of view of pattern this has involved a metamorphosis of the mathematical entities that underlie the theory: states are now waves and observables are operators, outcomes are apparently rooted in chance. The new vision is still systemic in nature and still based on difference and change, but the fundamental concepts are framed in an entirely new way and present to us a world that is very different from the mechanistic one that had preceded it. What are we to make of this new framework of understanding?

Quantum theory is paradoxical. Although its origins were simply to explain the curious and unexpected results of experiments at the atomic level, it has ended up questioning the very nature of our understanding of reality. It tells us that it is fantasy to believe that we can experience the world in a truly objective way, that uncertainty between the values of conjugate variables is an inevitability, and that the Cosmos is a dynamical process where evolution is wrought with determinism, randomness, continuity, and discreteness. It seems to embody indefiniteness yet it is extremely precise in its formulation and can be stunningly exact in its predictions.

And then there is the peculiarity that quantum theory almost defies normal language. Ordinary words, the language of our natural human experience of the world, simply fail to encompass it. Its language—observables, changes of state, chance, entanglement, spin, and uncertainty—is that of mathematics. Much of it lies beyond what we consider normal comprehension but it is expressed with exactitude in mathematical formulation and gives rise to experimentally verifiable predictions that are at times simply astounding (think of

Dirac's prediction of anti-matter, something that had not previously even been dreamed of). It seems impossible to intelligently engage in questions of the nature of reality or the significance of Bell's inequalities without some appreciation of this formalism.

These two chapters offer a glimpse into its internal elegance, its economy of expression, and its inherent beauty. Perhaps most beautiful of all is its revelation of the wondrous mix of chance and law that underlie the nature of the Cosmos: infinitely creative but at the same time perfectly guided by inherent patterning. Is not this intelligibility a miracle?

Pattern and experience are all about distinctions, but when we come to quantum theory we see that it is also about the *ambiguities* of distinguishing. We have seen that the nature of observation in quantum theory is such that it alters the state probabilistically, and observation results in new states that leaves previous states destroyed. So perception of difference is a matter both of chance and of destruction. The uncertainty principle holds that the observable differentiation between conjugate variables (like position and momentum) lies in reciprocal relationship.

We may say that this all happens at a scale far below those accessible to human senses and so not of importance. But this is false. Philosophical ideas about the nature of reality can sound remote from the important questions of our time. But they are not. The change of view from Nature as either an adversary or as a resource—ours to defeat or exploit—is far from one that sees human life as part of a larger system in which all of our actions reverberate. James Lovelock's *Gaia*, §8.4.1, is still an antidote to such narrow thinking.

CHAPTER 15

Afterword

> Every failure is a step to success. Every detection of what is false directs us towards what is true: every trial exhausts some tempting form of error. Not only so; but scarcely any attempt is entirely a failure; scarcely any theory, the result of steady thought, is altogether false; no tempting form of Error is without some latent charm derived from Truth.
>
> William Whewell, *Lectures on the History of Moral Philosophy in England, Lecture 7, (1852).*

15.1 Natural philosophy

Today any discovery in science is usually followed by the question of what it is good for, meaning how does it make our physical lives better, is it a possible cure for some form of cancer, does it increase longevity, does it give us faster and more enticing electronic gadgets, or better vaccines, less waste, more military security, etc. Science is seen as a driver of technology, and technology is seen as a means to "better life".

Science was not always seen in this way. In the early nineteenth century when the new scientific age was truly becoming a force in society, science was still seen as a part of natural philosophy, and its practitioners were natural philosophers. Even today a doctorate in science is commonly designated as a Ph.D., a doctor of philosophy. Science was a branch of philosophy, and its primary value lay in the ways in which it furthered our understanding of the world. Its practitioners were those who embraced it out of curiosity, out of a sense of excitement in this new age of discovery, and out of rivalry with kindred spirits. And perhaps more importantly, it was pursued only by those who were financially able to support themselves by other means. The "scientist" as the professional practitioner of natural philosophy did not exist. In fact the word "scientist" did not exist—and it was coined precisely on this point.

On June 24, 1833, at the meeting of the recently founded British Association for the Advancement of Science, Samuel Taylor Coleridge, the celebrated Romantic poet, stood up and argued that men digging in fossil pits or performing experiments with electrical apparatus could hardly be called natural philosophers, and he chastised the members of the association for referring to themselves as such. Chairing the meeting was William Whewell, the same Whewell who invented the word "consilience", and he rose to the occasion and, coining a new word, suggested that perhaps the members of the association should call themselves "scientists" [156]. So it gradually came about that scientists were no longer viewed as natural philosophers and were now universally known as scientists. Increasingly they became specialists in specific areas of science that were supported by governments, universities, pharmaceutical companies, and a vast assortment of technological industries. Nowadays they are rewarded for their discoveries by honors, prizes, distinctions, and by promotion to lucrative and powerful positions.[1]

The Pattern of Change is about modern science, but in many ways it points towards that older ideal of science, science as natural philosophy and the search for the relationships of things, one to another. We see those relationships as dynamical, as patterning, and as systems-based in nature. They arise out of the physical reality of Nature through the powers of the senses and emergence of mind as our brains incorporate and *enstructure* sensation, place, language, culture, and the process of life itself. It is in this sense that each of us comes to make the world. Each of us is an active participant in the making of the world that we find ourselves living in, by choice and circumstance.

In spite of the immense complexities that this entails, the idea of patterning really comes down to a few basic things: we always experience things within some context, what we experience are differences and change, and what we can do with that experience is to more appropriately conform to the nature of reality, making estimations of chance, magnitude, and scale. The processes of forming contexts and studying the relationships between distinguishable and measurable components of a context as they appear and disappear under change is what underlies pattern.

The paradox is that we come to see patterns themselves as suggesting some system, and then having framed the system we see the patterns as its manifestations. In this way we see that each engenders the other. This is put subtly in Moss Robert's translation of the last part of the twenty-fifth chapter of the *Daodejing*, where we see the Way (Dao), that mental appreciation of oneness, referred back to to the processes of becoming. (See also §8.2.1.)

> [Man] [w]ho is bound to follow the rule of earth,
> As earth must follow heaven's rule,

And heaven the rule of the Way itself;
And the moving Way is following
The self-momentum of all becoming.

Life itself emerges within the wholeness of things and then, whether it be through evolutionary adaptation, learning, or memory, has to discover how to survive and thrive in the world. Awareness or understanding of the context, the details that make up that context, and then how contexts fit together as ongoing processes are necessities of survival. This is pattern as we understand it.

What becomes clear from the very outset is that there are both reductionist components (know the parts and their details) and integrative components (know the coherence of the whole) to this. Without the analytic component there can be no deep understanding. How much could we know about the brain before we knew about neurons? The detail is important, but it is not enough. How much of the human brain, and beyond that the human mind, do we know because we know that it is a network of neurons? It is a common criticism that many modern scientists are specialists with little appreciation of wholes. To do anything new in modern science usually requires enormous specialization, yet that very specialization, so localized in nature, can be seen to recede from a global appreciation of how it all fits together. The linking together of principles to form a comprehensive theory, especially when it involves very different disciplines, is the most daunting challenge of all.

In the end, we think that it is this linking together, what Whewell had called *consilience* and classical Chinese subsumed into the word Li (see the Frontispiece), that is most inspiring. To Whewell consilience is what pointed to the truth of science. Newton's new physics and the mathematics that supported it was a prime example. Now, more than two centuries later, we are surrounded by new examples. Think of Maxwell's amazing theory of the electro-magnetic force that resulted, *after the fact*, in the complete unification of light, radio waves, infra-red, ultra-violet, and X-rays into a single continuum, or think of quantum mechanics, which has even been able to reveal the patterns underlying the fundamental material elements of our Universe, and their arrangement into the periodic table.

More potent is genetics and Darwin's theory of evolution, where suddenly the basis of life and the role of chance come to the fore, with all their disturbing inferences about human evolution. Yet we have learned that far from being simply a source of disorder and indirection, when linked with principles of invariance chance is a source of unbounded creativity and change. Even the most basic processes of life at the cellular level depend on the random motions of molecules in order to reach the receptors that can recognize and utilize them.

Still, even nowadays, evolution based on elements of pure chance is often hard to accept. If chance is taken to be a name that stands for a lack of true understanding or perception of some underlying causal processes, then that is acceptable. But for many people acausal non-teleological chance, especially in the creation of human life, is in direct conflict with their deeply held intuitions and beliefs. We have seen that even Einstein tried valiantly to uproot Bohr's insistence on the acausal nature of chance. But the great lesson is that intuition and the lenses of language and culture through which we see the world, though vital, often fail to be reliable guides when it comes to the workings of Nature.

As it stands today, quantum mechanics as understood by Bohr has withstood all attempts to show that at some hidden level it must be causal. In §8.5.4 we have questioned what ultimate meanings we might possibly give to chance and determinism, but even so, with pure acausal chance we have learned that there are laws, or principles, of physics that as far as we can tell are unchanging and present in all parts of our universe. In this way we have come to an idea of pattern in which chance and principle live together, and it is this dynamic pairing that manifests its endless flow of forms and creativity. Consilience is not lost with the incorporation of chance into our thinking. It is actually amplified in the most wonderful ways. It just isn't what we first thought.

An unexpected feature of all these developments has been the idea of events and event spaces. We know that the pattern systems we can talk about and understand are often seen as aspects of underlying systems that we cannot truly access with our senses. We cannot directly access the world of atoms, and even less the world of resonances and sub-atomic particles from which they arise. Similarly, whether we are talking about economics, sociology, or psychology, we usually cannot work at the micro-level of individual businesses, individual people, or the individual neurons that make up the brain.

The ultimate inaccessibility of the true constituents of Nature, if such there be, or of the minds of the individuals that make a society, is not just one of temporary ignorance. As parts of the whole ourselves, we should not expect to articulate its entirety. All rests on what aspects of distinction we can or wish to make, and what the lens of our culture has instilled. Here, we have formalized this, speaking of collections of states that represent aspects of alikeness and distinction. These alikeness/distinction-sets are what we have come to call events. It is at the level of events that probability is defined, and it is out of events and the probabilities that states will lie in them that we arrive at the definition of entropy.

We have seen that entropy means more than the revelation that everything eventually reverts to disorder. As much as it is a measure of disorder, it is a measure of order. Nature is profoundly coherent and that coherence can be, at least to some degree, expressed in terms of entropy. We can think of coherence as stored up entropy. It is entropy that suggests what it means for pattern systems to join together into larger pattern systems, and for the whole to be something other than the sum of its parts. Any interaction between pattern systems, no matter how seemingly insignificant, is an example of synthesis and integration. Interaction always involves inter-relationship and so some form of coherence, even if the outcome of that interaction is less than desirable. Entropy stands as both a qualitative and a quantitative aspect of that interaction.

As much as entropy refers to uncertainty associated with disorder, we have seen information as the removal of that uncertainty—the yin/yang of the same idea. We have also seen that information is expressed in terms of context, and so is based on pattern systems, just as often has been suggested. That being the case, the convertibility of information into many equivalent forms suggests that in the final analysis information is related to *structural and relational properties* of pattern systems, and so to classes of structurally equivalent pattern systems. This is the mathematical notion of isomorphism, and that is how we have come to describe it.

There seems to be a deep intuition that pattern is about repetition, or what we have also called "return", and this is a case in which intuition seems well founded. Repetition is about recognition, and recognition is the return of remembered experience. Repetition lies at the heart of the ubiquitous phenomenon of long-range order. Nature is extensive, and within it we see order at every imaginable scale. Long-range order is an expression of coherence, the similarity of conditions leading to similar outcomes of patterning, re-occurrence of what we have seen before. Still, it is easier to appreciate the general idea of long-range order than it is to define it in mathematical terms. We have talked about periodicity in crystals and wave phenomena, the almost periodicity arising from incommensurate periods, and even stochastic versions of long-range order arising from point processes. Still, it is not hard to imagine that we have much to learn.

In the end pattern is not about some "true" perception of what may or may not be Nature. Patterning is a process of grasping wholes or systems, gestalts, and it applies wherever we can establish a context and entities within it that fit the logical relationships of a pattern system. It applies as much to human patterns of behavior as it does to stones rolling down mountains or bees seeking the nectar of a flower.

It is always a two-way process. Models arise in the first place in attempting to isolate and abstract relationships between perceived identities. The mathematics expands on this and expresses what are often deep and hidden relationships implicit within them. These can then be referred back to the contexts in which they arose, or even to completely different contexts in which the same abstract principles can be found. Patterning applies both to the world we live in and to the internal mental worlds that we create. It amplifies our understanding. It may even lead to unexpected incongruences that are in conflict with what we expect.

There is no end to this. We explore. Sometimes there is enlightenment, and sometimes failure.

15.2 Wigner's question

In this respect mathematics has a long historical reputation of being based on truth beyond human boundaries. Whether this is justified or not, there are few disputes in the mathematical world about what has been proved by way of logical consequences of its axioms, and what has not. This is one of the reasons that the originators of the scientific revolution imagined that a Creator must have based the movements of the heavenly bodies on such truths. The brilliant success of Newton in forging this idea into a mathematical interpretation of the heavens based on a few simple empirical observations was the beginning of the modern mathematization of the world. It continues unabated.

There is a well-known paper by the famous twentieth-century mathematical-physicist Eugene Wigner (1902–1995) called *The unreasonable effectiveness of mathematics in the natural sciences* [50]. In this paper he muses about this success of mathematics in explaining the world.

He begins with two points. The first is that "mathematical concepts turn up in entirely unexpected connections" and that "they often permit an unexpectedly close and accurate description of the phenomena in these connections". The second is that "because we do not understand the reasons for their usefulness, we cannot know whether a theory formulated in terms of mathematical concepts is uniquely appropriate".

Our journey through this book has taken us far enough to have some appreciation for his thinking. The ability of mathematics to offer apparent insight into deep physical aspects of the world is striking—in fact, amazing. Often even just articulation of these insights is mathematical. We might take the point suggested by Nuñez in Chapter 1 that mathematics works because it was derived from our very experiences with the world. We should expect such things. Wigner is aware of this, but he sees that mathematics goes beyond what we can infer through some sort of metaphorical transfer via the senses. He makes the point that mathematics is primarily conceptual and it would quickly run out of interesting ideas if we only had the initial concepts given in the axioms of its foundations. We can see that for ourselves. Out of just the idea of sets we have created state spaces, operators upon them, partitions, measures that take values on subsets, concepts of change, chance, measures, observables, entropy, invariants, linear spaces, eigenspaces, and so on. Many of these have turned out to be useful in the science, but most of them were created out of the sheer imagination of mathematicians. This is particularly true for all the ideas around linear spaces and group theory which had been created in mathematics before suddenly they were discovered to be foundational to physicists in the creation of quantum theories. The complex numbers of quantum mechanics and the special unitary group SU(2) required to explain spin were known and explored by mathematicians long before they appeared in physics. These are aspects of the mind, Platonic in nature. In ‡Ch.16 we bring up this question again, for in the study of the prime numbers and their distribution, we enter a world that holds its own beauty and wonder, but cannot be seen as some metaphorical representation of anything we can experience directly.

Wigner concludes his essay with the words "The miracle of the appropriateness of the language of mathematics for the formulation of the laws of physics is a wonderful gift which we neither understand nor deserve". At the time of his paper in 1960, the role of DNA in genetics was hardly known and the understanding of the brain was far less advanced than it is now. But still, his questions remain valid. How is it that we were able to write this book about pattern and find that it is so deeply entrenched in modern mathematics?

Perhaps the answer is something that is staring us in the face. Perhaps mathematics works so well because it is through inter-relationship that we find explanation and meaning, and at its heart mathematics is the abstract study of relations. Often we hear the world spoken of as a field of dreams about which we can really know nothing. But we have seen in this book that in spite of the limitations of our understanding, the natural world is a very deeply structured unity. It is, as the Hua-yen philosophy would suggest, like Indra's net of jewels in which everything exists in relation to everything else, and *only* through relation to everything else. The miracle is that this unity exists and that it is intelligible to the human mind. Inevitably, mathematics or not, we make our own versions of the world through the sheer act of living in it. Its totality is a deep presiding coherence.

15.3 Belonging

Science is a methodology for gleaning knowledge of the physical world, not a revelation of absolute truths. Around this we see two quite different phenomena. First, an overwhelming tendency to presume that although our ideas in the past may have proven to be wrong, now our latest scientific knowledge is correct, even if still incomplete. The history of ideas teaches us to be very suspicious of this. It is unlikely that any picture of the world can truly represent it. We are creatures of Nature, not objective observers.

Second, there is the complex problem of what happens when science and belief are in conflict. Whewell's own experience with Darwin's ideas is a case in point. Publicly he said that the origin of man should stand outside the realm of science, but all the same he recognized the persuasive evidence that Darwin had so carefully collected and examined. Later in his life, in a letter to a friend, he wrote [156, Ch. 12]:

> I cannot see without some regrets the clear definite line, which used to mark the commencement of the human period of the earth's history, made obscure and doubtful .... It is true that the reconciliation of the scientific with the religious view is still possible, but it is not so clear and striking as it once

Figure 15.1: An artistic representation of the argument between reason and myth. On the left is Ptolemy (Claudius Ptolemaeus of Alexandria, the famous mathematician and scientist and writer of the *Almagest*); on the right Hermes-Trismegistus (a god of a Greek-Egyptian fusion of Hermes and Thoth), who represents hermetic thought—hidden, but divinely revealed, knowledge and myth. Above sits a judge, who by fate has been made absent, leaving it open for us to decide. So the debate continues! The silver plate, Byzantine from about 500 CE, is part of the collection of the Getty Villa. Photo R.V. Moody.

> was. But it is weakness to regret this; no doubt another generation will find some way of looking at the matter which will satisfy religious men.

Many people have had to contend with scientific evidence that is at odds with their traditional belief system. However scientific we are, and however guided by rational thought, we humans are story tellers and, inevitably, participants of myths of our time and culture. At the basis of quantum theory is both the realization and formalization that the observer and the observed always lie in a relationship, and that relationship is a unity that includes them both. In the end, we see that quantum theory too is a story about reality. Of course it has much in its favor, but we have said that of many of our previous stories. Heisenberg points out that even in making "rational" decisions we are not, and cannot ever be, fully rational, for we never truly have all the facts (which no doubt go back eons in time). "The decision finally takes place by pushing away all the arguments—both those that have been understood and others that might have come up through further deliberation—and by cutting off all further pondering. ... Even the most important decisions in life must always contain this inevitable element of irrationality" [77, p. 205].

There are no easy answers. Science has unseated our human indulgence in the idea that the World and Universe were created simply for us. Planet Earth, though just right for our kind of sun, is thought to be a relatively common kind of planet of a very normal star in a very normal galaxy in a universe that is immense beyond comprehension. Nature is thought to be without teleological intent, a creative mixture of chance and law indifferent to human concerns. We are but one of the many forms of life on Earth that survive only in mutual coexistence. We are gifted in so many ways, but also undone by tribalism, avarice, retribution, and the lust for power. Our faith in a loving and compassionate God that cares is tested every day as we see the world in endless suffering, both of our own and Nature's making, affecting good and bad with seeming equanimity.

"Nature treats us like straw dogs", says the *Daodejing* Ch. 5 —but, thankfully, it also says that "Nature is a sacred vessel", Ch. 29. The free play of randomness is wedded with the unmistakable and inconceivable coherence of it all. Modernity may struggle to find meaning but we can learn from both earlier cultures and brilliant voices within our own. There we can find inspiration, solace, even enlightenment, in fully embracing it.

> There is great beauty in the silent universe. There are manifest laws governing the four seasons without words. There is an intrinsic principle in the created things which is not expressed. The sage looks back to the beauty of the universe and penetrates into the intrinsic principle of created things. ...The spirit of the universe is subtle and informs all life. Things live and die and change their forms, without knowing the root from which they come. Abundantly it multiplies, eternally it stands by itself. The greatest reaches of space do not leave its confines, and the smallest down of a bird in autumn awaits its power to assume form (Zhuanzi, Ch, 22)[199, §6.1].

Such an understanding puts humanism into a more relative position, where human existence is understood as part of something far larger and more profound. It is a question of connection. We find it even more poignantly expressed in the famous speech of Chief Seathl in 1853:

> ...Every part of this soil is sacred in the estimation of my people. Every hillside, every valley, every plain and grove, has been hallowed by some sad or happy event in days long vanished. Even the rocks, which seem to be dumb and dead as they swelter in the sun along the silent shore, thrill with memories of stirring events connected with the lives of my people, and the very dust upon which you now stand responds more lovingly to their footsteps than yours, because it is rich with the blood of our ancestors and our bare feet are conscious of the sympathetic touch ...[119, Chief Seattle, p. 193].

Again, Robinson Jeffers (1887–1964), an American poet who lived in Carmel, on the rugged and wild Pacific coast just north of Big Sur in California, writes:

> I believe that the universe is one being, all its parts are different expressions of the same energy, and they are all in communication with each other, influencing each other, therefore parts of one organic whole. (This is physics, I believe, as well as religion.) The parts change and pass, or die, people and races and rocks and stars, none of them seems to me important in itself, but only the whole. This whole is in all its parts so beautiful, and is felt by me to be so intensely in earnest, that I am compelled to love it, and to think of it as divine. It seems to me that this whole alone is worthy of the deepest sort of love; and that here is peace, freedom, I might say a kind of salvation, in turning one's affection outwards towards this one God, rather than inwards on one's self, or on humanity, or on human imagination and abstractions ...[82, p. 11].

Nature is neither defined nor limited by our human ideas. Still, we are part of it and we can, and do, influence its changing manifestations on Earth. We can lose sight of its intrinsic beauty and mystery. We can deny climate change, pollution of land and sea, over-population, and the mass extinctions of species—perhaps our own too— that have been our companions for millennia, but their consequences will transpire anyway.

We conclude by going back to Steinbeck:

> The factors we have been considering as "answers" seem to be merely symbols or indices, relational aspects of things —of which they are integral parts —not to be considered in terms of causes and effects. The truest reason for anything's being so is that it is. This is actually and truly a reason, more valid and clearer than all the other separate reasons, or than any group of them short of the whole. Anything less than the whole forms part of the picture only, and the infinite whole is unknowable except by being it, by living into it [160, Steinbeck, *Sea of Cortez, Ch. 14*].

We may only partially understand, but we fully *belong*. That may be the most important truth of all.

> Not after listening to me, but after listening to the account (*logos*) one does wisely in agreeing that all things are one [136, Heraclitus, Fragment 50].

Appendix

Romanization of Chinese characters

There are numerous connections to Chinese thought and philosophy in this book. Ideally we would use Chinese characters exclusively since their romanizations are insufficient to identify the correct meaning. Many distinct characters have identical pronunciations and romanizations, so often we offer both the character and its romanization. To add to the complication there are two different romanizations in common use. The newer one, put forward by China in 1979 as the standard, is *pinyin*. The older one is the *Wade–Giles* romanization which has been in common use for over a hundred years and is still common in Taiwan. In quoting various sources where romanization occurs we follow whichever is given.

There are plenty of online sources that elaborate the pinyin–Wade–Giles relationship. In this book the most noticeable difference is that the Wade–Giles system uses the apostrophe in a special way that tends to lead to mispronunciation. Without it the sound is unaspirated and with it the sound is aspirated (so either voiced or unvoiced).

character	pinyin	Wade–Giles	Pronunciation	Meaning
道	dao	tao	dow	(the) Way
太	tai	t'ai	tie	great
易	yi	i	as in need	change
經	jing	ching	as in changing	classic (book)
氣	qi	ch'i	cheese	energy
極	ji	chi	as in jeep	extreme, as in roof ridge
德	de	te	duh	virtue, power

Thus the famous 道德經 *Tao Te Ching* is the Wade–Giles form of the pinyin *Daodejing* and the classic 易經 *Book of Changes*, the *I Ching*, is *Yijing* in pinyin. The popular meditative slow moving practice of *t'ai chi* (in the Wade–Giles system) is *taiji* in pinyin. In all of these the "j" is hard. 道, often translated as *Way*, is romanized as *dao* in pinyin and *tao* in Wade–Giles, with the pronunciation as in the English word *dow* in both cases.

道 is well known from many books with titles *The Tao of* .... In this book we use the words *Dao* and *Daodejing* in preference to the more familiar *Tao* and *Tao Te Ching*, partly because Tao has become so cheapened by its casual use everywhere and more importantly because the new romanizations are those that today's China prefers. When quoting other authors we use whichever romanizations they have given us.

Greek letters

α	alpha	β	beta	γ, Γ	gamma
δ, Δ	delta	ϵ, ε	epsilon	ζ	zeta
η, H	eta	θ, Θ	theta	ι	iota
κ	kappa	λ, Λ	lambda	μ	mu
ν	nu	ξ, Ξ	xi	o	omicron
π, Π	pi	ρ	rho	σ, Σ	sigma
τ	tau	υ, Υ	upsilon	ϕ, φ, Φ	phi
χ, X	chi	ψ, Ψ	psi	ω, Ω	omega

Missing capitals are the same as in the standard Roman alphabet, and so are omitted. We have left in H because it is the first letter of the Greek "entropy", from which it derives its use as a symbol for the measure of entropy.

Symbol Index

This index gives locations, by section and subsection numbers, where important symbols are introduced and further explained.

i	§11.5.1	$\overline{z}$	§11.5.2
$\lvert\cdot\rvert$	§11.5.2	$w_k, w_{\mathbf{k}}$	§11.6, §11.8
$c_k, c_{\mathbf{k}}$	§11.6, §11.7.2	$\widehat{f}$	§11.6.3
$\mathbb{T}$	§11.7	$\mathbf{k.x}$	§11.7.2
$\mathbf{1}_E$	§11.9.1	$\int_X f\,d\mathbf{m}$	§11.10, §11.10.1
av	§11.6.2, §11.10.3	$L^2(\mathscr{X}), L^2(X,\mathbf{m})$	§11.10.5, §19.4.1
Δ	§12.3.2, §19.2.6	φ	§12.4.2
$N(A,k)$	§12.5	$\mathscr{M}$	§12.8,§16.7
$h, \hbar$	§13.1	$\psi, \psi(x,t), \psi_t(x)$	§13.2.1
$\lvert\psi(x)\rvert^2$	§13.2.1	$\lVert\psi\rVert$	§13.2.1
$\mathscr{H}$	§13.2.1, §13.5.2	a_k, α_k	§13.2.2, §13.4
Δ_ψ	§13.3	$T(t)$	§13.3.1
Δt	§13.3.1	H, H^*	§13.3.1, §19.2.6
f, k	§13.3.2	p, E	§13.3.2,§18.1
$\langle\ \mid\ \rangle$	§13.3.3	c_i	§13.4
$\mathbb{E}$	§13.6	$\langle A\rangle_\psi$	§13.6
$\Delta_\psi A$	§13.6	$\mathbf{X}, \mathbf{P}$	§13.6.3
n, l, m	§14.2.1, §19.2.3	s, p, d, f, g	§14.2.1
m_s	§14.2.3	ψ_{spin}	§14.3.2
SU(2)	§14.3.2 ,§19.1	$\otimes$	§14.5
$E(A,B)$	§14.5	A_d, M_k	§16.7
$\pi(x), Li(x)$	§16.7.1	ln	§16.7.1
$\zeta(s)$	§16.7.4	$\mathscr{A}, \mathscr{A}^*$	§17.1.2
$\mathscr{L}$	§17.1.2	$\mathscr{Q}, q$	§17.1.3
$\square$	§17.1.3	L, R	§17.1.3
S, Q, A	§17.2.1	$\mathbb{M}, \mathbb{M}_H, \mathbb{M}_N$	§17.2.1
f, λ	§18.1	i, j, k	§19.1
$\mathbb{S}$	§19.1	E_n	§19.2.3
$\psi, \psi^\uparrow, \psi^\circ$	§19.2.4	Poly$[\mathbb{R}^3]$, A_k	§19.2.5
$\Delta f, \nabla f$	§19.2.6	V_X, V_E	§19.4.3
$\mathbf{m}^A, \mathbf{m}^A_\psi$	§19.4.4	$\mathscr{X}^A$	§19.4.4

Bibliography

[1] S. Addiss and S. Lombardo. *Tao Te Ching Lao-Tzu*. Shambhala, 2007.

[2] Stephen Addiss. *The art of haiku*. Shambhala, 2012.

[3] Roger T. Ames and David L. Hall. *Dao De Jing: A Philosophical Translation*. Ballantine Books, 2003.

[4] *Anthropic Principle*. 2024. URL: en.wikipedia.org/wiki/Anthropic_principle.

[5] Shundo Aoyama. *Zen Seeds*. Ed. by Patricia trans. Dai-En Bennage. Shambhala, 2019.

[6] Aristotle. *Physics*. Trans. by Robin Waterfield. Oxford Paperbacks, 1999.

[7] Asvaghosa. *The awakening of faith*. Trans. by Y.S. Hakeda. BDK English Tripitaka 63-IV. Numata Center for Buddhist Translation and Research, 2005.

[8] Ellen Baake. *Stepping-stones to modern mathematical biology*. (in preparation).

[9] M. Baake and U. Grimm. *Aperiodic Order, Vol. 1: A mathematical invitation*. Camb. U. Press, 2013.

[10] M. Baake and U. Grimm. *Aperiodic Order, Vol. 2: Crystallography and almost periodicity*. Camb. U. Press, 2017.

[11] M. Baake, H. Kösters, and R.V. Moody. "Diffraction theory of point processes: systems with clumping and repulsion". In: *J. Stat. Phys* 159 (2015), pp. 915–936.

[12] Michael Baake, Franz Gähler, and Lorenzo Sadun. "Dynamics and topology of the hat family of tilings". In: *arxiv.org/pdf/2305.05639* (). URL: arxiv.org/pdf/2305.05639.pdf.

[13] S.M. Baer and J. Rinzel. "Propagation of dendritic spikes mediated by excitable spines: a continuum theory". In: *Journal of Neurophysiology* 65(4) (1991), pp. 874–890.

[14] Sandrine Bailly. *Japan Season by Season*. Abrams, 2009.

[15] Gregory Bateson. *Mind and Nature: A necessary unity*. Bantam Books, 1980.

[16] C. Beeli. "Quasicrystal structures studied by high-resolution transmission electron microscopy". In: *Zeitschrift für Kristallographie* 215 (2000), pp. 606–617.

[17] J. S. Bell. "On the Einstein-Podolsky-Rosen Paradox". In: *Physics* (1964), pp. 1195–200.

[18] J. S. Bell. “On the problem of hidden variables in quantum mechanics”. In: *Reviews of Modern Physics* 38 (1966), pp. 447–452.

[19] Jeffrey Bennett. *A global warming primer*. Big Kid Science, 2016.

[20] J. M. Berg, J. L. Tymoczko, and L. Stryer. *Biochemistry*. 5th Ed. W H Freeman, 2000.

[21] M. Sassioli de Bianchi. “Theoretical and conceptual analysis of the celebrated 4π-symmetry neutron interferometry experiments”. In: *Foundations of Science* 22.3 (2017), pp. 627–653. arXiv: 1601.07053.

[22] Jonathan Birch, Alexandra Schnell, and Clayton Nicola. “Dimensions in animal consciousness”. In: *Trends in Cognitive Sciences* 24, No. 10 (2020), pp. 789–801.

[23] Daniel J. Boorstin. *The Discoverers*. Harry N. Abrams Inc., 1991.

[24] Robert Bringhurst. *Private communication*. 2014.

[25] Robert Bringhurst and Jan Zwicky. *Learning how to die: Wisdom in the age of climate crisis*. University of Regina Press, 2018.

[26] William Byers. *How mathematicians think*. Princeton University Press, 2007.

[27] C. S. Calude et al. “Experimental evidence of quantum randomness incomputability”. In: *Physical Review A* 82 (2010), p. 022102.

[28] Fritjof Capra. *The Tao of physics*. Flamingo. Fontana Paperbacks, 1983.

[29] Fritjof Capra and Pier Luigi Luisi. *The systems view of life: a unifying vision*. Camb. University Press, 2014.

[30] David G. Casagrande. “Information as verb: reconceptualizing information for cognitive and ecological models”. In: *Georgia Journal of Ecological Anthropology* 3 (1999).

[31] E. Castellani and J. Ismael. *Which Curie’s principle?* 2015. URL: philsci-archive.pitt.edu/11543/.

[32] Chung-yuan Chang. *Creativity and Taoism*. Singing Dragon, 2011.

[33] N. Chhabra, M. Aseri, and D. Padmanabhan. “A review of drug isomerism and its significance”. In: *Int. J. Appl. Basic Med. Res.* 3.1 (2013), pp. 16–18.

[34] Enrico Clementi. *Computational Aspects for Large Chemical Systems*. Vol. 19. Lecture Notes in Chemistry. Springer Berlin Heidelberg, 1980.

[35] Francis H. Cook. *Hua-yen Buddhism: The Jewel Net of Indra*. Pennsylvania State University Press, 1977.

[36] Steve Coutinho. *An introduction to Daoist philosophies*. Columbia University Press, 2014.

[37] Miguel Covarrubias. *Island of Bali*. Alfred A. Knopf, 1937.

[38] H. S. M. Coxeter. *Introduction to geometry*. 2nd Ed. John Wiley & Sons, 1969.

[39] Paul Davies. *The fifth miracle: the search for the origin and meaning of life*. Simon & Schuster, 1999.

[40] Peter Dear. *The Intelligibility of Nature: how science makes the world*. University of Chicago Press, 2006.

[41] *Dendritic Spine*. URL: en.wikipedia.org/wiki/Dendritic_spine.

[42] Ming-Dao Deng. *Decoding the Dàodéjīng*. In preparation.

[43] Ming-Dao Deng. *Lunar tao*. HarperCollins, 2013.

[44] Daniel Dennett. *Darwin's Dangerous Idea*. Simon & Schuster, 1996.

[45] P. Diaconis, S. Holmes, and R. Montgomery. *Dynamical bias in the coin toss*. 2007. URL: theoremoftheday.org/MathPhysics/Cointossing/TotDCoinTossing.pdf.

[46] Paul Dirac. *The principles of quantum mechanics*. 4th ed. Clarendon Press, 1982.

[47] Eihei Dogen. *Moon in a dewdrop: writings of zen master Dogen*. Ed. by Kazuaki Tanahashi. North Point Press (Farrar, Straus and Giroux), 1985.

[48] Patrick DuVal. *Homographies, quaternions, and rotations*. Mathematical Monographs, 1974.

[49] H.-D. Ebbinghaus. *Numbers*. Graduate Texts in Mathematics. Springer-Verlag, 1991.

[50] M. Eigen and R. Winkler. *Laws of the Game: How the principles of Nature govern chance*. Princeton University Press, 1981.

[51] A. Einstein, B. Podolski, and N. Rosen. "Can quantum-mechanical description of physical reality be considered complete?" In: *Physical Review* 47 (1935), pp. 777–780.

[52] G.C.R. Ellis-Davis. "Two-photon uncaging of glutamate". In: *Frontiers of Synaptic Neuroscience* 10 (2018).

[53] *Entropy*. 2024. URL: en.wikipedia.org/wiki/Entropy.

[54] Leonardo Pisano Fibonacci. *Fibonacci's Liber Abaci*. Ed. by translator L.E. Sigler. Sources and studies in the history of mathematics and physical sciences. Springer-Verlag, 2002.

[55] *Flavin adenine dinucleotide*. URL: wikipedia.org/wiki/Flavin_adenine_dinucleotide.

[56] Nancy Forbes and Basil Mahon. *Faraday, Maxwell, and the Electromagnetic Field*. Prometheus Books, 2019.

[57] Maya Frankfurt and Luine Victoria. "The evolving role of dendritic spines and memory". In: *Hormones and Behavior* 74 (2015), pp. 28–36.

[58] Olival Freire. *The quantum dissidents: Rebuilding the foundations of quantum mechanics (1950-1990)*. Springer-Verlag, 2015.

[59] J. M. Fuster. *Cortex and the mind*. Oxford University Press, 2003.

[60] Breen.Philip G. et al. *Newton vs the machine: solving the chaotic three-body problem using deep neural networks*. 2019. arXiv: 1910.07291v1.

[61] Galileo Galilei. *The essential Galileo*. Trans. by Maurice A. Finocchiaro. Hackett Publ. Co., 2008.

[62] Hans-Otto Georgii. *Stochastics*. de Gruyter, 2008.

[63] Ghirardi Giancarlo. *Sneaking a look at God's cards*. Princeton University Press, 2004.

[64] Raymond T. Gibbs. *The Cambridge Handbook of Metaphor and Thought*. Camb. University Press, 2008.

[65] Stuart Gillespie and Philip Hardie, eds. *The Cambridge Companion to Lucretius*. Camb. University Press, 2007.

[66] J. B. Gouéré. "Quasicrystals and almost periodicity". In: *Commun. Math. Phys.* 255 (2005), pp. 651–681.

[67] Timothy Gowers. *The Princeton Companion to Mathematics*. Princeton University Press, 2008.

[68] A. C. Graham. "Studies in Chinese Philosophy". In: University of New York Press, 1990. Chap. Being in Western Philosophy compared with shi/fei and yu/wu in Chinese Philosophy.

[69] Stephen Greenblatt. *The swerve: how the world became modern*. Norton, 2011.

[70] John Gribbin. *The Scientists*. Random House, 2004.

[71] B. Grünbaum and G. C. Shephard. *Tilings and Patterns*. W. H. Freeman, 1987.

[72] Ian Hacking. *The emergence of probability*. 2nd Ed. Camb. University Press, 2006.

[73] Brian C. Hall. *Quantum theory for mathematicians*. Graduate Texts in Mathematics. Springer: Springer, 2013.

[74] Brian Handwerk. "Animals dream too—here's what we know". In: *National Geographic* 2 (2022). URL: www.nationalgeographic.com/animals/article/animals-dream-too-heres-what-we-know.

[75] Stephen Hawking, ed. *The dreams that stuff is made of: the most astounding papers on quantum mechanics and how they shook the scientific world*. Running Press, 2011.

[76] R. Healy. "Change without change, and how to observe it in general relativity". In: *Synthese* 141 (2004), pp. 1–35.

[77] Werner Heisenberg. *Physics and Philosophy*. Torchbooks. Harper and Row, 1962.

[78] Herodotus. *The Histories*. Trans. by Tom Holland. Penguin Books Ltd., 2013.

[79] David Hinton. *China root*. Shambhala, 2020.

[80] David Hinton. *The blue-cliff record*. Shambhala Publication Inc., 2024.

[81] David Hinton. *The four Chinese classics*. Counterpoint, 2013.

[82] David Hinton. *Wild Mind, Wild Earth: our place in the sixth extinction*. Shambhala Publication Inc., 2022.

[83] Douglas Hofstadter. *Gödel, Escher, Bach: an eternal golden braid*. Basic Books, 1979.

[84] Bert Hölldobler and Edward O. Wilson. *The Ants*. Belknap Press, 1990.

[85] James Hollis. *The Middle Passage*. Studies in Jungian psychology by Jungian analysts. Inner City Books, 1993.

[86] F. C. Hoppensteadt and E. M. Izhikevich. *Weakly connected neural networks*. Applied Mathematical Sciences 126. Springer-Verlag New York, Inc: Springer-Verlag, 1997.

[87] Christian Huygens. *Horologium Oscillatorium ; Part One*. Trans. by Ian Bruce. 2013. URL: 17centurymaths.com/contents/huygens/horologiumpart1.pdf.

[88] Noboru Ikeya and J. R. Woodward. "Cellular autofluorescence is magnetic field sensitive". In: *Proceedings of the National Academy of Sciences of the United States* 118 (2021), p. 3.

[89] James. *Master organic chemistry*. URL: masterorganicchemistry.com/2017/06/07/valence-electrons/.

[90] Minford John. *I Ching*. Viking, 2014.

[91] Mark Johnson. "Philosophy's debt to metaphor". In: *The Cambridge handbook of metaphor and thought*. Ed. by Raymond W. Gibbs. Cambridge, New York: Camb. University Press, 2008.

[92] George Gheverghese Joseph. *The crest of the peacock: the roots of non-European mathematics*. 3rd. Princeton University Press, 2011.

[93] Renn Jürgen. "Einstein's invention of Brownian motion". In: *Ann. Phys (Leipzig)* 14 Supplement (2005), pp. 23–37.

[94] Charles H. Kahn. *The art and thought of Heraclitus*. Camb. University Press, 1981.

[95] H. Kawase, Y. Okata, and K. Ito. "Role of Huge Geometric Circular Structures in the Reproduction of a Marine Pufferfish". In: *Scientific Reports* 3.2106 (2013). URL: https://doi.org/10.1038/srep02106.

[96] K. F. Kelton et al. "First X-ray scattering studies on electrostaticaly levitated metallic liquids: demonstrated influence of local icosahedral order on the nucleation barrier". In: *Physical Review Letters* 90.19 (2003).

[97] J. Kepler. *Harmonices Mundi (The harmony of the world)*. Vol. Vol.2 (Book 2). Published at the expense of Gottfried Tampach. Johannes Planck, 1619.

[98] Donald E. Knuth. *The art of computer programming*. 3rd. Vol. 2. Addison-Wesley, 1998.

[99] A. Kolmogorov. *Foundations of the theory of probability*. Trans. by N. Morrison. translation of Grundbegriffe der Wahrscheinlichkeitrechnung, 1933. Chelsea Publ., 1950.

[100] S. Kondo. "An updated kernel-based Turing model for studying the mechanisms of biological pattern formation". In: *J. Theoretical Biology* (2017), pp. 41120–127.

[101] Manjit Kumar. *Quantum*. W. W. Norton and Co., 2008.

[102] D. R. Lande. "Development of the Binary Number System and the Foundations of Computer Science". In: *The Mathematics Enthusiast* 11(3) (2014).

[103] Peter Lax. *Functional analysis*. John Wiley & Sons, 2002.

[104] Beiying Liu and Feng Qin. "Use dependence of heat sensitivity of vanilloid receptor TRPV2". In: *Biophysical Journal* 110 (2016), pp. 1523–1537.

[105] G.E.R. Lloyd. *Greek Science*. Folio Society, 2012.

[106] Richard Semon (transl. Louis Simon). *The Mneme.* George Allen & Unwin Ltd., 1921.

[107] Philip Low. *The Cambridge Declaration on Consciousness.* URL: animalcognition.org/2015/03/25/the-declaration-of-nonhuman-animal-consciousness/.

[108] Lucretius. *On the nature of things.* Trans. by Martin Ferguson Smith. Hackett Publ. Co., 2012.

[109] Ma-Tsu. *The recorded sayings of Ma-Tsu.* Ed. by Julian Pas translators Lievens. Edwin Mellen Press, 1987.

[110] Nancy Maryboy and David Begay. *Sharing the skies.* Rio Nuevo Publ.

[111] H. R. Maturana and Francisco Varela. *Autopoiesis and cognition.* D. Reidel Publ. Co., 1980.

[112] B. Mazur and W. Stein. *Prime numbers and the Riemann hypothesis.* Camb. University Press, 2016.

[113] Merriam-Webster. *Merriam-Webster Online Dictionary.* URL: merriam-webster.com.

[114] Phillip Mitsis, ed. *The Oxford Handbook of Epicurus and Epicureanism.* Oxford University Press, 2020.

[115] R.V. Moody. "Meyer sets and diffraction". In: *Yves Meyer-Selecta.* Ed. by Aline Bonami, Stéphane Jaffard, and Stéphane Seuret. Vol. 22. Documents Mathématiques of the French Mathematical Society. 2024.

[116] V. B. Mountcastle. *The cerebral cortex.* Harvard University Press, 1998.

[117] D. Mumford and A. Desolneux. *Pattern theory.* A. K. Peters Ltd., 2010.

[118] M. G. Nadkarni. *Basic ergodic theory.* Birkhäuser, 1998.

[119] Kent Nerburn, ed. *The wisdom of the native Americans.* New World Library, 1999.

[120] Nathan Ng. *Large gaps between the zeros of the Riemann zeta function.* 2005. URL: /arxiv.org/pdf/math/0510530.

[121] *Nobel Prize for Physics.* 2022. URL: nytimes.com/2022/10/04/science/nobel-prize-physics-winner.html.

[122] Rafael Nuñez. "Conceptual metaphor, human cognition, and the nature of mathematics". In: *The Cambridge Handbook of Metaphor and Thought.* Ed. by R. W. Gibbs. Cambridge: Camb. University Press, 2008, pp. 339–362.

[123] David Orenstein. *A single memory is stored across many connected brain regions.* Tech. rep. Tonegawa Lab-Picower Institute. URL: news.mit.edu/2022/single-memory-stored-across-many-connected-brain-regions-0502.

[124] Abraham Pais. *Inward bound: Of matter and forces in the physical world.* Clarendon Press, 1986.

[125] Roger Penrose. "Remarks on Tiling". In: *The Mathematics of Long-Range Aperiodic Order.* Ed. by Robert V. Moody. Vol. 489. NATO Series C: Mathematical and Physical Sciences. Dordrecht, Boston, London: Kluwer Academic Publishers, 1997, pp. 467–497.

[126] Plato. *Plato Complete Works.* Ed. by J.M. Cooper. Hackett Publ. Co., 1997.

[127] T. D. Pollard et al. *Cell Biology*. 3rd Ed. Elsevier, 2016.

[128] *Polynesian Voyaging Society*. URL: hokulea.com (visited on 2023).

[129] Stephen W. Porges. *The pocket guide to the polyvagal theory*. W. W. Norton & Co., 2017.

[130] Ezra Pound. *Confucius*. New Directions Publ. Corp., 1951.

[131] D. E. Presti. *Foundational concepts in neuroscience: a brain-mind odyssey*. W.W. Norton & Co, 2016.

[132] Tony Prince. *Universal Enlightenment: An introduction to the teachings and practices of Hua-yen Buddhism*. 2nd Ed. Kindle Books, 2020.

[133] Hussain Rashed. *The Stone's representation theorems and compactification of a discrete space*. URL: arXiv.org,arXiv:2112.07821 (visited on 2022).

[134] Bernhard Riemann. "Über die Anzahl der Primzahlen unter einer gegebenen Grösse". In: *Monatsberichte der Berliner Akademie* (1879).

[135] Moss Roberts. *Dao De Jing: The book of the Way*. University of Calfifornia Press, 2001.

[136] T. M. Robinson. *Heraclitus: Fragments, a text and translation with commentary*. University of Toronto Press, 1987.

[137] Carlo Rovelli. *Helgoland: Making sense of the quantum revolution*. Riverhead Books, 2021.

[138] Carlo Rovelli. *The order of time*. Riverhead Books, 2018.

[139] Dheeraj Roy et al. "Brain-wide mapping reveals that engrams for a single memory are distributed across multiple brain regions". In: *Nature Communications* 13 (2022), p. 1799.

[140] A. Ryan James. "Leibniz' Binary System and Shao Yong's Yijing". In: *Philosophy East and West* 46.1 (1996). University of Hawai'i Press, pp. 59–90.

[141] Eric R. Scerri. *The periodic table: a very short introduction*. Oxford University Press, 2011.

[142] Michael Schirber. "Golden mystery solved". In: *Physical Review Letters* 10 (2017). URL: physics.aps.org/articles/v10/s3.

[143] Manfred Schroeder. *Fractals, chaos, power laws: Minutes from an infinite paradise*. WH. Freeman, 1991.

[144] Science Alert. *Birds have a mysterious 'quantum sense'*. URL: sciencealert.com/birds-have-a-quantum-sense-and-for-the-first-time-scientists-see-it-in-action.

[145] Richard Semon. *Die Mneme: Als Erhaltendes Prinzip im Wechsel des Organischen Geschehens*. 2nd ed. Classic Reprint Series. Forgotten Books, 1907.

[146] Marjorie Senechal. *Quasicrystals and Geometry*. Camb. University Press, 1995.

[147] A. R. Shakhnovich. *The brain and regulation of eye movement*. Plenum Press, 1977.

[148] C. E. Shannon. "A mathematical theory of communication". In: *The Bell System Technical Journal* 27 (1948), pp. 379–423.

[149] D. Shechtman et al. "Metalllic phase with long-range orientational order and no translational symmetry". In: *Phys. Rev. Lett.* 53 (1984), pp. 1951–1953.

[150] Josselyn A. Sheena, Stefan Köhler, and Frankland Paul W. "Heroes of the engram". In: *Journal of Neuroscience* 18 (2017), pp. 4647–4657.

[151] Michio Shinozaki, Brook A. Ziporyn, and David C. Earhart. *The threefold Lotus Sutra.* Kosei Publishing Company, 2019.

[152] Nina Simone. *Who Knows Where the Time Goes (Song).* RCA: from *The Essential Nina Simone,* 1993.

[153] Ya. G. Sinai. "Introduction to ergodic theory". In: Mathematical Notes, Princeton University Press. 1977.

[154] David Smith et al. "An aperiodic monotile". In: *arXiv.arXiv.2303.10798* (). URL: doi.org/10.48550/arXiv.2303.10798.

[155] Gary Snyder. *The Gary Snyder Reader.* Counterpoint, 1999.

[156] Laura J. Snyder. *The Philosophical Breakfast Club: Four Remarkable Friends Who Transformed Science and Changed the World.* Crown Publishing Group, 2011.

[157] D. M. Sparlin. *Physics 107.* URL: mst.edu/~sparlin/phys107/lecture/chap06.pdf.

[158] Saul Stahl. "The Evolution of the Normal Distribution". In: *Mathematics Magazine* 79.2 (2006).

[159] L. A. Steen. "The science of patterns". In: *Science* 240 (1988), pp. 611–616.

[160] J. Steinbeck and Edward F. Ricketts. *The sea of Cortez.* Penguin Books, 1941.

[161] Steven H. Strogatz. *Nonlinear dynamics and chaos.* Perseus Publ., 1994.

[162] D. T. Suzuki. *The Buddha of Infinite Light.* Boston and London: Shambhala Publications, 1997.

[163] Shunryu Suzuki. *Branching streams flow in the darkness.* University of Calfifornia Press, 1999.

[164] W. Swanson Larry et al. *The beautiful brain: The drawings of Santiago Ramón y Cajal.* Abrahams, 2017.

[165] L. Tarasov. *This amazingly symmetrical world.* Trans. by Alexander Repyev. MIR Publishers, 1986.

[166] Robert Temple. *The genius of China: 3000 years of science, discovery, & invention.* 3rd Ed. André Deutsch, Carlton Publishing Group.

[167] *Thalidomide: Molecule of the Week.* 2014. URL: acs.org/molecule-of-the-week/archive/t/thalidomide.html.

[168] *The International System of Units. 9th Ed. 2022.* URL: www.bipm.org/documents/20126/41483022/SI-Brochure-9.pdf/.

[169] W. J. Thompson. *Angular Momentum: an illustrated guide to rotational symmetries for physical systems.* Wiley, 1994.

[170] H. D. Thoreau. *The Journal 1837-1861.* Ed. by Damion Searls. New York Review of Books, 2009.

[171] *Time Crystal.* 2023. URL: en.wikipedia.org/wiki/Time_crystal.

[172] *Timeline of crystallography.* URL: wikipedia.org/wiki/Timeline_of_crystallography.

[173] A. M. Turing. *The chemical basis of morphogenesis.* B 237. Phil. Trans. Royal Society London, 1952.

[174] *Uncertainty principle.* URL: wikipedia.org/wiki/Uncertainty_principle (visited on 2024).

[175] B. L. van de Waerden. *Science Awakening.* Vol. 1. Oxford University Press, 1971.

[176] B. L. van de Waerden. *Science Awakening II: the birth of astronomy.* Noordhoff International Publ., 1974.

[177] G. Vantomme and J. Crassous. "Pasteur and chirality: A story of how serendipity favors the perpared minds." In: *Chirality* 33 (2021), pp. 597–601.

[178] Francisco Varela et al. *The embodied mind.* MIT Press, 1993.

[179] D.J. Velleman. "The fundamental theorem of algebra: a visual approach". In: *The Mathematical Intelligencer* 37 (2015), pp. 12–21.

[180] F. Verhulst. *Nonlinear differential equations and dynamical systems.* 2nd Ed. Springer-Verlag, 2000.

[181] Ernest Vinberg. *Linear representations of groups.* Birkhäuser, 1989.

[182] David Wade. *Crystal and Dragon.* Destiny Books, 1992.

[183] David Wade. *Li, Dynamic form in Nature.* Wooden Books Ltd, 2007.

[184] P. Walters. *An introduction to ergodic theory.* GTM 79, 2000.

[185] Burton Watson. *Zhuangzi.* Columbia University Press, 2003.

[186] J. Watson Andrew and E. Lovelock James. *Biological homeostasis of the global environment: the parable of Daisyworld.* 1983. URL: doi.org/10.1111/j.1600-0889.1983.tb00031.x.

[187] H. Weyl. *Symmetry.* Princeton U. Press, 1952.

[188] Rick Wicklin. *Analyzing the first 10 million digits of pi: randomness within structure.* URL: blogs.sas.com/content/iml/2015/03/12/digits-of-pi.html.

[189] Norbert Wiener. *Cybernetics.* 2nd Ed. Martino Publ., 1961.

[190] Frank Wilczek. *A beautiful question: Finding Nature's deep design.* Penguin Press, 2015.

[191] Frank Wilczek. *Quantum Time Crystals.* 2012. URL: arxiv.org/abs/1202.2539.

[192] Allan R. Willms, Petko M. Kitanov, and William F. Langford. "Huygens' clocks revisited". In: *Royal Society Open Science* 4 (2017).

[193] William Scott Wilson. *Tao Te Ching: a new translation.* Shambhala, 2010.

[194] William Scott Wilson. *The Unfettered Mind.* Shambhala Publications, 2012.

[195] Andrew J. Wood et al. "Daisy world: a review". In: *Reviews of Geophysics* 46.1 (2008).

[196] *WordSense Online Dictionary.* URL: wordsense.eu/world/.

[197] Guanghui Xie. *The composition of common Chinese characters*. Peking University Press, 1997.

[198] Ed Yong. *An Immense World*. Random House, 2022.

[199] Lin Yutang. *The wisdom of Laotse*. The Modern Library, 1948.

[200] Dmitrii Zabeleskii et al. "Structure-based insights into evolution of rhodopsins". In: *Nature Communications Biology* 4 (821) (2021).

[201] Brook Ziporyn. *Interpreting the Daodejing: "The Minimally Discernible Position"*. 2023. URL: voices.uchicago.edu/ziporyn.

[202] Paul G. Zolbrod. *Diné bahane': the Navajo creation story*. University of New Mexico Press, 1984.

[203] Jan Zwicky. *Personal communication*.

[204] Jan Zwicky. *The experience of meaning*. McGill-Queen's University Press, 2019.

Index